2D Nanomaterials

Scrivener Publishing
100 Cummings Center, Suite 541J
Beverly, MA 01915-6106

Publishers at Scrivener
Martin Scrivener (martin@scrivenerpublishing.com)
Phillip Carmical (pcarmical@scrivenerpublishing.com)

2D Nanomaterials

Synthesis, Properties and Applications

Edited by
Subhendu Chakroborty
Research Coordinator, IES University, Bhopal, India
and
Kaushik Pal
University Centre for Research and Development (UCRD), Chandigarh University, India

Scrivener Publishing

WILEY

This edition first published 2024 by John Wiley & Sons, Inc., 111 River Street, Hoboken, NJ 07030, USA and Scrivener Publishing LLC, 100 Cummings Center, Suite 541J, Beverly, MA 01915, USA

For more information about Scrivener publications please visit www.scrivenerpublishing.com.

Wiley Global Headquarters
111 River Street, Hoboken, NJ 07030, USA

For details of our global editorial offices, customer services, and more information about Wiley products visit us at www.wiley.com.

Library of Congress Cataloging-in-Publication Data

ISBN 978-1-394-16649-7

Cover image: Pixabay.Com
Cover design by Russell Richardson

Set in size of 11pt and Minion Pro by Manila Typesetting Company, Makati, Philippines

Printed in the USA

10 9 8 7 6 5 4 3 2 1

Contents

Preface

In recent years, 2D nanomaterials have emerged as a remarkable cornerstone in the field of advanced materials research, with their unique properties and versatile applications captivating the attention of scientists and engineers worldwide. This book is a testament to the ever-growing interest and importance of 2D nanomaterials in the realm of material science, nanotechnology, pharmaceuticals, and a myriad of engineering specializations. This book provides a comprehensive overview of the synthesis, modification, characterization, and application of 2D nanomaterials. Since the topic is vast, spanning a wide range of scientific disciplines and technological advancements, it is important to acknowledge that no single book can encompass the entirety of this field. However, this diverse collection of chapters covers essential themes and materials, focusing particularly on the synthesis strategies and potential applications of 2D nanomaterials.

The book is structured into three sections, each delving into different aspects of 2D nanomaterials. The first section explores the synthesis of these materials, providing an overview of both top-down and bottom-up strategies. Understanding the methods by which these materials can be synthesized is crucial for advancing their potential applications. Additionally, this section delves into the structural characterization of 2D nanomaterials, shedding light on their intricate compositions and properties.

The second section examines the diverse characteristics exhibited by 2D nanomaterials. From their magnetic and mechanical properties to their electrical, plasmonic, and optical behaviors, these materials possess an array of intriguing attributes that make them highly attractive for a wide range of applications. This section of the book provides a comprehensive understanding of these properties, enabling readers to appreciate the unique potential of 2D nanomaterials.

The final section focuses on the applications of 2D nanomaterials, highlighting their use in various fields such as energy, water purification, biomedical applications, multimodal tumor therapy, and supercapacitor technology. By showcasing the breadth of their applications, we hope to inspire readers to explore further and unlock the immense possibilities that lie within these materials.

The editors would like to express sincere gratitude to all the authors and co-authors who contributed their exceptional research to this book. Their expertise and dedication have enriched its content, ensuring its relevance and significance. Furthermore, we extend our heartfelt appreciation to the Wiley-Scrivener publishing team for their unwavering support throughout the challenging process of bringing this book to fruition. Their professionalism and commitment have been invaluable in making this project a reality.

As editors, we firmly believe that this book will serve as a valuable resource for students, researchers, and professionals in material science and related fields. It will spark new ideas, ignite curiosity, and pave the way for groundbreaking discoveries in the realm of 2D nanomaterials. We are confident that this book will find its rightful place in university and institute libraries across the globe and will be a source of inspiration for future research and innovation.

Once again, we extend our deepest gratitude to all who have contributed to this book. We are honored to present this collection of knowledge and discoveries, and we eagerly anticipate witnessing the impact it will have on the ever-evolving landscape of 2D nanomaterial research.

Subhendu Chakroborty
Ph.D.; MRSC
Kaushik Pal
Ph.D.; D.Sc. (Malaysia)
January 2024

Part I

SYNTHESIS OF 2D NANOMATERIALS

1

Top-Down Strategies Synthesis of 2D Nanomaterial

Pranjyan Dash[1] and Pradeep Kumar Panda[2]*

[1]*National Taipei University of Technology, Taipei, Taiwan*
[2]*Department of Packaging & Logistics, Yonsei University, Gangwon-do, South Korea*

Abstract
The nanotechnology field originated in the 21st century. Especially 2D-based nanomaterials have received a lot of attention since they are easily affordable, non-toxic, and have excellent electrical, optical, thermal, and mechanical properties. Moreover, it is simple to synthesize and can be applied to a wide range of applications. The present chapter mainly focuses on various types of synthesis methods for 2D nanomaterials using top-down strategies. Many top-down strategies have been developed to synthesize 2D nanomaterials, such as etching, mechanical milling, sputtering, and laser ablation. In all methods, we would introduce their synthesis parameters, advantages, disadvantages, and applications. Moreover, its characterization and toxicity were briefly introduced.

***Keywords*:** Nanomaterials, synthesis, 2D nanomaterials, top-down strategy, application of nanomaterials

1.1 Introduction

Nowadays, nanomaterials (NMs) are diverging materials among all research fields [1–4]. Nanomaterials are tiny-sized materials with external diameters up to 100 nm. The nanomaterials are primarily used to make the tubes, rods, and fibers. In addition, the nanoparticles found their physical existence in nature [5]. Nanomaterials possess different physical properties

**Corresponding author*: rkpanda277@gmail.com

Subhendu Chakroborty and Kaushik Pal (eds.) 2D Nanomaterials: Synthesis, Properties and Applications, (3–16) © 2024 Scrivener Publishing LLC

as well as chemical properties to form the bulk of their counterparts. As the size of the nanomaterials is too tiny, they cannot be seen with the naked eye. These nanomaterials are added to different materials, such as cloth, cement, and other materials. The tiny size of these materials also makes them useful in electronics, environmental remediation, and neutralizing toxins. The emergent properties of nanomaterials make them beneficial and impart great impacts in electronics, medicine, and other fields [6–8]. The chemical and physical properties of NMs highly depend on the surface atoms. As per the applications, NM size can be controlled by various techniques, such as modification of surface and micelle concentration [9, 10]. Dimensionally, NMs are divided into zero-dimensional (0D), one-dimensional (1D), two-dimensional (2D), and three dimensional (3D). In 0D, all three dimensions merge into a nanoscale range. A schematic diagram of dimension-based NMs is provided in Figure 1.1. In this category, nanospheres, quantum dots, and nanoclusters are included. In 1D, two dimensions merge into one. In this category, nanotubes and nanorods are included. In the 2D category, one dimension is at the nanoscale and the other dimension is outside [11]. In this category, nanofilms and nanolayers are included. 3D NMs are bulk NMs with diameters greater than the nanoscale (1–100 nm). The building blocks for 3D NMs are 0D, 1D, and 2D NMs. Core shells, nanowire bundles, nanotube bundles, and multi-nanolayers are included in this category [12].

Among these, more and more attention has been paid to two-dimensional (2D) NMs due to their unique properties, such as excellent

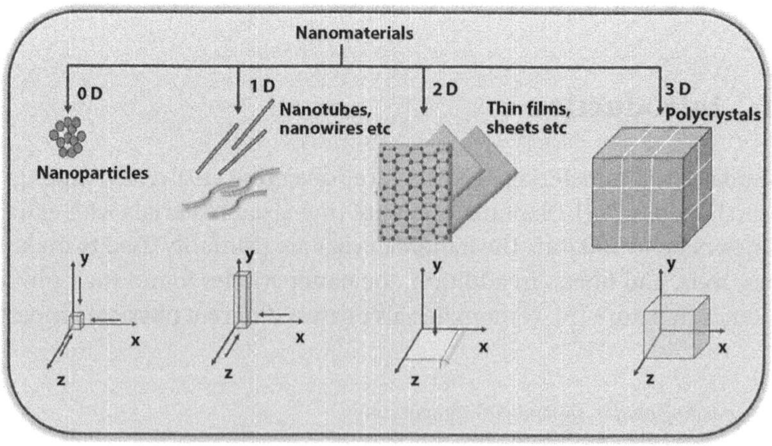

Figure 1.1 NM classification based on dimensions [13].

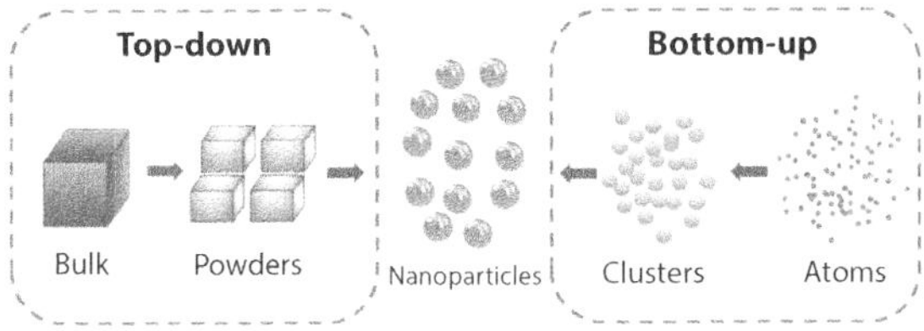

Figure 1.2 Nanoparticle's production strategies [23].

electrical, optical, thermal, and mechanical properties [14–17]. Many strategies have been developed to synthesize it for a specific application. These synthesis methods are broadly divided into two types of strategies: top-down and bottom-up approaches. Figure 1.2 displays an illustration of NM top-down and bottom-up strategies. The distinction between these two general classifications is based on the processes involved in the creation of nanometer-sized structures, and the choice of method depends on the specific requirements of the desired end product and the available techniques and technologies [18]. In the bottom-up approach, nanoscale materials are constructed from atomic or molecular precursors that are allowed to react, grow in size, or self-assemble into more complex structures [19]. By contrast, the top-down approach carves nanoscale structures by controlling the removal of materials from larger or bulk solids [20]. Each strategy has its own advantages and disadvantages [21]. In bottom-up strategies, nanoshells, ultrafine nanoparticles, and even nanotubes can be produced with a size of 1–20 nm. However, massive production is not possible, and synthesized nanomaterials need chemical purification. In a top-down strategy, nanomaterials can be produced massively, and purification is not needed. However, with this strategy, it is difficult to optimize the synthesized parameters [22]. This strategy is also not cost-effective.

In this chapter, we mainly emphasize the synthesis of 2D nanomaterials using a top-down strategy. In this strategy, we introduce the synthesis parameters, advantages, disadvantages, and applications for all methods. Eventually, the characterization and toxicity of 2D materials will be proposed.

1.2 Top-Down Strategy Synthesis Method

In a top-down strategy, bulk material is first converted into powder-based materials, which are then converted into nanomaterials [20]. There are mainly four methods available in this strategy (Figure 1.3), such as etching,

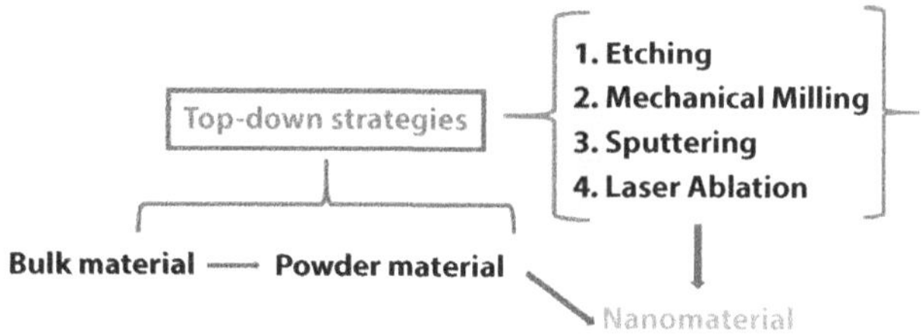

Figure 1.3 Top-down synthesis method of 2D nanomaterials.

mechanical milling, sputtering, and laser ablation. Each method's advantages, disadvantages, and own perspectives on applications are critically discussed. A summary of 2D NMs is listed in Table 1.1 using different top-down strategy methods.

Table 1.1 Synthesis of 2D NMs by different methods of top-down strategy.

NMs	Method	Parameters	Size	Applications	Ref.
GO	Mechanical milling	Ball size 5 nm; rotation speed 600 rpm, time (6, 16, and 24 h)	5 nm	Dye removal	[33]
GO	Mechanical milling	Ball size 3 mm, rotation speed 200 rpm, time 12 h	40 nm	Drug delivery	[34]
MoO_3	Laser ablation	Argon emission lines, Oxygen supplied	NA	Cancer therapy	[42]
GO-nano ribbon	Laser ablation	NA	NA	Optical	[43]
WS_2	Sputtering	Pulsing frequency of 20 kHz, DC power supply	>100	NA	[44]
MoS_2	Sputtering	Working distance 6 mm, a base pressure 2×10^{-6} Torr prior, 60 W	721	Supercapacitor	[45]

(*Continued*)

Table 1.1 Synthesis of 2D NMs by different methods of top-down strategy. (*Continued*)

NMs	Method	Parameters	Size	Applications	Ref.
MoO_2	Electro explosion	Surfactant, Thickness, temperatures, and current density	18.2	Antibacterial activity	[46]
MoS_2	Etching	Cl_2 plasma and Ar plasma	NA	Nanodevices	[47]
Hexagonal BN	Etching	O_2 and N_2 plasma (10 W); pressure: 1 Torr	NA	NA	[48]

1.2.1 Etching

In this method, the 2D NM surface is modified for enhanced physical and chemical properties. Mainly, this method of obtaining materials is applicable to the semiconductor industry. In this approach, a bulk material is treated with an etchant that selectively removes or dissolves certain layers or regions, leaving behind thin layers or flakes with desired 2D NM properties. This method has been used to synthesize 2D NMs such as graphene oxide (GO), transition metal dichalcogenides (TMDs), and boron nitride by selectively oxidizing or etching away layers from their respective bulk materials [24]. In this method, it is necessary to understand growth mechanisms [25]. Etching can be used to create patterns, structures, or features on the nanoscale, and it is a crucial step in the fabrication of various nanodevices and nanosystems [26]. Depending on the material and the desired outcome, several etching techniques can be used for 2D nanomaterials. Here are some commonly used etching techniques for 2D nanomaterials:

Wet etching: Wet etching involves the use of liquid chemicals to dissolve or remove material from a substrate selectively. For 2D nanomaterials, wet etching can be performed by immersing the substrate containing the nanomaterial in a chemical solution that selectively reacts with the material to be etched while leaving other parts of the substrate untouched [27]. Wet etching is relatively simple and can be used for a wide range of 2D NMs, including graphene, transition metal dichalcogenides (TMDs), MoS_2, and WS_2.

Dry etching: Dry etching, also known as plasma etching, involves using reactive gases and plasma to remove material from a substrate. Dry etching can be used for 2D NMs by exposing the substrate to a reactive gas, typically in a plasma chamber, which reacts with the material being etched and removed [28]. Dry etching offers higher precision and control over the etching process than wet etching, but it may require more complex equipment and processing conditions. Standard dry etching techniques for 2D NMs include reactive ion etching (RIE) and plasma etching.

Chemical vapor etching: Chemical vapor etching involves using reactive gases that are deposited onto the substrate as a vapor, which then reacts with the material to be etched. Chemical vapor etching can be used for 2D NMs by exposing the substrate to the reactive gases in a controlled environment, such as a vacuum chamber, and allowing the gases to react with the material to be etched [29]. Chemical vapor etching offers high control over the etching process and can be used for selective and precise material removal from 2D nanomaterials.

Atomic layer etching: atomic layer etching (ALE) is a specialized technique that offers precise control over the etching process at the atomic scale. ALE involves using sequential, self-limiting reactions to remove material from a substrate, layer by layer, selectively [30]. ALE can be used for 2D nanomaterials by controlling the exposure of the substrate to the reactive gases in a cyclic manner, which allows for highly controlled etching with atomic-level precision.

It is important to note that the choice of etching technique depends on the specific material properties of the 2D nanomaterial, the desired outcome, and the equipment and facilities available for the fabrication process. Careful consideration of the material properties, etching parameters, and safety precautions is necessary to ensure the successful etching of 2D NMs.

1.2.2 Mechanical Milling

Mechanical milling is the simplest top-down strategy. Mechanical milling, also known as mechanical alloying or ball milling, is a technique used to synthesize and process materials at the nanoscale through mechanical means. In this method, NMs are produced by the collision of rigid balls and bulk materials [31]. The whole system works under high pressure and is conducted through a sealed container. Usually, containers are made of ceramic and steel-based materials. The final product size depends upon many factors, such as ball size, milling time, milling duration, rotating speed, amount of bulky raw materials, and milling environment. The milling process is more effective when the ball-to-powder ratio is higher [32]. The milling process have many more advantages such as easy to handle, safety, high capacity, and long term maintained, even sometime no need solvent [32]. Despite this, disadvantages include disordering the crystal structure, requiring a lot of energy, the chance of contamination, and noise. It has been widely used to fabricate 2D nanomaterials, which exhibit unique properties due to their thickness being limited to the nanometer scale. The process of mechanical milling for 2D nanomaterials typically involves the following steps:

Material selection: Suitable precursor materials, typically in the form of powders, are chosen based on the desired properties of the final 2D nanomaterial.

Ball milling: The precursor materials are loaded into a ball mill and milling balls are usually made of a hard material such as stainless steel or tungsten carbide. The ball mill is then rotated at high speeds, causing the milling balls to collide with the precursor materials, leading to mechanical deformation, grinding, and mixing.

Milling parameter optimization: Various milling parameters, such as milling time, milling speed, and ball-to-powder ratio, are optimized to control the size, shape, and properties of the resulting 2D nanomaterials. These parameters can be adjusted to achieve the desired nanoscale features and properties of the 2D nanomaterial.

Post-milling treatment: Additional post-milling treatments such as annealing, doping, or functionalization may be employed to tailor the properties of the 2D nanomaterials further.

In this method, various types of 2D nanomaterials are synthesized and listed in Table 1.1 with different parameters. Recently, Mahmoud and his co-workers synthesized graphene oxide (GO) MPs through the ball milling method [33] The obtained NM size was 5 nm. Authors set different parameters such as rotating speed (600 rpm), milling time, ball size, etc. Further authors characterize it using various sophisticated techniques such as BET, UV, XRD, TEM, and Raman. Moreover, synthesized materials are applicable for methylene blue dye removal. Further, Caicedo *et al.* fabricated GO from graphite using this method. The obtained GO NM size was 40 nm in a 12 h processing time and is applicable in drug delivery systems. From the graphite precursor, GO was produced [34]. The main mechanism is the oxidation process.

1.2.3 Sputtering

Sputtering is a well-known synthesis method in the category of top-down strategies. Sputtering is a versatile and widely used technique for synthesizing 2D nanomaterials due to its ability to produce thin films with precise control over their thickness, composition, and properties [35]. It has been employed to synthesize various types of 2D NMs and 2D layered materials, with applications in areas such as electronics, optoelectronics, energy storage, and sensors [36]. Generally, in this method, thin films are fabricated on a substrate through bombardment of the target with energetic ions [37]. In practice, there are several types of sputtering, such as DC diodes, RF-diodes, magnetron diodes, and ion beam sputtering. In this method, thin films are commonly produced on the substrate. For the production films, an ultra-high vacuum system was needed [38]. The sputtering process for the synthesis of 2D nanomaterials typically involves the following steps:

Target preparation: A solid target material, which is the source of the atoms or ions to be deposited, is selected and prepared. The target material can be a pure element or a compound, depending on the desired composition of the 2D nanomaterial.

Substrate preparation: A substrate, which is a flat surface onto which the atoms or ions will be deposited to form the 2D nanomaterial, is prepared. The substrate can be made of various materials, such as silicon, glass, or a flexible polymer, depending on the intended application of the 2D nanomaterial.

Sputtering process: The target material is bombarded with high-energy ions, which dislodge atoms or ions from the target surface. These atoms or ions are then deposited onto the substrate surface, where they can form a thin film with a thickness in the nanometer range. The substrate is typically placed in close proximity to the target material, and the sputtering process is carried out in a vacuum chamber to minimize contamination and promote uniform deposition.

Control of deposition parameters: Various parameters, such as the sputtering power, gas pressure, and deposition time, can be adjusted to control the thickness, composition, and morphology of the deposited 2D nanomaterial. These parameters can be optimized to achieve the desired properties of the final nanomaterial.

Post-deposition treatments: After the deposition, the 2D nanomaterial may undergo additional treatments, such as annealing or etching, to further modify its properties, such as crystal structure, electrical conductivity, or surface morphology.

1.3 Laser Ablation

The laser concept was first introduced by Einstein. Laser ablation is a straightforward process. Using this process, lasers are the main source for converting bulk materials to NMs. Lasers are widely applicable in the medical field and industry [39, 40]. Laser ablation (LA) is a technique in which one atom or ion is removed from the substrate surface to form a nano/microstructure. Ablation is a term that refers to both the vaporization and melt ejection processes involved [41]. The obtained nano/microstructure pattern in polymers, metals, ceramics, and glasses. There are many influencing factors, such as wavelength, rate, duration, target material, and chemical composition. In this technique, low energy is needed for the formation of nanostructures. There are many benefits to using this method to produce nanostructures, such as safety, purity, reduced porosity, and simplicity. Due to the effects of laser irradiation, nanomaterial structures show different structures and mechanical properties. Moreover, in this method, the laser ablation rate can increase with the laser energy.

The basic principle of laser ablation for 2D nanomaterial synthesis involves irradiating a target material with a high-energy laser beam, causing the material to vaporize or melt. The vaporized or melted material then condenses or solidifies to form nanoscale particles, which can subsequently deposit on a substrate to form a thin film or accumulate to form flakes.

1.4 Characterizations and Toxicity of 2D Nanomaterials

2D nanomaterials are characterized through various techniques [47]. The X-ray diffraction pattern is mainly utilized to determine its conformational structure. It gives information about the crystal structure along with the phase. Transmission electron microscope (TEM), scanning electron microscope (SEM) and polarized optical microscope (OPM) are used to determine its morphology and size [49, 50]. Other fundamental properties, such as composition and element detection, are characterized through X-ray photoelectron microscopy (XPS), energy dispersive X-ray (EDX), etc. The optical properties are evaluated through ultraviolet–visible (UV–Vis) spectroscopy, Differential reflectance spectroscopy (DRS), and photoluminescence (PL) instruments.

The toxicity of 2D nanomaterials should be considered before applying them anywhere. Graphene is the most prominent material among all 2D nanomaterials. However, the toxic behavior of graphene is a challenging factor for biomedical applications. Recently, Strojny *et al.* studied *in vivo* the toxicity of graphene material based on its toxicity. However, mice died due to the toxicity of graphene [51]. The large surface area of 2D nanomaterials can lead to increased interactions with biological tissues and cells, potentially resulting in cellular uptake, oxidative stress, inflammation, and other adverse effects [15]. Some studies have shown that certain 2D nanomaterials, such as graphene and TMDs, can induce cytotoxicity (cell damage or death) *in vitro*, meaning in laboratory cell cultures. However, it is important to note that *in vitro* studies may not always accurately predict the behavior and effects of 2D nanomaterials in living organisms, and further research is needed to fully understand their toxicity *in vivo*, meaning in living organisms. In addition to cytotoxicity, other potential toxic effects of 2D nanomaterials may include genotoxicity (DNA damage), immunotoxicity (immune system disruption), neurotoxicity (nervous system damage), and ecotoxicity (environmental impacts). However, the toxicological properties of 2D nanomaterials can vary greatly depending on their specific characteristics and the biological system being studied. It is worth noting that research on the toxicity of 2D nanomaterials is also accompanied by efforts to mitigate or minimize potential risks. Strategies such as surface functionalization, coating, and encapsulation can be used to modify the properties of 2D nanomaterials to reduce their toxicity or enhance their biocompatibility. Safe handling, use, and disposal practices are also important in minimizing potential exposures and risks associated with 2D nanomaterials.

1.5 Conclusions

In summary, top-down synthetic strategies have been employed for synthesizing two-dimensional NMs. The top-down strategy includes etching, mechanical milling, sputtering, and laser ablation. Using top-down strategies, one can obtain high-quality and large-size two-dimensional crystals conveniently. Even so, this method is only appropriate for materials whose bulk crystals are layered. In the future, scientists should make more and more efforts to develop new synthetic technologies for high-quality 2D materials and the relevant applications.

References

1. Panda, P.K., Dash, P., Yang, J.M., Chang, Y.H., Development of chitosan, graphene oxide, and cerium oxide composite blended films: Structural, physical, and functional properties. *Cellulose*, 29, 2399, 2022.

2. Jayakumar, A., Mathew, S., Radoor, S., Kim, J.T., Rhim, J.W., Siengchin, S., Recent advances in two-dimensional nanomaterials: Properties, antimicrobial, and drug delivery application of nanocomposites. *Mater. Today Chem.*, 30, 101492, 2023.
3. Ahmad, I., Alayande, A.B., Jee, H., Wang, Z., Park, Y.J., Im, K.S., Nam, S.Y., Bae, T.H., Yang, E., Recent progress of MXene-based membranes for high-performance and efficient gas separation. *Diam. Relat. Mater.*, 135, 109883, 2023.
4. Marian, M., Berman, D., Rota, A., Jackson, R.L., Rosenkranz, A., Layered 2D nanomaterials to tailor friction and wear in machine elements—A review. *Adv. Mater. Interfaces*, 9, 3, 2101622, 2022.
5. Paramasivam, G., Palem, V.V., Sundaram, T., Sundaram, V., Kishore, S.C., Bellucci, S., Nanomaterials: Synthesis and applications in theranostics. *Nanomaterials*, 11, 12, 3228, 2021.
6. Lehner, R., Wang, X., Marsch, S., Hunziker, P., Intelligent nanomaterials for medicine: Carrier platforms and targeting strategies in the context of clinical application. *Nanomed. Nanotechnol. Biol. Med.*, 9, 6, 742, 2013.
7. Chakroborty, S. and Sahoo, S.K., Carbon-based nanomaterials in drug delivery systems, in: *Environmental Applications of Carbon Nanomaterials-Based Devices*, pp. 375–394, 2021.
8. Chakroborty, S., Bharadwaj, V., Sahoo, S.K., Sensing and biosensing with 2D nanosheets beyond grapheme, in: *Sensing and Biosensing with Optically Active Nanomaterials*, pp. 119–141, Elsevier, Amsterdam, Netherlands, 2022.
9. Zheng, Z., Cox, M., Li, B., Surface modification of hexagonal boron nitride nanomaterials: A review. *J. Mater. Sci.*, 53, 66, 2018.
10. Putro, J.N., Ismadji, S., Gunarto, C., Soetaredjo, F.E., Ju, Y.H., A study of anionic, cationic, and nonionic surfactants modified starch nanoparticles for hydrophobic drug loading and release. *J. Mol. Liq.*, 298, 112034, 2020.
11. Singh, V., Yadav, P., Mishra, V., Recent advances on classification, properties, synthesis, and characterization of nanomaterials, in: *Green Synthesis of Nanomaterials for Bioenergy Applications*, p. 83, 2020.
12. Kolahalam, L.A., Viswanath, I.K., Diwakar, B.S., Govindh, B., Reddy, V., Murthy, Y.L.N., Review on nanomaterials: Synthesis and applications. *Mater. Today Proc.*, 182182, 2019, 2019.
13. Abid, N., Khan, A.M., Shujait, S., Chaudhary, K., Ikram, M., Imran, M., Haider, J., Khan, M., Khan, Q., Maqbool, M., Synthesis of nanomaterials using various top-down and bottom-up approaches, influencing factors, advantages, and disadvantages: A review. *Adv. Colloid Interface Sci.*, 300, 102597, 2022.
14. Liu, B. and Zhou, K., Recent progress on graphene-analogous 2D nanomaterials: Properties, modeling and applications. *Prog. Mater. Sci.*, 100, 99, 2019.
15. Cheng, L., Wang, X., Gong, F., Liu, T., Liu, Z., 2D nanomaterials for cancer theranostic applications. *Adv. Mater.*, 32, 13, 1902333, 2020.

16. El-Khawaga, A.M., Zidan, A., El-Mageed, A.I., Preparation methods of different nanomaterials for various potential applications: A review. *J. Mol. Struct.*, 1281, 135148, 2023.
17. Soren, S., Chakroborty, S., Pal, K., Enhanced in tunning of photochemical and electrochemical responses of inorganic metal oxide nanoparticles via rGO frameworks (MO/rGO): A comprehensive review. *Mater. Sci. Eng. B*, 278, 115632, 2022.
18. Ghamsarizade, R., Ramezanzadeh, B., Mohammadloo, H.E., A review on recent advances in 2D-transition metal carbonitride-MXenes nano-sheets/polymer composites' electromagnetic shields, mechanical and thermal properties. *J. Taiwan Inst. Chem. Eng.*, 144, 104740, 2023.
19. Jamkhande, P.G., Ghule, N.W., Bamer, A.H., Kalaskar, M.G., Metal nanoparticles synthesis: An overview on methods of preparation, advantages and disadvantages, and applications. *J. Drug Deliv. Sci. Technol.*, 53, 101174. 2019.
20. Kiefer, V., Combining the top-down approach with bottom-up advantages in logic synthesis, in: *Proceedings Seventh Annual IEEE International ASIC Conference and Exhibit*, IEEE, 1994.
21. Biswas, A., Bayer, I.S., Biris, A.S., Wang, T., Dervishi, E., Faupel, F., Advances in top–down and bottom–up surface nanofabrication: Techniques, applications & future prospects. *Adv. Colloid Interface Sci.*, 170, 2, 2012.
22. Modan, E.M. and Plăiaşu, A.G., Advantages and disadvantages of chemical methods in the elaboration of nanomaterials. *Ann. "Dunarea Jos" Univ. of Galati. Fascicle IX. Metallurgy Mater. Sci.*, 43, 53, 2020.
23. Lu, H., Tang, S.Y., Yun, G., Li, H., Zhang, Y., Qiao, R., Li, W., Modular and integrated systems for nanoparticle and microparticle synthesis—A review. *Biosensors*, 10, 165, 2020.
24. Stehle, Y.Y., Sang, X., Unocic, R.R., Voylov, D., Jackson, R.K., Smirnov, S., Vlassiouk, I., Anisotropic etching of hexagonal boron nitride and graphene: Question of edge terminations. *Nano Lett.*, 17, 12, 7306, 2017.
25. Standaert, T.E.F.M., Schaepkens, M., Rueger, N.R., Sebel, P.G.M., Oehrlein, G.S., Cook, J.M., High density fluorocarbon etching of silicon in an inductively coupled plasma: Mechanism of etching through a thick steady state fluorocarbon layer. *J. Vac. Sci. Technol. A*, 16239, 1, 1998, 1998.
26. Samukawa, S., Ultimate top-down etching processes for future nanoscale devices: Advanced neutral-beam etching. *Int. J. Appl. Phys.*, 45, 2395, 2006.
27. Luo, J., Matios, E., Wang, H., Tao, X., Li, W., Interfacial structure design of MXene-based nanomaterials for electrochemical energy storage and conversion. *InfoMat*, 2, 6, 1057, 2020.
28. Lee, J.-M. and Kim, B.-I., Thermal dewetting of Pt thin film: Etch-masks for the fabrication of semiconductor nanostructures. *Mater. Sci. Eng. A*, 449, 769, 2007.
29. Ji, X., Kong, N., Wang, J., Li, W., Xiao, Y., Gan, S.T., Zhang, Y., Li, Y., Song, X., Xiong, Q., Shi, S., A novel top-down synthesis of ultrathin 2D boron

nanosheets for multimodal imaging-guided cancer therapy. *Adv. Mater.*, 30, 36, 1803031, 2018.
30. Tsai, Y., Li, Z., Hu., S., Recent progress of atomic layer technology in spintronics: Mechanism, materials and prospects. *Nanomaterials*, 12, 4, 661, 2022.
31. Chen, M., Guan, R., Yang, S., Hybrids of fullerenes and 2D nanomaterials. *Adv. Sci.*, 6, 1, 1800941, 2019.
32. Baig, N., Two-dimensional nanomaterials: A critical review of recent progress, properties, applications, and future directions. *Compos. Part A Appl. Sci. Manuf.*, 165, 107362, 2022.
33. Mahmoud, A.E.D., Stolle, A., Stelter, M., Sustainable synthesis of high-surface-area graphite oxide via dry ball milling. *ACS Sustain. Chem. Eng.*, 6, 5, 6358, 2018.
34. Caicedo, F.M.C., López, E.V., Agarwal, A., Drozd, V., Durygin, A., Hernandez, A.F., Wang, C., Synthesis of graphene oxide from graphite by ball milling. *Diam. Relat. Mater.*, 109, 108064, 2020.
35. Borca, B. and Bartha, C., Advances of nanoparticles and thin films. *Coatings*, 12, 8, 1138, 2022.
36. Gupta, D., Chauhan, V., Kumar, R., Sputter deposition of 2D MoS_2 thin films-a critical review from a surface and structural perspective. *Inorg. Chem. Commun.*, 144, 109848, 2022.
37. Maqbool, M., Ahmad, I., Richardson, H.H., Kordesch, M.E., Direct ultraviolet excitation of an amorphous AlN: Praseodymium phosphor by codoped Gd^{3+} cathodoluminescence. *Appl. Phys. Lett.*, 91, 19, 193511, 2007.
38. Jamnig, A., Pliatsikas, N., Konpan, M., Lu, J., Kehagias, T., Kotanidis, A.N., Kalfagiannis, N., Bellas, D.V., Lidorikis, E., Kovac, J., Abadias, G., 3D-to-2D morphology manipulation of sputter-deposited nanoscale silver films on weakly interacting substrates via selective nitrogen deployment for multifunctional metal contacts. *ACS Appl. Nano Mater.*, 3, 5, 4728, 2020.
39. Zeng, H., Du, X.W., Singh, S.C., Kulinich, S.A., Yang, S., He, J., Cai, W., Nanomaterials via laser ablation/irradiation in liquid: A review. *Adv. Funct. Mater.*, 22, 7, 1333, 2012.
40. Hong, S., Lee, H., Yeo, J., Ko, S.H., Digital selective laser methods for nanomaterials: From synthesis to processing. *Nano Today*, 11, 5, 547, 2016.
41. Ravi-Kumar, S., Lies, B., Zhang, X., Lyu, H., Qin, H., Laser ablation of polymers: A review. *Polym. Int.*, 68, 8, 1391, 2019.
42. Rahman, M., Pemmaraju, D.B., Murty, U.S., Phukan, S., Deshpande, U.P., Sathe, V., Kakati, M., A single-step plasma method for rapid production of 2D, ferromagnetic, surface vacancy-engineered MoO3–x nanomaterials, for photothermal ablation of cancer. *Nanotechnology*, 34, 19, 195601, 2023.
43. Dhongade, S., Koinkar, P., Furube, A., Sugano, S., Liquid exfoliation of graphene oxide nanoribbons using chemical assisted laser ablation. *Int. J. Mod. Phys. B*, 35, 2140009, 2021.

44. Villamayor, M.M.S., Lindblad, A., Johansson, F.O., Tran, T., Pham, N.H., Primetzhofer, D., Sorgenfrei, N.L., Giangrisotomi, E., Föhlisch, A., Lourenço, P., Bernard, R., Growth of two-dimensional WS_2 thin films by reactive sputtering. *Vacuum*, 188, 110205, 2021.
45. Ranjan, B., Sharma, G.K., Malik, G., Kumar, A., Kaur, D., *In-situ* sputtered 2D-MoS_2 nanoworms reinforced with molybdenum nitride towards enhanced Na-ion based supercapacitive electrodes. *Nanotechnology*, 32, 45, 455402, 2021.
46. Hashemzadeh, A., Ahmadi, R., Yarali, D., Sanaei, N., Synthesis of MoO_2 nanoparticles via the electro-explosion of wire (EEW) method. *Mater. Res. Express*, 6, 12, 1250d3, 2020.
47. Kim, K.S., Kim, K.H., Nam, Y., Jeon, J., Yim, S., Singh, E., Lee, J.Y., Lee, S.J., Jung, Y.S., Yeom, G.Y., Kim, D.W., Atomic layer etching mechanism of MoS_2 for nanodevices. *ACS Appl. Mater. Interfaces*, 9, 13, 11967, 2017.
48. Ma, Z., Prawoto, C., Ahmed, Z., Xiao, Y., Zhang, L., Zhou, C., Chan, M., Control of hexagonal boron nitride dielectric thickness by single layer etching. *J. Mater. Chem. C*, 7, 21, 6273, 2019.
49. Baig, N., Kammakakam, I., Falath, W., Nanomaterials: A review of synthesis methods, properties, recent progress, and challenges. *Mater. Adv.*, 2, 6, 1821, 2021.
50. Nicolosi, V., Chhowalla, M., Kanatzidis, M.G., Strano, M.S., Coleman, J.N., Liquid exfoliation of layered materials. *Science*, 340, 6139, 1226419, 2013.
51. Strojny, B., Kurantowicz, N., Sawosz, E., Grodzik, M., Jaworski, S., Kutwin, M., Wierzbicki, M., Hotowy, A., Lipińska, L., Chwalibog, A., Long term influence of carbon nanoparticles on health and liver status in rats. *Plos One*, 10, 12, e0144821, 2015.

2

Bottom-Up Strategies for Synthesis of 2D Nanomaterial

Nibedita Nath[1], Subhendu Chakroborty[2*] and Anita Routaray[3]

[1]*Department of Chemistry, D.S. Degree College, Sambalpur, Odisha, India*
[2]*Department of Basic Sciences, IITM, IES University, Bhopal, Madhya Pradesh, India*
[3]*Department of Chemistry, Ravenshaw University, Cuttack, Odisha, India*

Abstract
Two-dimensional nanomaterials (2D NMs) are a distinct class of nanomaterial that have gained popularity due to a wide range of promising applications and their intriguing physiochemical characteristics. The general top-down and bottom-up synthetic methodologies used on 2D NMs are briefly introduced in this chapter. Finally, we reviewed in detail the most recent work on the synthesis of different 2D NMs using the bottom-up strategy after discussing the various 2D NM synthetic methods.

Keywords: Nanomaterial, graphene, MXene, MOF, COF, top down, bottom up, nanocomposite

2.1 Introduction

Research and daily speech use nanomaterials (NMs). Its current and future fiscal promise is huge. Each of the three types of NMs—zero-dimensional (0D), one-dimensional (1D), and two-dimensional (2D)—has distinct properties [1]. Since Novoselov, Geim, and coworkers discovered graphene in 2004, physics, chemistry, material science, and nanoscience have studied ultrathin 2D nanoparticles [2]. This study has long-term fiscal benefits.

**Corresponding author*: subhendu.cy@gmail.com

Subhendu Chakroborty and Kaushik Pal (eds.) 2D Nanomaterials: Synthesis, Properties and Applications, (17–42) © 2024 Scrivener Publishing LLC

Since graphene's 2004 discovery, 2D materials have become a study center. Electron confinement in two dimensions makes the 2D feature unique and important for reaching previously unachievable physical, electrical, and chemical qualities. Due to its remarkable characteristics, including tremendous carrier mobility, the quantum hall effect, a large specific surface area, and exceptional optical, electrical, and thermal properties, graphene, with its thick crystalline carbon sheet, offers an excellent model. Due to its unique characteristics, graphene has been explored for use in high-speed electronics [3], optoelectronic devices [4], energy generation and storage [5–7], hybrid materials [8], sensors [9], and DNA sequencing [10, 11].

Research into graphene-like 2D NMs has grown since its discovery. Graphene-like 2D nanomaterials with various constituents include MXenes, black phosphorus (BP), layered metal oxides, hexagonal boron nitride (h-BN), and graphitic carbon nitride (g-C_3N_4) [12].

2.2 Types of 2D Nanomaterial

2.2.1 Graphene

Graphene is an allotrope of carbon with a 2D structure. Each atom in the hexagonally close-packed carbon network that makes up the substance forms a covalent link with three of its neighbors through the σ-bond. Around 1.42 Å degrees separate one carbon atom from the next. The graphite is formed when individual layers stack on top of one another using the van der Waals force, with the distance between each layer being around 3.35 Å [13]. A few elemental 2DNMs, such as silicene, germanene, stanene, phosphorene, arsenene, borophene, and others, are being explored in response to the exceptional properties of graphene [14–16]. Silicene and germanene are thought to just be single layers of Si and Ge that have undergone sp^2 hybridization to produce a 2D honeycomb lattice. First, theoretical hypotheses combining the structures of germanene and silicone were made [17, 18]. Contrary to graphene, the honeycomb lattices of silicene and germanene, are predicted to buckle and consist of two vertically displaced sublattices with, respectively, 0.4 and 0.7 buckling distances between the two A and B sublattices [18]. It is of great scientific interest to find out whether silicone may be produced on suitable substrates that might preserve a 2D honeycomb lattice to build a unique silicon allotrope with properties resembling those of graphene, as silicone is not typically seen in independent form [19].

2.2.2 MXenes

MXenes have the formula Mn + 1AXn (n = 1–3), where M = transition metals like Ti, V, Nb, or Cr, A = group IIIA or IVA elements like Al, Si, Sn, or In, and X = C or N [20]. MXenes are 2D-stacked transition metal carbides/nitride. The MAX phase has P63/mmc symmetry when M layers are almost hexagonally tightly packed and X atoms cover the octahedral locations. The element A is interspersed between Mn + 1Xn layers and metallically bonded to an element. Strong etching solutions, such as HF, can be used to selectively erode the A layers from the MAX phase, resulting in the formation of MXenes [21].

2.2.3 Black Phosphorus

Another 2D substance that has drawn attention more recently is black phosphorous (BP). A layered orthorhombic crystalline structure results from the bulk crystallization of BP. Individual layers begin to stack one on top of the other at a spacing of 5.3 [22] because of the van der Waals force. A BP monolayer, also known as phosphorene, is made up of a link between one P atom and another three P atoms within an "armchair" vertically staggered hexagonal lattice structure. The four P atoms are arranged as follows: three are in an identical plane and have bonds that are 0.2244 nm long, while the fourth occupies a parallel adjacent plane [23]. The hexagonal structure of BP with bond angles of 96.3° and 102.1° is visible in the top view in the z direction [22]. With such a carrier mobility of up to 1,000 cm^2 V1 s1, BP exhibits an adjustable bandgap that ranges from 0.33 eV for bulk to 1.5 eV for monolayer, according to the number of layers [24, 25]. BP may close the gap among graphene and TMDCs thanks to its natural direct bandgap and respectable carrier mobility.

2.2.4 Hexagonal Boron Nitride

The bulk type of hexagonal boron nitride (H-BN) that resembles graphite in its layered structure. A hexagonal arrangement of equal numbers of N and B atoms makes up its structure [26]. Each layer contains covalently coupled B and N atoms, and the van der Waals force causes the layers to stack on top of one another to form the bulk crystal. The interlayer distances and lattice constants of bulk H-BN are similar to those of graphite (3.30–3.33 Å). Similar to graphite, the bulk form of H-BN may be exfoliated into a single or few atomic layers. Carbon, nitrogen, and boron can all be used to create doped or alloyed layers with tunable electronic characteristics [27]. Exfoliated thin-layer H-BN has drawn a lot of interest as an insulating substrate, encapsulating material, and dielectric [28–30].

2.2.5 Transition Metal Dichalcogenides

Transition metal dichalcogenides (TMDCs) are composed of two distinct layers containing chalcogen atoms (X) sandwiched within two hexagonal layers of metal atoms (M) and have an MX_2 stoichiometry. Based on the chalcogen (such as S, Se, or Te) and transition metal (often Mo, W, Nb, Re, Ni, or V), TMDCs can be divided into more than 40 different groups. Every layer in layered structures usually possesses a thickness of 6–7 A° and is made up of two layers of chalcogen atoms sandwiched between a layer of metallic atoms packed hexagonally. Contrary to the mostly covalent nature of the intralayer MdX bonds, which enable the crystal to easily break across the layer surface, the sandwich layers are joined by a negligible van der Waals force [31]. Some of the most studied TMDCs include MoS_2, $MoSe_2$, WS_2, and WSe_2, and they all have tunable bandgaps that switch from indirect to direct inside the monolayers [32]. Additionally, strange properties are seen, such as the valley Hall effect [33], valley polarization [34], and superconductivity [35].

2.2.6 Graphitic Carbon Nitride

g-C_3N_4 is a different counterpart of graphite with a van der Waals layer structure. A framework of N-substitute graphite composed of graphitic planes that undergo sp^2 hybridization between nitrogen and carbon atoms to produce conjugated graphitic planes makes up the g-C3N4 crystal structure. The distance between two layers of g-C_3N_4 is d143.26, which is 3% closer than the gap between layers of crystalline graphite (d143.35 A°). This decreased interlayer distance could be explained by shifting electron locations and strengthening the bonds within layers because of the substitution of nitrogen atoms. For g-C3N4, there are two distinct structural hypotheses: the initial one is a condensed tri-s-triazine structure linked by planar tertiary –NH_2 groups, while the other is a condensed tri-s-triazine structure linked by wider regular gaps in the lattice [36].

2.2.7 MOF and COF

MOFs are a subclass of crystalline, porous coordination polymers made up of organic ligands and metal-containing molecules [37]. The benefits of

MOF for gas separation, catalysis, and biomedicine include its huge surface area, tailorable function and structure, and ultrahigh porosity [38–40]. In addition to the initial benefits, 2D MOFs have the advantage of 2D materials as we previously described, hence broadening their application fields [41, 42]. The term "COF" refers to a different kind of organic crystalline porous polymer that assembles covalently bonded organic parts into extended structures with periodic frameworks with ordered holes [43, 44]. Among them, planar construction elements eventually lead to an appearance of a layered structure, which results in 2D COFs. Due to its adaptable surface chemistry, low mass densities, and changeable pore sizes and shapes, this type of material has gained popularity recently [45, 46].

2.3 Synthesis Strategies

It has taken years of work to fully utilize and perfect the synthesis techniques for the effective creation of a variety of 2D NMs with controlled, large-scale, and ultrathin layers. For property modification and certain purposes, it makes sense. Two synthetic methodologies top-down and bottom-up approaches are frequently employed Figure 2.1.

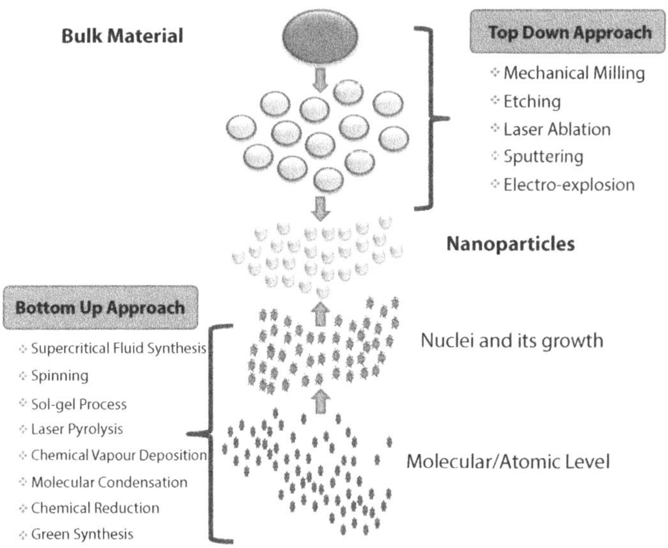

Figure 2.1 Synthesis techniques of NMs [47].

2.3.1 Top-Down

Through eliminating the van der Waals interaction within the layers that are stacked on top of the multilayered bulk crystals, top-down approaches enable the manufacture of one- or few-layer 2D NMs. Mechanical compression, the exfoliation process, and nanolithography are a few examples of top-down strategies. Top-down separation of bulk components creates nanostructured materials. Laser ablation, etching, mechanical milling, sputtering, and electro-explosion are a few examples of top-down techniques.

2.3.1.1 Mechanical Milling

Mechanical milling is a useful technique for producing nanomaterials from bulk materials. This process easily produces blends of different phases and creates nanocomposites [48]. Mechanical milling is used to create a variety of nanocomposite materials, like aluminum, nickel, copper, magnesium, and wear-resistant spray coatings [49]. A brand-new nanomaterial called ball-milled carbon nanomaterials has the potential to satisfy the needs for environmental cleanup and solve energy-related problems [50].

2.3.1.2 Electrospinning

One of the easiest top-down techniques for creating nanomaterials is electrospinning. It is frequently employed to create nanofibers out of a range of substances, primarily polymers [51]. The use of coaxial electrospinning was a significant advancement in the field of electrospinning. In this type of electrospinning, the spinneret is made up of two coaxial capillaries. In these capillaries, core-shell nanoarchitectures can be constructed using double-viscous liquids, with this liquid acting as the shell and a non-viscous liquid as the core. Hybrid, inorganic, organic, and core-shell hollow polymer structures have all been made using this technique [52].

2.3.1.3 Lithography

Lithography uses a focused electron/light beam to create nanoarchitectures in a practical way. Mask- and maskless-lithography are the two primary subcategories of lithography [53]. In masked nanolithography, a particular mask or template is used to transfer nanopatterns over a substantial surface area. Examples of maskless lithography include scanner probe lithography, focused ion beam lithography, and electron beam lithography.

2.3.1.4 *Sputtering*

Sputtering involves influencing solid surfaces via high-energy particles like plasma or gas to create nanomaterials. This method is very useful for creating thin nanomaterial sheets [54]. It is an exciting process since it is less costly as compared to electron-beam lithography, which produces nanomaterials that are identical to the target material with fewer impurities [55].

2.3.1.5 *The Arc Discharge Method*

Many different nanostructured materials can be produced using the arc discharge method. It is well known for producing amorphous spherical carbon nanoparticles, carbon nanotubes, fullerenes, carbon nanohorns, and few-layer graphene [56]. Techniques for manufacturing arc discharge nanostructures are efficient. Graphene's properties can vary depending on the conditions of manufacturing [57].

2.3.1.6 *Laser Ablation*

Under the intense radiation from a laser used in the laser ablation method, the initial material or precursor evaporates, producing nanoparticles. It may be said that employing laser ablation to make noble metal nanoparticles is environmentally beneficial because neither stabilizing agents nor other chemicals are needed [58]. Numerous types of nanomaterials can be produced using this technique [59–63].

2.3.2 Bottom-Up Method

2.3.2.1 *Chemical Vapor Deposition*

Techniques like chemical vapor deposition (CVD) were crucial for the creation of a nanomaterial made of carbon. During CVD, a chemical reaction involving vapor-phase precursors results in the formation of a thin coating over the surface of the substrate [64]. Precursors are useful for CVD if they have enough volatility, great evaporative stability, low cost, a nontoxic nature, outstanding chemical purity, and a long shelf life. Additionally, no pollutants must be left behind after it disintegrates [64]. For instance, the CVD technique for creating carbon nanotubes involves heating a substrate to a high temperature in an oven. Then, a precursor gas containing carbon (such as hydrocarbons) is gradually introduced to the system. The carbon atoms are released after the gas decomposes at high temperatures, and as

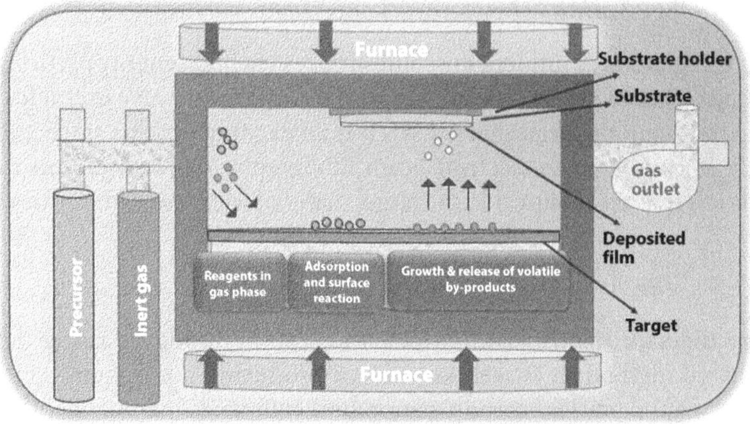

Figure 2.2 CVD procedure schematic diagram [68].

they recombine, they create carbon nanotubes on a substrate [65]. The form and nature of the nanomaterial produced, however, are greatly influenced by the catalyst selected. A well-known method for producing excellent-quality nanomaterials is the production of 2D nanomaterials using CVD [66, 67]. The CVD procedure is schematically depicted in Figure 2.2.

2.3.2.2 Sol–Gel Method

To create nanomaterials, a wet chemical sol–gel method is often utilized (Figure 2.3). Numerous superior metal-oxide-based nanomaterials may be created using this technique. This process is characterized as a sol–gel approach since the liquid precursor is changed into a sol during the production of the metal-oxide nanostructures, and the sol is then changed into a network-like structure known as a gel [69]. The final product made by the sol–gel process is influenced by the kind of precursor, rate of hydrolysis, aging time, pH, and molar ratio of the precursor to water [70]. One of the numerous advantages of the sol–gel method, which is also advantageous economically, is the homogeneous quality of the material formed. Other benefits are the lower operating temperatures, the technology's ease of use in producing complex nanostructures and composites, etc. [71].

2.3.2.3 Solvothermal and Hydrothermal Methods

The hydrothermal process is the most popular and frequently used process for producing nanomaterials [73, 74]. Using an aqueous hydrothermal

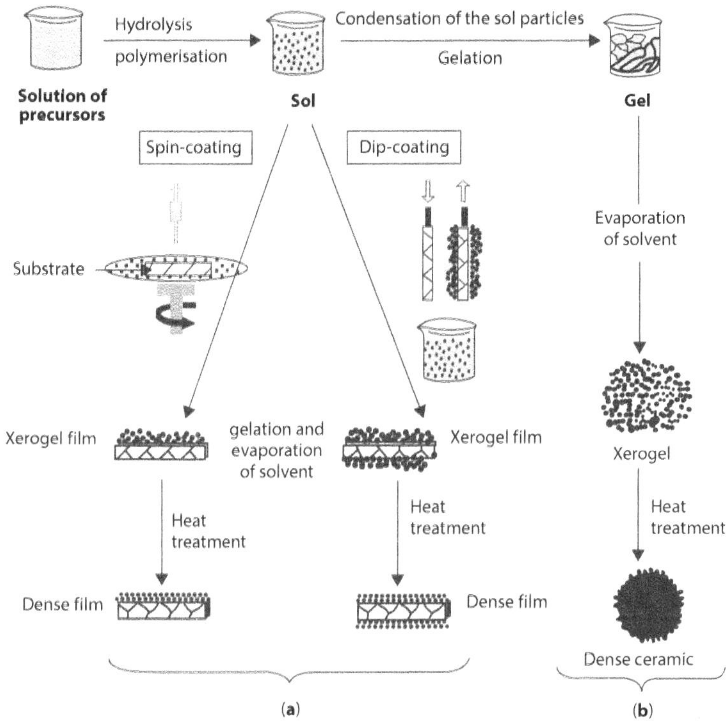

Figure 2.3 Shows two examples of sol–gel synthesis [72].

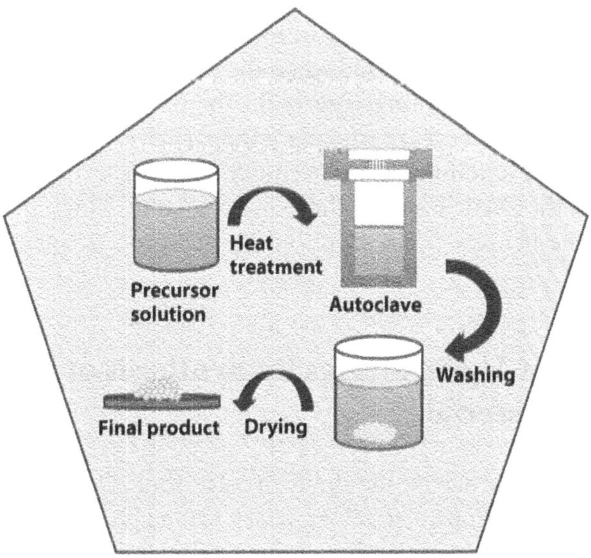

Figure 2.4 Hydrothermal technique schematic view [68].

approach, a heterogeneous reaction produces nanostructured materials in the hydrothermal process. The use of a non-aqueous medium is the only distinction. Most frequently, closed systems are employed to carry out solvothermal and hydrothermal treatments [75]. The microwave-assisted hydrothermal approach, which combines the advantages of the hydrothermal and microwave processes, has recently drawn a lot of interest in engineering nanomaterials [76]. Fascinating and practical techniques for creating different nano-geometries for materials, including nanosheets, nanorods, and nanowires, include hydrothermal and solvothermal processes [77–79]. The hydrothermal mechanism is shown in Figure 2.4.

2.3.2.4 Soft and Hard Template and Reverse Micelle Methods

To make nanoporous materials, both soft and hard template techniques are used. A straightforward traditional method for creating nanomaterials is the soft template approach. Due to its simple implementation, versatility to fit experimental conditions, and capability to manufacture materials with a range of morphologies, the soft template approach has been deemed effective [80]. The creation of nanoporous materials involves the employment of a variety of soft templates, such as flexible organic compounds, cationic, anionic, and non-ionic surfactants [81].

A helpful way for creating nanomaterials with specific shapes is the reverse micelle method. The usual micelles produced by an oil-in-water emulsion have tails that are hydrophobic oriented at a core with trapped oil droplets inside of them. However, in the instance of a water-in-oil emulsion, reversal micelles are created when the hydrophilic heads are directed towards the water-containing core [82]. The reverse micelles work as a nanoreactor in their core, producing nanoparticles. For the synthesis of nanomaterials, it functions as a water reservoir. By altering the water/surfactant ratio, the particle size of those nanoreactors, which in turn affects the size of the nanoparticles produced by this process, may be changed [83].

2.4 Bottom-Up Strategies for Synthesis of 2D Nanomaterial

It is especially crucial to have the ability to create thin 2D nanomaterials with a variety of compositions, sizes, thicknesses, crystal phases, and surface features for the ongoing investigation of their distinct physical, chemical, and electrical properties as well as the wide variety of possible applications [84].

The many bottom-up methods for producing 2D NMs are covered in this chapter. Alagh *et al.* were the first to disclose the straightforward fabrication of 2D layered WS_2 nanosheets stacked atop 1D WS_2 nanostructures utilizing the aerosol-aided CVD method (AA-CVD) and H_2-free atmospheric pressure CVD. Utilizing AACVD Figure 2.5a and tungsten hexacarbonyl W(CO)6 (50 mg, purity 97%), which has been dissolved in a 3:1 mixture of acetone and methanol, WO_3 nanoneedles (NNs) are first produced. Figure 2.5b shows how the second stage of the procedure involved sulfurizing sheets of previously created WO3 nanomaterial to create WS_2 [85].

It is still challenging to precisely control the shape and quality of 2D materials using CVD technologies because of our poor understanding of the underlying growth principles of these materials. The single-domain net growth rate (SD-NGR), which is a crucial factor in determining the configuration as well as the quality of 2D-material crystals, is further shown using the notion of 2D diffusion-limited aggregation (2D-DLA). A number of improved single-crystal 2D materials having fractal morphologies are created by carefully altering SD-NGR, allowing for the first time precise control of 2D-material CVD growth [86]. Using inexpensive glass substrates, a CVD method was used for the direct development of a 2D material-based ReS_2@N-doped Graphene heterostructure made up of vertically erected ReS_2 nanowalled cores & NG conformal-coated shells shown in Figure 2.6a. SEM images of a bare ReS_2 sheet and a ReS_2@NG heterostructure are

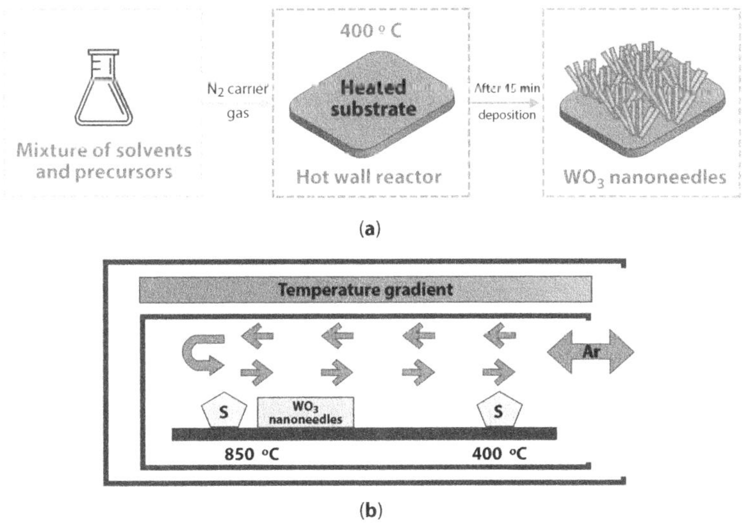

Figure 2.5 Shows the manufacture of WO_3 nanomaterials using AACVD in (a) and WS_2 nanomaterials using CVD in (b).

shown in Figures 2.6b, c. It is possible to clearly see the vertically erected nanostructure, which may provide excellent surface utilization and simple mass transfer in LiPS adsorption as well as electrocatalysis [87]. An improved CVD reactor with graphene and cold walls. The technology's benefits include the simple design of the reactor, the quick deposition time, and the low-cost production of graphene due to the high-efficiency heating provided by a cold-wall CVD system. When compared to a traditional CVD system, that represents an energy reduction of more than 90% [88]. *Ex situ* and *in situ* CVD were both used to produce graphene and hBN heterostructures with in-plane and vertical stacking. It is anticipated that these perfectly interwoven hybrid atomically layered graphene with hBN heterostructures will display fascinating electrical characteristics and uses due to the extremely small lattice misfit and the similar crystal structure. Using *in situ* CVD or a combination of *in situ* and *ex situ* CVD, the crisp interface of in-plane graphene having hBN heterostructures with tailored structures and characteristics may be produced [89].

According to Ji *et al.*, graphene was created utilizing the CVD method with ethanol as the carbon source. Characterizations of the surface and structure

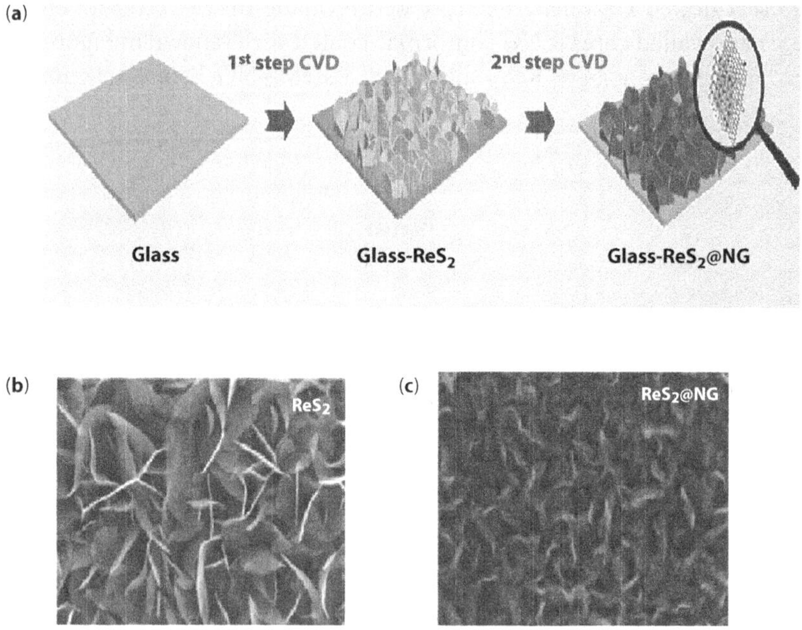

Figure 2.6 Schematic representation of the two-step CVD procedure used to produce ReS_2@NG heterostructures *in situ* is shown in (a). Vertically erected ReS_2 and ReS_2@NG nanosheets are depicted in (b, c) SEM pictures.

of CVD-grown films point to the production of double-layer graphene [90]. Sol–gel synthesis of NMs is the most popular bottom-up method due to its ease of use. Using the sol–gel method and heating at 500 °C (in which x 14 = 0, 5, 10, 20, 30 mol%), we created thin films of CeO_2-xZrO_2. Figure 2.7 shows the X-ray patterns for thin films with a long signal accumulation time at the location (where the strongest reflex (111) is located) along the angular range of 2Theta = 27–31o (b). According to a monotonic widening of the reflex (111), as there is more ZrO_2 in the system, a solid CeO_2-xZrO_2 solution develops into the film structure similarly to powders. It should be emphasized, too, that for thin films with x > 510, the reflex intensity was so weak that it was difficult to determine its location quantitatively and draw the appropriate conclusions [91]. A unique two-step surfactant sol–gel technique is employed to generate Fe_3C/few-layered graphene core/shell nanoparticles embedded inside a carbon matrix, with the hydrolysis, polycondensation, and drying processes all occurring in the same pot [92]. Wang *et al.* created MXene/amarid nanofiber (ANF) nanocomposite papers on a large scale using a blade-coating approach and a sol–gel conversion step [93]. The creation of novel adsorbents based upon anionic silica precursors, cellulose nanofiber (CNF), as well as graphene oxide (GO), has been generated through an ionic reaction accompanied by a sol–gel technique [94]. A molybdenum oxide/reduced graphene oxide (MoO_3/rGO) nanocomposite was created by employing the sol–gel method, accompanied by the ultrasound method, in the presence of a zwitterionic surfactant, myristylsulfobetaine (SB3-14), at the critical micelle concentration. Li4Ti_5O_{12}@C nanocomposite (SG-T@C)

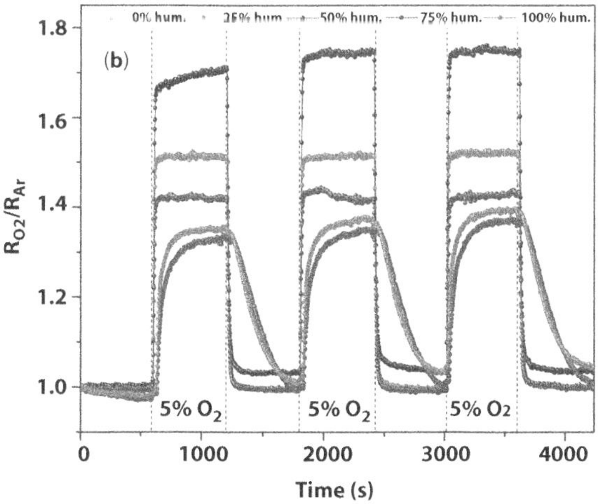

Figure 2.7 Demonstrates thin films of CeO_2-xZrO_2 with their X-ray patterns. (b), in which x = 0, 5, 10, 20, 30, and 50 mol%.

with sulfur-doped graphene modifications was initially made using a sol–gel approach supplemented by a solid-state procedure. The $Li_4Ti_5O_{12}$@C nanocrystals (T@C) are uniformly distributed throughout the 2D sulfur-doped grapheme in this composite [95].

Both hydrothermal and solvothermal techniques are frequently used in the synthesis of different nanomaterials because of their easy setup and operation [96–98]. Typically, the synthesis is carried out above ambient pressure and temperature in Teflon-lined stainless-steel autoclaves utilizing hydrothermal or solvothermal methods. Nanosheets made of Rh [99, 100], Co nanoplatelets [101], nanosheets of Co [102], Ni nanoplatelets [101, 103], and nanobelts of Ru nanoplates as well as Cu nanoplatelets [104].

Li *et al.* reported the formation of Co nanoplatelets using a solution of water including CoCl2, sodium dodecyl sulfate, sodium hydroxide, and sodium hypophosphite monohydrate over 24 h at 110 °C. This technique may also be employed to make both nickel and copper nanoplatelets by simply altering the temperature and duration of the reaction [101]. A simple process for polymerizing aniline utilizing a soft template and the hydrothermal method to build well-aligned PANI nanostructures on reduced graphene oxide nanosheets (SPANI/S-rGO), where naphthalene sulfonic acid is employed as a dopant for both the soft template and the PANI [105]. With the use of a sodium hydroxide soft template, Lertthanaphol *et al.* reported the heterostructure of copper-doped sodium dititanate nanosheets with graphene oxide (CTGN) after a one-step hydrothermal method. The template's function in this scenario is to aid in the production of the nanosheets, resulting in the 2D-2D heterostructure of nanosheets and graphene [106].

The solvothermal procedure makes use of nonaqueous solvents like benzyl alcohol, formamide, ethylene glycol, and DMF. Ag, Rh, Ir, and Ru are the monometallic nanosheets that have so far been created using this process [107–112]. For instance, Li *et al.* produced single-atomic-layered Rh nanosheets from $Rh(acac)_3$ at 180 °C for 8 h while using PVP, formaldehyde, and benzyl alcohol. The edge length of these nanosheets was approximately 500–600 nm. By using a two-step solvothermal process, PtCu nanocones and ultrathin PtCu alloy nanosheets having a thickness of 4–6 atomic layers were synthesized under controlled conditions, according to Wang *et al.* [113].

To build the 2D-2D nanochannel membranes depicted in Figure 2.8, Yu *et al.* coupled rGO with mesoporous TiO_2 nanosheets using a solvothermal-induced assembly approach. Materials made of 2D TiO_2 and 2D-2D rGO-TiO_2 are first described to explore their potential characteristics. By using SEM and TEM analysis, the shape and microscopic structure of the

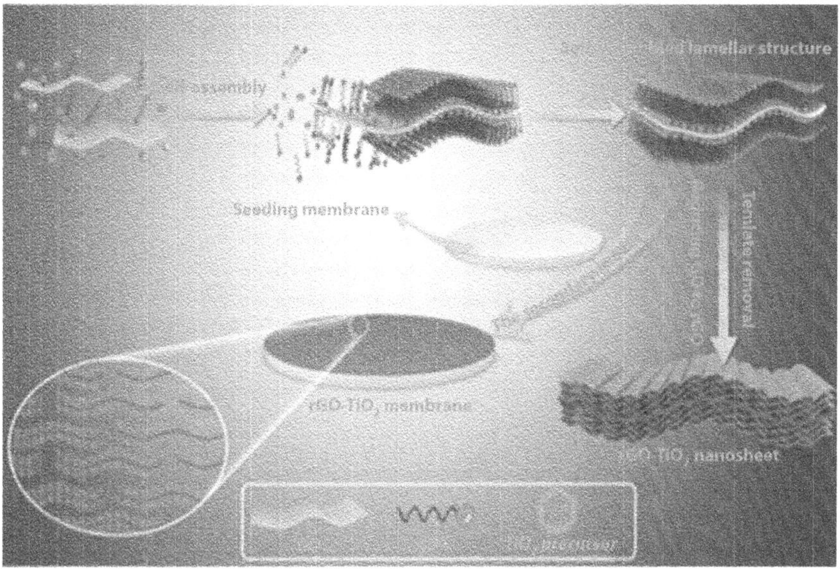

Figure 2.8 rGO-TiO_2 membrane self-assembly is shown schematically.

generated bare TiO_2 and rGO-TiO2 nanosheets have been intuitively evaluated. In the SEM image Figure 2.9a, a typical "sheet-like" structure for TiO2 with restacking is readily visible. The generated nanostructures were layer-defined nanosheets, and a TEM image of the monodisperse TiO2 materials showed that their edges curled up due to increased surface tension (Figure 9c). When interacting with rGO sheets, the sample of rGO-TiO2 shows a characteristic laminar structure on a micrometer-lateral scale (Figure 2.9b). According to interfacial contact, the sheet-like TiO2 or rGO nanosheets were incorporated in the 2D-2D rGO-TiO2 composite edge region in the TEM picture of rGO-TiO2 nanosheets shown in Figure 2.9d [114].

The ultrathin iron-cobalt oxide nanosheets (FeCo-ON 1.2 nm) are one example of an amorphous metal oxide nanosheet (M-ONS) that the Zhuang research group reported using as sacrificial templates to create MOF-74 nanosheets (M-MNS) *via* restricted ligand coordination. It is presumable that under hydrothermal circumstances, acidic ligand solutions could achieve the regulated leaching of metal ions from MONS and result in the loading of metal ions on the M-ONS surface, facilitating coordination reactions. The presence of more metal ions close to a metal oxide surface may cause MOF crystals to grow constrainedly into a 2D shape [115].

By using a soft-template technique, graphene was sandwiched between mesoporous silica to produce unfolded graphene/SiO_2 nanosheets (rGOSs) [116]. Sulfated lanthanum oxy-carbonate nanorods (S-$La_2O_2(CO_3)$ NRs)

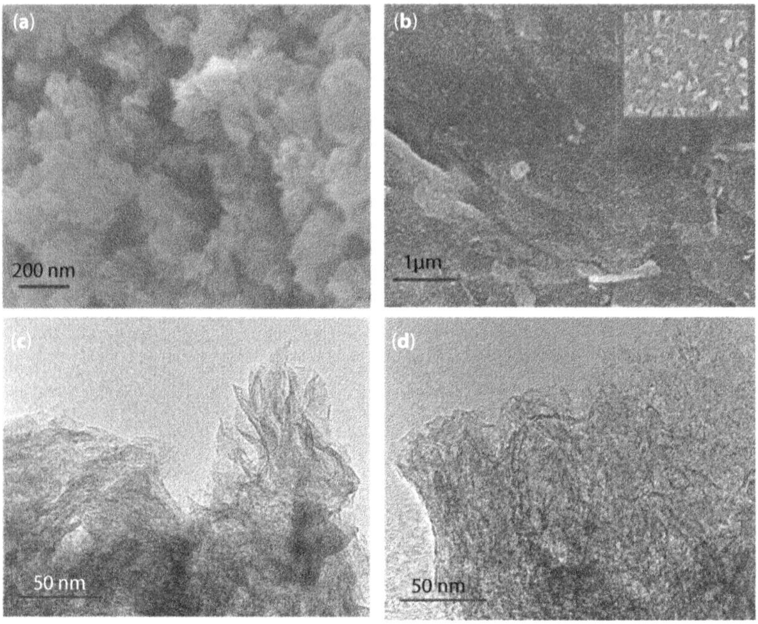

Figure 2.9 Shows TEM pictures of 2D TiO_2 (c) and 2D-2D rGO-TiO_2 (d) nanosheets, as well as SEM images of blank TiO_2 (a) and 2D-2D rGO-TiO_2 nanosheets (b).

were made using the reverse micelle microemulsion method, and they were then utilized to modify graphene oxide nanosheets to generate a graphene oxide-sulfated lanthanum oxy-carbonate (GO-S-$La_2O_2(CO_3)$) nanocomposite [117]. Cu-MOF was developed using a 1:1 molar ratio of $Cu(NO_3)_2{\bullet}3H_2O$ and 2,6-pyridine dicarboxylic acid using the ultrasonic-aided reverse micelle method (UARM) [118]. To create a variety of cobalt-based substances that have various anions and a similar ultrathin 2D structure, a self-assembly-triggered reverse micelle templating approach is used [119].

2.5 Conclusion and Outlook

The study of 2D NMs has seen rapid advancement in the last ten years, from basic research to the creation of cutting-edge technologies. This chapter covered the recent advancements in 2D NM synthesis using bottom-up methods. There are numerous synthetic techniques that can be used to create these 2D NMs, each of which has benefits and drawbacks. 2D NMs with various structural characteristics, including size, thickness, crystallinity, and surface properties, can be created using a variety of synthetic techniques. For the design of next-generation technology, the study of 2D NMs, particularly their sophisticated synthesis techniques, has received significant attention in recent years. This chapter covers a variety of industrial criteria that have not yet been met by the 2D NMs existing production yield, rate, and quantity. A regulated synthesis strategy is required to be created for 2D NMs because a material's structural features typically affect its chemical and physical characteristics and consequently, their suitability for a given application. Another difficulty is the creation of efficient characterization techniques for comprehending the growth mechanisms of 2D NMs.

Moreover, research must concentrate on enhancing the stability and longevity of 2D NMs, as these properties have a significant influence on prospective applications. The main cause of this instability is the present synthesis methods and procedures. Utilizing novel technologies for controlled synthesis methods is necessary to solve instabilities, including structure deterioration and change, irreversible aggregation, etc., for the purpose of stabilizing 2D NMs. In this area, there is still plenty to be discovered. Moreover, despite the development of numerous synthesis techniques, there are still no efficient means for achieving mass manufacture of controlled quality, which delays the use of most 2DNMs in practical applications.

References

1. Nath, N., Kumar, A., Chakroborty, S., Soren, S., Barik, A., Pal, K., de Souza Jr., F.G., Carbon nanostructure embedded novel sensor implementation for detection of aromatic volatile organic compounds: An organized review. *ACS Omega*, 8, 5, 4436–4452, 2023.
2. Novoselov, K.S., Geim, A.K., Morozov, S.V., Jiang, D.E., Zhang, Y., Dubonos, S.V., Grigorieva, I.V., Firsov, A.A., Electric field effect in atomically thin carbon films. *Science*, 306, 5696, 666–669, 2004.
3. Lin, Y.M., Dimitrakopoulos, C., Jenkins, K.A., Farmer, D.B., Chiu, H.Y., Grill, A., Avouris, P., 100-GHz transistors from wafer-scale epitaxial graphene. *Science*, 327, 5966, 662–662, 2010.
4. Liu, M., Yin, X., Ulin-Avila, E., Geng, B., Zentgraf, T., Ju, L., Wang, F., Zhang, X., A graphene-based broadband optical modulator. *Nature*, 474, 7349, 64–67, 2011.
5. Wu, Z.S., Parvez, K., Feng, X., Mullen, K., Graphene-based in-plane micro-supercapacitors with high power and energy densities. *Nat. Commun.*, 4, 2487, 2013.

6. Wu, Z.S., Parvez, K., Winter, A., Vieker, H., Liu, X., Han, S., Turchanin, A., Feng, X., Müllen, K., Layer-by-layer assembled heteroatom-doped graphene films with ultrahigh volumetric capacitance and rate capability for micro-supercapacitors. *Adv. Mater.*, 26, 26, 4552–4558, 2014.
7. Zhu, Y., Murali, S., Stoller, M.D., Ganesh, K.J., Cai, W., Ferreira, P.J., Pirkle, A., Wallace, R.M., Cychosz, K.A., Thommes, M., Su, D., Carbon-based supercapacitors produced by activation of graphene. *Science*, 332, 6037, 1537–1541, 2011.
8. Wu, Z.S., Zheng, Y., Zheng, S., Wang, S., Sun, C., Parvez, K., Ikeda, T., Bao, X., Müllen, K., Feng, X., Stacked-layer heterostructure films of 2D thiophene nanosheets and graphene for high-rate all-solid-state pseudocapacitors with enhanced volumetric capacitance. *Adv. Mater.*, 29, 3, 1602960, 2017.
9. Deng, M., Yang, X., Silke, M., Qiu, W., Xu, M., Borghs, G., Chen, H., Electrochemical deposition of polypyrrole/graphene oxide composite on microelectrodes towards tuning the electrochemical properties of neural probes. *Sens. Actuators B. Chem.*, 158, 1, 176–184, 2011.
10. Xu, M., Fujita, D., Hanagata, N., Perspectives and challenges of emerging single-molecule DNA sequencing technologies. *Small*, 5, 23, 2638–2649, 2009.
11. Garaj, S., Hubbard, W., Reina, A., Kong, J., Branton, D., Golovchenko, J.A., Graphene as a subnanometre trans-electrode membrane. *Nature*, 467, 7312, 190–193, 2010.
12. Parvez, K., Two-dimensional nanomaterials: Crystal structure and synthesis, in: *Biomedical Applications of Graphene and 2D Nanomaterials*, pp. 1–25, Elsevier, 2019.
13. Lu, L., Yang, Q., Xu, Q., Sun, Y., Tang, S., Tang, X., Liang, H., Yu, Y., Two-dimensional materials beyond graphene for the detection and removal of antibiotics: A critical review. *Crit. Rev. Environ. Sci. Technol.*, 52, 19, 3493–3524, 2022.
14. Kong, X., Liu, Q., Zhang, C., Peng, Z., Chen, Q., Elemental two-dimensional nanosheets beyond graphene. *Chem. Soc. Rev.*, 46, 8, 2127–2157, 2017.
15. Zhang, Y., Rubio, A., Le Lay, G., Emergent elemental two-dimensional materials beyond graphene. *J. Phys. D Appl. Phys.*, 50, 5, 05300450, 2017.
16. Tsai, H.S. and Liang, J.H., Production and potential applications of elemental two-dimensional materials beyond graphene. *Chem. Nano Mater.*, 3, 9, 604–613, 2017.
17. Takeda, K. and Shiraishi, K., Theoretical possibility of stage corrugation in Si and Ge analogs of graphite. *Phys. Rev. B*, 50, 20, 14916, 1994.
18. Cahangirov, S., Topsakal, M., Aktürk, E., Şahin, H., Ciraci, S., Two-and one-dimensional honeycomb structures of silicon and germanium. *Phys. Rev. Lett.*, 102, 23, 236804, 2009.
19. Dimoulas, A., Silicene and germanene: Silicon and germanium in the "flatland". *Microelectron. Eng.*, 131, 68–78, 2015.

20. Dhakal, C., Aryal, S., Sakidja, R., Ching, W.Y., Approximate lattice thermal conductivity of MAX phases at high temperature. *J. Eur. Ceram. Soc.*, 35, 12, 3203–3212, 2015.
21. Naguib, M., Mochalin, V.N., Barsoum, M.W., Gogotsi, Y., 25th anniversary article: MXenes: A new family of two-dimensional materials. *Adv. Mater.*, 26, 7, 992–1005, 2014.
22. Ling, X., Wang, H., Huang, S., Xia, F., Dresselhaus, M.S., The renaissance of black phosphorus. *Proc. Natl. Acad. Sci.*, 112, 15, 4523–4530, 2015.
23. Zhang, X., Xie, H., Liu, Z., Tan, C., Luo, Z., Li, H., Lin, J., Sun, L., Chen, W., Xu, Z., Xie, L., Black phosphorus quantum dots. *Angew. Chem. Int. Ed.*, 54, 12, 3653–3657, 2015.
24. Qiao, J., Kong, X., Hu, Z.X., Yang, F., Ji, W., High-mobility transport anisotropy and linear dichroism in few-layer black phosphorus. *Nat. Commun.*, 5, 1, 4475, 2014.
25. Liu, H., Neal, A.T., Zhu, Z., Luo, Z., Xu, X., Tománek, D., Ye, P.D., Phosphorene: An unexplored 2D semiconductor with a high hole mobility. *ACS Nano*, 8, 4, 4033–4041, 2014.
26. Ba, K., Jiang, W., Cheng, J., Bao, J., Xuan, N., Sun, Y., Liu, B., Xie, A., Wu, S., Sun, Z., Chemical and bandgap engineering in monolayer hexagonal boron nitride. *Sci. Rep.*, 7, 1, 45584, 2017.
27. Ci, L., Song, L., Jin, C., Jariwala, D., Wu, D., Li, Y., Srivastava, A., Wang, Z.F., Storr, K., Balicas, L., Liu, F., Atomic layers of hybridized boron nitride and graphene domains. *Nat. Mater.*, 9, 5, 430–435, 2010.
28. Dean, C.R., Young, A.F., Meric, I., Lee, C., Wang, L., Sorgenfrei, S., Watanabe, K., Taniguchi, T., Kim, P., Shepard, K.L., Hone, J., Boron nitride substrates for high-quality graphene electronics. *Nat. Nanotechnol.*, 5, 10, 722–726, 2010.
29. Mayorov, A.S., Gorbachev, R.V., Morozov, S.V., Britnell, L., Jalil, R., Ponomarenko, L.A., Blake, P., Novoselov, K.S., Watanabe, K., Taniguchi, T., Geim, A.K., Micrometer-scale ballistic transport in encapsulated graphene at room temperature. *Nano Lett.*, 11, 6, 2396–2399, 2011.
30. Jang, S.K., Youn, J., Song, Y.J., Lee, S., Synthesis and characterization of hexagonal boron nitride as a gate dielectric. *Sci. Rep.*, 6, 1, 30449, 2016.
31. Zhang, Y.J., Yoshida, M., Suzuki, R., Iwasa, Y., 2D crystals of transition metal dichalcogenide and their iontronic functionalities. *2D Mater.*, 2, 4, 044004, 2015.
32. Mak, K.F., Lee, C., Hone, J., Shan, J., Heinz, T.F., Atomically thin MoS2: a new direct-gap semiconductor. *Phys. Rev. Lett.*, 105, 13, 136805, 2010.
33. Mak, K.F., McGill, K.L., Park, J., McEuen, P.L., The valley Hall effect in MoS2 transistors. *Science*, 344, 6191, 1489–1492, 2014.
34. Zeng, H., Dai, J., Yao, W., Xiao, D., Cui, X., Valley polarization in MoS2 monolayers by optical pumping. *Nat. Nanotechnol.*, 7, 8, 490–493, 2012.

35. Ye, J.T., Zhang, Y.J., Akashi, R., Bahramy, M.S., Arita, R., Iwasa, Y., Superconducting dome in a gate-tuned band insulator. *Science*, 338, 6111, 1193–1196, 2012.
36. Zheng, Y., Liu, J., Liang, J., Jaroniec, M., Qiao, S.Z., Graphitic carbon nitride materials: Controllable synthesis and applications in fuel cells and photocatalysis. *Energy Environ. Sci.*, 5, 5, 6717–6731, 2012.
37. Furukawa, H., Cordova, K.E., O'Keeffe, M., Yaghi, O.M., The chemistry and applications of metal-organic frameworks. *Science*, 341, 6149, 1230444, 2013.
38. Peng, Y., Li, Y., Ban, Y., Yang, W., Two-dimensional metal–organic framework nanosheets for membrane-based gas separation. *Angew. Chem. Int. Ed.*, 129, 33, 9889–9893, 2017.
39. Kang, Y.S., Lu, Y., Chen, K., Zhao, Y., Wang, P., Sun, W.Y., Metal–organic frameworks with catalytic centers: From synthesis to catalytic application. *Coord. Chem. Rev.*, 378, 262–280, 2019.
40. Simon-Yarza, T., Mielcarek, A., Couvreur, P., Serre, C., Nanoparticles of metal-organic frameworks: On the road to *in vivo* efficacy in biomedicine. *Adv. Mater.*, 30, 37, 1707365, 2018.
41. Zhao, M., Huang, Y., Peng, Y., Huang, Z., Ma, Q., Zhang, H., Two-dimensional metal–organic framework nanosheets: Synthesis and applications. *Chem. Soc. Rev.*, 47, 16, 6267–6295, 2018.
42. Zang, Y., Pei, F., Huang, J., Fu, Z., Xu, G., Fang, X., Large-area preparation of crack-free crystalline microporous conductive membrane to upgrade high energy lithium–sulfur batteries. *Adv. Energy Mater.*, 8, 31, 1802052, 2018.
43. Diercks, C.S. and Yaghi, O.M., The atom, the molecule, and the covalent organic framework. *Science*, 355, 6328, eaal1585, 2017.
44. Jin, E., Asada, M., Xu, Q., Dalapati, S., Addicoat, M.A., Brady, M.A., Xu, H., Nakamura, T., Heine, T., Chen, Q., Jiang, D., Two-dimensional sp2 carbon–conjugated covalent organic frameworks. *Science*, 357, 6352, 673–676, 2017.
45. Li, G., Zhang, K., Tsuru, T., Two-dimensional covalent organic framework (COF) membranes fabricated via the assembly of exfoliated COF nanosheets. *ACS Appl. Mater. Interfaces*, 9, 10, 8433–8436, 2017.
46. Chen, X., Li, Y., Wang, L., Xu, Y., Nie, A., Li, Q., Wu, F., Sun, W., Zhang, X., Vajtai, R., Ajayan, P.M., High-lithium-affinity chemically exfoliated 2D covalent organic frameworks. *Adv. Mater.*, 31, 29, 1901640, 2019.
47. Khanna, P., Kaur, A., Goyal, D., Algae-based metallic nanoparticles: Synthesis, characterization and applications. *J. Microbiol. Methods*, 163, 105656, 2019.
48. Zhuang, S., Lee, E.S., Lei, L., Nunna, B.B., Kuang, L., Zhang, W., Synthesis of nitrogen-doped graphene catalyst by high-energy wet ball milling for electrochemical systems. *Int. J. Energy Res.*, 40, 15, 2136–2149, 2016.
49. Yadav, T.P., Yadav, R.M., Singh, D.P., Mechanical milling: A top down approach for the synthesis of nanomaterials and nanocomposites. *Nanosci. Nanotechnol.*, 2, 3, 22–48, 2012.

50. Lyu, H., Gao, B., He, F., Ding, C., Tang, J., Crittenden, J.C., Ball-milled carbon nanomaterials for energy and environmental applications. *ACS Sustain. Chem. Eng.*, 5, 11, 9568–9585, 2017.
51. Ostermann, R., Cravillon, J., Weidmann, C., Wiebcke, M., Smarsly, B.M., Metal–organic framework nanofibers via electrospinning. *Chem. Commun.*, 47, 1, 442–444, 2011.
52. Kumar, P.S., Sundaramurthy, J., Sundarrajan, S., Babu, V.J., Singh, G., Allakhverdiev, S.I., Ramakrishna, S., Hierarchical electrospun nanofibers for energy harvesting, production and environmental remediation. *Energy Environ. Sci.*, 7, 10, 3192–3222, 2014.
53. Pimpin, A. and Srituravanich, W., Review on micro-and nanolithography techniques and their applications. *Eng. J.*, 16, 1, 37–56, 2012.
54. Ayyub, P., Chandra, R., Taneja, P., Sharma, A.K., Pinto, R., Synthesis of nanocrystalline material by sputtering and laser ablation at low temperatures. *Appl. Phys. A Mater. Sci. Process.*, 73, 67–73, 2001.
55. Nie, M., Sun, K., Meng, D.D., Formation of metal nanoparticles by short-distance sputter deposition in a reactive ion etching chamber. *J. Appl. Phys.*, 106, 5, 054314, 2009.
56. Zhang, D., Ye, K., Yao, Y., Liang, F., Qu, T., Ma, W., Yang, B., Dai, Y., Watanabe, T., Controllable synthesis of carbon nanomaterials by direct current arc discharge from the inner wall of the chamber. *Carbon*, 142, 278–284, 2019.
57. Wu, Z.S., Ren, W., Gao, L., Zhao, J., Chen, Z., Liu, B., Tang, D., Yu, B., Jiang, C., Cheng, H.M., Synthesis of graphene sheets with high electrical conductivity and good thermal stability by hydrogen arc discharge exfoliation. *ACS Nano*, 3, 2, 411–417, 2009.
58. Amendola, V. and Meneghetti, M., Laser ablation synthesis in solution and size manipulation of noble metal nanoparticles. *Phys. Chem. Chem. Phys.*, 11, 20, 3805–3821, 2009.
59. Zhang, J., Chaker, M., Ma, D., Pulsed laser ablation based synthesis of colloidal metal nanoparticles for catalytic applications. *J. Colloid Interface Sci.*, 489, 138–149, 2017.
60. Ismail, R.A., Mohsin, M.H., Ali, A.K., Hassoon, K.I., Erten-Ela, S., Preparation and characterization of carbon nanotubes by pulsed laser ablation in water for optoelectronic application. *Phys. E*, 119, 113997, 2020.
61. Chrzanowska, J., Hoffman, J., Małolepszy, A., Mazurkiewicz, M., Kowalewski, T.A., Szymanski, Z., Stobinski, L., Synthesis of carbon nanotubes by the laser ablation method: Effect of laser wavelength. *Phys. Status Solidi, 252(8)*, 1860-1867, 2015, 2015.
62. Duque, J.S., Madrigal, B.M., Riascos, H., Avila, Y.P., Colloidal metal oxide nanoparticles prepared by laser ablation technique and their antibacterial test. *Colloids Interfaces*, 3, 1, 25, 2019.
63. Su, S.S. and Chang, I., *Commercialization of Nanotechnologies–A Case Study Approach*, pp. 15–29, Springer International Publishing, Cham, 2018.

64. Jones, A.C. and Hitchman, M.L., Overview of chemical vapour deposition, in: *Chemical Vapour Deposition: Precursors, Processes and Applications*, vol. 1, pp. 1–36, 2009.
65. Shah, K.A. and Tali, B.A., Materials science in semiconductor processing synthesis of carbon nanotubes by catalytic chemical vapour deposition: a review on carbon sources, catalysts and substrates. *Mater. Sci. Semicond. Process.*, 41, 67–82, 2016.
66. Machac, P., Cichon, S., Lapcak, L., Fekete, L., Graphene prepared by chemical vapour deposition process. *Graphene Technol.*, 5, 9–17, 2020.
67. Wu, Q., Wongwiriyapan, W., Park, J.H., Park, S., Jung, S.J., Jeong, T., Lee, S., Lee, Y.H., Song, Y.J., *In situ* chemical vapor deposition of graphene and hexagonal boron nitride heterostructures. *Curr. Appl. Phys.*, 16, 9, 1175–1191, 2016.
68. Abid, N., Khan, A.M., Shujait, S., Chaudhary, K., Ikram, M., Imran, M., Haider, J., Khan, M., Khan, Q., Maqbool, M., Synthesis of nanomaterials using various top-down and bottom-up approaches, influencing factors, advantages, and disadvantages: A review. *Adv. Colloid Interface Sci.*, 300, 102597, 2022.
69. Danks, A.E., Hall, S.R., Schnepp, Z.J.M.H., The evolution of 'sol–gel' chemistry as a technique for materials synthesis. *Mater. Horiz.*, 3, 2, 91–112, 2016.
70. de Coelho Escobar, C. and dos Santos, J.H.Z., Effect of the sol–gel route on the textural characteristics of silica imprinted with Rhodamine B. *J. Sep. Sci.*, 37, 7, 868–875, 2014.
71. Parashar, M., Shukla, V.K., Singh, R., Metal oxides nanoparticles via sol–gel method: A review on synthesis, characterization and applications. *J. Mater. Sci. Mater. Electron.*, 31, 3729–3749, 2020.
72. Znaidi, L., Sol–gel-deposited ZnO thin films: A review. *Mater. Sci. Eng. B*, 174, 1-3, 18–30, 2010.
73. Wu, X., Lu, G.Q.M., Wang, L., Shell-in-shell TiO2 hollow spheres synthesized by one-pot hydrothermal method for dye-sensitized solar cell application. *Energy Environ. Sci.*, 4, 9, 3565–3572, 2011.
74. Cao, S., Zhao, C., Han, T., Peng, L., Hydrothermal synthesis, characterization and gas sensing properties of the WO3 nanofibers. *Mater. Lett.*, 169, 17–20, 2016.
75. Chen, A. and Holt-Hindle, P., Platinum-based nanostructured materials: Synthesis, properties, and applications. *Chem. Rev.*, 110, 6, 3767–3804, 2010.
76. Meng, L.Y., Wang, B., Ma, M.G., Lin, K.L., The progress of microwave-assisted hydrothermal method in the synthesis of functional nanomaterials. *Mater. Today Chem.*, 1, 63–83, 2016.
77. Dong, Y., Du, X.Q., Liang, P., Man, X.L., One-pot solvothermal method to fabricate 1D-VS4 nanowires as anode materials for lithium ion batteries. *Inorg. Chem. Commun.*, 115, 107883, 2020.

78. Jiang, Y., Peng, Z., Zhang, S., Li, F., Liu, Z., Zhang, J., Liu, Y., Wang, K., Facile *in-situ* Solvothermal Method to synthesize double shell ZnIn2S4 nanosheets/TiO2 hollow nanosphere with enhanced photocatalytic activities. *Ceram. Int.*, 44, 6, 6115–6126, 2018.
79. Chai, B., Xu, M., Yan, J., Ren, Z., Remarkably enhanced photocatalytic hydrogen evolution over MoS2 nanosheets loaded on uniform CdS nanospheres. *Appl. Surf. Sci.*, 430, 523–530, 2018.
80. Liu, Y., Goebl, J., Yin, Y., Templated synthesis of nanostructured materials. *Chem. Soc. Rev.*, 42, 7, 2610–2653, 2013.
81. Li, W. and Zhao, D., An overview of the synthesis of ordered mesoporous materials. *Chem. Commun.*, 49, 10, 943–946, 2013.
82. Malik, M.A., Wani, M.Y., Hashim, M.A., Microemulsion method: A novel route to synthesize organic and inorganic nanomaterials: 1st nano update. *Arab. J. Chem.*, 5, 4, 397–417, 2012.
83. Shiri, M.S.Z., Henderson, W., Mucalo, M.R., A review of the lesser-studied microemulsion-based synthesis methodologies used for preparing nanoparticle systems of the noble metals, Os, Re, Ir and Rh. *Materials*, 12, 12, 1896, 2019.
84. Butler, S.Z., Hollen, S.M., Cao, L., Cui, Y., Gupta, J.A., Gutiérrez, H.R., Heinz, T.F., Hong, S.S., Huang, J., Ismach, A.F., Johnston-Halperin, E., Progress, challenges, and opportunities in two-dimensional materials beyond graphene. *ACS Nano*, 7, 4, 2898–2926, 2013.
85. Alagh, A., Annanouch, F.E., Umek, P., Bittencourt, C., Sierra-Castillo, A., Haye, E., Colomer, J.F., Llobet, E., CVD growth of self-assembled 2D and 1D WS2 nanomaterials for the ultrasensitive detection of NO2. *Sens. Actuators B Chem.*, 326, 128813, 2021.
86. Li, J., Chen, M., Zhang, C., Dong, H., Lin, W., Zhuang, P., Wen, Y., Tian, B., Cai, W., Zhang, X., Fractal-theory-based control of the shape and quality of CVD-grown 2D materials. *Adv. Mater.*, 31, 35, 1902431, 2019.
87. Wei, N., Cai, J., Wang, R., Wang, M., Lv, W., Ci, H., Sun, J., Liu, Z., Elevated polysulfide regulation by an ultralight all-CVD-built ReS2@ N-Doped graphene heterostructure interlayer for lithium–sulfur batteries. *Nano Energy*, 66, 104190, 2019.
88. Machac, P., Cichon, S., Lapcak, L., Fekete, L., Graphene prepared by chemical vapour deposition process. *Graphene Technol.*, 5, 9–17, 2020.
89. Wu, Q., Wongwiriyapan, W., Park, J.H., Park, S., Jung, S.J., Jeong, T., Lee, S., Lee, Y.H., Song, Y.J., *In situ* chemical vapor deposition of graphene and hexagonal boron nitride heterostructures. *Curr. Appl. Phys.*, 16, 9, 1175–1191, 2016.
90. Ji, D., Wen, X., Foller, T., You, Y., Wang, F., Joshi, R., Chemical vapour deposition of graphene for durable anticorrosive coating on copper. *Nanomaterials*, 10, 12, 2511, 2020.

91. Mokrushin, A.S., Simonenko, E.P., Simonenko, N.P., Bukunov, K.A., Sevastyanov, V.G., Kuznetsov, N.T., Gas-sensing properties of nanostructured CeO2-xZrO2 thin films obtained by the sol-gel method. *J. Alloys Compd.*, 773, 1023–1032, 2019.
92. Castellano-Soria, A., López-Sánchez, J., Granados-Miralles, C., Varela, M., Navarro, E., González, C., Marín, P., Novel one-pot sol-gel synthesis route of Fe3C/few-layered graphene core/shell nanoparticles embedded in a carbon matrix. *J. Alloys Compd.*, 902, 163662, 2022.
93. Liu, S., Pan, X., Liu, H., Two-dimensional nanomaterials for photothermal therapy. *Angew. Chem.*, 132, 15, 5943–5953, 2020.
94. Fishman, Z.S., He, Y., Yang, K.R., Lounsbury, A.W., Zhu, J., Tran, T.M., Zimmerman, J.B., Batista, V.S., Pfefferle, L.D., Hard templating ultrathin polycrystalline hematite nanosheets: Effect of nano-dimension on CO 2 to CO conversion via the reverse water-gas shift reaction. *Nanoscale*, 9, 35, 12984–12995, 2017.
95. Savic, S., Vojisavljevic, K., Nešić, M.P., Zivojevic, K., Mladenovic, M., Knezevic, N., Hard template synthesis of nanomaterials based on mesoporous silica. *Metall. Mater. Eng.*, 24, 4, 225–241, 2018.
96. Feng, S. and Xu, R., New materials in hydrothermal synthesis. *Acc. Chem. Res.*, 34, 239–247, 2001.
97. Lai, J., Niu, W., Luque, R., Xu, G., Solvothermal synthesis of metal nanocrystals and their applications. *Nano Today*, 10, 240–267, 2015.
98. Xu, B. and Wang, X., Solvothermal synthesis of monodisperse nanocrystals. *Dalton Trans.*, 41, 4719–4725, 2012.
99. Jiang, Y., Su, J., Yang, Y., Jia, Y., Chen, Q., Xie, Z., Zheng, L., A facile surfactant-free synthesis of rh flower-like nanostructures constructed from ultrathin nanosheets and their enhanced catalytic properties. *Nano Res.*, 9, 849–856, 2016.
100. Bai, J., Xu, G.R., Xing, S.H., Zeng, J.H., Jiang, J.X., Chen, Y., Hydrothermal synthesis and catalytic application of ultrathin rhodium nanosheet nanoassemblies. *ACS Appl. Mater. Interfaces*, 8, 33635–33641, 2016.
101. Xu, R., Xie, T., Zhao, Y., Li, Y., Single-crystal metal nanoplatelets: Cobalt, nickel, copper, and silver. *Cryst. Growth Des.*, 7, 1904–1911, 2007.
102. Gao, S., Lin, Y., Jiao, X., Sun, Y., Luo, Q., Zhang, W., Li, D., Yang, J., Xie, Y., Partially oxidized atomic cobalt layers for carbon dioxide electroreduction to liquid fuel. *Nature*, 529, 68–71, 2016.
103. Liu, Z.P., Li, S., Yang, Y., Peng, S., Hu, Z.K., Qian, Y.T., Complex- surfactant-assisted hydrothermal route to ferromagnetic nickel nanobelts. *Adv. Mater.*, 15, 1946–1948, 2003.
104. Yin, A.X., Liu, W.C., Ke, J., Zhu, W., Gu, J., Zhang, Y.W., Yan, C.H., Ru nanocrystals with shape-dependent surface-enhanced raman spectra and catalytic properties: Controlled synthesis and DFT calculations. *J. Am. Chem. Soc.*, 134, 20479–20489, 2012.

105. Macherla, N., Singh, K., Nerella, M., Kumari, K., Reddy Lekkala, R.G., Improved performance of flexible supercapacitor using naphthalene sulfonic acid-doped polyaniline/sulfur-doped reduced graphene oxide nanocomposites. *Int. J. Energy Res.*, 46, 5, 6529–6542, 2022.
106. Lertthanaphol, N., Prawiset, N., Soontornapaluk, P., Kitjanukit, N., Neamsung, W., Pienutsa, N., Srinives, S., Soft template-assisted copper-doped sodium dititanate nanosheet/graphene oxide heterostructure for photoreduction of carbon dioxide to liquid fuels. *RSC Adv.*, 12, 37, 24362–24373, 2022.
107. Zhang, Q., Yang, Y., Li, J., Iurilli, R., Xie, S., Qin, D., Citrate-free synthesis of silver nanoplates and the mechanistic study. *ACS Appl. Mater. Interfaces*, 5, 6333–6345, 2013.
108. Yang, Y., Matsubara, S., Xiong, L., Hayakawa, T., Nogami, M., Solvothermal synthesis of multiple shapes of silver nanoparticles and their SERS properties. *J. Phys. Chem. C*, 111, 9095–9104, 2007.
109. Duan, H., Yan, N., Yu, R., Chang, C.R., Zhou, G., Hu, H.S., Rong, H., Niu, Z., Mao, J., Asakura, H., Tanaka, T., Ultrathin rhodium nanosheets. *Nat. Commun.*, 5, 3093, 2014.
110. Hou, C., Zhu, J., Liu, C., Wang, X., Kuang, Q., Zheng, L., Formaldehyde-assisted synthesis of ultrathin rh nanosheets for applications in co oxidation. *CrystEngComm*, 15, 6127–6130, 2013.
111. Pi, Y., Zhang, N., Guo, S., Guo, J., Huang, X., Ultrathin laminar Ir superstructure as highly efficient oxygen evolution electrocatalyst in broad pH range. *Nano Lett.*, 16, 4424–4430, 2016.
112. Kong, X., Xu, K., Zhang, C., Dai, J., Norooz Oliaee, S., Li, L., Zeng, X., Wu, C., Peng, Z., Free-standing two-dimensional Ru nanosheets with high activity toward water splitting. *ACS Catal.*, 6, 1487–1492, 2016.
113. Saleem, F., Zhang, Z., Xu, B., Xu, X., He, P., Wang, X., Ultrathin Pt–Cu nanosheets and nanocones. *J. Am. Chem. Soc.*, 135, 18304–18307, 2013.
114. Yu, J., Zhang, Y., Chen, J., Cui, L., Jing, W., Solvothermal-induced assembly of 2D-2D rGO-TiO2 nanocomposite for the construction of nanochannel membrane. *J. Membr. Sci.*, 600, 117870, 2020.
115. Zhuang, L., Ge, L., Liu, H., Jiang, Z., Jia, Y., Li, Z., Yang, D., Hocking, R.K., Li, M., Zhang, L., Wang, X., A surfactant-free and scalable general strategy for synthesizing ultrathin two-dimensional metal–organic framework nanosheets for the oxygen evolution reaction. *Angew. Chem.*, 131, 38, 13699–13706, 2019.
116. Wu, T., Yang, Y., Sun, W., Yang, Z., Wang, L., Wang, J., Liu, G., Unfolding graphene nanosheets towards high barrier performance of epoxy/graphene nanocomposite coating. *Compos. A Appl. Sci. Manuf.*, 153, 106732, 2022.
117. Eftekhari, M., Gheibi, M., Monhemi, H., Tabrizi, M.G., Akhondi, M., Graphene oxide-sulfated lanthanum oxy-carbonate nanocomposite as an adsorbent for the removal of malachite green from water samples with application of statistical optimization and machine learning computations. *Adv. Powder Technol.*, 33, 6, 103577, 2022.

118. Akhavan-Sigari, R., Zeraati, M., Moghaddam-Manesh, M., Kazemzadeh, P., Hosseinzadegan, S., Chauhan, N.P.S., Sargazi, G., Porous Cu-MOF nanostructures with anticancer properties prepared by a controllable ultrasound-assisted reverse micelle synthesis of Cu-MOF. *BMC Chem.*, 16, 1, 10, 2022.
119. Guo, X., Duan, M., Zhang, J., Xi, B., Li, M., Yin, R., Xiong, S., A general self-assembly induced strategy for synthesizing 2D ultrathin cobalt-based compounds toward optimizing hydrogen evolution catalysis. *Adv. Funct. Mater.*, 32, 2209397, 2022.

3

Unveiling the Intricacies: Characterization Techniques for 2D Nanomaterials

Siba Soren[1], Subhendu Chakroborty[2*], Rudra N. Purusottam[3†] and Amiya Ranjan Panda[4]

[1]*Department of Chemistry, Govt. Women's College, Baripada, Odisha, India*
[2]*Department of Basic Sciences, IITM, IES University, Bhopal, India*
[3]*Department of Chemistry, University of Miami, Coral Gables, FL, USA*
[4]*Kabi Samrat Upendra Bhanja (KSUB) College, Bhanjanagar, Ganjam, Odisha, India*

Abstract

Thin layers that may have a thickness of at least one atomic layer make up two-dimensional (2D) nanomaterials. These nanomaterials have significantly more atoms on their surfaces than bulk materials because of their high aspect ratio (surface-area-to-volume ratio). The presence of more surface atoms alters the behavior of 2D nanomaterials because they serve a different purpose than the interior atoms. A great variety of characterization techniques for 2D materials are highlighted in this chapter. The advantages of each technique are discussed, including those of X-ray diffraction, scanning electron microscopy, transmission electron microscopy, atomic force microscopy, X-ray photoelectron spectroscopy, and Raman spectroscopy. These techniques allow for the investigation of the 2D materials' flake size and shape, number of layers, conductivity, morphology, and chemical composition.

Keywords: 2D nanomaterials, XRD, optical microscope, raman spectroscopy

**Corresponding author*: subhendu.cy@gmail.com
†*Corresponding author*: rxp1097@miami.edu

Subhendu Chakroborty and Kaushik Pal (eds.) 2D Nanomaterials: Synthesis, Properties and Applications, (43–58) © 2024 Scrivener Publishing LLC

3.1 Introduction

The planar nature and single layer atomic thickness of two-dimensional (2D) nanomaterials have become versatile promising materials for various applications. The best option for 2D materials is the hierarchy of layers stacked together by van der Waals interactions. Each layer was bonded by a covalent bond between their atoms. Owing to their special hierarchical nature, 2D nanomaterials exhibit unique interfacial, elastic, surface, and fracture properties. In this race, honeycomb lattice graphene became the first 2D research material. Extensive studies on the physical and chemical properties of 2D nanomaterials began with the discovery of ultrathin two-dimensional (2D) graphene sheets from graphite in 2004 by Novoselov *et al.* [1, 2]. One can find examples of 2D nanomaterials, such as graphitic carbon nitride (g-C3N4) and metal dichalcogenides (transition metal sulfides, such as MoS2, TiS2, TaS2, WS2, and MoSe2) [3–7], hexagonal boron nitride (h-BN) [8], graphitic carbon nitride (g-C3N4) [9], binary group III–V materials [10], layered metal oxides, and layered double hydroxides, are shown in Figure 3.1 [11, 12]. Non-layered varieties of 2D materials have also recently been investigated, including metals [13–16], metal-organic frameworks [17, 18], polymers [19, 20], MXenes [21], black phosphorus [22], and silicone [23] (Figure 3.1).

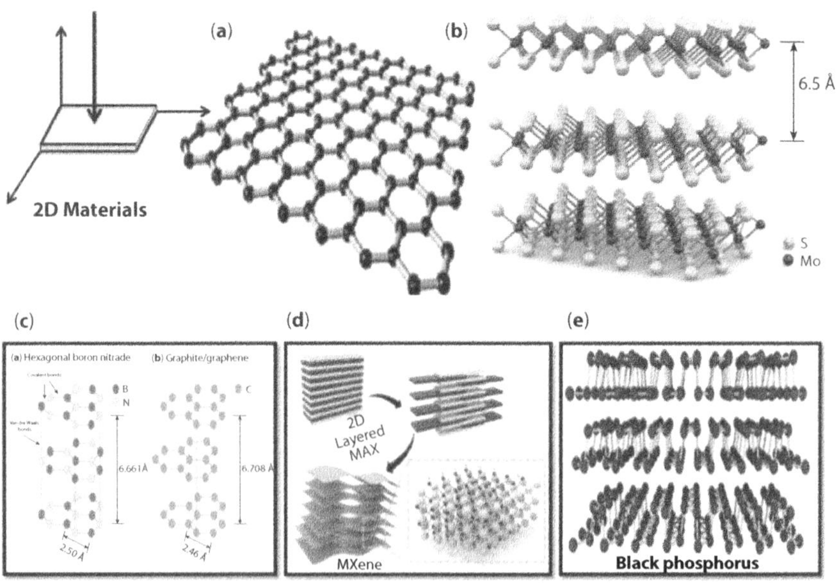

Figure 3.1 2D layered structures of (a) graphene [2], (b) MoS_2 [3], (c) boron nitride [8], (d) MXene [21], and (e) black phosphorus [22]. Figure reprinted with permission.

3.2 Characterization Techniques

To understand two-dimensional nanomaterials in terms of their crystallinity, size, composition, thickness, atomic oxidation state, etc., powerful characterization techniques are needed. Characterization can aid in the application of 2D nanomaterials in various fields. To date, the differentiation of two-dimensional nanomaterials has required the use of a wide variety of sophisticated characterization methods.

The probe-based analytical methods mentioned above, such as AFM and STM, may provide details on the orientation, smoothness, number of layers, grain size, depth profile, and dielectric characteristics of 2D materials. The thickness, number of layers, high-resolution imaging, surface characteristics, composition, and shape can all be ascertained using electron diffraction-based methods, including TEM, SEM, and energy-dispersive X-ray spectroscopy (EDS). Information on crystallinity, interlayer distance, interface information, composition, chemical bonding, layer interaction, and defects was obtained using X-ray-based methods such as XRD and XPS. Owing to the limitations of each technique, it is necessary to combine all characterization techniques to provide a complete understanding results of 2D-nanomaterials. In this chapter, we discuss some important typical characterization techniques (Figure 3.2) that have been widely used to characterize 2D nanomaterials.

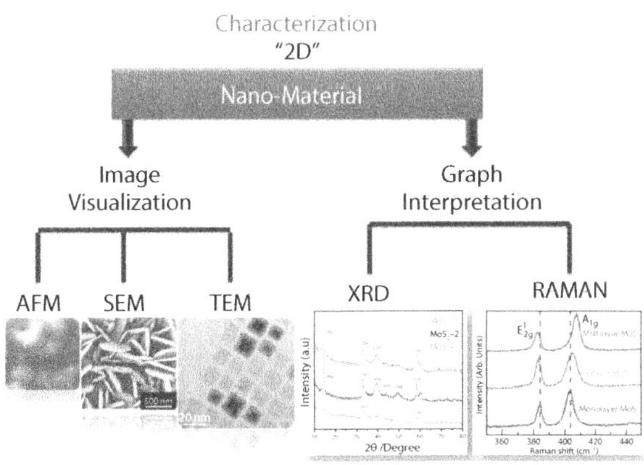

Figure 3.2 Some important typical characterization techniques: AFM [24], SEM [25], TEM [26], XRD [27], and RAMAN [28]. Figure reprinted with permission.

3.2.1 XRD

The most common characterization technique used to evaluate the phase structure of nanoscale materials is X-ray diffraction (XRD). It is a non-contact and nondestructive technique used to identify the crystalline or amorphous nature of nanomaterials. For crystalline nanomaterials, X-ray diffractograms provide a wealth of information on phase composition and crystallite size. In the case of 2D nanomaterials, XRD provides three types of structural information: the interplane distance of the lattices, thickness, and number of layers. The diffraction of incident X-rays from nanomaterials can be interpreted using Bragg's equation, $n\lambda = 2d \sin\theta$, which gives ideas about the positive integer (n), wavelength of incident X-ray (λ), interplanar distance of lattice (d), and diffraction angle(θ), respectively. The d-spacing or interlayer distance is the main concept of 2D layered nanomaterials. A simple relationship exists between the d-spacing of the interlayer and the related observed peak position [29]. This relation was clearly observed when the XRD characterization technique with a wavelength of 0.15418 nm (Cu Kα) was employed for the synthesis of ultrathin MoS_2 nanosheets at different temperatures (140–220 °C). In the XRD graph in Figure 3.3a, the first diffraction peak (002) of MoS_2 (S220 = synthesized at high temperature 220 °C) was observed at $2\theta = 14.38°$, while there was a shift the (002) peak at $2\theta = 9.308°$ for the low-temperature synthesis of MoS_2 nanosheets. With taking of these (002) plane for both types of MoS_2, the interlayer distanced was calculated and found to be 0.62 nm for high temperature synthesis of MoS_2 nanostructures and 0.95 nm for the rest of MoS_2 synthesis at low temperature. Another significant change in the XRD

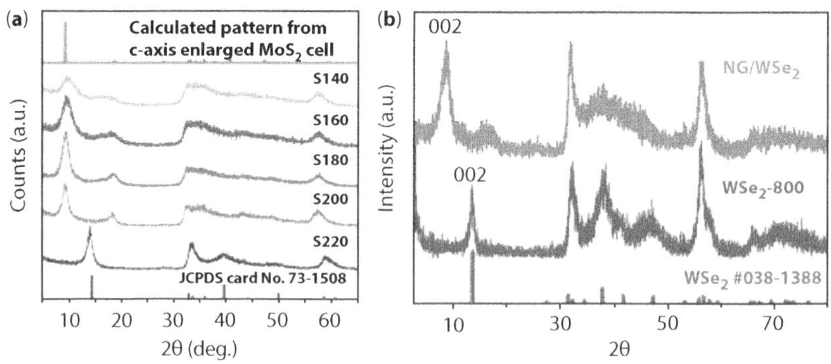

Figure 3.3 (a) XRD patterns of MoS_2 obtained at various temperatures. Figure reprinted with permission [29] and (b) XRD patterns of WSe_2-800 and NG/WSe_2 samples. Figure reprinted with permission [30].

peaks was observed in the 2D WeS_2 nanomaterials, as shown in Figure 3.3b [30]. Except for the (002) family planes, which were dramatically displaced from 13.6° to 8.5°, all the XRD peaks of N-doped graphene/WeS_2 (NG/WSe_2) corresponded to those of WSe_2-800. In the case of NG/WSe_2, this change indicates an increase in interlayer distance.

In addition to the interlayer distance, the XRD graph can also predict the average number of stacked layers using Debye function analysis (DFA). It uses the Debye scattering equation (DSE) to calculate a full diffraction profile from fundamental physical principles [31].

$$I(s) = P(s)\sum_i \sum_j f_i(s) f_j(s) \frac{\mathrm{Sin}(sr)_{ij}}{sr_{ij}}$$

where,

$s = 4\pi \cdot \sin(\theta)/\lambda$ is the scattering vector,

P(s) are the correction functions,

$f_i(s)$ and $f_j(s)$ are the scattering factors of atoms i and j,

r_{ij} is the distance between atoms,

θ and λ denote the diffraction angle and the X-ray wavelength, respectively.

The total X-ray scattering from model nanoparticles can be directly calculated using computer analysis of the Debye scattering equation. X-ray scattering from even tiny molecules, atomic clusters, and crystallites that are only a few nanometers in size can be mimicked using the DFA approach [32–39]. An increasing number of researchers are using the DFA approach to analyze the structural properties of nanocrystals and nanostructures. Pakharukova *et al.* discussed about the possibilities of the DFA approach for determining the average size and stacking number of supported MoS_2 crystallites [40].

3.2.2 SEM and TEM

SEM and TEM characterizations are the most versatile imaging characterizations for nanomaterials, which can provide a visualization image for all types of materials at the nano level. Under these characterizations, the morphology, topology, and detailed surface as well as inner layers of all types of dimensional (0D, 1D, 2D, and 3D) nanomaterials can be easily determined. The morphology, topography, composition, orientation of grains, crystallographic information etc. of nanomaterials can be visualized through SEM by the results of back-scattered electrons from the sample, whereas in TEM, the electrons are replaced by photons and images are viewed on a screen instead of an eyepiece. TEM characterization is a powerful image magnification technique that collects information from the inner layer of nanostructures and provides good spatial resolution. The bright-field and dark-field images provided by TEM provide details regarding the precise particle size. Overall, both instruments are helpful for characterizing all types of materials, ranging from 0-D to 3-D nanomaterials.

3.2.3 Optical Microscope

Optical microscopes have been established as rapid information providers such as in the imaging of the shape and thickness of 2D nanomaterials [41, 42]. These images are the result of the interference of the reflected light from various interfaces. The reflectance from the 2D material depends on the refractive indices of the corresponding material, number of layers (thickness) of the 2D material, and thickness of the underlying substrate or dielectric SiO_2 layer. The layer-identification mechanism is illustrated in Figure 3.4. For 2D-layered materials with a high refractive index, an accurate number of layers up to 15-layer can be identified from contrast imaging. In the case of a low refractive index, there is a lack of image contrast; hence, it is difficult to accurately calculate the number of layers.

From the optical microscopic image of 2D graphene nanosheets underlying a dielectric SiO_2 film with a thickness of 90 or 300 nm, its thickness was calculated using theoretical calculations [43–47]. It has been possible to calculate the graphene nanosheet layers from a single layer (1L) to fifteen

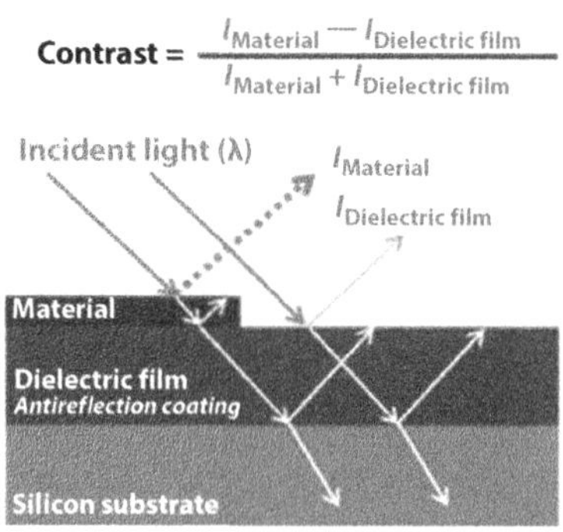

Figure 3.4 Schematic showing optical transmission and reflection at the interconnections of a layered system. Figure reprinted with permission [41].

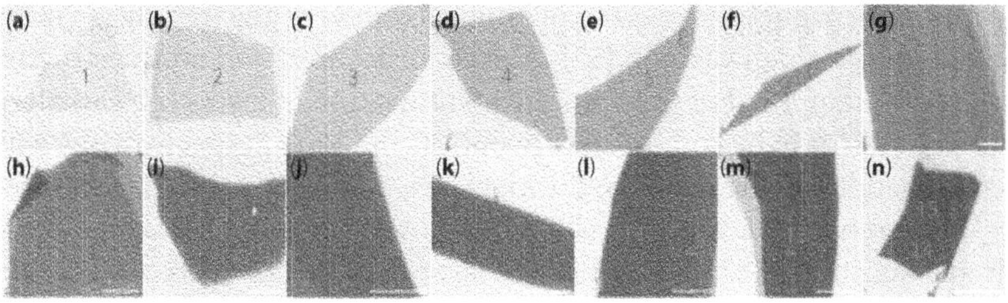

Figure 3.5 (a-n) 1L–15L graphene nanosheets on 300 nm SiO_2/Si are shown in color optical pictures. Figure reprinted with permission [42].

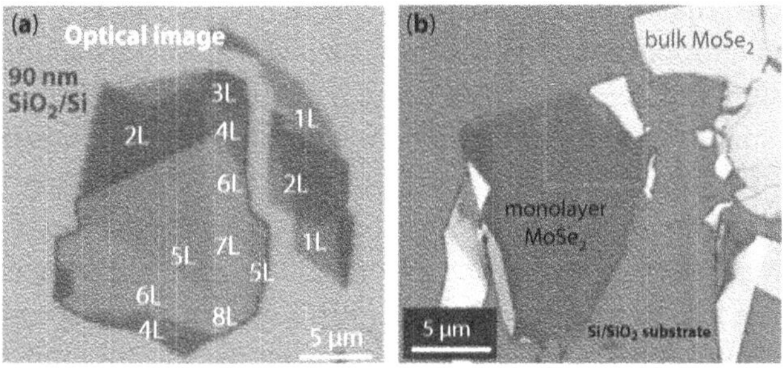

Figure 3.6 (a) MoS_2 flakes on 90 nm SiO_2. Figure reprinted with permission [42]. (b) Optical micrograph of exfoliated $MoSe_2$ flakes on SiO_2/Si substrate. Figure reprinted with permission [50].

layers (15L) using optical contrast with 300 nm dielectric SiO_2/Si (Figure 3.5) [42]. Accordingly, most studies have been conducted to quickly determine the thickness of the graphene sheet, and it has been discovered that SiO2 substrates between 90 and 280 nm thick are the most suitable for supporting graphene when illuminated with either white or green light [48]. Thus, many 2D nanomaterials have been investigated on SiO_2 substrates with specific thicknesses [42, 49]. The different numbers of layers of 2D MoS_2 can be distinctly visualized in Figure 3.6 on a 90 nm SiO_2 substrate.

3.2.4 AFM

The non-optical high-resolution imaging technique, atomic force microscopy (AFM), for nanomaterials was first demonstrated by Binnig, Quate, and Gerber in 1985 [51]. AFM can be termed as scanning probe microscopy (SPM). AFM contains a cantilever probe that provides a sharp tip to scan the sample surface and measure the topography, friction, height, magnetic, optical, and magnetic properties of samples with very high resolution in air, liquids, or ultrahigh vacuum [52]. AFM is an accurate and non-destructive analysis of samples. For the accurate identification of layers of 2D materials, AFM needs to be combined with optical and Raman spectroscopy with careful handling. The apparent thickness of a 2D nanomaterial is measured by interference with the large chemical contrast between 2D nanomaterials and substrates [53, 54]. For single-layer thicknesses, there are discrepancies between the theoretical and measured values [55]. AFM can easily distinguish morphological changes on the surfaces of 2D nanomaterials [56]. High-spatial-resolution investigations of material conductivity may be performed using conductive atomic force microscopy (CAFM), an AFM method that integrates current sensing. In contrast to CAFM, which measures the topography by detecting the cantilever's deflection using an optical system, current is sensed using a current-to-voltage preamplifier. [57]. The spatial mapping of graphene conductivity on various surfaces has been successfully demonstrated using CAFM [58]. Owing to the significant variations in the tip-sample contact resistance, epitaxial graphene domains were easily distinguished from the nearby silicon carbide (SiC) surface using a high imaging contrast [59]. Kelvin probe force microscopy (KPFM), a non-contact variation of atomic force microscopy, also offers information on the thickness of 2D materials through contactless electric mapping with the distribution of electric potential and charge, as well as work function at the nanoscale [60]. The work function of graphene has been shown to vary with the number of layers [61]. High-resolution KPFM can detect differences in the work function of folds, standing collapses, line defects, and grain boundaries in graphene. The varied substrate-induced doping levels may explain the surface potential discrepancies between single layers and bilayers [62].

3.2.5 XPS

In 1978, Fadley wrote postulates on the basic concept of XPS theory with measurements of film thickness [63]. However, many of these postulates have not been fully utilized for the measurement of film thickness by some research groups who are actively working with XPS characterization. In 2000, Cumpson introduced a ground-breaking visual method called a "thickogram" to streamline the interpretation of XPS data for film thickness measurements [64]. This graphical approach for thickness measurement covers a limited thickness range of materials with some errors. Hence, this thickogram concept became obsolete. In addition to this thickogram proposal, quantitative analysis of angular-dependent peak intensities by

XPS has been used to quantify the thickness of 2D materials [65–70]. The thickness of a few layers of 2D-graphene grown on a SiC substrate was successfully estimated using XPS [70, 71]. The general approach for XPS is to measure the thickness by comparing the peak intensities originating from substrate and an overlayer film [72]. Apart from thickness measurement, XPS can easily identify the elemental composition within a material. XPS can measure the surface within a 10 nm depth and gives the binding energy peak that arises from the electronic configuration. Therefore, this method cannot be used to qualitatively analyze the elemental composition, but it may also be used to quantitatively determine the chemical composition of a material [73].

3.2.6 RAMAN

The most extreme characterization technique is the RAMAN technique, where Raman vibrational spectroscopy is observed to be related to the geometric structure and bonding of molecules. Raman sensitivity to geometric structure of layered nanomaterials is extremely useful for 2D nanomaterial characterization. Sir Chandrasekhara Venkata Raman, an Indian physicist, discovered the light-scattering effect in 1928. He was the first person to carefully observe the change in optical frequency, which resulted from the interaction between the incident light and molecular motion, as follows:

(1) Rayleigh scattering having no change in frequency of scattered light and
(2) Raman shift produced by scattering of light with two different frequencies: (a) low frequency (stroke line) and (b) high frequency (anti-stroke line). The anti-stroke lines are the result of the in-plane and out-of-plane vibrations of molecules (Figure 3.7b), while the low-frequency stroke lines provide information about the crystal structure, uniformity, layer number, etc. [74].

In 2D materials, there are two kinds of Raman modes are exhibits related: intralayer and interlayer modes [76]. Many studies have reported Raman spectra to identify of the thickness of 2D layered materials [77–90]. Figures 3.8(a,b) show the comparison between monolayer MoS_2 and particulate MoS_2 collected from two different studies [91, 92]. In their comparison, the difference in the Raman peak positions of E^1_{2g} and A^1_g directly provided information about the number of layers. Raman peak position of E^1_{2g} and A^1_g for MoS_2 particles is at 378.3 cm^{-1} and 404.7

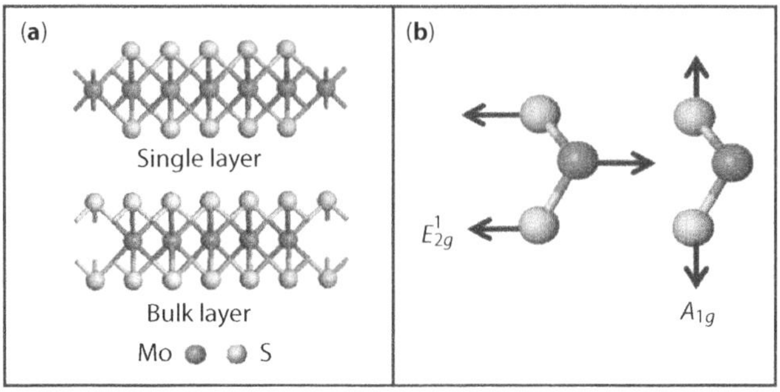

Figure 3.7 (a) Schematic model of MoS_2 single layer and bulk layer, and (b) in-plane phonone E^1_{2g} and out-of-plane phonon A_{1g} for a single layer of MoS_2. Figure reprinted with permission [75].

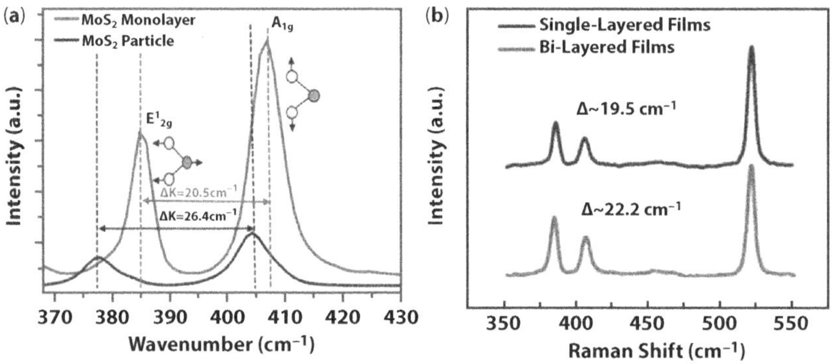

Figure 3.8 RAMAN intensity of MoS_2. The figure was reprinted with permission [91, 92].

cm^{-1}, respectively, whereas these peaks can be observed at 385.5 cm^{-1} and 405 cm^{-1} for the MoS_2 monolayer. From the peak position, one can observe the peak differences (ΔK) of the E^1_{2g} and A^1_g modes, and it was found that the ΔK value for particle-like MoS_2 was 26.4 cm^{-1}, while that of monolayer MoS_2 was 20.5 cm^{-1}. The ΔK values increased by 22 cm^{-1}, 23.5 cm^{-1}, and 25 cm^{-1} for the second, third, and fourth layers of MoS_2, respectively [91]. Another study reported that the ΔK value for monolayer MoS_2 is ~19.5 cm^{-1}, and that for a bilayer of MoS_2 is ~22.2 cm^{-1} [92]. From these observations the conclusion may be ascribed to the value of RAMAN peak differences increasing monotonically with the increasing number of layers. In addition, the narrow nature of the two phonon nodes (A^1_g and E^1_{2g}) gives the crystalline quality of 2D materials [93, 94].

3.3 Conclusion

Two-dimensional nanomaterials with an average size of thin layers and density defects at the atomic level behave differently than bulk materials. Most 2D nanomaterials are used in both electronic and biomedical applications. Therefore, it is important to characterize them based on the number of layers, layer size, and atomic concentration present in the layers. Here, we discuss the basic characterization techniques of XRD, SEM, TEM, AFM, XPS, and optical imaging for the identification of 2D nanomaterials. In recent years, *in-situ* characterization techniques such as *in-situ* transmission electron microscopy, X-ray photoelectron microscope, and Raman spectroscopy have been developed to accurately study the growth mechanism of 2D nanomaterials with potential relevance.

References

1. Novoselov, *et al.*, Electric field effect in atomically thin carbon films. *Science*, 306, 5696, 666–669, 2004, DOI: 10.1126/science.1102896.
2. Das, *et al.*, Graphene and graphene-like materials in biomass conversion: Paving the way to the future. *J. Mater. Chem. A*, 5, 25131–25143, 2017, DOI: 10.1.39/C7TA09418C.
3. Yadav, *et al.*, 2D MoS_2-based nanomaterials for therapeutic, bioimaging, and biosensing applications. *Small*, 15, 1, e1803706–e1803739, 2018, DOI: 10.1002/smll.201803706.
4. Huang, *et al.*, Metal dichalcogenide nanosheets: Preparation, properties and applications. *Chem. Soc. Rev.*, 42, 1934–1946, 2013, DOI.10.1039/c2cs35387c.
5. Chhowalla, *et al.*, Two-dimensional transition metal dichalcogenide (TMD) nanosheets. *Chem. Soc. Rev.*, 44, 9, 2584–2586, 2015.
6. Zhang, *et al.*, Ultrathin two-dimensional nanomaterials. *ACS Nano*, 9, 10, 9451–9469, 2015, DOI: 10.1021/acsnano.5b05040.
7. Tan, *et al.*, Two-dimensional transition metal dichalcogenide nanosheet-based composites. *Chem. Soc. Rev.*, 44, 9, 2713–2731, 2015, DOI:https://doi.org/10.1039/C4CS00182F.
8. Kim, *et al.*, A new horizon for hexagonal boron nitride film. *J. Korean Phys. Soc.*, 64, 10, 1605–1616, 2014.
9. Ong, *et al.*, Graphitic carbon nitride (g-C3N4)-based photocatalysts for artificial photosynthesis and environmental remediation: Are we a step closer to achieving sustainability? *Chem. Rev.*, 116, 7159–7329, 2016.
10. Xu, *et al.*, Graphene-like two-dimensional materials. *Chem. Rev.*, 113, 3766–3798, 2013.

11. Osada, *et al.*, Exfoliated oxidenanosheets: New solution to nanoelectronics. *J. Mater. Chem.*, 19, 2503–2511, 2009.
12. Wang, *et al.*, Recent advances in the synthesis and application of layered double hydroxide (LDH) nanosheets. *Chem. Rev.*, 112, 7, 4124–4155, 2012.
13. Huang, *et al.*, Freestanding palladium nanosheets with plasmonic and catalytic properties. *Nat. Nanotechnol.*, 6, 1, 28–32, 2011.
14. Huang, *et al.*, Synthesis of hexagonal close-packed gold nanostructures. *Nat. Commun.*, 2, 292–297, 2011, doi: 10.1038/ncomms1291.
15. Duan, *et al.*, Ultrathin rhodium nanosheets. *Nat. Commun.*, 5, 3093–3100, 2014.
16. Fan, *et al.*, Thin metal nanostructures: Synthesis, properties and applications. *Chem. Sci.*, 6, 1, 95–111, 2015, DOI: 10.1039/C4SC02571G.
17. Peng, *et al.*, Metal-organic framework nanosheets as building blocks for molecular sieving membranes. *Science*, 346, 6215, 1356–1359, 2014, DOI: 10.1126/science.125422.
18. Colson, *et al.*, Oriented 2D covalent organic framework thin films on single-layer graphene. *Science*, 332, 6026, 228–231, 2011.
19. Kory, *et al.*, Gm-scale synthesis of two-dimensional polymer crystals and their structure analysis by X-ray diffraction. *Nat. Chem.*, 6, 9, 779–789, 2014.
20. Kissel, *et al.*, A nanoporous two-dimensional polymer by single-crystal-to-single-crystal photopolymerization. *Nat. Chem.*, 6, 9, 774–778, 2014, DOI: 10.1038/nchem.2008.
21. Hwang, *et al.*, An emerging two-dimensional layered material for removal of radioactive pollutants. *Chem. Eng. J.*, 397, 125428–125493, 2020, https://doi.org/10.1016/j.cej.2020.125428.
22. Liu, *et al.*, Semiconducting black phosphorus: Synthesis, transport properties and electronic applications. *Chem. Soc. Rev.*, 44, 2732–2743, 2015, https://doi.org/10.1039/c4cs00257a.
23. Lalmi, *et al.*, Epitaxial growth of a silicene sheet. *Appl. Phys. Lett.*, 97, 223109–223111, 2010, https://doi.org/10.1063/1.3524215.
24. Gopalakrishnan, *et al.*, Removal of heavy metal ions from pharma-effluents using graphene-oxide nanosorbents and study of their adsorption kinetics. *J. Ind. Eng. Chem.*, 30, 14–19, 2015.
25. Vasudevan, *et al.*, Tailoring electrical conductivity of two dimensional nanomaterials using plasma for edge electronics: A mini review. *Front. Chem. Sci. Eng.*, 13, 13, 427–443, 2019.
26. Yalcin, *et al.*, Core–shell reconfiguration through thermal annealing in FexO/CoFe2O4 ordered 2D nanocrystal arrays. *Nanotechnology*, 25, 5, 055601–055610, 2014, DOI:10.1088/0957-4484/25/5/055601.
27. Guo, *et al.*, The novel and facile preparation of multilayer MoS_2 crystals by a chelation-assisted sol–gel method and their electrochemical performance. *RSC Adv.*, 7, 15, 9009–9014, 2017.

28. Pierucci, *et al.*, Large area molybdenum disulphide- epitaxial graphene vertical Van der Waals heterostructures. *Sci. Rep.*, 6, 26656–26667, 2016, DOI: 10.1038/srep26656.
29. Xie, *et al.*, Controllable disorder engineering in oxygen-incorporated MoS2 ultrathin nanosheets for efficient hydrogen evolution. *J. Am. Chem. Soc.*, 135, 47, 17881–17888, 2013.
30. Zhang, *et al.*, Robust lithium–sulfur batteries enabled by highly conductive WSe_2-based superlattices with tunable interlayer space. *Adv. Funct. Mater.*, 32, 24, 1–13, 2022.
31. Debye, Zerstreuung von Röntgenstrahlen. *Ann. Phys.*, 351, 6, 809–823, 1915, https://doi.org/10.1002/andp.19153510606.
32. Hall, *et al.*, Debye function analysis of structure in diffraction from nanometer-sized particles. *J. Appl. Phys.*, 87, 4, 1666–1675, 2000.
33. Yatsenko, *et al.*, (Diffraction analysis of nanopowders) – a software for structural analysis of nanosized powders. *Z. Kristallogr.-Cryst. Mater.*, 233, 1, 61–66, 2018.
34. Lee, *et al.*, Morphological determination of face-centered-cubic metallic nanoparticles by X-ray diffraction. *J. Colloid Interface Sci.*, 369, 1, 129–133, 2012.
35. Cernuto, *et al.*, Size and shape dependence of the photocatalytic activity of TiO_2 nanocrystals: A total scattering debye function study. *J. Am. Chem. Soc.*, 133, 9, 3114–3119, 2011.
36. Sławiński, *et al.*, Structural arrangement in close-packed cobalt polytypes. *Cryst. Growth Des.*, 18, 4, 2316–2325, 2018.
37. Bertolotti, *et al.*, Coherent nanotwins and dynamic disorder in cesium lead halide perovskite nanocrystals. *ACS Nano*, 11, 4, 3819–3831, 2017.
38. Beyerlein, *et al.*, A review of debye function analysis. *Powder Diffr.*, 28, S2, S2–S10, 2013, https://doi.org/10.1017/S0885715613001218.
39. Gelisio, *et al.*, 100 years of Debye's scattering equation. *Acta Crystallogr.*, A72, 6, 608–620, 2016, https://doi.org/10.1107/S2053273316014881.
40. Pakharukova, *et al.*, A total scattering debye function analysis: Effective approach for structural studies of supported MoS2-based hydrotreating catalysts. *Ind. Eng. Chem. Res.*, 59, 23, 10914–10922, 2020.
41. Jung, *et al.*, Simple approach for high-contrast optical imaging and characterization of graphene-based sheets. *Nano Lett.*, 7, 12, 3569–3575, 2007, https://doi.org/10.1021/nl0714177.
42. Li, *et al.*, Rapid and reliable thickness identification of two-dimensional nanosheets using optical microscopy. *ACS Nano*, 7, 11, 10344–10353, 2013, DOI: 10.1021/nn4047474.
43. Gao, *et al.*, Total color difference for rapid and accurate identification of graphene. *ACS Nano*, 2, 8, 1625–1633, 2008, https://doi.org/10.1021/nn800307s.

44. Roddaro, *et al.*, Optical visibility of graphene: Interference colors of ultrathin graphite on SiO_2. *Nano Lett.*, 7, 9, 2707–2710, 2007, https://doi.org/10.1021/nl071158l.
45. Nolen, *et al.*, High-throughput large-area automated identification and quality control of graphene and few-layer graphene films. *ACS Nano*, 5, 2, 914–922, 2011, https://doi.org/10.1021/nn102107b.
46. Blake, *et al.*, Making graphene visible. *Appl. Phys. Lett.*, 91, 063124–063127, 2007, https://doi.org/10.1063/1.2768624.
47. Casiraghi, *et al.*, Rayleigh imaging of graphene and graphene layers. *Nano Lett.*, 7, 9, 2711–2717, 2007.
48. Shearer, *et al.*, Accurate thickness measurement of graphene. *Nanotechnology*, 27, 125704–12514, 2016, DOI: 10.1088/0957-4484/27/12/125704.
49. Benameur, *et al.*, Visibility of dichalcogenide nanolayers. *Nanotechnology*, 22, 125706–125711, 2011.
50. Toth, *et al.*, Structural features dictate the photoelectrochemical activities of two-dimensional MoSe2 and WSe2 nanostructures. *J. Phys. Chem. C*, 125, 7701–7710, 2021.
51. Binnig, *et al.*, Atomic force microscope. *Phys. Rev. Lett.*, 56, 930–933, 1986.
52. Marti, *et al.*, Atomic force microscopy of liquid-covered surfaces: Atomic resolution images. *Appl. Phys. Lett.*, 51, 484–486, 1987.
53. Novoselov, *et al.*, Two-dimensional atomic crystals. *Proc. Natl. Acad. Sci. U.S.A.*, 102, 30, 10451–10453, 2005, https://doi.org/10.1073/pnas.0502848102.
54. Ferrari, *et al.*, Raman spectrum of graphene and graphene layers. *Phys. Rev. Lett.*, 97, 18, 187401–187404, 2006, https://doi.org/10.1103/PhysRevLett.97.187401.
55. Xu, *et al.*, Graphene visualizes the first water adlayers on mica at ambient conditions. *Science*, 329, 5996, 1188–91, 2010.
56. Wood, *et al.*, Effective passivation of exfoliated black phosphorus transistors against ambient degradation. *Nano Lett.*, 14, 12, 6964–6970, 2014.
57. Lanza, *Conductive atomic force microscopy*, p. 400, Wiley-VCH, Berlin, Germany, 2017, ISBN 978-3-527-34091-0.
58. Fisichella, *et al.*, Current transport in graphene/AlGaN/GaN vertical heterostructures probed at nanoscale. *Nanoscale*, 6, 8671–8680, 2014.
59. Kellar, *et al.*, Identifying and characterizing epitaxial graphene domains on partially graphitized SiC(0001) surfaces using scanning probe microscopy. *Appl. Phys. Lett.*, 96, 14, 143103–143116, 2010.
60. Liscio, *et al.*, Nanoscale quantitative measurement of the potential of charged nanostructures by electrostatic and kelvin probe force microscopy: Unraveling electronic processes in complex materials. *Acc. Chem. Res.*, 43, 541–550, 2010.
61. Panchal, *et al.*, Standardization of surface potential measurements of graphene domains. *Sci. Rep.*, 3, 2597, 1–8, 2013.
62. Ziegler, *et al.*, Variations in the work function of doped single- and few-layer graphene assessed by Kelvin probe force microscopy and density functional

theory. *Phys. Rev. B*, 83, 23, 235434–235440, 2011, https://doi.org/10.1103/PhysRevB.83.235434.
63. Fadley, *et al.*, Basic concepts of x-ray photoelectron spectroscopy electron spectroscopy theory, in: *Techniques and Applications*, vol. 2, pp. 1–156, Academic Press, London, 1978.
64. Cumpson, *et al.*, The Thickogram: A method for easy film thickness measurement in XPS. *Surf. Interface Anal.*, 29, 6, 403–6, 2000.
65. Fraser, *et al.*, Surface sensitivity and angular dependence of X-ray photoelectron spectra. *Surf. Sci.*, 36, 2, 661–674, 1973.
66. Fadley, Instrumentation for surface studies: XPS angular distributions. *J. Electron Spectrosc.*, 5, 725–754, 1974.
67. Baird, *et al.*, Direct observation of surface-profile effects on X-ray-photoelectron angular distribution. *Chem. Phys. Lett.*, 34, 1, 49–54, 1975, https://doi.org/10.1016/0009-2614(75)80198-9.
68. Fadley, Comments in Faraday Discussion no. 60, in: *Progress in Solid State Chemistry*, vol. f 1, Part 3, G.A. Samorjai and J.-O. McCaldin (Eds.), p. 19760, Pregamon Press, London, 1975.
69. Hill, *et al.*, Properties of oxidized silicon as determined by angular-dependent X-ray photoelectronspectroscopy. *Chem. Phys. Lett.*, 44, 225–231, 1976.
70. Briggs, *et al.*, Practical surface analysis, in: *Auger and X-Ray Photoelectron Spectroscopy*, 2, pp. 1–18, John Wiley and Sons, Chichester, 1990.
71. Biedermann, *et al.*, Insights into few-layer epitaxial graphene growth on 4H-SiC(000(ī) substrates from STM studies. *Phys. Rev. B*, 79, 12, 125411–125420, 2009, DOI:https://doi.org/10.1103/PhysRevB.79.125411.
72. Prakash, *et al.*, AFM study of ridges in few-layer epitaxial graphene grown on the carbon-face of 4HSiC(000(ī). *Carbon*, 48, 2383–93, 2010, http://dx.doi.org/10.1016/j.carbon.2010.02.026.
73. Khaled, Biomedical applications of graphene and 2D nanomaterials, in: *Micro and Nano Technologies*, 1st, Chapter: Chapter 2, pp. 27–41, Elsevier, Amsterdam, Netherlands, 2019.
74. Brien, *et al.*, Low wavenumber Raman spectroscopy of highly crystalline $MoSe_2$ grown by chemical vapor deposition. *Phys. Status Solidi B*, 252, 11, 2385–2389, 2015, https://doi.org/10.1002/pssb.201552225.
75. Su, *et al.*, Dependence of coupling of quasi 2-D MoS_2 with substrate on substrate type, probed by temperature dependent Raman scattering. *Nanoscale*, 6, 4920–4927, 2014.
76. Tan, *et al.*, The shear mode of multilayer graphene. *Nat. Mater.*, 11, 294–300, 2012.
77. Yu, *et al.*, Controlled scalable synthesis of uniform, high-quality monolayer and few-layer MoS_2 films. *Sci. Rep.*, 3, 1866–1871, 2013, https://doi.org/10.1038/srep01866.
78. Zhao, *et al.*, Interlayer breathing and shear modes in few-trilayer MoS_2 and WSe_2. *Nano Lett.*, 13, 13, 1007–1015, 2013.

79. Sahin, *et al.*, Anomalous Raman spectra and thickness-dependent electronic properties of WSe_2. *Phys. Rev. B*, 87, 16, 165409–65414, 2013, DOI:https://doi.org/10.1103/PhysRevB.87.165409.
80. Zhao, *et al.*, Lattice dynamics in mono- and few-layer sheets of WS_2 and WSe_2. *Nanoscale*, 5, 20, 9677–9683, 2013, DOI: https://doi.org/10.1039/C3NR03052K.
81. Lee, *et al.*, Anomalous lattice vibrations of single- and few-layer MoS_2. *ACS Nano*, 4, 5, 2695–2700, 2010, https://doi.org/10.1021/nn1003937.
82. Berkdemir, *et al.*, Identification of individual and few layers of WS_2 using Raman Spectroscopy. *Sci. Rep.*, 3, 1755–1762, 2013, DOI: 10.1038/srep01755.
83. Tan, *et al.*, The shear mode of multilayer graphene. *Nat. Mater.*, 5, 11(4), 294–300, 2012, DOI: 10.1038/nmat3245.
84. Gomez, *et al.*, Isolation and characterization of few-layer black phosphorus. *2D Mater.*, 1, 2, 025001–025020, 2014, DOI 10.1088/2053-1583/1/2/025001.
85. Late, *et al.*, Rapid characterization of ultrathin layers of chalcogenides on SiO_2/Si substrates. *Adv. Funct. Mater.*, 22, 9, 1894–1905, 2012, https://doi.org/10.1002/adfm.201102913.
86. Ferreira, *et al.*, Evolution of the Raman spectra from single-, few-, and many-layer graphene with increasing disorder. *Phys. Rev. B*, 82, 12, 125429–12537, 2010, https://doi.org/10.1103/PhysRevB.82.125429.
87. Wang, *et al.*, Integrated circuits based on bilayer MoS_2 transistors. *Nano Lett.*, 12, 9, 4674–4680, 2012, https://doi.org/10.1021/nl302015v.
88. Duan, *et al.*, Lateral epitaxial growth of two-dimensional layered semiconductor heterojunctions. *Nat. Nanotechnol.*, 9, 12, 1024–30, 2014, doi: 10.1038/nnano.2014.222.
89. Late, *et al.*, Thermal expansion, anharmonicity and temperature-dependent Raman spectra of single- and few-layer $MoSe_2$ and WSe_2. *ChemPhysChem*, 15, 8, 1592–1598, 2014, doi: 10.1002/cphc.201400020.
90. Tan, (Ed.), *Raman Spectroscopy of Two-Dimensional Materials*, vol. 276, pp. 1–269, Springer Series in Materials Science, United States, 2009, https://doi.org/10.1007/978-981-13-1828-3.
91. Gnanasekar, *et al.*, Promoter-free synthesis of monolayer MoS_2 by chemical vapour deposition. *CrystEngComm*, 20, 2, 1–10, 2018, https://doi.org/10.1039/C8CE00576A.
92. Najmaei, *et al.*, Vapour phase growth and grain boundary structure of molybdenum disulphide atomic layers. *Nat. Mater.*, 12, 754–759, 2013, https://doi.org/10.1038/nmat3673.
93. Molina-Sánchez, *et al.*, Phonons in single-layer and few-layer MoS_2 and WS_2. *Phys. Rev. B: Condens. Matter Mater. Phys.*, 84, 1–8, 2011.
94. Lee, *et al.*, Anomalous lattice vibrations of single- and few-layer MoS_2. *ACS Nano*, 4, 5, 2695–2700, 2010.

Part II

PROPERTIES OF 2D NANOMATERIALS

4

Crystal Structure, Magnetic and Mechanical Properties of 2D Nanomaterials

Nandini Roy

Department of Physics, B. Borooah College, Guwahati, Assam, India

Abstract

2D nanomaterials have been receiving increasing interest from the basic physical sciences to material engineering owing to their variant properties and diverging fields of uses in sensors, electronics, optoelectronics, energy storage devices such as solar cells, lithium batteries, composites, etc. Graphene is the pioneer of a family of 2D materials that has triggered active research. In addition to graphene, many other members, such as boron nitride nanosheets (BNNS), MXenes, and 2D oxide/hydroxide materials, black phosphorous (BP), transition metal dichalcogenides (TMDC), and mono-elemental 2D semiconductors, have joined the family of 2D nanomaterials.

2D nanomaterials have a very low thickness as well as macroscale/nanoscale dimensions, making them the thinnest of all the nanomaterials. They have strong in-plane bonds and weak interlayered van der Waals (vdW) forces. The upper- and lower layer neighboring atoms are arranged in a different manner. They are configured in a honeycomb structure.

Another aspect of 2D materials is the ease of production of 2D magnetic materials. On decreasing the thickness of 2D nanomaterials to a few sheets, they exhibit a magnetic ground state irrespective of either their antiferromagnetic or ferromagnetic states. The exceptional properties of 2D nanomaterials and their potential applications are largely dependent on their crystal structures. The crystal structures determine the inherent optical, electrical, magnetic, and mechanical properties. Therefore, an in-depth study of the crystal structures of 2D nanomaterials is of great importance. In this chapter, an attempt has been made to summarize the

Email: nandini.nit.10@gmail.com

Subhendu Chakroborty and Kaushik Pal (eds.) 2D Nanomaterials: Synthesis, Properties and Applications, (61–72) © 2024 Scrivener Publishing LLC

crystal structures of various 2D nanomaterials and their correlation to the magnetic and mechanical properties of these materials.

Keywords: 2D nanomaterials, crystal structure, magnetic properties, mechanical properties

4.1 Introduction

2D nanomaterials have attracted great attention since the discovery of the first 2D material, graphene, in 2004. Due to their unique properties such as optical transparency, stability, high thermal conductivity, mechanical strength, large specific surface area, etc., 2D materials have introduced a new era of research [42]. These properties of a family of 2D materials have been used in various device technological applications such as memory devices [36], sensors, hydrogen storage devices [30, 31], solar cells [48], graphene-based gas and biosensors [41–45], and many others [46, 47]. These 2D materials have interesting crystal structures as well as structure-dependent properties. They possess extraordinary mechanical as well as magnetic properties. In this chapter, it has been tried to present a few of the basic structures of 2D materials along with their mechanical and magnetic properties.

4.2 Structure of 2D Materials

2D materials can broadly be classified as layered and nonlayer-structured materials. The interatomic connection in every plane of a layered 2D material is due to strong chemical bonding between the atoms, whereas the interlayer stacking to form the bulk structure is formed through the van der Waal interaction. One of the best examples of stacking layers is graphite. In graphite, several layers of graphene are stacked upon one another to give the complete bulk structure. There are several other layered materials, such as black phosphorous, transition metal oxides (TMOs), transition metal dichalcogenides (TMDCs), h-BN, and g-C_3N_4.

4.2.1 Graphene

Graphene is an allotrope of carbon with a hexagonal lattice where each vertex is formed by a single carbon atom with sp^2 hybridization. The structure forms a benzene ring where one unpaired electron is donated by every atom. In a hexagonal structure of graphene, each lattice has three σ bonds.

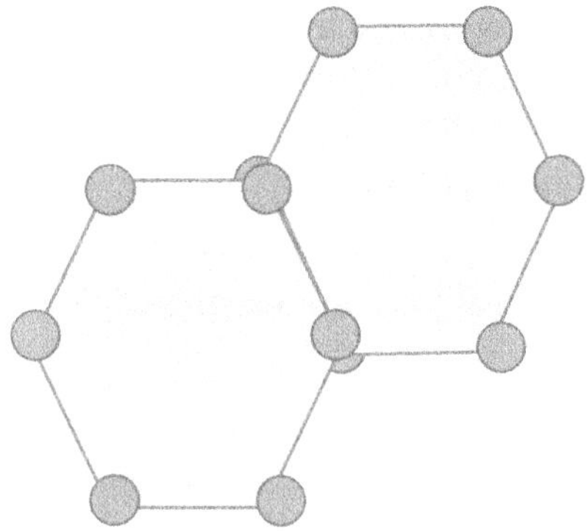

Figure 4.1 Carbon–carbon bonding in graphene.

The carbon–carbon bond as shown in Figure 4.1 is about 0.142 nm long. There is a π bond located vertically in the lattice plane. The sp^2 orbital hybridization and the tightly bonded carbon atoms are the reasons for the mechanical stability of graphene. Graphene on its own has a stable structure [1–11]. It is the basic structure of graphite, fullerene, and carbon nanotubes.

Despite its thickness, which is as low as one millionth of the diameter of a human hair, graphene is quite stable structurally. The interconnection of carbon atoms is considerably high in graphene. Graphene nanoribbon is one of its strong forms, where lateral charge movement brings the energy barrier near the central point. Thus, the energy barrier can be increased by decreasing the width of the ribbon [12]. This property of graphene makes it a promising material for graphene-based electronic devices.

4.2.2 Black Phosphorous

Black phosphorous (BP) is one of the latest interests in the family of 2D nanomaterials. It is the strongest of all the allotropes of phosphorous. On pulverization, it matches graphite. Like graphene, BP is orthorhombic in structure, where a honeycomb lattice is formed by the phosphorous atoms.

Unlike other 2D materials, BP is not atomically flat; it is formed by puckered double layers. In each single layer, every phosphorous atom that is sp^3 hybridized is bonded covalently with adjacent phosphorous atoms. The bond length is 2.18 Å. Out of the three adjacent phosphorous atoms, two are in the plane of the layer. The angle between these two atoms is 98°. The third atom makes an angle of 103° with the layers [13]. This arrangement of atoms is responsible for different anisotropic physical properties such as optical, mechanical, thermoelectronic, and electrical conductance.

The ABA stacking order is formed by the adjacent layers, which interact by weak van der Waals interactions alone in the z direction. The layers are separated by a distance of 5 Å. It is due to this layered structure; mechanical or liquid exfoliation can be employed on bulk BP to obtain a single-layer film.

4.2.3 Transition Metal Dichalcogenide (TMDC)

TMDCs themselves form a wide family of 2D nanomaterials. TMDCs can be found in different structural phases like 1T, 2H, and 3R, where the alphabets represent trigonal, hexagonal, and rhombohedral lattices, respectively, and the numerical values in the coefficient of the letters represent the number of X-M-X units in the unit cell. In the 2H phase, the adjacent layers form an ABA stacking order, whereas in the 1T phase, the layers form an ABC stacking order. The mechanical stability of these phases of TMDCs depends on the ratio of transition metal and chalcogen components in a particular phase. These may also form a metastable phase [14, 15].

For example, in the case of group VI transition metals (Mo, W), 2H is the thermodynamically most stable phase. Again, at room temperature (RT), orthorhombic 1Td is the stable phase in bulk WTe_2. The 1T and 1T' (distorted) phases are more stable than the 2H phase in WTe_2 at RT. Phase transformation in $MoTe_2$ from the 2H and 1T' phases occurs due to the similar cohesive energy difference [16–20]. The coordination of transition metal with its d-electron is the basis of the electronic structure of TMDCs. These bondings are responsible for the wide range of magnetic and electronic properties of TMDCs. In its 1T and 1H phases, it is the gap between bonding and antibonding bands of M–X bonds where the non-bonding d bands are situated. In TMDCs, the transition metal centers are organized in an octahedral structure (D_{3d}). These metal centers ultimately form the degenerate orbitals $d_z^{22} \cdot {}^{2}_{x-y}(e_g)$ and $d_{yz,xz,xy}(t_{2g})$ which are collectively called the d-electron of TMDCs. Metal centers in TMDCs are again organized in trigonal prismatic coordination (D_{3h}). These are, as in the above case, degenerated to $d^2_{z,} d^{22}_{x-y,\ \ xy}$, and $d_{yz,\ xz,\ xy}$ states. Here, the first two groups of orbitals have a significant gap of around 1 eV. When the nonbonding d orbitals are filled up by group IV–X species, a variety of electronic properties are shown by TMDCs. If metallic TMDCs are formed, e.g., 1T-ReS_2, 2H-$NbSe_2$, when d-orbitals are partially filled, a complete fill-up of d-bands leads to the formation of semiconductor TMDCs, e.g., 1T-HfS_2 and 2H-MoS_2. Chalcogens do not have much effect on TMDCs' electronic structure; rather, its stable phases depend on the transition metal d-electron count [21], which is the basis of the variety of electronic structures in pyrites. Studies show that pyrites can be insulators, e.g., NiS_2, semiconductors, e.g., FeS_2, metals, e.g., CoS_2, or superconductors, e.g., CuS_2 [22], depending on the progressive increase in the ant bonding d electron.

4.3 Magnetic 2D Materials

The discovery of long-range ferromagnetism and anti-ferromagnetism in 2D crystals has opened a spark to further explore the magnetic properties of 2D materials [23–27]. Many researchers have discovered 2D nanomaterials with improved properties [28–37]. Thus, the study of 2D crystals is a wide-ranging platform to explore the origin of magnetic properties in lower dimensions.

Many theories have been revealed about suppressed long-range ordering in 2D materials, and it has again been shown by studies that this magnetic ordering can be stabilized by magnetic anisotropy driven by atomic position-dependent spin-orbit coupling [38, 39]. Thus, structural changes in the crystal cause a large difference in magnetic ordering in 2D material.

In 2017, two ferromagnetic insulators, $Cr_2Ge_2Te_6$ and CrI_3, were reported with low critical temperatures [29, 40]. In 2018, 2D Fe_3GeTe_2 was a new finding whose trilayer has the highest critical temperature (T_c = room temperature) [23]. Many other 2D magneto crystals in different stable phases have been reported in the last few years.

$Cr_2Ge_2Te_6$ is an example of a honeycomb 2D lattice. In $Cr_2Ge_2Te_6$, each Cr atom occupies a major part of the octahedral sites formed by X anions, whereas the Ge-Ge dimers fill up the remaining empty sites along the out-of-plane axis. Replacement of the Ge atom by other atoms such as Si or Sn forms other promising FM semiconductors [41, 42]. Another group of honeycomb-structure 2D materials is layered transition metal trihalides, MX_3 [43, 44]. Monolayer CrI_3, $CrCl_3$, and $CrBr_3$ are promising members of this family.

The Fe_3GeTe_2 monolayer again possesses a triangular lattice structure. It is a recent emerging 2D FM material where ABABA stacking order is observed in Fe and Te atoms [48]. Many other 2D materials with different crystal structures are now being reported by many researchers.

4.4 Origin of Magnetization in 2D Materials

Partially filled 3D orbitals of transition metals are the main source of magnetization in 2D magnets [49]. Macroscopically, it is the intra- and interlayer interactions among the local spins of the magnetic ions that are responsible for the magnetism. A ferromagnetic state is one where all these local spins are aligned in a particular direction. The opposite or random alignment of local spins results in a net zero magnetization, which can form different sublattices. The energy required to align this magnetic moment from an easy axis to a hard axis can be found in anisotropy energy. This magnetocrystalline anisotropy energy has a vital role in the long-range ordering of the magnetic materials. Orientation of magnetic moment in a magneto crystal directs different categories of crystals; e.g., magnetic moment can be oriented randomly (isotropic) or in a plane (easy plane) or can be aligned along a particular axis, and according to the way of orientation, they are named as Heisenberg, XY, and Ising type magnets, respectively.

A magnetocrystalline anisotropy energy (MAE) study is used to determine the energy required for the alignment of the direction of magnetic moments along the hard axis from the easy axis. This energy has an important role in stabilizing of the long-range spin order in this family of materials. The magnetocrystalline anisotropy energy, MAE, arises due to the spin-orbit coupling being dominated by the interaction between the crystal field and magnetic moments. This shows a strong correlation between the structure and magnetic nature of a 2D magneto crystal.

4.5 Mechanical Properties of 2D Nanomaterials

Recently, nanomaterials having good mechanical properties have attracted the attention of researchers due to the large-scale utilization of matrices in advanced technological systems where efficiency of the device, durability, and environmental compatibility of the material are the main requirements [50]. Two-dimensional nanomaterials can be an exceptional option because of their large modulus of elasticity, greater strength, and very low friction. Significant changes in mechanical properties are also observed with the addition of a very small amount of 2D nanomaterials.

2D nanomaterials have unique mechanical properties due to their small thickness, large surface area, and absence of long-range order in comparison to bulk materials. Some of these properties include high tensile strength, elasticity, toughness, plasticity, hardness, fatigue resistance, etc. The hardness of 2D materials depends on the nature of the bonds between the atoms, with materials like graphene having a high hardness. The strong covalent bonds between the atoms in 2D materials result in high tensile strength; graphene is one of the strongest materials known, with a tensile strength of 130 GPa [51]. Black phosphorus has a tensile strength of around

90 GPa, making it one of the strongest 2D materials [53]. 2D materials have high elasticity, which allows them to be stretched and bent without breaking. Like a thin membrane, they can be deformed in-plane or out-of-plane by stretching and bending. Due to this, the elastic properties of this group of materials have both in-plane and bending moduli. Graphene has excellent elastic properties, with a Young's modulus of 1 TPa. Black phosphorus has a Young's modulus of around 100 GPa, making it a relatively stiff material. Beyond the elasticity of 2D materials, their mechanical properties depend strongly on the presence of defect states. A few common defect states noticed in 2D materials are vacancies, dislocations, grain boundaries, and crack-like flaws. These defect states are developed during the deformation. Two mechanical properties that are most important among all are strength and toughness, which describe the reason for failure in terms of the energy and stress of 2D materials [52].

The 2D nature of these materials makes them relatively brittle, so they are not as tough as their bulk counterparts. Due to their thinness and large surface area, 2D materials have high plasticity and can be easily deformed. 2D materials are highly resistant to fatigue, making them suitable for use in high-stress environments. The mechanical properties of 2D nanomaterials are highly anisotropic, meaning that they vary depending on the orientation of the crystal structure.

Nanosheets of boron nitride (BN) are structurally similar to graphene and are composed of atomically thin hexagonal boron nitride (hBN). They possess many properties that make them different from other 2D materials. These BN nanosheets are capable of sustaining at much higher temperatures compared to other materials, and hence they are the best candidates to reinforce the ceramic and metal matrix composites. In addition to enhanced thermal stability, BN nanosheets also have high electrical insulation and optical transparency, and due to this, they can also be used in polymer composites. Monolayers of BN nanosheets have other parameters that are very important in respect of mechanical strength, including Young's modulus of ~0.865 ± 0.073 TPa and fracture strength of ~70.5 ± 5.5 GPa [54]. One of the major interesting properties of BN is that it has a strength that is independent of its number of layers, i.e., with increasing the number of layers up to 9L, it is found to have a similar strength as that of 1L BN, which is found to be absent in graphene also [44–57]. BN nanosheets possess strong interlayer interaction as well as thermal stability, which makes them an ideal candidate for applications in mechanical reinforcement.

The 2D materials family and its members other than graphene have expanded largely, and in all, this family of materials provides great opportunities for different applications. Their strong mechanics and mechanical properties are key parameters to be used in many fields of applications, including large-scale manufacturing and integration [58]. Mechanical properties are influenced by the presence of various defects in 2D materials. Defect engineering may explore many parameters regarding the toughness and strength of 2D materials. Tailoring the mechanical properties of 2D materials from both continuum and atomistic points of view is possible by answering a few fundamental questions [59, 60]. Theoretical consideration beyond van der Waal interactions is necessary for the interfacial properties of 2D materials. Other than these, a detailed study of the effects of surface roughness and capillary bridging can lead to exploring many other areas of information. Adhesion and friction at the interface can be unified by developing a mixed-mode traction-separation relationship [60, 61]. Mechanochemical interactions that lead to wear of the 2D materials can be unraveled by further studying the matter theoretically as well as experimentally.

4.6 Conclusion

In this chapter, it has been tried to summarize the different properties of various 2D materials. From the above discussion, it can be concluded that due to their extraordinary properties, researchers are trying to give more emphasis to these materials. They possess unique magnetic as well as mechanical properties, which encourage the researchers to work in this field more and more. The unique structural properties of this large family of materials can be used to bring about a revolutionary change in technologies for a better tomorrow.

References

1. Avouris, P. and Dimitrakopoulos, C., Graphene: Synthesis and applications. *Mater. Today*, 15, 3, 86–97, 2012.
2. Tozzini, V. and Pellegrini, V., Prospects for hydrogen storage in graphene, Phys. *Chem. Chem. Phys.*, 15, 1, 80–89, 2013.
3. Verma, A. and Parashar, A., The effect of STW defects on the mechanical properties and fracture toughness of pristine and hydrogenated grapheme. *Phys. Chem. Chem. Phys.*, 19, 24, 16023–16037, 2017.
4. Sun, Y., Wu, Q., Shi, G., Graphene based new energy materials. *Energy Environ. Sci.*, 4, 4, 1113, 2011.

5. Li, M., Feng, Y.Y., Liu, E.Z., Qin, C.Q., Feng, W., Azobenzene/graphene hybrid for high-density solar thermal storage by optimizing molecular structure. *Sci. China Technol. Sci.*, 59, 9, 1383–1390, 2016.
6. Sakhaee-Pour, A., Ahmadian, M.T., Vafai, A., Potential application of single-layered graphene sheet as strain sensor. *Solid State Commun.*, 147, 7-8, 336–340, 2008.
7. Choi, W., Lahiri, I., Seelaboyina, R., Kang, Y.S., Synthesis of graphene and its applications: A review. *Crit. Rev. Solid State Mater. Sci.*, 35, 1, 52–71, 2010.
8. Kuila, T., Bose, S., Khanra, P., Mishra, A.K., Kim, N.H., Lee, J.H., Recent advances ingraphenebased biosensors. *Biosens. Bioelectron.*, 26, 12, 4637–4648, 2011.
9. Artiles, M.S., Rout, C.S., Fisher, T.S., Graphene based hybrid materials and devices for biosensing. *Adv. Drug Deliv. Rev.*, 63, 14-15, 1352–1360, 2011.
10. Pumera, M., Graphene in biosensing. *Mater. Today*, 14, 7-8, 308–315, 2011.
11. Ebrahimi, S., Montazeri, A., Rafii-Tabar, H., Molecular dynamics study of the interfacial mechanical properties of the graphene-collagen biological nanocomposite. *Comput. Mater. Sci.*, 69, 29–39, 2013.
12. Son, Y.W., Cohen, M.L., Louie, S.G., Energy gaps in grapheme nanoribbons. *Phys. Rev. Lett., 97*, 21, 216803, 2006.
13. Gomez, A.C., Vicarelli, L., Prada, E., Island, J.O., Acharya, K.L.N., Blanter, S., II, Groenendijk, D.J., Buscema, M., Steele, G.A., Alvarez, J.V., Zandbergen, H.W., Palacios, J.J., van der Zant, H.S.J., Isolation and characterization of few-layer black phosphorus. *2D Mater.*, 1, 025001, 2014.
14. Nan, H., Zhou, R., Gu, X., Xiao, S., Ostrikov, K.K., Recent advances in plasma modification of 2D transition metal dichalcogenides. *Nanoscale*, 11, 19202–19213, 2019.
15. Manzeli, S., Ovchinnikov, D., Pasquier, D., Yazyev, O.V., Kis, A., 2D transition metal dichalcogenides. *Nat. Rev. Mater.*, 2, 17033, 2017.
16. Lin, Y.C., Dumcenco, D.O., Huang, Y.S., Suenaga, K., Atomic mechanism of the semiconducting-to-metallic phase transition in single-layered MoS2. *Nat. Nanotechnol.*, 9, 5, 391–6, 2014.
17. Voiry, D., Fullon, R., Yang, J., de Carvalho Castro e Silva, C., Kappera, R., Bozkurt, I., Kaplan, D., Lagos, M.J., Batson, P.E., Gupta, G., Mohite, A.D., Dong, L., Er, D., Shenoy, V.B., Asefa, T., Chhowalla, M., The role of electronic coupling between substrate and 2D MoS2 nanosheets in electrocatalytic production of hydrogen. *Nat. Mater.*, 15, 1003–1009, 2016.
18. Duerloo, K.A.N., Li, Y., Reed, E.J., Structural phase transitions in two dimensional Mo- and W-dichalcogenide monolayers. *Nat. Commun.*, 5, 4214, 2014.
19. Wang, Q.H., Kalantar Zadeh, K., Kis, A., Coleman, J.N., Strano, M.S., Electronics and optoelectronics of two-dimensional transition metal dichalcogenides. *Nat. Nanotechnol.*, 7, 699–712, 2012.
20. Chae, S., Chae, S.S., Choi, M., Min Park, H., Chang, H., Lee, J.-O., Il Lee, T., Blocking of the 1T-to-2H phase transformation of chemically exfoliated

transition metal disulfides by using a lattice lock. *Nano Energy*, 56, 65–73, 2019.

21. Chhowalla, M., Shin, H.S., Eda, G., Li, L.J., Loh, K.P., Zhang, H., The chemistry of two-dimensional layered transition metal dichalcogenide nanosheets. *Nat. Chem.*, 5, 263–275, 2013.
22. Ogawa, S., Magnetic properties of 3d transition-metal dichalcogenides with the pyrite structure. *J. Appl. Phys.*, 50, 2308–2311, 1979.
23. Deng, Y., Yu, Y., Song, Y., Zhang, J., Wang, N.Z., Sun, Z., Yi, Y., Wu, Y.Z., Wu, S., Zhu, J., Wang, J., Chen, X.H., Zhang, Y., Gate-tunable room-temperature ferromagnetism in two dimensional Fe3GeTe2. *Nature*, 563, 94–99, 2018.
24. Bonilla, M., Kolekar, S., Ma, Y., Diaz, H.C., Kalappattil, V., Das, R., Eggers, T., Gutierrez, H.R., Phan, M.H., Batzill, M., Strong room-temperature ferromagnetism in VSe2 monolayers on van der Waals substrates. *Nat. Nanotechnol.*, 13, 289–293, 2018.
25. O'Hara, D.J., Zhu, T., Trout, A.H., Ahmed, A.S., Luo, Y.K., Lee, C.H., Brenner, M.R., Rajan, S., Gupta, J.A., McComb, D.W., Kawakami, R.K., Room temperature intrinsic ferromagnetism in epitaxial manganese selenide films in the monolayer limit. *Nano Lett.*, 18, 3125–3131, 2018.
26. Gong, Y., Guo, J., Li, J., Zhu, K., Liao, M., Liu, X., Zhang, Q., Gu, L., Tang, L., Feng, X., Zhang, D., Li, W., Song, C., Wang, L., Yu, P., Chen, X., Wang, Y., Yao, H., Duan, W., Xu, Y., Zhang, S.-C., Ma, X., Xue, Q.-K., He, K., Experimental realization of an intrinsic magnetic topological insulator. *Chin. Phys. Lett.*, 36, 076801, 2019.
27. Li, J., Li, Y., Du, S., Wang, Z., Gu, B.L., Zhang, S.C., He, K., Duan, W., Xu, Y., Intrinsic magnetic topological insulators in van der Waals layered MnBi2Te4-family materials. *Sci. Adv.*, 5, eaaw5685, 2019.
28. Xu, R., Liu, B., Zou, X., Cheng, H.-M., Chemical vapor deposition growth and applications of two-dimensional materials and their heterostructures. *ACS Appl. Mater. Interfaces*, 9, 38796–38801, 2017.
29. Gong, C., Li, L., Li, Z., Ji, H., Stern, A., Xia, Y., Cao, T., Bao, W., Wang, C., Wang, Y., Qiu, Z.Q., Cava, R.J., Louie, S.G., Xia, J., Zhang, X., Discovery of intrinsic ferromagnetism in two-dimensional van der Waals crystals. *Nature*, 546, 265–269, 2017.
30. Torelli, D., Thygesen, K.S., Olsen, T., High throughput computational screening for 2D ferromagnetic materials: The critical role of anisotropy and local correlations. *2D Mater.*, 6, 045018, 2019.
31. McGuire, M., Crystal and magnetic structures in layered, transition metal dihalides and trihalides. *Crystals*, 7, 121, 2017.
32. Lu, S., Zhou, Q., Guo, Y., Zhang, Y., Wu, Y., Wang, J., Coupling a crystal graph multilayer descriptor to active learning for rapid discovery of 2D ferromagnetic semiconductors/half-metals/metals. *Adv. Mater.*, 32, 2002658, 2020.
33. Mounet, N., Gibertini, M., Schwaller, P., Campi, D., Merkys, A., Marrazzo, A., Sohier, T., Castelli, I.E., Cepellotti, A., Pizzi, G., Marzari, N., Two-dimensional

materials from high-throughput computational exfoliation of experimentally known compounds. *Nat. Nanotechnol.*, 13, 246–252, 2018.
34. Liu, H., Sun, J.-T., Liu, M., Meng, S., Screening magnetic two-dimensional atomic crystals with nontrivial electronic topology. *J. Phys. Chem. Lett.*, 9, 6709–6715, 2018.
35. Zhu, Y., Kong, X., Rhone, T.D., Guo, H., Systematic search for two-dimensional ferromagnetic materials. *Phys. Rev. Mater.*, 2, 081001, 2018.
36. Li, X. and Yang, J., First-principles design of spintronics materials. *Natl. Sci. Rev.*, 3, 365–381, 2016.
37. Zhang, S.Q., Zou, X.L., Cheng, H.M., Mechanical-electro-magnetic coupling in strained bilayer CrI_3. *Sci. China Technol. Sci.*, 63, 1265–1271, 2020.
38. Zhong, T., Li, X., Wu, M., Liu, J.-M., Rev., Room-temperature multiferroicity and diversified magnetoelectric couplings in 2D materials. *Nat. Sci. Rev.*, 7, 373-380, 2019.
39. Huang, B., Clark, G., Klein, D.R., MacNeill, D., Navarro-Moratalla, E., Seyler, K.L., Wilson, N., McGuire, M.A., Cobden, D.H., Xiao, D., Yao, W., Jarillo-Herrero, P., Xu, X., Electrical control of 2D magnetism in bilayer CrI3. *Nat. Nanotechnol.*, 13, 544–548, 2018.
40. Gong, C., Li, L., Li, Z., Ji, H., Stern, A., Xia, Y., Cao, T., Bao, W., Wang, C., Wang, Y., Qiu, Z.Q., Cava, R.J., Louie, S.G., Xia, J., Zhang, X., Discovery of intrinsic ferromagnetism in two-dimensional van der Waals crystals. *Nature*, 546, 265–269, 2017.
41. Chakroborty, S., Bharadwaj, V., Sahoo, S.K., Sensing and biosensing with 2D nanosheets beyond graphene, in: *Sensing and Biosensing with Optically Active Nanomaterials*, pp. 119–141, Elsevier, 2021.
42. Soren, S., Chakroborty, S., Pradhan, L., Chandra, P., Sahu, J., Parhi, P., Hydrothermal synthesis of graphene modified SnO nanocomposite for oxygen reduction reaction. *Mater. Today Proc.*, 57, 1, 72–76, 2022.
43. Hillebrecht, H., Schmidt, P.J., Rotter, H.W., Thiele, G., Zönnchen, P., Bengel, H., Cantow, H.J., Magonov, S.N., Whangbo, M.H., Structural and scanning microscopy studies of layered compounds MCl3 (M = Mo, Ru, Cr) and MOCl2 (M = V, Nb, Mo, Ru, Os). *J. Alloys Compd.*, 246, 70–79, 1997.
44. Klemm, W. and Krose, E., The crystal structures of ScCl3, TiCl3 and VCl3. *Z. Anorg. Allg. Chem.*, 253, 218–225, 1947.
45. Panda, P., Pal, K., Chakroborty, S., Smart advancements of key challenges in graphene-assembly glucose sensor technologies: A mini review. *Mater. Lett.*, 303, 130508, 2021.
46. Soren, S., Chakroborty, S., Pal, K., Enhanced in tunning of photochemical and electrochemical responses of inorganic bi-metal oxide nanoparticles via rGO frameworks (MO/rGO): A comprehensive review. *Mater. Sci. Eng. B*, 278, 115632, 2022.
47. Chakroborty, S. and Panda, P., Nanovaccinology against infectious diseases, in: *Nanovaccinology Outbreak As Targeted Therapeutics*, Wiley-VCH, Verlag GmbH, 2022.

48. Deiseroth, H.-J., Aleksandrov, K., Reiner, C., Kienle, L., Kremer, R.K., Fe3GeTe2 and Ni3GeTe2–Two new layered transition-metal compounds: Crystal structures, HRTEM investigations, and magnetic and electrical properties. *Eur. J. Inorg. Chem.*, 2006, 8, 1561–1567, 2006.
49. Mak, K.F., Shan, J., Ralph, D.C., Probing and controlling magnetic states in 2D layered magnetic materials. *Nat. Rev. Phys.*, 1, 646–661, 2019.
50. Ji, Z., Zhang, L., Xie, G., Xu, W., Guo, D., Luo, J., Prakash, B., Mechanical and tribological properties of nanocomposites incorporated with two-dimensional materials. *Friction*, 8, 813–846, 2020.
51. Ranjbartoreh, A.R., Wang, B., Shen, X., Wang, G., Advanced mechanical properties of graphene paper. *J. Appl. Phys.*, 109, 1, 014306, 2011.
52. Imani Yengejeh, S., Wen, W., Wang, Y., Mechanical properties of lateral transition metal dichalcogenide heterostructures. *Front. Phys.*, 16, 1–7, 2021.
53. Wei, Q. and Peng, X., Superior mechanical flexibility of phosphorene and few-layer black phosphorus. *Appl. Phys. Lett.*, 104, 25, 251915, 2014.
54. Senturk, A.E., Outstanding thermo-mechanical properties of graphene-like B3C3 and C3N3. *Appl. Phys. A*, 126, 8, 584, 2020.
55. Baimova, J.A., Liu, B., Dmitriev, S.V., Zhou, K., Mechanical properties and structures of bulk nanomaterials based on carbon nanopolymorphs. *Phys. Status Solidi (RRL)–Rapid Res. Lett.*, 8, 4, 336–340, 2014.
56. Andrew, R.C., Mapasha, R.E., Ukpong, A.M., Chetty, N., Mechanical properties of graphene and boronitrene. *Phys. Rev. B*, 85, 12, 125428, 2012.
57. Falin, A., Cai, Q., Santos, E.J., Scullion, D., Qian, D., Zhang, R., Li, L.H., Mechanical properties of atomically thin boron nitride and the role of interlayer interactions. *Nat. Commun.*, 8, 1, 15815, 2017.
58. Papageorgiou, D.G., Kinloch, I.A., Young, R.J., Mechanical properties of graphene and graphene-based nanocomposites. *Prog. Mater. Sci.*, 90, 75–127, 2017.
59. Liu, Y., Xie, B., Zhang, Z., Zheng, Q., Xu, Z., Mechanical properties of graphene papers. *J. Mech. Phys. Solids*, 60, 4, 591–605, 2012.
60. Akinwande, D., Brennan, C.J., Bunch, J.S., Egberts, P., Felts, J.R., Gao, H., Zhu, Y., A review on mechanics and mechanical properties of 2D materials—Graphene and beyond. *Extreme Mech. Lett.*, 13, 42–77, 2017.
61. Jiang, H., Zheng, L., Liu, Z., Wang, X., Two-dimensional materials: From mechanical properties to flexible mechanical sensors. *InfoMat*, 2, 6, 1077–1094, 2020.

5

Electrical, Plasmonic, and Optical Properties of 2D Nanomaterials

Ankita Subhrasmita Gadtya and Srikanta Moharana*

School of Applied Sciences, Centurion University of Technology and Management, Bhubaneswar, Odisha, India

Abstract
Two-dimensional (2D) nanomaterials have received tremendous attention in recent times because of their optical, electrical, and plasmonic properties, which have many uses in optoelectronics. These uses span from light-emitting diodes and optical sensors to photovoltaic solar cells and photo-electrochemical cells. Metal nanomaterials may generate a wide range of fascinating hybrid nanostructures with exceptional plasmonic capabilities, thanks to the 2D shape and atomically thin thickness that characterize 2D materials. Furthermore, the hybrid nanostructures display unprecedented optical capabilities that are the result of the interaction between plasmonic optical effects and the particular physicochemical features of 2D materials. These capabilities emerge from the interplay between these two factors. Because of the synergistic relationship between plasmonic optical effects and the odd physicochemical features of 2D materials, hybrid systems made up of graphene or other 2D materials and plasmonic metal nanostructures offer exceptional optical properties. These qualities are a result of the fact that plasmonic optical effects enhance the peculiar physicochemical properties of 2D materials.

***Keywords*:** 2D nanomaterial, electrical, optical, plasmonic properties, biomedical, energy

**Corresponding author*: srikantanit@gmail.com

Subhendu Chakroborty and Kaushik Pal (eds.) 2D Nanomaterials: Synthesis, Properties and Applications, (73–102) © 2024 Scrivener Publishing LLC

5.1 Introduction

Chemical compounds or materials that are produced and used at a very small scale, i.e., 1–100 nm in at least one dimension, are known as nanomaterials (NMs). These nanoparticles can be found in a variety of states, including dispersed forms, suspensions, colloids, and agglomerates, all of which are dependent on their chemistry and electromagnetic properties. For example, magnetic NPs tend to aggregate until their surfaces are functionalized. Nanomaterials' surface and internal structures are frequently made up of multiple layers or phases containing multiple components, all essential for achieving the required functionality. Atoms can arrange themselves in different ways in lattice structures depending on external factors like temperature or pressure, which can result in distinct crystal forms for some materials with the same quantities of included components. However, the increased potential energy at the surface and the high reactivity of NMs are contributing factors to their instability. Exposure to air or light causes quick deterioration. Due to their large surface-to-volume ratio and the ability to expand the number of active surfaces by modifying the nanocrystal structure, nanomaterials have long been viewed as promising heterogeneous catalysts. As the particle size is reduced, the contribution of surface atoms to the overall characteristics of the nanomaterial increases, and this has a significant effect. The size and shape of NMs can affect a variety of their physical characteristics. Properties including melting point, fluorescence, electrical conductivity, magnetic permeability, and chemical reactivity vary as a function of particle size when the particle size is reduced to the nanoscale. These characteristics, however, are not primarily based on the size of the materials. It is becoming obvious that the structure of the materials also plays an equally vital role in influencing the properties of nanoparticles as nanotechnology develops toward applications in electronics and medicine. There are many advantages of NMs, such as: (a) adsorption and absorption of different molecules are higher and faster; (b) they work better when employed as catalysts, gas sensors, etc. because they have a high surface-to-volume ratio; (c) they also possess a high percentage of atoms or molecules on the surface, which again leads to unique properties; (d) metal nanoparticles provide superior scattering capabilities; and many more [1–5]. NMs may be found in a variety of states, including disseminated, suspended, colloidal, and agglomerated, all of which are dependent on their chemistry and electromagnetic properties. For instance, magnetic nanoparticles (NPs) tend to aggregate until their surfaces are functionalized. As researchers discovered that a substance's size affects its physicochemical qualities, including its physiochemical, electric, mechanical strength, and optoelectronic performances, they realized the significance of NMs. Due to thei enhanced properties in different sectors, NMs have a variety of applications. For example, NMs are used in energy storage applications through photovoltaic cells, light-emitting diodes, batteries, etc. [6–8]. Some of the many medical applications of NMs include bio-imaging, drug/gene delivery, live-cell tracking, anti-cancer medicines, and many more [9, 10]. Water purification is also one of the major applications of nanomaterials by removing heavy metals in the agricultural sector, among many others [11–14].

NMs may be classified into several subsets according to a wide range of criteria. Dimensions, shape, condition, and chemical composition are often used to classify NMs. At least one defining characteristic is that at least one of their dimensions is between 1 and 100 nm. So, there are several kinds of NMs depending on their shape, size, chemical composition, and other characteristics. Nanomaterials may range from those made of carbon to those made of metal, semiconductors, polymers, or lipid. The skeletal constituent of carbon-based NMs is carbon alone. Carbon nanotubes (CNTs) and fullerenes are commonly found in carbon-based NMs. Divalent and trivalent metal ions serve as the building blocks for metal-based NMs. They have a high surface area and better adsorption capabilities. Gold nanoparticles and silver nanoparticles are the main examples. Metallic and non-metallic characteristics can be found in semiconductor NMs. They display wide band gaps by modifying them and are used to improve the physiochemical properties of the semiconductor. Some examples are ZnS, CdTe, GaP, GaN, etc. Liposomes, nanostructured lipid carriers (NLCs), and solid lipid nanoparticles (SLNs) are all examples of lipid-based NMs employed in drug delivery [15, 16]. Based on their dimensionality, NMs fall into one of four broad classes. Nanomaterials may exist in a variety of dimensions, including zero, one, two, and three. Figure 5.1 displays several instances of NMs and a categorization scheme for them.

(i) Zero-dimensional NMs: The 0D carbon-based nanomaterials stand out for their unrivaled electrical and optoelectronic properties, low toxicity, and high quantum yield. There are many different types of 0D NMs, but the most well-known ones include graphene quantum dots (GQD), carbon quantum dots (CQD), fullerenes, nano-diamonds, magnetic nanoparticles (MNs), gold, and silver nanoparticles [17, 18].

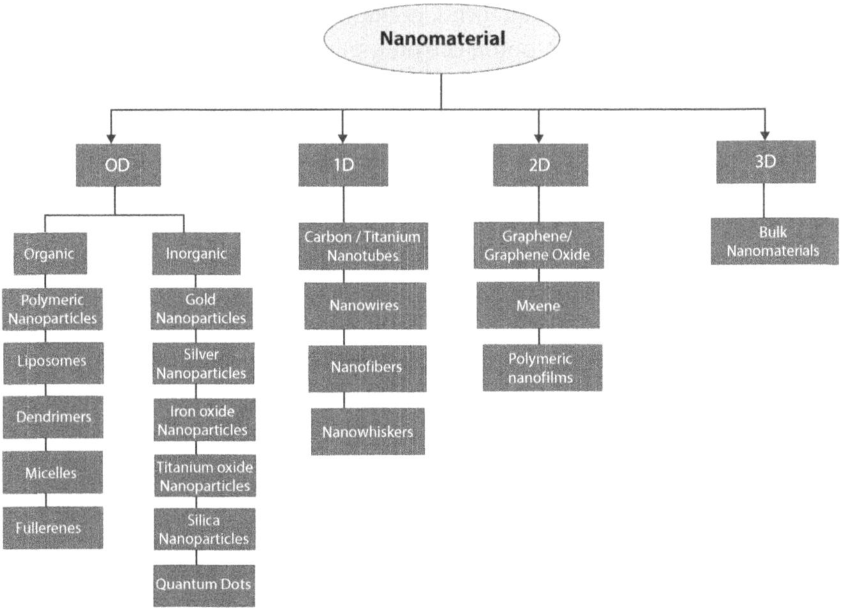

Figure 5.1 Classification of nanomaterials based on 0D, 2D. and 3D-dimensional.

(ii) One-dimensional NMs: The superior surface areas, additional surface-active sites, and improved permeability of 1D nanomaterials, such as nanowires (NWs), nanofibers, nanorods, nanobelts, and nanotubes (NTs), can considerably boost the energy density, power density, and excellent cycle presentation for energy storage [19, 20].

(iii) Two-dimensional NMs: 2D NMs are made up of thin layers, which may have a thickness of at least one atomic layer. A few examples of 2D NMs are graphene and graphene-based materials, transition metal oxides and dichalcogenides, hexagonal boron nitrides, etc. [21].

(iv) Three-dimensional NMs: The 3D NMs span all dimensions of the nanoscale. Over 100 nm, these materials have three arbitrary dimensions. Bundles of nanotubes, nanowires, core shells, and quantum dots are common examples of 3D NMs.

These nanomaterials can be synthesized *via* two approaches, i.e., the top-down approach and the bottom-up approach. The top-down approach includes mechanical grinding, ball milling, electric arc deposition, chemical etching, sputtering, and many more physical techniques. Whereas the bottom-up approach involves aggregation, pyrolysis, chemical vapor deposition (CVD), flame synthesis, spinning, etc. With the top-down method, larger, externally controlled devices are used to direct the assembly of nanoscale devices. The top-down method frequently makes use of conventional workshop or microfabrication techniques. Molecules organize themselves into more complex assemblies, such as atom-by-atom, molecule-by-molecule, and cluster-by-cluster, using the bottom-up method (e.g., the growth of a crystal). The bottom-up method is demonstrated by the colloid dispersion method of nanoparticle synthesis. When compared to nanostructures made using the top-down method, bottom-up methods typically result in fewer flaws. An approach that uses both methods is referred to as a hybrid approach. But recently, several research teams have suggested using biological systems to create NMs, which leads to enhanced polydispersity, dimensions, and stability. This biological synthesis includes synthesis through plant extracts, yeast, biological particles, etc. By permitting the creation of nanoparticles at physiological pH, temperature, and pressure while at the same time incurring minimal expense, biological techniques of NM synthesis would help to eliminate harsh processing conditions. The physical and chemical approaches are quite expensive [22–26].

In this typical chapter, we will cover the potential uses of 2D nanomaterials in the biomedical and energy sectors, along with the electrical, optical, and plasmonic properties of these materials. A clear analysis of the most recent discoveries is presented, and a brief discussion of the possible applications of these findings is included.

5.2 Overview of Two-Dimensional Nanomaterials (2D NMs)

Two-dimensional nanomaterials (NMs) with ultrathin structures and transverse dimensions bigger than 100 nm are a relatively new kind of nanomaterial and their thickness is under 5 nm. Because of their one-of-a-kind shapes, 2D NMs tend to have large surfaces and anisotropic physical and chemical properties [27, 28]. When it comes to their physiochemical and biological characteristics, NMs' atomic configurations are crucial. 2D materials are distinguished from other types of NMs by the fact that one of their dimensions has a thickness of only a few atomic layers. The size and thickness of 2D NMs at the macro and nanoscales suggest that they are the smallest NMs possible. Despite the strong in-plane connections between these NMs, the van der Waals force between the layers is rather moderate [21]. Moreover, various 2D materials like graphene, hexagonal boron nitride (hBN), and metal dichalcogenides (MX_2) have garnered significant interest among academic and industrial researchers in recent years due to their favorable properties and numerous applications in a variety of fields, including electronics, optoelectronics, catalysts, energy storage facilities, sensors, solar cells, lithium batteries, composites, etc. [29–34]. Ultrathin 2D nanomaterials are appealing for use in fundamental condensed matter research and electronic device applications because of their compelling electrical characteristics and lack of interlayer interactions, especially in the case of single-layer nanosheets. Due to their atomic thickness, 2D NMs provide the highest mechanical flexibility and optical transparency, making them attractive for the fabrication of exceptionally flexible and transparent electrical and optoelectronic devices. Their vast lateral breadth and very thin thickness provide a very high specific surface area, making them perfect for surface-active applications [35]. When stretched in-plane, 2D NMs are rigid, but when bent or folded out-of-plane, they exhibit a rippled or wrinkled appearance. The local electronic structure can be modulated by mechanical distortion of the 2D lattice, which also has a direct impact on chemical reactivity. Surface topography has an impact on the microscale interactions of 2D nanomaterials, which can be exploited to promote cellular adherence to surfaces [36–39].

2D NM has been categorized by Gupta *et al.* based on stability. They are the 2D oxides, 2D chalcogenides, and graphene families. There are various types of graphene, including graphene itself, graphene oxide (GO), fluorographene, etc., which are completely stable at room temperature in ambient environments. The 2D chalcogenides have metallic dichalcogenides, layered semiconductors, and semiconducting dichalcogenides like TaS_2, $NbSe_2$, $MoTe_2$, $MoSe_2$, etc. The semiconducting dichalcogenides are stable in air but unstable in inert atmospheric conditions. The 2D oxides comprise micas, layered Cu oxides, hydroxides, perovskite types, etc. This layered 2D material has a variety of good electrical, mechanical, optical, and thermal properties and may one day replace silicon in nanoelectromechanical systems (NEMS), photonics, and other electronic and photonic applications. Most scientists see using 2D materials in this way in practical applications [40, 41]. Different types of 2D nanomaterials (TMDs) include graphene, graphene oxide (GO), reduced graphene oxide (rGO), clay materials, transition metal oxides (TMOs), and transition metal dichalcogenides. Black phosphorus, graphitic C_3N_4, hexagonal boron nitride (hBN), graphitic C_3N_4, boron nanosheets (BNs), antimonene (AM), and elemental monolayers such as arsenate, germanene, silicene, and antimonide are examples

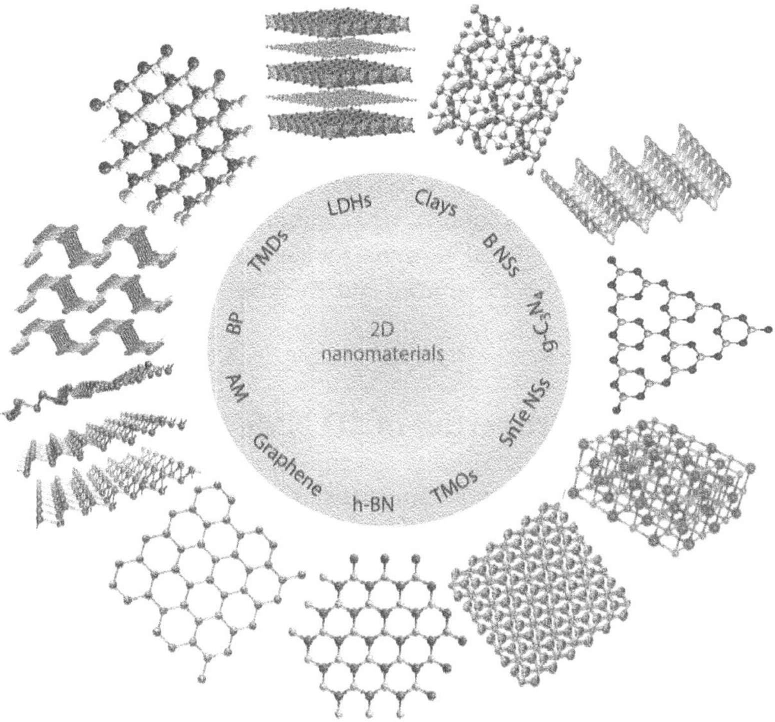

Figure 5.2 Carbon-based nanomaterials, silicate clays, LDHs, TMDs, TMOs, BP, g-C_3N_4, h-BN, AM, BNSs, and SnTe NSs are all depicted in this simplified diagram of 2D nanomaterials [42]. Adapted with permission from Ref [42] Copyright (2019) (Elsevier).

of other 2D materials. 2D nanomaterials typically consist of one or more atomically thin layers (Figure 5.2) [42].

However, the atomic arrangements of 2D nanomaterials have a significant impact on their biological and physicochemical properties. They are distinct from other materials because their vertical thickness is only a few atoms. Graphene is a carbon allotrope in which hexagonal carbon atoms are bonded together by strong sp^2 covalent interactions [43]. The presence of oxygenated groups on the surface of GO makes it more chemically active than graphene, despite both materials being comprised of hexagonal carbon atoms. This simultaneously increases the variety of potential interactions occurring on the GO surface. Several methods exist for the chemical or thermal reduction of GO, which can result in the production of rGO, a material with properties somewhere between graphene and GO [42]. When compared to graphene, reduced graphene oxide (rGO) exhibits a larger concentration of structural flaws because it maintains some oxygen functional groups after reduction. However, the thickness of 2D nanomaterials (silicate clays, LDHs, TMDs, and TMOs) is composed of stable single-crystal units. This is because 2D nanomaterials have a thickness of only a few hundred atoms. For instance, laponite is a synthetic clay mineral that has a structure that is very similar to that of natural clays. The structure of laponite consists of layered units of tetrahedral silica sheets that enclose an octahedral layer that contains Mg2+ cations. Na+ 0.7[($Si_8Mg_{5.5}$-$Li_{0.3}$) $O_{20}(OH)_4]^{-0.7}$ is the empirical formula for laponite, and its disk-shaped individual crystals have a diameter and thickness of about 25 nm and 0.92 nm [44].

5.3 Electrical Properties of 2D NMs

The electrical characteristics of nanoparticles have a significant impact on the mobility of charge carriers. When a material is shrunk to nanoscale dimensions, it inevitably experiences the quantum-size effect and the quantum confinement effect. Because electron energy is quantized, conductive metal nanoparticles may become nonconductive at very specific voltages. The conductivity and resistivity properties fall under the electrical properties [45, 46]. In the presence and absence of certain gases, the electrical properties of 2D NMs vary. Electric properties are intimately related to bandgaps and the technical parameters that change their values. The highly selective and sensitive sensors GO and rGO, which have zero bandgaps, are created without changing their electric characteristics. High electric conductivity and low Johnson electric noise characterize these materials. Electric conductivities, on the other hand, alter dramatically because of the change in carrier concentration brought on by gases. The bandgap is altered by dopants in 2D materials. The bandgap space is directly correlated with the electric conductivity of 2D materials due to semiconducting characteristics, dopants, defects, and functionalization. The electric conductivity of the sensing material will be high if the bandgap is narrow. Because the band gap energy of NMs increases as their particle sizes drop, especially for semiconductor nanoparticles, their electrical conductivity is often lower than that of bulk materials [47–49]. Self-assembled monolayers (SAMs) can alter the electrical properties of 2D NMs. The electrical properties are altered when 2D NMs are applied on top of SAMs [50]. Due to their thin layers' atomic-thick structure, 2D NMs have recently been employed for gas sensors owing to their superior surface activity and huge surface-to-volume ratios, which enable them to absorb enormous amounts of gas molecules. Additionally, because 2D NMs can exist in more than one crystalline structure, there is a greater chance that they will exhibit either a semiconducting phase or a semi-metallic phase, allowing for the band gap to be tuned. They are also highly sensitive to a wide range of chemicals, and their thickness alters the material's physical and chemical properties [51–54]. Gas sensing properties on graphene can be differentiated through chemical doping with foreign elements thanks to an improved reversible reaction. To explore novel sensing characteristics, ternary or multi-component 2D layers can be manufactured and studied. Due to the increased gas penetration *via* the pores, porous architectures can significantly improve the sensing properties of nanosheet materials for structural purposes. Also, by increasing the density of edge sites, nanopores can produce active reaction sites. In terms of sensor fabrication, integrating nanosheet materials on a flexible substrate will be a cutting-edge method for use in wearable real-time sensors. Although several 2D NMs have been combined with plastic substrates to demonstrate flexible gas sensors, the bulk of 2D NMs have been tested on rigid substrates due to the challenges involved in producing them on flexible substrates. Moreover, the sensitivity and dependability of the 2D NMs' sensing capability on flexible sensors need to be increased under mechanical stress [55]. Kanan and his co-workers discussed the use of 2D inorganic NMs in surface-enhanced Raman spectroscopy (SERS) chemical sensor applications. Due to their intriguing morphology (a lot of edge locations for hot spot production) and substantial surface area, dendritic nanostructures have recently been used in SERS sensing [56]. Figures 5.3(a–f) presents the scanning electron microscopy (SEM) image of gold "nanoflowers: made of dendritic nanoparticles that exhibit signal amplification for the detection of rhodamine 6G, and the utilization of Ag/Au bimetallic dendritic structures created on Si substrates for the SERS detection of crystal violet. Dendritic Ag/Au and Ag/Pd bimetallic nanostructures have been created through multistage galvanic replacement reaction (MGRR) for use as SERS substrates [56–63]. The development of field-effect transistors (FETs) with improved ON/OFF ratios, conductivity, carrier mobility, and lower power consumption has been a primary focus of research into 2D materials-based semiconductor devices. Since typical electrical characterization methods might yield incorrect findings when used with ultra-thin 2D layered materials, understanding the electrical characteristics of such devices is crucial. In a bulk semiconductor, for instance, the electrical conductivity at ambient temperature is proportional to the density of charge carriers. However, because of their atomic thinness, 2D materials cannot undergo the usual implanted substitutional doping. The generation of electron and hole carriers in 2D materials requires special procedures including charge transfer

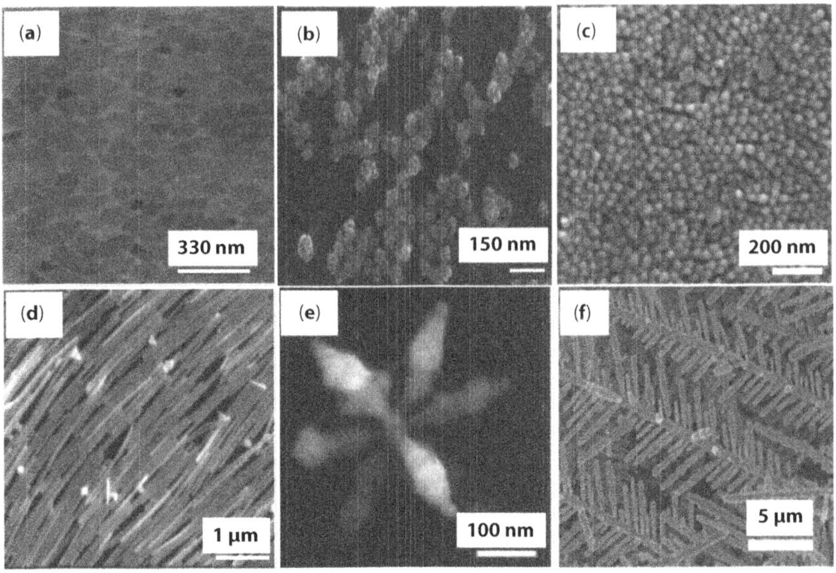

Figure 5.3 Nanostructures of Au and Ag used in SERS substrate manufacturing, as seen by scanning electron microscopy. A gold nanoprism (a), silver nanoflower (b), gold nanorod (c), silver nanowire (d), silver nanostar (e), and gold nanodendrite (f) [58–63]. Adapted with permission from Ref. [58–63].

doping, which are not necessary in ordinary semiconductor materials. Also, in few-layer materials, charge density drops sharply with depth [64].

Using conventional electron beam lithography and subsequent deposition of 3 nm of Cr and 15 nm of Au, a total of twelve two-terminal devices with Ti_3C_2Tx Mxene channels were successfully constructed. In Figure 5.4a, a monolayer Ti_3C_2Tx flake is shown connecting the source (S) and drain (D) electrodes on top of a Si/SiO_2 substrate, illustrating the device architecture. In the electrical tests, the substantially p-doped Si substrate functioned as the bottom gate (G) electrode. In Figure 5.4b, it is observed that the Cr/Au electrodes are highlighted in yellow in a scanning electron microscopy (SEM) picture of a typical Ti_3C_2Tx device. Typically, the length of the channels in the manufactured devices was around 5 mm. To reduce the impact of surface adsorbates, electrical characterization of the devices was carried out in a vacuum at a residual pressure of [23] 106 Torr after 2 days of evacuation [27]. Figure 5.4c depicts the relationships between the IDS and VDS for gates set to 40 V, 0 V, and 40 V. There is only a little change in the electrical conductivity of Ti_3C_2Tx as a function of gate voltage, as shown by the linearity of the IDS-VDS relationships and their

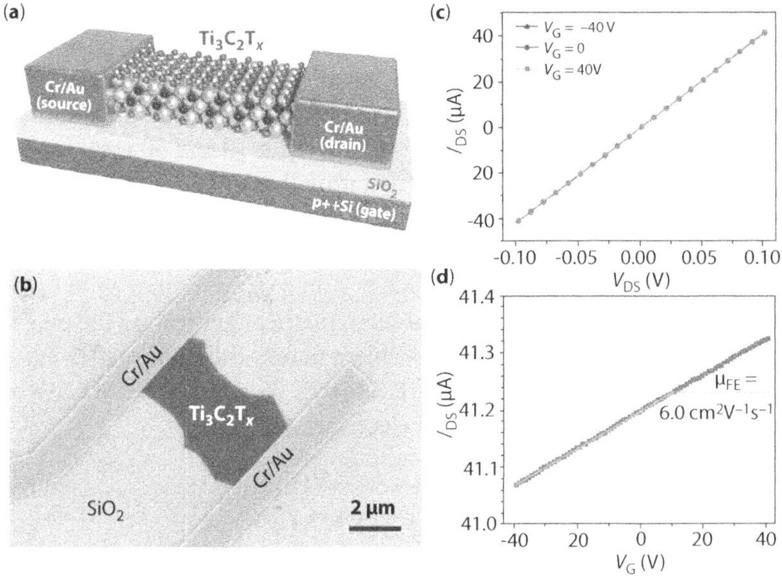

Figure 5.4 (a) Ti_3C_2Tx channel layout for a two-terminal device; (b) SEM picture in false colors showing a monolayer Ti3C2Tx MXene flake being used as the basis for a two-terminal device. To make it easier to see, the Cr/Au electrodes are shaded a bright yellow. (c) Two-terminal IDS-VDS curves for the Ti3C2Tx device illustrated in (b) at 40, 0, and 40 V gate voltages; and (d) The transport properties of a monolayer Ti3C2Tx FET with two terminals. VDS = 0.1 V [65]. Adapted with permission from Ref. [65]. Open access under a CC BY 4.0 license. Copyright © 2021.

close coincidence. The transfer characteristics recorded between 40 V and +40 V applied to the gate electrode are displayed in Figure 5.4d. The greatest value of mFE for Ti_3C_2Tx MXene devices was found to be 6 $cm^2\ V^{-1}\ s^{-1}$ for four devices. $Ti_3C_2T_x$ flakes synthesized using the MILD technique have an average value of μFE = 5.2 G 0.9 $cm^2\ V^{-1}\ s^{-1}$; this is a significant increase from our earlier report of mFE = 2.6 $cm^2\ V^{-1}\ s^{-1}$ [65].

In terms of electrical conductivity, the Ti_3C_2Tx film spray coated onto a PET sheet performed the best (1.2×10^6 S m^{-1}). When combined with other conductive fillers, graphene-based coatings (used by 39%) often produce higher conductivities compared to MXene-based coatings (12%). After being blade coated onto a PET sheet and sintered with copper nanoparticles (Cu NPs) and silver nitrate ($AgNO_3$), the most conductive graphene-based coating measured 8×10^5 S m^{-1}. Similarly, rGO improved with PEDOT; PSS was employed in the most electrically conductive textile coating tested (6.4×10^4 S m^{-1}) [93]. The combination of silver nanowires

(AgNWs) and polydopamine (PDA) on rGO resulted in the most electrically conductive foam (59 S m^{-1}) [66]. Matrix conductivity performance and dispersion stability can both be negatively affected by MXenes' propensity to oxidize. Conductivity values dropped to 7% and 40% of the initial conductivity, respectively, in a PVA matrix at 10 and 50 wt.% loadings, as shown by Habib *et al.* [67], demonstrating that Ti_3C_2Tx oxidizes quicker in liquid media, a significant fall in performance. The long-term stability of these MXene nanocomposites is, however, seldom considered in studies. Factors like humidity, UV exposure, voltage, and frequency of use may impact the rate and quantity of oxidation, further aggravating the problem. To comprehend the scope of the problem across different matrix systems and flake surface chemistries, future research on conductive MXene coatings should incorporate long-term electrical stability.

5.4 Optical Properties of 2D NMs

The optical properties of a material are strongly affected by its electronic band structure and screening. All the optical characteristics, such as reflectance, transmission, absorption, and emission, of a nanomaterial are determined by its electronic structure, which changes greatly for various morphologies since it is highly reliant on surface atoms. Increased gaps between energy levels (the quantum effect) and surface plasmon resonance play important roles in how nanomaterials' optical characteristics vary with size [68]. Gregorenko *et al.* [69] have given a great example of graphene plasmonics. They have discussed the optical conductivity of graphene, which is unaffected by any characteristics of the substance. The absorption of monolayer graphene, $\pi\alpha$, (around 2.3%), is exclusively determined by the fine structure constant and is unrelated to optical wavelength. As light shines on graphene, the excited electron-hole pairs cause rapid heating of the carriers, which can last for a few picoseconds at a temperature higher than that of the lattice. After that, slower scattering between charge carriers and acoustic phonons brings the heated electrons and lattice to equilibrium. The states close to the edges of the conduction and valence bands were filled when the optical excitation intensity was strong [70]. Oscillator intensity between localized d-states with substantial spatial overlap and dipole transitions with large joint densities of states on the transition metal atoms is responsible for the extraordinary optical absorption at visible light in Transition Metal Dichalcogenides (TMD) monolayers. Recently, 2D-TMDs have been suggested for use in photovoltaics and photocatalysis because of their very strong optical absorption in the visible frequency range.

Because the number of hot carriers created exceeded the density of carriers already present in graphene. In terms of its optical properties, black phosphorus (BP) is a prototypical 2D nanomaterial. A direct bandgap exists in BP, and this bandgap can connect large bandgap TMDs with zero bandgap graphenes. Moreover, the sample thickness can change the band gap of BP, which facilitates fine-tuning the BP absorption spectra. In the hexagonal structure of BP, there are two in-plane directions known as zigzag and armchair directions. It has been determined that, regardless of thickness, the armchair direction is preferable for polarized light absorption [71, 72].

Excitonic effects dominate the optical response in 2D monolayer semiconductors. The reduction in dimensionality and weak dielectric screening lead to excitons even at room temperature [73, 74]. Yang and his co-workers have performed an experiment about the excitonic effects on the optical response of graphene and bilayer graphene. Using many-electron effects, we have undertaken first-principles calculations on the quasiparticle energies and optical characteristics of single and bilayer graphene and graphite. In comparison to the independent-particle model, the optical absorption spectra in the energy range around a van Hove singularity are significantly altered by resonant excitonic processes in graphene and bilayer graphene. Lastly, in accordance with experimental results, we have demonstrated that excitonic effects do not alter the absorbance in the infrared range from that calculated inside the single-particle image [75]. One of the most anisotropic layer compounds is hexagonal boron nitride (h-BN), which is a fascinating quasi-two-dimensional insulator comparable to semimetallic graphite [76, 77].

5.5 Plasmonic Properties of 2D NMs

In the emerging field of plasmonics, researchers are investigating the materials' nanoscale light manipulation capabilities. Surface plasmon resonance is a phenomenon through which plasmonic materials can obtain exceptional optical properties by exploiting the light-matter interaction. Surface plasmons may be used to confine electromagnetic fields to incredibly small areas, and their behavior can be modulated in a broad variety of ways. The poor performance characteristics of Ag, Au, and Al metals (radiative losses, high energy dissipation, and restricted tunability) led to the failure of earlier attempts to use these metals as plasmonic materials. As a solution to these problems, a family of two-dimensional (2D) materials is proposed due to their significant light–matter interaction phenomena leading to efficient quantum confinement effects, making them ideal for plasmonic applications. Semiconductors, conductive oxides, and dielectrics are just a few of the numerous classes of materials that have been investigated as possible plasmonic materials due to their exceptional plasmonic properties. The better features of 2D materials, such as their malleability in terms of bandgap and electron transport, have piqued more interest in plasmonic applications [35, 78]. Atomically thin 2D nanomaterial sheets have electrical structures and optical features that differ from their bulk morphologies due to the quantum confinement phenomenon. The enhanced electrical and optical characteristics of 2D nanomaterials make their plasmonic features more alluring. Because of the strong coupling impact shown by the excitonic layered transition metal dichalcogenides (LTMDCs)-plasmonic structures hybrid system and metallic graphene's support for SP mode in the infrared regime, the plasmonic capabilities of 2D nanomaterials are enhanced in exciting ways [79]. However, MXenes are a novel type of 2D plasmonic material that was created in 2011 by combining nitrides and carbides. They are more flexible than graphene, have superior electromagnetic properties, and can be folded and stretched without breaking, making them ideal for a wide range of applications in fields including energy storage, supercapacitors, photonics, and plasmonics. The metal or semiconductor properties of MXenes can be tuned by tinkering with their conductivity by surface termination, composition, doping, or mixing [80]. It has been observed that h-BN has focused on its usage as a substrate for thin-layered materials; however, recent studies have shown that it also holds promise as

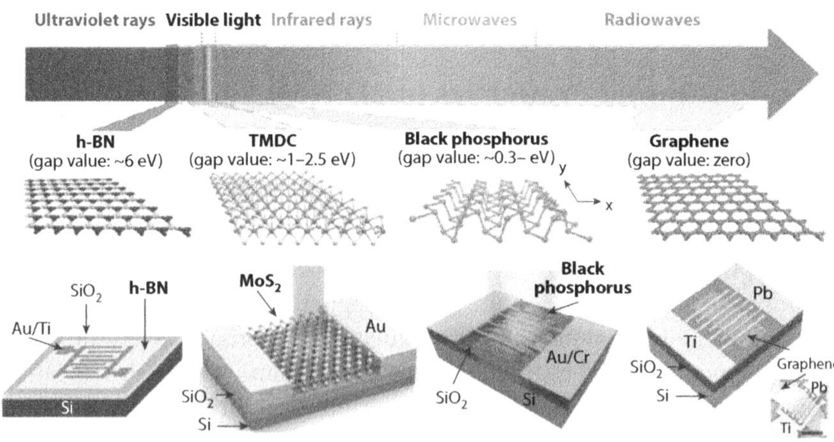

Figure 5.5 Optoelectronic devices based on 2D plasmonic materials [82]. Adapted with permission from Ref. [82]. Copyright (2019) (Royal Society of Chemistry).

an active plasmonic material. Because it protects graphene from the environment and increases its electrical mobility, it is an excellent encapsulant for plasmonic applications [81]. Surface plasmon modeling in plasmonic applications has recently been attributed to near-field IR optical imaging and IR microscopy of graphene. Figure 5.5 provides an overview of devices based on 2D plasmonic materials [82]. Interestingly, a 2D arrangement of three nanoparticles in an equilateral fashion shows negligible sensitivity to the in-plane polarization angle of the incoming light. For symmetric tetramers in three dimensions (3D), this planar symmetry may be extended to full 3D isotropy [83].

5.6 Recent Applications of 2D NMs

A variety of technologies, from the biomedical to the LED to the solar cell, energy storage, batteries, and lubricant additives, can benefit from categorizing 2D nanomaterials (2D NMs) based on their electrical, optical, and plasmonic properties. Since 2D graphene-based photonic and optoelectronic devices have many potential uses, including in sensing, communication, and imaging technologies, they have garnered a lot of attention. The few applications that have been covered in these chapters are as follows:

5.6.1 2D NMs for BioMedical Application

Nanomaterials based on 2D structures have garnered a great deal of attention in recent years, particularly in the biomedical research community. These structures will also be utilized in foreseeable pharmaceutical treatments. These nanomaterials have a huge range of unique features (physical, chemical, electrical, and optical) with planar structures and promising applications such as drug delivery, biosensing, bioimaging, regenerative medicine, and additive manufacturing. However, 2D nanomaterials are also used in regenerative medicine owing to their superior biocompatibility and positive interactions with the biological environment. For instance, nanoclay and layered double hydroxides (LDHs). They can be utilized to encapsulate various drugs for controlled release because of their layered architecture. Regenerating load-bearing tissues and incorporating antibacterial characteristics into mechanically weak scaffolds has greatly benefited from the use of MXenes and TMDs as reinforcing materials [84, 85]. Most recently, composite scaffolds comprising hydrothermally produced MoS_2, a kind of TMD, have been used for the photothermal treatment of akermanite (AKT, $Ca_2MgSi_2O_7$) for the regeneration of ablated bone tissue. Runt-related transcription factor 2 (RunX2), osteopontin (OPN), and osteocalcin (OCN) expression were higher in AKT-MoS2 composite scaffolds than in pure AKT scaffolds [86]. Similar to this, poly-propylene fumarate (PPF) composites had their compressive and flexural moduli increased by using WS2 nanotubes as reinforcing materials in bone tissue engineering [87]. To improve cell adhesion and spread, Marinucci *et al.* have used Ti-based TMO nanosheets. Cells from rat bone marrow (rBMSCs) transplanted on Ti-based nanosheets showed higher levels of osteogenic differentiation compared to TCPS controls. By increasing a-integrin receptor-mediated cell adhesion, implants coated with Ti nanosheets show enhanced osteointegration with surrounding tissues [88]. In another experiment by Qui *et al.*, a nanocomposite aerogel was prepared with graphene and poly(*N*-isopropyl acrylamide) (PNIPAM) with a highly interconnected architecture. In comparison to PNIPAM hydrogels, the electrical conductivity, and thermo-responsiveness of nanocomposite hydrogels were also markedly enhanced by the inclusion of graphene. Graphene has been demonstrated to hasten the adhesion, multiplication, and differentiation of human mesenchymal stem cells (hMSCs) toward an osteogenic cell destiny. It is also able to boost the mechanical strength and stiffness of hydrogel scaffolds without impairing cytocompatibility [89]. GO is also a good candidate as a nanocarrier for stimuli-responsive nanocomposites for drug administration due to its wide surface area, aromatic structure, and functional groups. Weaver *et al.* created electrically responsive GO–PPy nanocomposites for regulated drug delivery in a recent study. In comparison to pure PPy, the GO–PPy nanocomposites were able to trap and release two times as much of the anti-inflammatory chemical dexamethasone. Additionally, the addition of GO made the nanocomposites more sensitive to electrical stimulation and enabled linear release kinetics across 400 stimulation cycles without any detectable drug release when stimulation was not present. By altering the GO's sonication period, it was also demonstrated that drug loading and release rates could be adjusted [90]. Silicate nanoparticles are currently being investigated for use in musculoskeletal tissue engineering. Even in the absence of exogenous growth factors like bone morphogenetic protein 2 (BMP2), silicate nanoparticles can stimulate the differentiation of stem cells into osteoblasts and enhance the creation of type I collagen [91].

2D NMs are often utilized in the treatment of cancer because of their superior physiochemical properties, such as their particular targeting abilities, therapeutic binding capacity, and sustained drug release. Due to their capacity to bind anionic medicines, layered double hydroxides (LDHs) are intriguing 2D NMs for the therapeutic administration of anticancer treatments. Drug release is managed within the tumor microenvironment, and drug loading efficiency is enhanced using rare earth elements. One group doped LDHs with Mn^{2+} and Fe^{3+} ions to improve MRI contrast for their cancer imaging and medicine delivery platform [92]. Anti-cancer drugs like doxorubicin and methotrexate may be released and transported into the tumor microenvironment in a pH-dependent manner because of the unique surface charges on nanoclay. Moreover, ligands like polyethylene glycol-linked (PEG-FA) may be attached to these GMs to facilitate targeted delivery to cancer cells that have an abundance of folic acid receptors.

Two-dimensional nanomaterials (NMs) excel in biosensing and bioimaging because they are biocompatible and may be widely dispersed in an aqueous environment. Transition metal-based 2D NMs, such as TMDs and TMOs, are promising candidates for biosensing and bioimaging because of their excellent biocompatibility and fluorescence quenching capabilities. Electrochemical biosensors have made use of 2D NMs because of their high electrical conductivity, fast transfer kinetics, and sizable surface areas for functionalization. Some examples are the two-dimensional layered materials phosphorene, MoS2, and WS2. Biosensors based on MoS2 have been employed for a variety of applications, including the detection of glucose and the selective detection of dopamine in ascorbic acid and uric acid combinations. Biomolecules can be immobilized on transducers coated with LDH without sacrificing activity, stability, or access to the analyte [93–95]. Mg/Al LDHs were able to have enhanced green fluorescent protein (EGFP) adsorb onto their surfaces after being built up in layers. The immobilized EGFP exhibited different fluorescence responses based on the pH of its surrounding environment in the different layers of the LDH. Small biomolecules like protoporphyrin might also be detected [84, 96]. Therefore, there is a degree of reserved optimism regarding the use of 2D nanomaterials in biomedical applications, particularly those requiring more permanent treatment techniques for tissue regeneration. However, during the past decade, there has been a dramatic increase in translational research employing 2D nanomaterials, as shown in Figures 5.6(a,b). In this chapter, we have looked at the emerging field of 2D nanomaterials for use in medicine (Figure 5.6c). Due to the abundance of high-quality evaluations of graphene-based nanomaterials for healthcare applications [84].

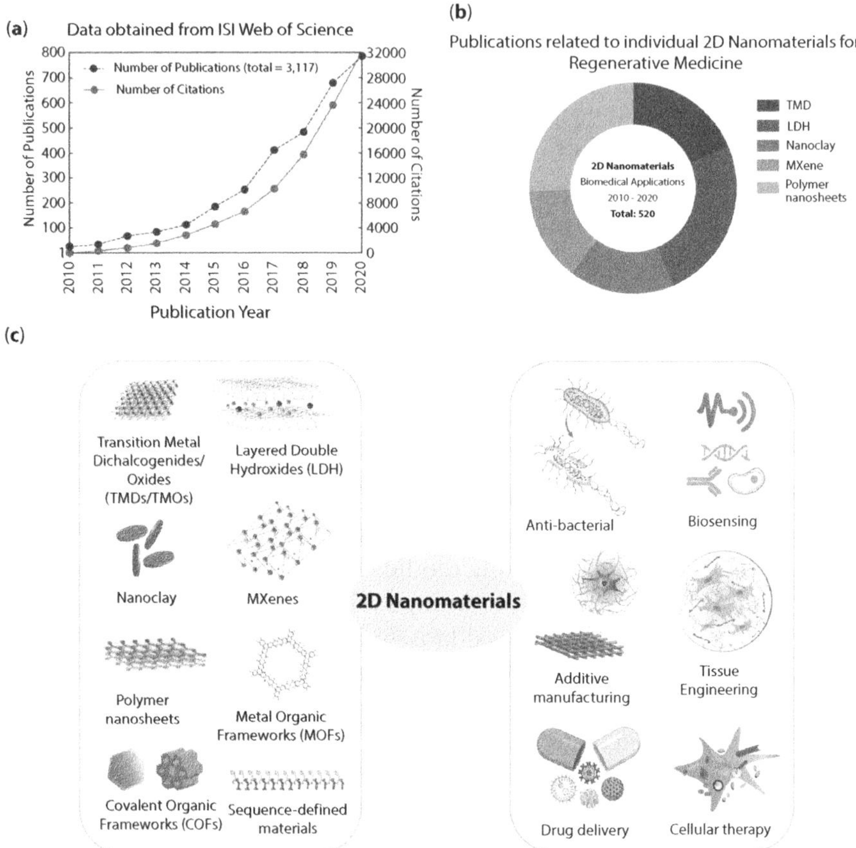

Figure 5.6 Biomedical applications of 2D nanomaterials. (a) The number of papers related to 2D nanomaterials has increased exponentially during the previous decade (2010–2020); (b) What have certain 2D nanomaterials contributed to the field of biomedicine during the past decade? And (c) Two-dimensional nanomaterials with a wide range of potential medicinal uses [84]. Adapted with permission from Ref. [84] Copyright (2021) (Elsevier).

5.6.2 2D NMs in the Field of Energy

Due to the unique geometry of 2D NMs, creating a flaw or adding doping typically requires meticulous atomic engineering. The interlayer space, a unique engineering option for modifying the energy storage and catalytic capabilities of 2D nanomaterials, is a particular structural feature. The interlayer space can be expanded to accommodate more active or storage sites on the layers, or it can be employed by itself to create a controlled microenvironment for desired reactions. It is possible to artificially create an interlayer gap even for materials that are not layered. When properly planned and built, these 2D nanofluidic channels function as excellent ion transport paths that significantly improve energy storage systems based on these materials. On a larger scale, 2D nanomaterials can be designed into porous structures with simple mass and charge transfer for use in catalytic and energy storage applications. The porous or holey structures may eventually develop into more sophisticated composites. As a result, they act as a strong material foundation on which cutting-edge material systems can be constructed. The development of 2D heterostructures and their widespread use in the field of electronic materials have sparked interest in using this idea for energy storage and catalytic applications. While many of the concepts in this direction are still in the conceptual stage, they have the potential to greatly enhance state-of-the-art technology [97]. Without binders or other additives, 2D nanosheets can provide flexible electrodes, but they also offer the advantage of a substantially higher packing density and volumetric performance compared to their 1D counterpartS. The liquid phase exfoliation (LPE) procedure makes it possible to create 2D inks for printing devices in a variety of solvents. In-plane high electronic conductivity is a feature of 2D NMs. The development of 2D heterostructures, which allow the combination of highly conducting and high-energy density 2D materials, is a further method for enabling fast transit of electrons and ions. One of the most well-liked flexible and integrated energy storage technologies is in-plane micro-supercapacitors built of carbide-derived carbons and 2D NMs, such as graphene, Mxene, metal oxides, and conductive MOFs [98]. A new method to include pseudocapacitive effects and create possible hybrid electrode materials is the chemical functionalization of 2D NMs. Reduced graphene oxide (RGO) is one of the most promising possibilities among the functionalized 2D NMs because of the surface redox interaction between the oxygen functional groups on the RGO and the electrolyte ions [99]. Chen *et al.* have shown that aqueous electrolytes can more easily penetrate graphene, which results in pseudocapacitive effects. The lowering of oxygen functional groups corresponds to an enhancement in RGO's electrochemical performance. Moreover, RGO can demonstrate that in 1 m H_2SO_4 and $BMIPF_6$, respectively, maximum capacitance values of 348 and 158 F g^{-1} were attained [100]. High-performance pseudo capacitors can benefit from the ultrahigh-specific surface areas of 2D nanosheets, particularly metal hydroxides or oxides. When ions intercalate into the tunnels or layers of a redox-active material with a Faradaic charge transfer and no crystallographic phase shift, intercalation pseudocapacitances of 2D nanosheets result [101, 102]. Some of the benefits of using 2D materials are illustrated in Figure 5.7 [103].

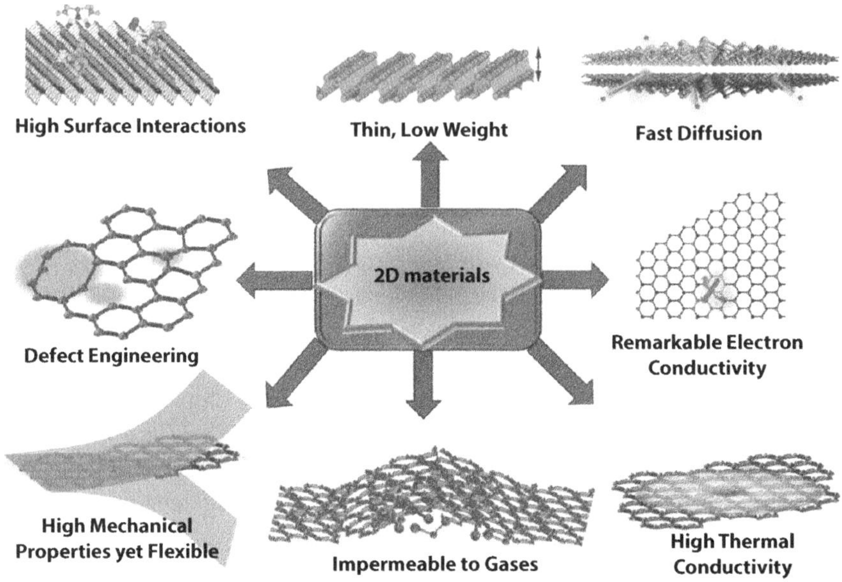

Figure 5.7 The advantages of 2D materials are discussed. Incorporating 2D materials into energy storage systems can introduce novel features [103]. Adapted with permission from Ref. [103] Copyright (2020) (American Chemical Society).

5.6.3 2D NMs as Lubricant Additive

Lubricants help reduce wear and friction. Additives to lubricants are a proven way to boost performance, and many of the newest, highest-quality lubricants on the market already consist of both base oil and an additive. Base oil, the primary component of a lubricant, is responsible for its primary characteristics. Two common solid lubricants are graphite and MoS2, both of which have layered architectures. When exposed to shear strain, the neighboring layers of these structures readily slide across one another because of the weak Van der Waals forces that hold them together. Since 2D NMs have much larger specific surface areas, they may cover more substrate surface area when absorbed, reducing the chance of direct contact and the amount of friction between rubbing surfaces. Experimental evidence shows that the 2D NMs may enhance the performance of base lubricants by decreasing their viscosity [104–108]. Graphene's large surface area and layered structure allow it to readily penetrate the interface between a tribo-contact pair, preventing the pair's rough surfaces from coming into direct contact with one another and thereby reducing wear and friction. The capacity of graphene to reduce energy loss and enhance the toughness and reliability of mechanical assemblies is proved when even a small quantity of graphene is put between rubbing surfaces [109, 110]. Elomaa *et al.* have studied GO's interactions with diamond-like carbon and stainless steel when added to water. They found that the COF (clean water) was 57% lower than in the control group [111]. Kinoshita *et al.* discovered an extremely low COF of around 0.05 at tungsten carbide on stainless steel contact, even after 60,000 cycles of rubbing testing with no evident wear. Due to graphene and its derivatives' tremendous ability to decrease friction and minimize wear, they are extensively studied as additives in both oil-based and water-based lubricating media [112].

5.7 Challenges and Prospective

Research into 2D materials proceeded in the same vein when graphene was discovered, which quickly led to the demonstration of many electrical devices and novel charge transport phenomena. Although our understanding of how electrons pass through particular 2D materials is improving, there are still several significant roadblocks that must be overcome before this knowledge can be applied to wafer size. From a materials science point of view, strong wafer-scale growth processes that provide uniform and predictable thickness face significant challenges. The chemistry of 2D materials has its own unique set of challenges and opportunities when contrasted with the chemistry of standard bulk semiconductors. With the lone electron pair on chalcogen atoms in TMDCs and phosphorus atoms in BP offering opportunities for Lewis' acid/base chemistry and radical-based chemistry, the huge surface area to volume ratio of 2D materials implies that chemical functionalization can radically modify properties. In addition, the lone electron pair on chalcogen atoms in TMDCs and phosphorus atoms in BP give chances for Lewis' acid–base chemistry.

5.8 Conclusion

Two-dimensional nanomaterials with outstanding electrical, optical, and plasmonic capabilities have become more widely used recently for a range of medical applications, including medication delivery, energy, sensors, and additive manufacturing. However, plasmonics is a concept in condensed matter physics that describes surface plasmon resonance, where surface plasmons are collective electron oscillations imited by the dielectric-metal contact.

When light interacts with a plasmonic surface, remarkable plasmonic characteristics are revealed. The unique plasmonic effects of semiconductors, metals, and 2D nanomaterials are all attributable to their unique confinement situations. Nanomaterials and their structures are the basis of surface plasmons. Within a decade, extensive basic research, and early industrial uses of two-dimensional (2D) materials have revolutionized several fields of modern science and industry. In this chapter, we present a detailed overview of 2D materials, their characteristics, and their applications in the context of electrochemistry and allied scientific domains. Furthermore, the general characteristics of 2D nanomaterials as well as their electrical, optical, and plasmonic qualities are discussed.

Acknowledgments

The authors gratefully acknowledge the support provided by Centurion University of Technology and Management, Odisha, India, for carrying out the present research work.

References

1. Saleh, T.A., Nanomaterials: Classification, properties, and environmental toxicities. *Environ. Technol. Innov.*, 20, 101067, 2020.
2. Baig, N., Kammakakam, I., Falath, W., Nanomaterials: A review of synthesis methods, properties, recent progress, and challenges. *Mater. Adv.*, 2, 6, 1821–1871, 2021.
3. Wu, Q., Miao, W.S., Zhang, Y.D., Gao, H.J., Hui, D., Mechanical properties of nanomaterials: A review. *Nanotechnol. Rev.*, 9, 1, 259–273, 2020.
4. Pomerantseva, E., Bonaccorso, F., Feng, X., Cui, Y., Gogotsi, Y., Energy storage: The future enabled by nanomaterials. *Science*, 366, 6468, eaan8285, 2019.
5. Yaqoob, A.A., Ahmad, H., Parveen, T., Ahmad, A., Oves, M., Ismail, I.M., Mohamad Ibrahim, M.N., Recent advances in metal decorated nanomaterials and their various biological applications: A review. *Front. Chem.*, 8, 341, 2020.
6. Gajanan, K. and Tijare, S.N., Applications of nanomaterials. *Mater. Today Proc.*, 5, 1, 1093–1096, 2018.
7. Pomerantseva, E., Bonaccorso, F., Feng, X., Cui, Y., Gogotsi, Y., Energy storage: The future enabled by nanomaterials. *Science*, 366, 6468, eaan8285, 2019.
8. Ni, J., and Li, Y., Carbon nanomaterials in different dimensions for electrochemical energy storage. *Adv. Energy Mater.*, 6, 17, 1600278, 2016.

9. Xia, Y., Nanomaterials at work in biomedical research. *Nat. Mater.*, 7, 10, 758–760, 2008.
10. Siddique, S. and Chow, J.C., Application of nanomaterials in biomedical imaging and cancer therapy. *Nanomaterials*, 10, 9, 1700, 2020.
11. Sannino, D., Rizzo, L., Vaiano, V., Progress in nanomaterials applications for water purification, in: *Nanotechnologies for Environmental Remediation: Applications and Implications*, pp. 1–24, 2017.
12. Singh, K.K., Singh, A., Rai, S., A study on nanomaterials for water purification. *Mater. Today Proc.*, 51, 1157–1163, 2022.
13. Dwivedi, S., Saquib, Q., Al-Khedhairy, A.A., Musarrat, J., Understanding the role of nanomaterials in agriculture, in: *Microbial Inoculants in Sustainable Agricultural Productivity: Vol. 2: Functional Applications*, pp. 271–288, 2016.
14. Kaphle, A., Navya, P.N., Umapathi, A., Daima, H.K., Nanomaterials for agriculture, food and environment: Applications, toxicity and regulation. *Environ. Chem. Lett.*, 16, 43–58, 2018.
15. Kolahalam, L.A., Viswanath, I.K., Diwakar, B.S., Govindh, B., Reddy, V., Murthy, Y.L.N., Review on nanomaterials: Synthesis and applications. *Mater. Today Proc.*, 18, 2182–2190, 2019.
16. Stone, V., Nowack, B., Baun, A., van den Brink, N., von der Kammer, F., Dusinska, M., Fernandes, T.F., Nanomaterials for environmental studies: classification, reference material issues, and strategies for physico-chemical characterisation. *Sci. Total Environ.*, 408, 7, 1745–1754, 2010.
17. Wang, Z., Hu, T., Liang, R., Wei, M., Application of zero-dimensional nanomaterials in biosensing. *Front. Chem.*, 8, 320, 2020.
18. Raj, A.M. and Balachandran, M., Coal-based fluorescent zero-dimensional carbon nanomaterials: A short review. *Energy Fuels*, 34, 11, 13291–13306, 2020.
19. Chen, C., Fan, Y., Gu, J., Wu, L., Passerini, S., Mai, L., One-dimensional nanomaterials for energy storage. *J. Phys. D Appl. Phys.*, 51, 11, 113002, 2018.
20. Gong, S. and Cheng, W., One-dimensional nanomaterials for soft electronics. *Adv. Electron. Mater.*, 3, 3, 1600314, 2017.
21. Rafiei-Sarmazdeh, Z., Zahedi-Dizaji, S.M., Kang, A.K., Two-dimensional nanomaterials, in: *Nanostructures*, IntechOpen, 2019.
22. Varghese, R.J., Parani, S., Thomas, S., Oluwafemi, O.S., Wu, J., Introduction to nanomaterials: Synthesis and applications, in: *Nanomaterials for Solar Cell Applications*, pp. 75–95, Elsevier, 2019.
23. Tulinski, M. and Jurczyk, M., Nanomaterials synthesis methods, in: *Metrology and Standardization of Nanotechnology: Protocols and Industrial Innovations*, pp. 75–98, 2017.
24. Mageswari, A., Srinivasan, R., Subramanian, P., Ramesh, N., Gothandam, K.M., Nanomaterials: Classification, biological synthesis and characterization. *Nanosci. Food Agric.*, 3, 31–71, 2016.

25. Huston, M., DeBella, M., DiBella, M., Gupta, A., Green synthesis of nanomaterials. *Nanomaterials*, 11, 8, 2130, 2021.
26. Aarthye, P. and Sureshkumar, M., Green synthesis of nanomaterials: An overview. *Mater. Today Proc.*, 47, 907–913, 2021.
27. Miró, P., Audiffred, M., Heine, T., An atlas of two-dimensional materials. *Chem. Soc. Rev.*, 43, 18, 6537–6554, 2014.
28. Fang, Z., Xing, Q., Fernandez, D., Zhang, X., Yu, G., A mini review on two-dimensional nanomaterial assembly. *Nano Res.*, 13, 5, 1179–1190, 2020.
29. Sun, Y., Gao, S., Lei, F., Xiao, C., Xie, Y., Ultrathin two-dimensional inorganic materials: New opportunities for solid state nanochemistry. *Acc. Chem. Res.*, 48, 1, 3–12, 2015.
30. Sun, Y., Gao, S., Lei, F., Xie, Y., Atomically-thin two-dimensional sheets for understanding active sites in catalysis. *Chem. Soc. Rev.*, 44, 3, 623–636, 2015.
31. Zhu, Y., Peng, L., Fang, Z., Yan, C., Zhang, X., Yu, G., Structural engineering of 2D nanomaterials for energy storage and catalysis. *Adv. Mater.*, 30, 15, 1706347, 2018.
32. Yu, X., Yun, S., Yeon, J.S., Bhattacharya, P., Wang, L., Lee, S.W., Park, H.S., Emergent pseudocapacitance of 2D nanomaterials. *Adv. Energy Mater.*, 8, 13, 1702930, 2018.
33. Hu, T., Mei, X., Wang, Y., Weng, X., Liang, R., Wei, M., Two-dimensional nanomaterials: Fascinating materials in biomedical field. *Sci. Bull.*, 64, 22, 1707–1727, 2019.
34. Haque, F., Daeneke, T., Kalantar-Zadeh, K., Ou, J.Z., Two-dimensional transition metal oxide and chalcogenide-based photocatalysts. *Nanomicro Lett.*, 10, 2, 1–27, 2018.
35. Geim, A.K. and Novoselov, K.S., The rise of graphene. *Nat. Mater.*, 6, 3, 183–191, 2007.
36. Mu, P., Zhou, G., Chen, C.L., 2D nanomaterials assembled from sequence-defined molecules. *Nano-Struct. Nano-Objects*, 15, 153–166, 2018.
37. Guo, Y., Xu, K., Wu, C., Zhao, J., Xie, Y., Surface chemical-modification for engineering the intrinsic physical properties of inorganic two-dimensional nanomaterials. *Chem. Soc. Rev.*, 44, 3, 637–646, 2015.
38. Lim, J.Y. and Donahue, H.J., Cell sensing and response to micro- and nano-structured surfaces produced by chemical and topographic patterning. *Tissue Eng.*, 13, 8, 1879–1891, 2007.
39. Bettinger, C.J., Langer, R., Borenstein, J.T., Engineering substrate topography at the micro-and nanoscale to control cell function. *Angew. Chem. Int. Ed.*, 48, 30, 5406–5415, 2009.
40. Gupta, A., Sakthivel, T., Seal, S., Recent development in 2D materials beyond graphene. *Prog. Mater. Sci.*, 73, 44–126, 2015.
41. Geim, A.K. and Grigorieva, I.V., Van der Waals heterostructures. *Nature*, 499, 7459, 419–425, 2013.

42. Hu, T., Mei, X., Wang, Y., Weng, X., Liang, R., Wei, M., Two-dimensional nanomaterials: Fascinating materials in biomedical field. *Sci. Bull.*, 64, 22, 1707–1727, 2019.
43. Huang, X., Tan, C., Yin, Z., Zhang, H., 25th Anniversary article: Hybrid nanostructures based on two-dimensional nanomaterials. *Adv. Mater.*, 26, 14, 2185–2204, 2014.
44. Tomás, H., Alves, C.S., Rodrigues, J., Laponite®: A key nanoplatform for biomedical applications? *Nanomed. Nanotechnol. Biol. Med.*, 14, 7, 2407–2420, 2018.
45. Ramalingam, G., Kathirgamanathan, P., Ravi, G., Elangovan, T., Manivannan, N., Kasinathan, K., Quantum confinement effect of 2D nanomaterials, in: *Quantum Dots-Fundamental and Applications*, IntechOpen, 2020.
46. Hasan, M.A.M., Wang, Y., Bowen, C.R., Yang, Y., 2D nanomaterials for effective energy scavenging. *Nano-Micro Lett.*, 13, 1–41, 2021.
47. Qu, J.F., Benz, S.P., Rogalla, H., Tew, W.L., White, D.R., Zhou, K.L., Johnson noise thermometry. *Meas. Sci. Technol.*, 30, 11, 112001, 2019.
48. Sett, A., Rana, T., Rajaji, U., Sha, R., Liu, T.Y., Bhattacharyya, T.K., Emergence of two-dimensional nanomaterials-based breath sensors for non-invasive detection of diseases. *Sens. Actuators A Phys.*, 338, 113507, 2022.
49. Zheng, T., Zhou, Y., Feng, E., Tian, Y., Surface-enhanced Raman scattering on 2D nanomaterials: Recent developments and applications. *Chin. J. Chem.*, 39, 3, 745–756, 2021.
50. Lee, W.H. and Park, Y.D., Tuning electrical properties of 2d materials by self-assembled monolayers. *Adv. Mater. Interfaces*, 5, 1, 1700316, 2018.
51. Meng, Z., Stolz, R.M., Mendecki, L., Mirica, K.A., Electrically-transduced chemical sensors based on two-dimensional nanomaterials. *Chem. Rev.*, 119, 1, 478–598, 2019.
52. Varghese, S.S., Varghese, S.H., Swaminathan, S., Singh, K.K., Mittal, V., Two-dimensional materials for sensing: graphene and beyond. *Electronics*, 4, 3, 651–687, 2015.
53. Yang, W., Gan, L., Li, H., Zhai, T., Two-dimensional layered nanomaterials for gas-sensing applications. *Inorg. Chem. Front.*, 3, 4, 433–451, 2016.
54. Tang, X., Du, A., Kou, L., Gas sensing and capturing based on two-dimensional layered materials: Overview from theoretical perspective. *Wiley Interdiscip. Rev. Comput. Mol. Sci.*, 8, 4, e1361, 2018.
55. Choi, S.J. and Kim, I.D., Recent developments in 2D nanomaterials for chemiresistive-type gas sensors. *Electron. Mater. Lett.*, 14, 221–260, 2018.
56. Karthick Kannan, P., Shankar, P., Blackman, C., Chung, C.H., Recent advances in 2D inorganic nanomaterials for SERS sensing. *Adv. Mater.*, 31, 34, 1803432, 2019.
57. Yi, Z., Chen, S., Chen, Y., Luo, J., Wu, W., Yi, Y., Tang, Y., Preparation of dendritic Ag/Au bimetallic nanostructures and their application in surface-enhanced Raman scattering. *Thin Solid Films*, 520, 7, 2701–2707, 2012.

58. Xu, J.Y., Wang, J., Kong, L.T., Zheng, G.C., Guo, Z., Liu, J.H., SERS detection of explosive agent by macrocyclic compound functionalized triangular gold nanoprisms. *J. Raman Spectrosc.*, 42, 9, 1728–1735, 2011.
59. Zhen, S., Wu, T., Huang, X., Li, Y., Huang, C., Facile synthesis of gold nanoflowers as SERS substrates and their morphological transformation induced by iodide ions. *Sci. China Chem.*, 59, 1045–1050, 2016.
60. Saute, B., Premasiri, R., Ziegler, L., Narayanan, R., Gold nanorods as surface enhanced Raman spectroscopy substrates for sensitive and selective detection of ultra-low levels of dithiocarbamate pesticides. *Analyst*, 137, 21, 5082–5087, 2012.
61. Tao, A., Kim, F., Hess, C., Goldberger, J., He, R., Sun, Y., Yang, P., Langmuir–Blodgett silver nanowire monolayers for molecular sensing using surface-enhanced Raman spectroscopy. *Nano Lett.*, 3, 9, 1229–1233, 2003.
62. Garcia-Leis, A., Garcia-Ramos, J.V., Sanchez-Cortes, S., Silver nanostars with high SERS performance. *J. Phys. Chem. C*, 117, 15, 7791–7795, 2013.
63. Dao, T.C., Luong, T.Q.N., Cao, T.A., Kieu, N.M., Application of silver nanodendrites deposited on silicon in SERS technique for the trace analysis of paraquat. *Adv. Nat. Sci. Nanosci. Nanotechnol.*, 7, 1, 015007, 2016.
64. Mitta, S.B., Choi, M.S., Nipane, A., Ali, F., Kim, C., Teherani, J.T., Yoo, W.J., Electrical characterization of 2D materials-based field-effect transistors. *2D Mater.*, 8, 1, 012002, 2020.
65. Lipatov, A., Goad, A., Loes, M.J., Vorobeva, N.S., Abourahma, J., Gogotsi, Y., Sinitskii, A., High electrical conductivity and breakdown current density of individual monolayer Ti3C2Tx MXene flakes. *Matter*, 4, 4, 1413–1427, 2021.
66. Orts Mercadillo, V., Chan, K.C., Caironi, M., Athanassiou, A., Kinloch, I.A., Bissett, M., Cataldi, P., Electrically conductive 2D material coatings for flexible and stretchable electronics: A comparative review of graphenes and MXenes. *Adv. Funct. Mater.*, 32, 38, 2204772, 2022.
67. Habib, T., Zhao, X., Shah, S.A., Chen, Y., Sun, W., An, H., Green, M.J., Oxidation stability of Ti3C2Tx MXene nanosheets in solvents and composite films. *NPJ 2D Mater. Appl.*, 3, 1, 8, 2019.
68. Bernardi, M., Ataca, C., Palomo, M., Grossman, J.C., Optical and electronic properties of two-dimensional layered materials. *Nanophotonics*, 6, 2, 479–493, 2017.
69. Grigorenko, A.N., Polini, M., Novoselov, K.S., Graphene plasmonics. *Nat. Photonics*, 6, 11, 749–758, 2012.
70. Falkovsky, L.A., Optical properties of graphene. *J. Phys. Conf. Ser.*, IOP Publishing, 129, 1, 012004, October 2008.
71. Wang, X. and Lan, S., Optical properties of black phosphorus. *Adv. Opt. Photonics*, 8, 4, 618–655, 2016.
72. Chen, X., Ponraj, J.S., Fan, D., Zhang, H., An overview of the optical properties and applications of black phosphorus. *Nanoscale*, 12, 6, 3513–3534, 2020.

73. Singh, S., Singh, P.K., Umar, A., Lohia, P., Albargi, H., Castañeda, L., Dwivedi, D.K., 2D nanomaterial-based surface plasmon resonance sensors for biosensing applications. *Micromachines*, 11, 8, 779, 2020.
74. Woodward, R., II and Kelleher, E.J., 2D saturable absorbers for fibre lasers. *Appl. Sci.*, 5, 4, 1440–1456, 2015.
75. Yang, L., Deslippe, J., Park, C.H., Cohen, M.L., Louie, S.G., Excitonic effects on the optical response of graphene and bilayer graphene. *Phys. Rev. Lett.*, 103, 18, 186802, 2009.
76. Arnaud, B., Lebègue, S., Rabiller, P., Alouani, M., Huge excitonic effects in layered hexagonal boron nitride. *Phys. Rev. Lett.*, 96, 2, 026402, 2006.
77. Wirtz, L., Marini, A., Grüning, M., Attaccalite, C., Kresse, G., Rubio, A., Comment on "huge excitonic effects in layered hexagonal boron nitride". *Phys. Rev. Lett.*, 100, 18, 189701, 2008.
78. Novoselov, K.S., Morozov, S.V., Mohinddin, T.M.G., Ponomarenko, L.A., Elias, D.C., Yang, R., Geim, A.K., Electronic properties of graphene. *Phys. Status Solidi (B)*, 244, 11, 4106–4111, 2007.
79. Li, Y., Li, Z., Chi, C., Shan, H., Zheng, L., Fang, Z., Plasmonics of 2D nanomaterials: properties and applications. *Adv. Sci.*, 4, 8, 1600430, 2017.
80. Dillon, A.D., Ghidiu, M.J., Krick, A.L., Griggs, J., May, S.J., Gogotsi, Y., Fafarman, A.T., Highly conductive optical quality solution-processed films of 2D titanium carbide. *Adv. Funct. Mater.*, 26, 23, 4162–4168, 2016.
81. Iqbal, M.A., Malik, M., Shahid, W., Ahmad, W., Min-Dianey, K.A., Pham, P.V., Plasmonic 2D materials: Overview, advancements, future prospects and functional applications, in: *21st Century Nanostructured Materials-Physics, Chemistry, Classification, and Emerging Applications in Industry, Biomedical, and Agriculture*, 2021.
82. Wang, H., Li, S., Ai, R., Huang, H., Shao, L., Wang, J., Plasmonically enabled two-dimensional material-based optoelectronic devices. *Nanoscale*, 12, 15, 8095–8108, 2020.
83. Borah, R. and Verbruggen, S.W., Coupled plasmon modes in 2D gold nanoparticle clusters and their effect on local temperature control. *J. Phys. Chem. C*, 123, 50, 30594–30603, 2019.
84. Murali, A., Lokhande, G., Deo, K.A., Brokesh, A., Gaharwar, A.K., Emerging 2D nanomaterials for biomedical applications. *Mater. Today*, 50, 276–302, 2021.
85. Wang, X., Li, T., Ma, H., Zhai, D., Jiang, C., Chang, J., Wu, C., A 3D-printed scaffold with MoS2 nanosheets for tumor therapy and tissue regeneration. *NPG Asia Mater.*, 9, 4, e376–e376, 2017.
86. Lalwani, G., Henslee, A.M., Farshid, B., Parmar, P., Lin, L., Qin, Y.X., Sitharaman, B., Tungsten disulfide nanotubes reinforced biodegradable polymers for bone tissue engineering. *Acta Biomater.*, 9, 9, 8365–8373, 2013.
87. Marinucci, L., Balloni, S., Becchetti, E., Belcastro, S., Guerra, M., Calvitti, M., Locci, P., Effect of titanium surface roughness on human osteoblast

proliferation and gene expression in vitro. *Int. J. Oral. Maxillofac. Implants*, 21, 5, 719–725, 2006.

88. Bai, H., Li, C., Wang, X., Shi, G., A pH-sensitive graphene oxide composite hydrogel. *Chem. Commun.*, 46, 14, 2376–2378, 2010.
89. Weaver, C.L., LaRosa, J.M., Luo, X., Cui, X.T., Electrically controlled drug delivery from graphene oxide nanocomposite films. *ACS Nano*, 8, 2, 1834–1843, 2014.
90. Kerativitayanan, P., Carrow, J.K., Gaharwar, A.K., Nanomaterials for engineering stem cell responses. *Adv. Healthc. Mater.*, 4, 11, 1600–1627, 2015.
91. Huang, G., Zhang, K.L., Chen, S., Li, S.H., Wang, L.L., Wang, L.P., Yang, H.H., Manganese-iron layered double hydroxide: A theranostic nanoplatform with pH-responsive MRI contrast enhancement and drug release. *J. Mater. Chem. B*, 5, 20, 3629–3633, 2017.
92. Wen, W., Song, Y., Yan, X., Zhu, C., Du, D., Wang, S., Lin, Y., Recent advances in emerging 2D nanomaterials for biosensing and bioimaging applications. *Mater. Today*, 21, 2, 164–177, 2018.
93. Su, S., Sun, Q., Gu, X., Xu, Y., Shen, J., Zhu, D., Wang, L., Two-dimensional nanomaterials for biosensing applications. *TrAC Trends Anal. Chem.*, 119, 115610, 2019.
94. Yadav, V., Roy, S., Singh, P., Khan, Z., Jaiswal, A., 2D MoS2-based nanomaterials for therapeutic, bioimaging, and biosensing applications. *Small*, 15, 1, 1803706, 2019.
95. Zhang, P., Hu, Y., Ma, R., Li, L., Lu, J., Enhanced green fluorescence protein/layered double hydroxide composite ultrathin films: Bio-hybrid assembly and potential application as a fluorescent biosensor. *J. Mater. Chem. B*, 5, 1, 160–166, 2017.
96. Han, J., Wei, M., Evans, D.G., Duan, X., Hierarchical layered double hydroxide materials, in: *Advanced Hierarchical Nanostructured Materials*, pp. 231–266, 2014.
97. Zhu, Y., Peng, L., Fang, Z., Yan, C., Zhang, X., Yu, G., Structural engineering of 2D nanomaterials for energy storage and catalysis. *Adv. Mater.*, 30, 15, 1706347, 2018.
98. Pomerantseva, E., Bonaccorso, F., Feng, X., Cui, Y., Gogotsi, Y., Energy storage: The future enabled by nanomaterials. *Science*, 366, 6468, eaan8285, 2019.
99. Yu, X., Yun, S., Yeon, J.S., Bhattacharya, P., Wang, L., Lee, S.W., Park, H.S., Emergent pseudocapacitance of 2D nanomaterials. *Adv. Energy Mater.*, 8, 13, 1702930, 2018.
100. Chen, Y., Zhang, X., Zhang, D., Yu, P., Ma, Y., High performance supercapacitors based on reduced graphene oxide in aqueous and ionic liquid electrolytes. *Carbon*, 49, 2, 573–580, 2011.
101. Hasan, M.A.M., Wang, Y., Bowen, C.R., Yang, Y., 2D nanomaterials for effective energy scavenging. *Nanomicro Lett.*, 13, 1–41, 2021.

102. Feng, F., Wu, J., Wu, C., Xie, Y., Regulating the electrical behaviors of 2D inorganic nanomaterials for energy applications. *Small*, 11, 6, 654–666, 2015.
103. Rojaee, R. and Shahbazian-Yassar, R., Two-dimensional materials to address the lithium battery challenges. *ACS Nano*, 14, 3, 2628–2658, 2020.
104. Xiao, H. and Liu, S., 2D nanomaterials as lubricant additive: A review. *Mater. Des.*, 135, 319–332, 2017.
105. Wen, P., Lei, Y., Li, W., Fan, M., Two-dimension layered nanomaterial as lubricant additives: Covalent organic frameworks beyond oxide graphene and reduced oxide graphene. *Tribol. Int.*, 143, 106051, 2020.
106. Zhao, J., Huang, Y., Li, Y., Gao, T., Dou, Z., Mao, J., Luo, J., Superhigh-exfoliation graphene with a unique two-dimensional (2D) microstructure for lubrication application. *Appl. Surf. Sci.*, 513, 145608, 2020.
107. Liu, W., Qiao, X., Liu, S., Chen, P., A review of nanomaterials with different dimensions as lubricant additives. *Nanomaterials*, 12, 21, 3780, 2022.
108. Guo, W., Yin, J., Qiu, H., Guo, Y., Wu, H., Xue, M., Friction of low-dimensional nanomaterial systems. *Friction*, 2, 209–225, 2014.
109. Liu, Y., Ge, X., Li, J., Graphene lubrication. *Appl. Mater. Today*, 20, 100662, 2020.
110. Berman, D., Erdemir, A., Sumant, A.V., Graphene: A new emerging lubricant. *Mater. Today*, 17, 1, 31–42, 2014.
111. Elomaa, O., Singh, V.K., Iyer, A., Hakala, T.J., Koskinen, J., Graphene oxide in water lubrication on diamond-like carbon vs. stainless steel high-load contacts. *Diam. Relat. Mater.*, 52, 43–48, 2015.
112. Kinoshita, H., Nishina, Y., Alias, A.A., Fujii, M., Tribological properties of monolayer graphene oxide sheets as water-based lubricant additives. *Carbon*, 66, 720–723, 2014.

Part III

APPLICATION OF 2D NANOMATERIALS

6

Challenges Surrounding 2D Nanomaterials and Their Application to Photocatalytic Industrial Wastewater Treatment

Anchit Modi[1]* and N. K. Gaur[2]

[1]*Department of Basic Sciences, IITM, IES University, Bhopal, India*
[2]*Department of Physics, Barkatullah University, Bhopal, India*

Abstract
Using 2D nanomaterials in photocatalytic industrial wastewater treatment shows great potential for mitigating water pollution. However, significant challenges must be addressed to harness these materials' capabilities in practical applications effectively. One major challenge lies in the synthesis and scalable production of 2D nanomaterials with the desired properties. It is crucial to develop reliable and cost-effective methods for large-scale production, while ensuring the stability and durability of these materials under harsh operating conditions to maintain their photocatalytic activity over extended periods. Additionally, the efficient immobilization and integration of 2D nanomaterials into photocatalytic systems pose further challenges. Strong adhesion, prevention of aggregation, and enhancement of interfacial contact with the support matrices are necessary to maximize the photocatalytic efficiency of these materials.

Understanding the interactions between pollutants and 2D nanomaterials and developing strategies to mitigate these effects is vital for efficient wastewater treatment. The choice of nanomaterials, combinations, and catalyst loading significantly affects the efficiency of pollutant degradation. Moreover, the photocatalytic treatment of industrial wastewater involves diverse organic and inorganic contaminants, which can interfere with the photocatalytic process. To enable industrial adoption of photocatalytic processes, scalability and cost-effectiveness are key considerations. This entails developing continuous-flow systems, integrating them with the existing wastewater treatment infrastructure, and implementing cost-effective catalyst recovery methods. Another critical aspect is the selection and

**Corresponding author*: anchitmodi87@gmail.com

Subhendu Chakroborty and Kaushik Pal (eds.) 2D Nanomaterials: Synthesis, Properties and Applications, (105–120) © 2024 Scrivener Publishing LLC

optimization of suitable catalyst compositions. By addressing these challenges, we can pave the way for successful application of 2D nanomaterials in photocatalytic industrial wastewater treatment, leading to the development of efficient and sustainable wastewater treatment technologies.

Keywords: 2D nanomaterials, wastewater treatment, photocatalysis, graphene

6.1 Introduction

Industrial wastewater treatment is a pressing environmental concern due to its complex composition and high levels of pollutants. Conventional treatment methods often struggle to remove these contaminants effectively. However, the promising approach of photocatalysis, which utilizes semiconductor materials to harness solar energy and degrade organic pollutants, has gained significant interest. In recent years, the application of two-dimensional (2D) nanomaterials in photocatalytic industrial wastewater treatment has emerged as a fascinating field of research [1–3]. Two-dimensional nanomaterials exhibit exceptional physicochemical properties, which make them highly desirable for various applications, including photocatalysis. For instance, graphene, composed of a single layer of carbon atoms arranged in a hexagonal lattice, has been extensively studied due to its remarkable electronic and mechanical properties. Additionally, transition metal dichalcogenides (TMDs) and black phosphorus (BP) are other noteworthy 2D nanomaterials with favorable properties suitable for photocatalytic applications. This chapter provides a comprehensive overview of the challenges associated with 2D nanomaterials and their potential applications [4–6].

The high surface-to-volume ratio, excellent electrical conductivity, and tunable bandgaps of these nanomaterials enable efficient solar-energy utilization in photocatalytic reactions. Their large surface area provides ample active chemical reaction sites, which enhances their catalytic activity. Moreover, their distinctive electronic properties facilitate efficient charge separation, minimize recombination, and improve overall photocatalytic efficiency [7]. Notably, the tunable bandgaps of 2D nanomaterials allow them to absorb a broader spectrum of light, enabling photocatalytic reactions under both UV and visible-light irradiation [8].

Despite their promising properties, the use of 2D nanomaterials for photocatalytic industrial wastewater treatment presents several challenges. One significant challenge lies in the synthesis and characterization of these materials. Precise control over their size, morphology, and composition is vital for optimizing their photocatalytic performance. Various techniques including chemical vapor deposition, solution-based methods, and mechanical exfoliation have been employed for their synthesis. However, achieving large-scale production with uniform properties remains a challenge. Additionally, addressing the stability and recyclability of 2D nanomaterials under harsh wastewater treatment conditions is crucial to ensure their long-term effectiveness [9, 10]. Nevertheless, overcoming these challenges can unlock the substantial potential of 2D nanomaterials in improving the efficiency and effectiveness of industrial wastewater treatment. Their enhanced light absorption, charge separation, and catalytic activity contribute to the more efficient degradation of various organic pollutants in industrial wastewater. Furthermore, the versatility and tunability of these nanomaterials offers opportunities for tailored solutions to address different types of contaminants [5, 11–13]. For example, specific modifications and functionalization can enhance their selectivity towards particular pollutants or improve their stability in the presence of inhibitory substances.

Moreover, combining 2D nanomaterials with other advanced technologies such as nanocomposites, heterostructures, and co-catalysts can enhance their photocatalytic performance. These synergistic approaches improve light harvesting, prolong the lifetime of charge carriers, and enhance the overall efficiency of the photocatalytic reactions. By addressing the challenges surrounding 2D nanomaterials and exploring their potential applications in photocatalytic industrial wastewater treatment, this chapter contributes to the development of innovative and sustainable solutions to address the pressing environmental concerns associated with industrial wastewater pollution. Continued research and advancements in the synthesis, characterization, and application of 2D nanomaterials hold great promise for photocatalytic wastewater treatment, paving the way for more efficient and environmentally friendly industrial processes.

6.2 Photocatalysis for Industrial Wastewater Treatment

Industrial wastewater treatment is a critical process that aims to mitigate the environmental impact of various industries. While traditional methods, such as chemical treatment and filtration, have been widely used, they often have limitations in terms of effectiveness and cost. In recent years, photocatalysis has emerged as a promising technology for industrial wastewater treatment [11, 12]. This section explores the principles of photocatalysis and its applications in the treatment of industrial wastewater.

6.2.1 Principles of Photocatalysis

Photocatalysis harnesses light energy and a photocatalyst to initiate chemical reactions. The critical component of photocatalysis is the photocatalyst, typically a semiconductor material such as titanium dioxide (TiO_2). When a photocatalyst is illuminated with light energy, it generates electron-hole pairs, which can participate in various chemical reactions. As a mediator, the photocatalyst facilitates the conversion of harmful pollutants into less toxic substances through oxidation or reduction reactions. The photocatalytic process involves two main steps: light absorption and a catalytic reaction. The photocatalyst absorbs photons with energies higher than its bandgap during light absorption, generating electron-hole pairs [1, 5]. These charge carriers can then react redoxally with pollutants in the wastewater.

6.2.2 Photocatalytic Processes for Industrial Wastewater Treatment

Photocatalysis involves several processes that can be used for industrial wastewater treatment. Direct photodegradation is the most used process, in which the photocatalyst directly breaks down pollutants upon light absorption. This process is highly effective for degrading organic compounds such as dyes and pesticides. Another photocatalytic process involves the generation of reactive oxygen species (ROS). When the photocatalyst interacts with light, it can produce ROS such as hydroxyl radicals (•OH). These highly reactive species can oxidize and degrade organic pollutants in wastewater, thereby providing an efficient means of treatment [14–16]. Furthermore, photocatalysis can be combined with other processes such as adsorption and membrane filtration to enhance the overall treatment efficiency. By immobilizing the photocatalyst onto an adsorbent or integrating it into a membrane, wastewater can be pre-treated to remove larger particles or adsorbable compounds before undergoing photocatalytic degradation.

6.2.3 Advantages and Limitations of Photocatalysis

Photocatalysis offers several advantages for industrial wastewater treatment, making it an attractive option in various industries. First, it is a sustainable and environmentally friendly process as it utilizes light energy, a renewable resource, to drive reactions. Unlike chemical treatment, photocatalysis does not generate additional waste streams, thereby reducing the overall environmental impact. Moreover, photocatalysis has been proven to be highly effective in degrading various organic pollutants in industrial wastewater. It has shown promising results for the treatment of dyes, phenols, pesticides, and various recalcitrant organic compounds. The process can be optimized by controlling factors such as the photocatalyst type, dosage, light intensity, and pH, allowing for tailored treatment solutions. Photocatalysis offers the advantage of non-specific treatments [15, 17]. Unlike conventional treatment techniques that target specific pollutants, photocatalysis can simultaneously degrade multiple contaminants without the need for individual process adjustments. This versatility makes it suitable for treating complex industrial wastewater streams.

Despite its numerous advantages, photocatalysis has certain limitations that must be considered. One of the main limitations of this study was its dependence on light availability. The intensity and wavelength of the light source influence the efficiency of the photocatalytic reactions. Therefore, the process may be less effective in treating wastewater with low light penetration, such as highly concentrated or dark-colored wastewater. Another limitation is the potential for catalyst deactivation and fouling. Over time, the surface of the photocatalyst may become fouled or deactivated because of the deposition of pollutants or other contaminants [18]. Regular cleaning or replacement of the photocatalyst may be necessary to maintain the optimal performance.

In additional, the scalability and cost-effectiveness of photocatalysis for large-scale industrial applications require further investigation. Factors such as the synthesis and immobilization of photocatalysts and the energy requirements of light sources should be considered to ensure the viability of this technology in industrial wastewater treatment. Photocatalysis has emerged as a promising technology for the treatment of industrial wastewater. Its principles revolve around generating electron-hole pairs through light absorption by a photocatalyst, leading to the degradation of pollutants through oxidation or reduction reactions. Photocatalytic processes, such as direct photodegradation and ROS generation, offer an effective means of treating organic compounds in wastewater [15, 19]. The advantages of photocatalysis include sustainability, versatility, and nonspecificity. However, limitations such as dependence on light availability and catalyst deactivation must be addressed for practical implementation. Further research and development efforts are required to optimize the process, improve scalability, and assess its cost-effectiveness for large-scale industrial applications. By overcoming these challenges, photocatalysis can achieve an efficient and sustainable industrial wastewater treatment.

6.3 2D Nanomaterials in Photocatalysis

Integrating two-dimensional (2D) nanomaterials has emerged as a promising strategy for enhancing photocatalytic performance. Because of their unique properties and characteristics, these materials offer new opportunities for efficient and sustainable wastewater treatment. This section provides an overview of 2D nanomaterials, their types used in photocatalysis, and their key roles in enhancing photocatalytic performance.

6.3.1 Introduction to 2D Nanomaterials and Types Used in Photocatalysis

2D nanomaterials are materials characterized by their atomic or molecular thicknesses and large lateral dimensions. Their ultrathin nature and high surface-area-to-volume ratio contributed to their exceptional properties. These materials exhibit unique electronic, optical, and catalytic properties, making them highly attractive for various applications, including photocatalysis. Graphene, transition metal dichalcogenides (TMDs) such as molybdenum disulfide (MoS2), tungsten diselenide (WSe2), black phosphorus (BP), and hexagonal boron nitride (h-BN) are some of the commonly utilized 2D nanomaterials in photocatalysis. Graphene, composed of a single layer of carbon atoms arranged in a honeycomb lattice, possesses excellent electrical conductivity and high electron mobility [4, 6, 8, 20–24]. TMDs offer tunable bandgaps and light-matter solid interactions, enhancing light absorption capabilities. BP exhibits a high carrier mobility and remarkable catalytic properties. H-BN, which is known for its excellent thermal and chemical stability, is suitable for use in harsh photocatalytic conditions. The unique properties of these 2D nanomaterials make them well suited for enhancing the efficiency of photocatalytic processes.

6.3.2 Key Properties and Characteristics of 2D Nanomaterials

2D nanomaterials possess several vital properties and characteristics that contribute to their enhanced photocatalytic performance. These properties include a large specific surface area, high light absorption capacity, efficient charge separation and transfer, and excellent catalytic activity. The large specific surface area provides more active sites for chemical reactions, allowing increased contact between the photocatalyst and pollutants. This enhances the photocatalytic degradation process. Additionally, the high light absorption capacity of 2D nanomaterials enables the efficient utilization of light energy across a broader spectrum, leading to improved photocatalytic efficiency. The efficient charge separation and transfer capabilities of 2D nanomaterials prevent the recombination of electron-hole pairs, prolong the lifetime of charge carriers, and increase the chances of redox reactions with pollutants [3, 12]. Furthermore, their excellent catalytic activity facilitates the generation of reactive species and promotes contaminant degradation. These properties make 2D nanomaterials highly promising for enhancing photocatalytic performance in wastewater treatment applications.

6.3.3 Role of 2D Nanomaterials in Enhancing Photocatalytic Performance

Incorporating 2D nanomaterials into photocatalytic systems offers significant benefits for enhancing the overall wastewater treatment performance. Their unique properties, such as large specific surface area, high light absorption capacity, and efficient charge separation and transfer, contribute to improved photocatalytic efficiency. These nanomaterials provide more active sites for chemical reactions, increasing the contact between the photocatalyst and pollutant molecules. Their high light absorption capacity allows for efficient utilization of a broader range of the solar spectrum, leading to enhanced photocatalytic efficiency. Additionally, the efficient charge separation and transfer properties of 2D nanomaterials minimize the recombination of electron-hole pairs, prolong the charge carrier lifetime, and increase the chances of redox reactions with pollutants [25, 28]. Moreover, 2D nanomaterials can act as co-catalysts, promoting specific responses and improving the overall catalytic activity of photocatalysts.

Furthermore, the excellent thermal and chemical stabilities of 2D nanomaterials ensure their durability and long-term performance under harsh photocatalytic conditions. This stability reduces the need for frequent catalyst replacement, contributing to the cost-effectiveness of the process. Additionally, the tunable bandgaps of specific 2D nanomaterials allow for the efficient utilization of different wavelengths of light, enabling tailored photocatalytic systems that can target particular pollutants or optimize the utilization of available light sources [21, 24]. Overall, integrating 2D nanomaterials in photocatalysis holds great promise for developing more efficient and sustainable wastewater treatment solutions, benefiting both environmental and industrial processes.

6.4 Challenges in Utilizing 2D Nanomaterials for Photocatalytic Wastewater Treatment

6.4.1 Synthesis and Fabrication Challenges

The synthesis and fabrication of 2D nanomaterials presents significant challenges for their utilization in photocatalytic wastewater treatment. The development of scalable and cost-effective synthesis methods that can consistently produce high-quality 2D nanomaterials remains challenging. Achieving precise control over the size, shape, and composition of these materials is crucial for optimizing their properties and performance. Additionally, integrating these nanomaterials into practical photocatalytic systems requires careful consideration of factors such as stability, compatibility with the reaction environment, and efficient immobilization techniques [28, 29].

6.4.2 Stability and Degradation Issues

The stability of 2D nanomaterials under photocatalytic conditions is critical. Ensuring the long-term strength and durability of these nanomaterials is essential for their practical application in wastewater treatment. Exposure to harsh environments, such as high temperatures, corrosive solutions, and intense light irradiation, can lead to the degradation and loss of catalytic activity over time. Strategies such as surface passivation, protective coatings, and structural modifications have been explored to address these stability and degradation issues [30].

6.4.3 Efficiency and Selectivity Considerations

The efficiency and selectivity of photocatalytic reactions using 2D nanomaterials require careful optimization. Although these materials offer enhanced photocatalytic performance, the efficiency of charge separation, transfer, and utilization of light energy can still be improved. Additionally, achieving high selectivity for the degradation of specific pollutants while minimizing the degradation of harmless compounds is challenging [30, 31]. The development of strategies to enhance the efficiency and selectivity of photocatalytic reactions using 2D nanomaterials is an active area of research.

6.4.4 Scalability and Cost-Effectiveness Challenges

The scalability and cost-effectiveness of 2D nanomaterials for photocatalytic wastewater treatment are essential for practical implementation. Large-scale synthesis methods, cost-effective production techniques, and the efficient utilization of these nanomaterials in industrial-scale treatment systems require further exploration. In addition, the cost of raw materials, energy requirements for light sources, and catalyst recovery and recycling methods must be optimized to ensure the economic viability of using 2D nanomaterials in large-scale wastewater treatment processes. Addressing these challenges is crucial for harnessing the full potential of 2D nanomaterials in photocatalytic wastewater treatment [32]. Continued research and development efforts are necessary to overcome these obstacles and enable the practical implementation of these materials for efficient and sustainable wastewater treatment [33].

6.5 Strategies to Overcome Challenges

6.5.1 Improvement of Synthesis and Fabrication Techniques

Efforts should focus on advancing the synthesis and fabrication techniques for 2D nanomaterials to overcome the challenges related to their production. This includes exploring innovative methods that enable precise control over size, shape, and composition. The development of scalable and cost-effective synthesis routes, such as solution-based approaches or chemical vapor deposition, can facilitate the large-scale production of high-quality 2D nanomaterials. Additionally, the integration of these nanomaterials into practical photocatalytic systems should be optimized through improved immobilization techniques and compatibility with the reaction environment [3, 9].

6.5.2 Enhancement of Stability and Durability

Research should aim to improve the long-term stability and durability of 2D nanomaterials under photocatalytic conditions to address the stability and degradation issues. Surface passivation techniques, protective coatings, and structural modifications can enhance resistance to harsh environments. Developing novel approaches, such as hybrid composites or core-shell structures, can provide additional stability and protection to nanomaterials, ensuring their prolonged catalytic activity [7, 10, 11].

6.5.3 Optimization of Photocatalytic Performance

Efforts should be directed toward optimizing the photocatalytic performance of 2D nanomaterials by improving their efficiency and selectivity. Strategies include engineering the electronic structure of nanomaterials, introducing doping or heterostructure formation, and tailoring the bandgap for efficient light absorption. Enhancing the charge separation and transfer efficiency by designing appropriate heterojunctions or introducing co-catalysts can further improve the photocatalytic activity. Additionally, the development of selective catalysts or modification techniques can enable the targeted degradation of specific pollutants, thereby enhancing the selectivity of the photocatalytic process [7, 9].

6.5.4 Economical and Scalable Production Methods

To address the scalability and cost-effectiveness challenges, research should focus on developing economical and scalable production methods for 2D nanomaterials. This includes exploring alternative raw materials, improving the energy efficiency of the synthesis process, and developing efficient catalyst recovery and recycling techniques. Optimizing large-scale photocatalytic systems, considering factors such as reactor design, light sources, and integration with existing wastewater treatment processes, can enhance the economic viability of utilizing 2D nanomaterials for industrial applications. By implementing these strategies and continuing research and development efforts, the challenges associated with utilizing 2D nanomaterials for photocatalytic wastewater treatment can be effectively addressed. These advancements will contribute to the practical implementation of these materials, enabling more efficient, sustainable, and cost-effective solutions for industrial wastewater treatment [28–30].

6.6 Case Studies and Applications

6.6.1 Examples of Successful Applications of 2D Nanomaterials

The numerous successful applications of 2D nanomaterials in various fields have demonstrated their potential. For instance, graphene-based nanomaterials have been employed as photocatalysts to degrade organic pollutants in wastewater, exhibiting excellent performance and high degradation rates. Transition metal dichalcogenides (TMDs) have been utilized for hydrogen production *via* photocatalytic water splitting, offering promising alternatives to traditional catalysts. In addition, black phosphorus-based nanomaterials have shown efficient pollutant removal and degradation. These examples highlight the versatility and effectiveness of 2D nanomaterials for various photocatalytic applications [3–7].

6.6.2 Case Studies in Photocatalytic Industrial Wastewater Treatment

Several case studies have demonstrated the application of 2D nanomaterials in the photocatalytic treatment of industrial wastewater. For instance, the use of graphene-based nanomaterials as photocatalysts has been shown to effectively degrade dyes and organic pollutants in textile wastewater. Transition metal dichalcogenides (TMDs) have been applied in the treatment of pharmaceutical wastewater, demonstrating their ability to remove persistent organic compounds. Black phosphorus-based nanomaterials have been employed in the treatment of industrial effluents containing heavy metals, demonstrating their potential for efficient pollutant removal. These case studies highlight the successful utilization of 2D nanomaterials to address specific challenges and treat various types of industrial wastewater [15].

6.6.3 Lessons Learned and Future Prospects

Case studies and applications of 2D nanomaterials in photocatalytic wastewater treatment provide valuable insights and lessons for future research and development. They emphasized the importance of tailoring nanomaterials for specific pollutants and optimizing photocatalytic systems to achieve the desired treatment outcomes. Furthermore, these studies highlight the need for interdisciplinary collaboration among scientists, engineers, and industry experts to address technical and practical challenges associated with large-scale implementation. Prospects include exploring the synergistic effects of combining multiple types of 2D nanomaterials, developing multifunctional materials, and integrating them with other treatment technologies to achieve enhanced wastewater treatment efficiency. Moreover, the development of standardized protocols, regulations, and guidelines for using 2D nanomaterials in wastewater treatment is crucial for their safe and sustainable application [26, 27]. The successful application and case studies of 2D nanomaterials in photocatalytic wastewater treatment show their potential to address industrial wastewater challenges. By leveraging their unique properties, optimizing photocatalytic systems, and addressing scalability and cost-effectiveness considerations, nanomaterials can contribute to more efficient and sustainable solutions. Future research and collaboration will pave the way for further advancements and open new opportunities for the widespread adoption of 2D nanomaterials in industrial wastewater treatment.

6.7 Conclusion

In this comprehensive exploration of 2D nanomaterials in photocatalysis for industrial wastewater treatment, we covered several essential aspects. We discuss the principles of photocatalysis, highlighting the role of photocatalysts in initiating chemical reactions through light absorption. The advantages and limitations of photocatalysis as a wastewater treatment method were examined, emphasizing its sustainability, versatility, and non-specificity, while considering challenges such as light availability and catalyst deactivation. We then investigated the significance of 2D nanomaterials in enhancing the photocatalytic performance, emphasizing their large surface area, high light absorption capacity, efficient charge separation and transfer, and catalytic activity. Furthermore, we examined the types of 2D nanomaterials commonly used in photocatalysis, such as graphene, TMDs, black phosphorus, and h-BN, highlighting their unique properties for wastewater treatment. We also discuss the challenges of utilizing 2D nanomaterials, including synthesis and fabrication issues, stability concerns, efficiency considerations, scalability, and cost-effectiveness.

The field of 2D nanomaterials in photocatalysis for industrial wastewater treatment is evolving rapidly, and several emerging trends and future directions have been identified. These include improving synthesis and fabrication techniques to enhance the quality and scalability of 2D nanomaterials. Efforts to improve the stability and durability of these materials under harsh photocatalytic conditions are also crucial for their long-term performance. Optimization of photocatalytic performance through the design of tailored systems and exploration of multifunctional nanomaterials will contribute to more efficient treatment processes. Developing economical and scalable production methods for 2D nanomaterials will facilitate their widespread adoption in industrial wastewater treatment.

In conclusion, integrating 2D nanomaterials into photocatalysis holds great promise for addressing the challenges of industrial wastewater treatment. Their unique properties and characteristics contribute to enhanced photocatalytic performance, enabling the degradation of organic pollutants and the removal of contaminants. Despite existing challenges, strategies can be implemented to overcome these hurdles, including advancements in synthesis techniques, stability improvement, optimization of photocatalytic performance, and cost-effective production methods. The case studies and applications discussed highlight the successful utilization of 2D nanomaterials for the treatment of various types of industrial wastewater. As research and development efforts continue, emerging trends and future directions indicate a promising path for further advancements and widespread adoption of 2D nanomaterials in industrial wastewater treatment. By harnessing the potential of 2D nanomaterials, we can achieve more efficient and sustainable solutions for mitigating the environmental impacts of industrial activities.

References

1. Lazar, M.A., Varghese, S., Nair, S.S., Photocatalytic water treatment by titanium dioxide: Recent updates. *Catalysts*, 2, 4, 572–601, 2012. https://doi.org/10.3390/catal2040572.
2. Chen, K., Wang, Y., Lu, Y., Treatment of polluted water for reclamation using photocatalysis and constructed wetlands. *Catal. Today*, 175, 1, 276–282, 2011. https://doi.org/10.1016/j.cattod.2011.06.013.
3. Priyadharshini, S.D., Manikandan, S., Kiruthiga, R., Rednam, U., Babu, P.S., Subbaiya, R., Karmegam, N., Kim, W., Govarthanan, M., Graphene oxide-based nanomaterials for the treatment of pollutants in the aquatic environment: Recent trends and perspectives-a review. *Environ. Pollut.*, 306, 119377, 2022. https://doi.org/10.1016/j.envpol.2022.119377.
4. Arul, N.S. and Nithya, V.D., *Two Dimensional Transition Metal Dichalcogenides*, Springer Singapore, 2019, https://link.springer.com/book/10.1007/978-981-13-9045-6.
5. Elbanna, O., Zhu, M., Fujitsuka, M., Majima, T., Black phosphorus sensitized TiO2 mesocrystal photocatalyst for hydrogen evolution with visible and near-infrared light irradiation. *ACS Catal.*, 9, 4, 3618–3626, 2019. https://doi.org/10.1021/acscatal.8b05081.
6. Li, S., Zhang, Y., Huang, H., Black phosphorus-based heterostructures for photocatalysis and photoelectrochemical water splitting. *J. Energy Chem.*, 67, 745–779, 2022. https://doi.org/10.1016/j.jechem.2021.11.023.
7. Ganguly, P., Harb, M., Cao, Z., Cavallo, L., Breen, A., Dervin, S., Dionysiou, D.D., Pillai, S.C., 2D nanomaterials for photocatalytic hydrogen production. *ACS Energy Lett.*, 4, 7, 1687–1709, 2019. https://doi.org/10.1021/acsenergylett.9b00940.
8. Khan, M., Assal, M.E., Tahir, M.N., Khan, M., Ashraf, M., Hatshan, M.R., Khan, M., Varala, R., Badawi, N.M., Adil, S.F., Graphene/inorganic nanocomposites: Evolving photocatalysts for solar energy conversion for environmental remediation. *J. Saudi Chem. Soc.*, 26, 6, 101544, 2022. https://doi.org/10.1016/j.jscs.2022.101544.
9. Zeng, M., Chen, M., Huang, D., Lei, S., Zhang, X., Wang, L., Cheng, Z., Engineered two-dimensional nanomaterials: An emerging paradigm for water purification and monitoring. *Mater. Horiz.*, 8, 758–802, 2021. https://doi.org/10.1039/D0MH01358G.

10. Chen, Y., Fan, Z., Zhang, Z., Niu, W., Li, C., Yang, N., Two-dimensional metal nanomaterials: Synthesis, properties, and applications. *Chem. Rev.*, 118, 13, 6409–6455, 2018. https://doi.org/10.1021/acs.chemrev.7b00727.
11. Jeong, G.H., Sasikala, S.P., Yun, T., Lee, G.Y., Lee, W.J., Kim, S.O., Nanoscale assembly of 2D materials for energy and environmental applications. *Adv. Mater.*, 32, 35, 1907006, 2020. https://doi.org/10.1002/adma.201907006.
12. Rasool, K., Pandey, R.P., Rasheed, P.A., Buczek, S., Gogotsi, Y., Mahmoud, K.A., Water treatment and environmental remediation applications of two-dimensional metal carbides (MXenes). *Mater. Today*, 30, 80–102, 2019. https://doi.org/10.1016/j.mattod.2019.05.017.
13. Fatima, J., Shah, A.N., Tahir, M.B., Mehmood, T., Shah, A.A., Tanveer, M., Nazir, R., Latief, B., Alansi, S., Tunable 2D nanomaterials; Their key roles and mechanisms in water purification and monitoring. *Front. Environ. Sci.*, 10, 766743, 2022. https://doi.org/10.3389/fenvs.2022.766743.
14. Mohsin, M., Ishaq, T., Bhatti, I.A., Maryam, A., Melaibari, A.A., Abu-Hamdeh, N.H., Semiconductor nanomaterial photocatalysts for water-splitting hydrogen production: The holy grail of converting solar energy to fuel. *Nanomaterials*, 13, 3, 546, 2023. https://doi.org/10.3390/nano13030546.
15. Acar, C., Dincer, I., Zamfirescu, C., A review on selected heterogeneous photocatalysts for hydrogen production. *Int. J. Energy Res.*, 38, 1903–1920, 2014. https://doi.org/10.1002/er.3211.
16. Fajrina, N. and Tahir, M., A critical review in strategies to improve photocatalytic water splitting towards hydrogen production. *Int. J. Hydrog. Energy*, 44, 540–577, 2019. https://doi.org/10.1016/j.ijhydene.2018.10.200.
17. Goveas, L.C., Nayak, S., Vinayagam, R., Selvaraj, R., Pugazhendhi, A., Recent advances of nano-photocatalysts involved in hydrogen production by water splitting. *Fuel*, 348, 128460, 2023. https://doi.org/10.1016/j.fuel.2023.128460.
18. Ahmad, H., Kamarudin, S.K., Minggu, L.J., Kassim, M., Hydrogen from photo-catalytic water splitting process: A review. *Renew. Sust. Energ. Rev.*, 43, 599–610, 2015. https://doi.org/10.1016/j.rser.2014.10.101.
19. Ortega Méndez, J.A., López, C.R., Melián, E.P., Díaz, O.G., Rodríguez, J.M.D., Hevia, D.F., Macías, M., Production of hydrogen by water photo-splitting over commercial and synthesised Au/TiO2 catalysts. *Appl. Catal. B*, 147, 5, 439–452, 2014. https://doi.org/10.1016/j.apcatb.2013.09.029.
20. Rafiq, M., Hu, X., Ye, Z., Qayum, A., Xia, H., Hu, L., Lu, F., Chu, P.K., Recent advances in structural engineering of 2D hexagonal boron nitride electrocatalysts. *Nano Energy*, 91, 106661, 2022. https://doi.org/10.1016/j.nanoen.2021.106661.
21. Ling, X., Wang, H., Huang, S., Dresselhaus, M.S., The renaissance of black Phosphorus, 112, 15, 4523–4530, 2015. https://doi.org/10.1073/pnas.1416581112.
22. Chhowalla, M., Shin, H.S., Eda, G., Li, L., Loh, K.P., Zhang, H., The chemistry of two-dimensional layered transition metal dichalcogenide nanosheets. *Nat. Chem.*, 5, 263–275, 2013. https://doi.org/10.1038/nchem.1589.

23. Laursen, L.B., Kegnæs, S., Dahla, S., Chorkendorff, I., Molybdenum sulfides efficient and viable materials for electro-and photoelectrocatalytic hydrogen evolution. *Energy Environ. Sci.*, 5, 5577–5591, 2012. https://doi.org/10.1039/c2ee02618j.
24. Li, T. and Galli, G., Electronic properties of MoS2 nanoparticles. *J. Phys. Chem. C*, 111, 16192–16196, 2007. https://doi.org/10.1021/jp075424v.
25. Novoselov, K.S., Mishchenko, A., Carvalho, A., Neto, A.H.C., 2D materials and van der Waals heterostructures. *Science*, 353, 6298, aac9439, 2014. DOI: 10.1126/science.aac9439.
26. Ida, S. and Ishihara, T., Recent progress in two-dimensional oxide photocatalysts for water splitting. *J. Phys. Chem. Lett.*, 5, 15, 2533–2542, 2014. https://doi.org/10.1021/jz5010957.
27. Saleem, Z., Pervaiz, E., Yousaf, M.U., Khan Niazi, M.B., Two-dimensional materials and composites as potential water splitting photocatalysts: A review. *Catalysts*, 10, 464, 2020. https://doi.org/10.3390/catal10050464.
28. Ismail, A.A. and Bahnemann, D.W., Photochemical splitting of water for hydrogen production by photocatalysis: A review. *Sol. Energy Mater. Sol. Cells*, 128, 85–101, 2014. https://doi.org/10.1016/j.solmat.2014.04.037.
29. Shi, Y., Li, H., Li, L., Recent advances in controlled synthesis of two-dimensional transition metal dichalcogenides via vapor deposition techniques. *Chem. Soc. Rev.*, 44, 2744–2756, 2015. https://doi.org/10.1039/C4CS00256C.
30. Zhang, L., Wang, N., Li, Y., Design, synthesis, and application of some two-dimensional materials. *Chem. Sci.*, 14, 5266–5290, 2023. https://doi.org/10.1039/D3SC00487B.
31. Zhang, J., Chena, Y., Wang, X., Two-dimensional covalent carbon nitride nanosheets: synthesis, functionalization, and applications. *Energy Environ. Sci.*, 8, 3092–3108, 2015. https://doi.org/10.1039/C5EE01895A.
32. Wang, T., Zhang, X., Mei, L., Ma, D., Liao, Y., Zu, Y., Xu, P., Yin, W., Gu, Z., A two-step gas/liquid strategy for the production of N-doped defect-rich transition metal dichalcogenide nanosheets and their antibacterial applications. *Nanoscale*, 12, 8415–8424, 2020. https://doi.org/10.1039/D0NR00192A.
33. Huan, Y., Shi, J., Zou, X., Gong, Y., Xie, C., Yang, Z., Zhang, Z., Gao, Y., Shi, Y., Li, M., Yang, P., Jiang, S., Hong, M., Gu, L., Zhang, Q., Yan, X., Zhang, Y., Scalable production of two-dimensional metallic transition metal dichalcogenide nanosheet powders using NaCl templates toward electrocatalytic applications. *J. Am. Chem. Soc.*, 141, 47, 18694–18703, 2019. https://doi.org/10.1021/jacs.9b06044.

7

Application of 2D Nanomaterials for Energy Storage

Tulasi Barik[1]* and Subhendu Chakroborty[2]

[1]*Department of Chemistry, SRICT-ISR, UPL University of Sustainable Technology, Gujarat, India*
[2]*Department of Basic Sciences, IITM, IES University, Bhopal, Madhya Pradesh, India*

Abstract
2D nanomaterials have recently attracted considerable attention in energy storage technology. Supercapacitors and batteries are two recent advancements in energy storage technology that have significantly assisted mankind in meeting their energy demands. 2D nanomaterials must possess different features, including a large surface area, outstanding electronic conductivity, exceptional electrochemical properties, and superior thermal stabilities for batteries and supercapacitors to work well. This chapter provides a concise summary of the use of 2D nanomaterials, such as graphene analogs, MXenes, chalcogenides, and transition metal oxides/hydrides, in batteries and supercapacitors. The relationship between the composition of 2D nanomaterials and their electrochemical properties depends entirely on the storage performance of the supercapacitors and batteries.

Keywords: 2D nanomaterial, graphene, transition metal oxides, MXenes, chalcogenides, supercapacitor, batteries

7.1 Introduction

The biggest challenge faced by human beings in the twenty-first century was the energy crisis. The growing need for energy has led to the improvement of energy-storing devices to justify power supply for several applications, ranging from household uses to grids of stationary storage and transport systems. Numerous studies have been conducted to improve energy generation,

**Corresponding author*: bariktulsi693@gmail.com

Subhendu Chakroborty and Kaushik Pal (eds.) 2D Nanomaterials: Synthesis, Properties and Applications, (121–154) © 2024 Scrivener Publishing LLC

transportation, and storage. Electrochemical reactions that add energy to a system are essential for achieving this goal. These redox processes are caused by the difference in the electrochemical potential between the cathode and anode. Control of the redox reaction is crucial for achieving a long cycle life and good electrochemical performance without any side reactions. Therefore, selecting appropriate elements or compounds is crucial. 2D materials have drawn more interest in this sector as materials that would provide higher energy storage capacities. Researchers and scientists have been motivated to design new 2D nanomaterials since the discovery of graphene. Owing to its desirable characteristics, including high surface area, exceptional thermal conductivity, superior physical properties (high Young's modulus) and distinctive structural properties, this family of materials facilitates rapid surface redox reactions and reduces the ion diffusion distance [1–4].

The structural design of 2D nanomaterials, having planar, buckling, and puckered structures, causes facile ion transport, resulting in high ion conductivity [5]. Many synthetic techniques, including chemical vapor deposition (CVD), organic conversion, liquid-phase and mechanical exfoliation, wet chemical processes, and the hot-injection method, have been used to design various types of 2D nanomaterials. These materials have been classified as transition metal oxides, chalcogenides, graphene and graphene analogs, elemental 2D materials, and MXene materials [6–9]. The shapes and characteristics of 2D materials can be easily changed and expanded by controlling and adjusting the surface atoms. The extraordinary behavior of 2D nanomaterials is due to the absence of van der Waals interactions in 2D nanomaterials, higher specific surface area, electron confinement, and direct bandgap in 2D nanomaterial. The ionic conduction of 2D nanomaterials has been significantly improved by altering their chemical composition, which is responsible for boosting the electrochemical performance of energy storage devices. Several 2D materials have been introduced as interesting options for electrodes and electrode additives to improve the electrochemical properties of electrode devices. The development of numerous 2D nanomaterials is outlined in this chapter along with their applications in electrochemical energy storage devices.

7.2 2D Nanomaterials for Application of Lithium Ion Batteries

The energy issue has risen to the forefront owing to global industrialization. The development of alternative energy sources is crucial to avoid energy issues. Secondary batteries, particularly lithium-ion batteries (LIBs), have been developed as an integral part of the energy industry as portable energy storage and supply devices. The transformation of chemical energy into electrical energy occurs in energy storage devices called LIBs. In as LIB, an electrolyte wedged between two electrodes engages in a variety of electrochemical processes to provide energy for the system. Li-ions are intercalated and deintercalated between anodes and cathodes in LIBs. More specifically, when charging through the electrolyte, the Li-ion de-intercalates from the cathodes and intercalates into the anodes. Li-ions, on the other hand, switch from the anode to cathode during discharge. During the operation of a battery with electrically conductive current collectors, the resultant electrons are transmitted. Despite the expansion of LIB research, significant material-related challenges have made it difficult for commercial applications that require high energy density and performance to use LIBs. Traditional electrode materials perform poorly electrochemically, which results from the substantial irreversible volume and compositional changes that high-capacity electrode materials go through anode and cathode materials are essential components to develop a high capacity of LIB.

Despite significant attempts to develop high-capacity batteries, these materials still experience severe issues. To prevent short circuits between the cathode and anode and extend cycle stability, separators and suitable electrolytes are essential in terms of safety. Another essential component of the LIBs is the current collector. To achieve consistent electrochemical characteristics, the extremely conductive nature of electrode materials is essential. Significant attempts have been made to increase the electrochemical enactment of devices by improving ionic conduction and changing the chemical composition [10]. To improve the electrochemical properties of electrode devices, several 2D materials have been introduced as interesting electrode materials or electrode additives [11]. The extensive surface area of the electrodes offers many reactive active sites, which cause lithium ions to intercalate easily, leading to a high theoretical capacitance. The greater the availability of active materials, the more frequently the electrical properties of the 2D materials can be changed. As current collectors, 2D materials are essential for increasing the lifetime and energy density of LIB. The ionic conductivity and electrochemical stability can be increased using 2D nanomaterials as electrolytic additives. 2D nanomaterials as electrode materials can improve the electrochemical stability and prevent cracking owing to their outstanding mechanical properties [12–16]. This section discusses the increased electrochemical performance of 2D nanomaterials, such as MXene materials, chalcogenides, graphene analogs, elemental 2D materials, and transition metal oxides/hydrides in LIBs.

The unsatisfactory electrochemical properties of typical electrode materials have led to the development and use of 2D nanomaterials as electrode batteries. Graphene and its equivalents are frequently used as electrodes in LIBs [17–26]. Owing to its effective electrical conductivity, which increases its cyclability and substantial reversible capacity, graphene is an ideal 2D material for use as a lithium-ion battery anode, with a notable performance of >600 mAh/g [27, 28]. Wang *et al.* [29] designed N-doped graphene (GN), with the high electrical conductivity of N doping, has higher lithium storage efficiency in comparison to pure graphene. The structural defects of N-doped graphene (GN) cause quicker Li^+ diffusivity, with a higher capacity of 10–30% than that of pure graphene. In the first cycle at 0.2 °C, this GN electrode had exceptional discharge capacities of 1,284 mA h/g and 432 mA h/g at 50 °C.

The lightweight interlayers of Li-S batteries were designed using graphene and functionalized boron nitride nanosheets [30]. When coated with functionalized boron nitride nanosheets/graphene (FBN/G) film, Li-S batteries show an increase in storage capacities ranging from 382 to 560 mA h g^{-1} at 3 °C because of the interaction between the positive functional groups of FBN and negatively charged polysulfide compounds. Moreover, the rate of cyclic capacity degradation decreased every cycle from 0.032% to 0.0037%. After 1,000 cycles, a degradation capacity rate of 0.0067% per cycle at 1 °C was crucial. Lei *et al.* used an easy and environmentally friendly approach to synthesize a substantial amount of boron carbon nitride (BCN) nanosheets. After 5,000 cycles, the steady and improved storage capacity of BCN nanosheets in lithium batteries is ~100 mA h g^{-1} at a current density of 2 A g^{-1} because of their large surface area and 2–6 atomic layers [31].

Phosphorene, silicene, and germanene, which have planar, buckled, and puckered structures, play a substantial role in the introduction of ions and electron transport channels. Owing to the diverse atomic structural designs of these graphene analog nanomaterials, they have been used as high-capacity anode materials [32–34]. With binding energies of 3.26 eV, 2.6 eV, 4.03 eV, 2.92 eV, and for Li, Mg, Ca, and Na, respectively, borophene has been explored using DFT calculations as an ideal candidate for rechargeable battery anode materials. The investigation showed that the 2D material borophene can store Ca, Na, Mg, and Li ions with capacities of 800, 1380, and 1860 mAh/g, respectively [35].

Recently, phosphorene, a highly anisotropic substance, was synthesized using black phosphorous and exploited as an LIB anode material. Theoretical calculations (DFT) of phosphorene were performed to investigate its diffusion barrier and binding energy. This work reveals a metallic transition with good electrical conductivity, low diffusion rate of 0.8 eV and 0.68 eV along the armchair and zigzag routes, respectively, and high average voltage of 2.4 eV. Owing to its anisotropic properties, this material is regarded as a suitable electrode material. Further analysis was carried out using DFT calculations to explore the pristine and defective phosphorene. The evidence shows that the defect improves the ability of phosphorene to bind energy and acts as a diffusion barrier. In comparison to pure phosphorene, the predicted binding energy improved to 1 eV, whereas the calculated diffusion barrier was 0.13 eV, which was lower than that of graphene and molybdenum dioxide [36, 37]. In addition to graphene, the most prevalent monoelemental analogs have been employed as LIB electrodes.

Chen *et al.* [38] reported Layer Black Phosphorus nanosheet by using a sustainable gas-phase transformation process supported by a mineralizer. They provided examples of mixing these BP nanosheets with graphene sheets that are extremely conductive to create flexible, high performance LIB electrodes. These electrodes had an outstanding rate and durable efficiency, with a significant capacity of 501 mAh $g^{\times 1}$. The molecular beam epitaxy deposition technique and solid-state reaction technique have both been used to successfully produce silicene, a honeycomb-shaped single-layer-thick silicon nanosheet. Because of its chemical stability, metal–silicene interactions, and charge-carrier mobility, it has served as an anode for LIBs. Both theoretical and practical results have demonstrated the electrochemical performance of silicene [39].

During the lithiation process, alloy Sb experiences noticeable expansion and compositional disintegration, which significantly reduces the battery cycling stability in LIB [40]. DFT was used to analyze Li atom adsorption and diffusion in a 2D graphene/antimonene (G/Sb) heterostructure. This heterostructure has strong thermodynamic and dynamic stability and can offer rapid electron transit through the electrode during the lithiation/delithiation processes, with a 0.06 eV band gap, which is much lower than that of monolayer Sb (1.24 eV). The calculated higher Li diffusion coefficient and volume expansion at 300 K, as compared to bulk, monolayer, and bilayer materials, were studied to improve the charge and discharge performance of the G/Sb heterostructure. This heterostructure is used as a potential flexible and wearable anode in LIBs which has been confirmed by charge analysis and density of state calculations.

Transition metal oxides (TMOs) have been prepared to improve the capacity of LIBs as electrodes in LIBs by diminishing their capacity decay. According to numerous reports, TMO materials with short diffusion routes and an abundance of intercalation sites exhibit quick charging and discharging rates, marginal volume expansion during Li^+ intercalation, and high specific capacities [41–43]. Shen *et al.* [44] reported a unique MnO_2/carbon nanotube (CNT) (MnO_2/CNTs) by mixing MnO_2 nanosheets with CNTs *via* a simple, high-yield, one-step procedure. This putty-like substance can be rolled into flexible films or easily molded into various shapes. This layered structure provides good electronic conductivity in addition to interlayer spaces for quick and reversible lithium-ion insertion/extraction and can be reused as an anode material with a specific capacity of 796 mAh g^{-1} and a current density of 500 mA g^{-1} in LIBs. Jia *et al.* [45] reported oxygen-deficient $SnO_{2-\delta}$/C composite (Figure 7.1) has been designed by carbon coating of porous SnO_2 nanoflakes and this meticulous nanoarchitecture composite is used as LIBs anode with large reversible specific capacity of 1,378.6 mA h g^{-1} having current density of 0.1 A g^{-1} with coulombic efficiency of approximately 74.3% because the carbon layer increases the conductivity of SnO_2 and significantly reduces nucleation and volume alteration during the charging/discharge processes.

The liquid exfoliation technique was used to design few-layer V_2O_5 nanosheets. Owing to their substantial reversible capacity, large coulombic efficiency, and steady cyclability, these nanosheets are employed as LIB cathode materials with an enhanced capability of 117 mA h g^{-1} for charge and discharge at 50 °C. These materials can be used in high-power batteries for electric vehicle applications owing to their superior electrochemical properties. They are also appealing for quick charging [46]. Current research on generation energy storage systems looks forward to using transition metal hydroxides in LIB applications as anode materials instead of commercial graphite.

Zhu *et al.* [47] designed a 2D ultrathin nanosheet of -$Ni(OH)_2$ having a large specific surface area of 189.57 m^2 g^{-1} which causes high electrochemical activity and excellent reversibility towards lithium. These materials

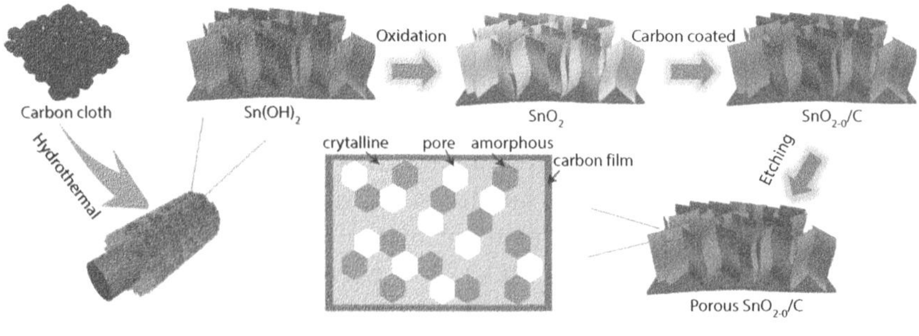

Figure 7.1 Schematic representation of the synthesis and assembly technique for porous $SnO_{2-\delta}$/C nanosheet arrays [45] (reprinted with permission by copyright 2018, Elsevier).

are employed as LIB anodes because of their reversible conversion-type electrochemical behavior, which results in outstanding cycling longevity. The ultrathin 2D-$Ni(OH)_2$ nanostructure was responsible for the longer lithium storage performance owing to the effective surface electrochemical reaction. Using a self-assembly solution-based molecular method, Mei *et al.* designed nanosheets of heterostructured bismuth oxide (BOHNs). These materials were used as LIB anode materials with an initial discharging capacity 647.6 mAh g^{-1}. Improved capacities of more than 200 mAh g^{-1} have been observed even after 260 cycles owing to more surface/edge-active sites and assorted α- and β-phases [48]. Several types of 2D materials group of TMO/TMH materials have been developed by different research groups for LIB applications [49–59].

New ultrathin 2D materials are being investigated because of the rapid advancement of graphene in LIBs. Owing to their semiconductor qualities with large bandgaps and abundance in nature, lamellar structures of single layers of transition metal dichalcogenides (TMDs) have received considerable attention. New theoretical and technical approaches have been used to develop inorganic 2D nanomaterials with the MX_2 chemical formula (where M = Ta, Mo, Nb, W, and X = Se, S, etc.). These 2D materials have been demonstrated to be potential anode materials for LIBs [60, 61]. 2D materials, such as MoS_2 and VS_2, have been employed as anode materials in LIBs and have high charging capacities compared to graphene, which has high charge capabilities of 466 mAh/g and 335 mAh/g and low energy barriers of 0.22 eV and 0.25 eV, respectively. Theoretical calculations show the remarkable binding interaction and diffusion of Li on MoS_2 accelerates the charging/discharging processes and increases the high-power density [62, 63].

Im *et al.* [64] reported a photolysis reaction using a gas-phase laser technique to design unique $SnSe_x$ and $GeSe_x$ nanocrystals (where x = 2 – 1). After 70 cycles, these materials exhibited good reversible capacities of 400–800 mA h/g, which is nearly equal to the theoretical capacity of $Li_{4.4}Ge$ and $Li_{4.4}Sn$ for LIBs. These materials exhibit phase conversion of Sn and Ge, with a cubic-tetragonal structure, which has been studied using X-ray diffraction and Raman spectroscopy, and results in an increase in the lithiation and delithiation processes. The first-principle calculation shows that the enhanced cycling proficiency of Sn derivatives is due to the small expansion of volume in the cubic Sn phase. Sen *et al.* synthesised MoS_2 nanowalls by sintering at 700 °C with carbon and MoS2 [65], These materials have achieved excellent electrochemical activity with a capacity of 880 mAh g^{-1} at a current density of 100 mA g^{-1} due to the ample space between two layers of thin MoS_2, which facilitates diffusion of the electrolyte, and the thickness of the 2D-wall in the nano range, which causes diffusion of lithium-ion the MoS_2 matrix.

Chen *et al.* [66] used a one-step hydrothermal approach was used by to design a new 2D heterostructure ($SnSe_2$/graphene) using $SnSe_2$ nanosheets supported on graphene (Figure 7.2). This hybrid heterostructure can stop the tin agglomeration formed during the electrochemical processes because graphene and $SnSe_2$ are chemically bonded to one another. This 2D heterostructure can operate normally in a specified temperature range −30 to 70 °C with a capacity of 490.9 mA h g^{-1} and a current rate of 0.1 C. After 1,500 cycles, this material also had greater longevity, with a capacity preservation of 59.3% at 0.5 C.

Another 2D nanomaterial, MXene, is composed of carbides and nitrides of the transition metals. Its high electrical conductivity and mechanical properties make it suitable for use as an anode or cathode in LIBs. The structural composition of MXenes might vary according to the synthesis and processing methods used, which in turn affects their storage capacities for LIBs. According to theoretical calculations, the theoretical storage capacity of terminated MXene sheets is lower in Li-ion batteries than that of bare MXene sheets. Recent studies have suggested that heat treatment is an effective method for altering the surface chemistry of MXene sheets. Ti_2CT_x's capacitive performance is considerably enhanced by heating in various atmospheres. Some of the compounds in this family,

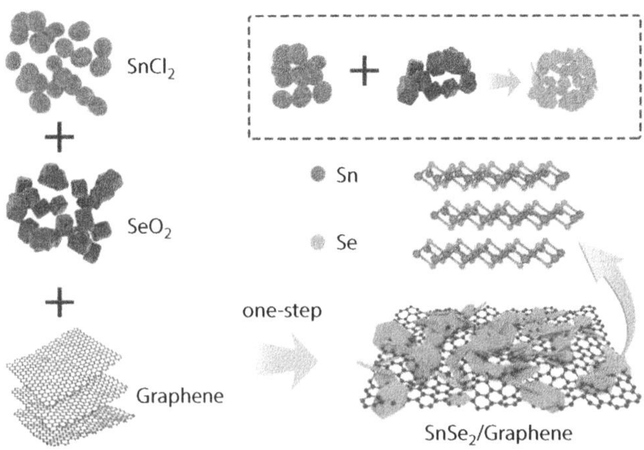

Figure 7.2 Schematic representation of the synthesis and assembly technique for 2D/2D SnSe2/graphene heterostructure [66] (reprinted with permission by copyright 2020, Elsevier).

M_2C (M = Sc, Ti, V, and Cr), have extraordinary discharge capacities of greater than 400 mAh/g [67, 68].

Ahmed *et al.* [69] observed TiO_2 nanocrystals on the surface of MXene sheets by after oxidizing Ti_2CT_x MXene with hydrogen peroxide (H_2O_2) solution at room temperature. The performance of Li-ion batteries can be significantly improved by changing the surface chemistry of the material. After 50 charge and discharge cycles, H_2O_2 treated MXene exhibited discharge capacities of 389 mAh g^{-1}, 337 mAh g^{-1}, and 297 mAh g^{-1} at current densities of 100, 500, and 1,000 mA/g, respectively. Furthermore, MXene treated with H_2O_2 demonstrated a pronounced capacity of 150 mAh g^{-1} with an exceptional rate capability at a high current density of 5,000 mA/g. When the MXene family-Nb_2CT_x was delaminated, an amine was used instead of DMSO, yielding a flexible CNT/MXene composite. These electrodes have a significant capacity of 400 mAh g^{-1} at 0.5 °C in Li-ion batteries, which is more than their multilayered counterparts [70].

7.3 Application of 2D Nanomaterials in Sodium Ion Batteries

Although LIBs are currently the most widely used energy storage technology, they may not be used as energy storage systems on a large scale because of their high cost and other related issues [71–74]. Most the world's lithium resources are also concentrated in a small number of areas, including China, South America, Australasia, North America, and Russia [75]. Other nations would have to endure LIBs, including the aforementioned, as well as the unsteady supply that drives up the price of lithium. Low-cost sodium-ion batteries (SIBs) have become increasingly attractive for use in massive electrical energy storage systems. Sodium ions are desired as charge carriers, which may be simply and clearly explained by the quantity of the material.

Because of their similar chemical properties and placement in the same group of periodic tables as lithium, sodium-ion batteries present a competitive challenge to the next wave of innovative rechargeable battery electrode materials. These batteries have large resource reserves and are inexpensive. The electrolytic equivalent of Li^+/Li is three times that of Na^+/Na. However, as time progressed, the gap in the potential reversible capacity became less important. In addition, owing to their relatively large size, sodium ions have the potential to increase the versatility of material design. An additional advantage of higher ionic radii is that they have a weaker solvation energy in polar solvents, according to methodical and speculative investigations of Li^+, Na^+, and Mg^{2+} as well as other aprotic polar solvents. The relatively low desolvation energy is a key discovery for the development of high-power batteries because it significantly affects the insertion of alkali ions at the interface of electrolyte. Finally, the high conductivity of the Na^+-based electrolyte contributed to the improved performance of the batteries. However, SIBs have some drawbacks when used as batteries, including the fact that sodium has a higher molecular mass than lithium, which results in a lower theoretical energy density than LIBs, and that the radius of sodium is greater than that of lithium, which makes it more challenging for sodium ions to intercalate and deintercalate during charging and discharging [76, 77]. The theoretical volumetric and gravimetric reversible capacities of the Li and Na electrodes decreased. If the invention of the material is realized and current research advances to solve it, this issue may be resolved by one day. To increase the performance of SIBs, reasonable electrode material design and structure optimization are crucial. Numerous attempts have been made to expand the use of Na storage materials for SIBs with high capacities and rapid diffusion kinetics. 2D nanomaterials have been used as prospective candidates for electrodes in SIBs because of their wide surface area that promotes ionic adsorption and flexible interlayer spacing, which makes it easier for the transfer of Na ions and the intercalation of electrons. Owing to their strong electronic conductivity and good structural stability, two-dimensional materials have a huge capacity for NIBs. When used as electrodes, 2D materials have a large surface area and a wide-open structure, which is responsible for increasing the interaction between the electrolytes and the active materials and also reducing Na-ion diffusion which causes a higher theoretical capacity.

Large Na ions can be easily incorporated into 2D materials because they are stacked by weak van der Waals interactions and have considerable mechanical flexibility. While sodiation and desodiation occur, these materials undergo volume changes. It is crucial to understand how 2D materials in NIBs maximize battery capacity by enhancing the charge storage capacity. Current research into the sodiation and desodiation mechanisms of 2D materials includes a variety of approaches. Currently, work is being done to change the composition of 2D materials to include many active reactive sites to improve their electrochemical properties. Consequently, sodium ions can intercalate more readily, increasing the capacity of NIBs [78–80]. This section will cover the increased electrochemical performance and various compositions of assembling 2D materials such as elemental 2D chalcogenides, transition metal oxides/hydrides, graphene analogs, and MXene materials in NIBs. First-principle calculations have established that borophene makes an effective anode material for NIBs with high energy storage.

The highest storage capacity for sodium ion batteries is 1,240 mAh/g, which was calculated by adsorbing sodium ions at various concentrations on borophene [81]. Sun *et al.* [82] fabricated a black phosphorous/grapheme hybrid by alternating phosphorene layers with graphene layers as a sandwich-structure. After 100 cycles, the capacity of this phosphorene/graphene structure was 2,440 mA h/g at a current density of 50 mA/g, with a retention capacity of 83%. Physical vapor deposition (PVD) was used by Liu *et al.* for the uniform and dense deposition of red phosphorus nanodots on sheets of reduced graphene oxide, resulting in new P@RGO [83]. This P@RGO has significant performance as flexible anode material used in SIB having charge capacities of 1165.4, 510.6, and 135.3 mAh/g at current densities of 159.4, 31 878.9, and 47 818.3 mA g^{-1}, respectively, at high charge and discharge rates. Poly(2,5-Dimercapto-1,3,4-thiadiazole) (PDMcT) was carbonized on nanosheets of graphene oxide according to Xu *et al.* [84], producing sheets of N/S co-doped graphene (PDMcT/RGO). These sheets were used as anodes for SIBs with a capacity of 240 mA h g^{-1} at a current density of 500 mA/g and the rate of performance increased to 144 mA h g^{-1} at a current density of 10 A g^{-1}. Moreover, it exhibited a strong cycle stability at 153 mAh/g over 5,000 cycles at 5,000 mA/g^{1}.

A layered black phosphorous/graphene composite (BP/rGO) was described by Liu *et al.* [85] using a straightforward and scalable pressurization method at ambient temperature. The BP/rGO anode was employed at various current densities to attain specific charge capacities of 1,460.1, 1,401.2, 1,377.6, 1,339.7, 1,277.8, 1,123.78, and 720.8 mAh/g. The BP/rGO anodes with charge capacities of 1,250 and 640 mAh/g at current densities of 1 and 40 A/g over 500 cycles, respectively, stabilized the long-cycling performance, indicating a substantial enhancement in the performance of NIB anodes. Nanosheets of reduced graphene oxide (rGO) with an extraordinary capacity of 140 mAh g^{-1} at 0.2 °C (40 mA/g) and long-term cycling endurance durability of more than 1,000 cycles were first proposed by Wang *et al.* [86] in 2013. They also have an incredible and reversible rate capability that provides a capacity of 95.6 mA hg^{-1} at a noticeably high current rate of 5 °C (1,000 mA/g). The sodium ion diffusion length may be shortened by the nanosheet structure, which also has outstanding performance.

According to Yang *et al.* [87] N-rich carbon nanosheets and a mixture of H_2S and Ar gas were used to design S-doped N-rich carbon nanosheets (S-N/C). This (S-N/C) nanosheet is used as an SIB anode with a high rate and a large capacity due to the substitution of pyrrolic-N for S while retaining pyridinic-N and the development of strong C-S-C-bonds, which increased the distance between the interlayers and surface area and provided additional active sites to produce more reactions of reversible sodium storage. Consequently, the porous N,S-co-doped carbon nanosheets used as SIB anodes showed high capacity and outstanding rate performance. Because the electrolyte passes through the holes and the diffusion of Na-ion and distance can be reduced, porous 2D materials provide good electrodes for NIBs because their electrochemical performance is enhanced [88–90]. Liu *et al.* [91] reported an easy electrostatic and chemical reduction procedure to design a double-decker-like structure of graphene-supported Sb nanocomposite Sb@RGO@Sb by using ultra-fine Sb nanoparticles symmetrically embedded on an RGO surface. The Sb@RGO@Sb composite has a high sodium storage capacity owing to Sb encapsulation in the matrix, which induces strain and stress without pulverizing it, shortening the paths of electronic and ionic transport, and limiting the loss of electrical contact. Even after 200 cycles, it still had a capacity of approximately 430 mAh/g when used as an SIB anode material, and it had a good rate performance of 330 mAh/g at a current density of 5 A/g.

Table 7.1 2D transition metal sulfide nanoflakes as electrodes in SIBs.

Materials	Cyclic performance	Ref
VS_2 nanoflakes	700 mAh g^{-1} after 100 cycles at 0.1 A g^{-1}	[98]
2D $MoSe_2$/graphene nanocomposites	324 mAh g^{-1} after 1500 cycles at 3.2 A g^{-1}	[99]
MoS_2 nanoflakes aligned vertically on carbon paper	230 mAh g^{-1} after 100 cycles at 0.08 A g^{-1}	[100]
Hierarchical flower-like VS_2 nanoflakes	352 mAh g^{-1} after 700 cycles at 2 A g^{-1}	[101]
Self-supported VG/$MoSe_2$/N–C sandwiched arrays	298 mAh g^{-1} after 1,000 cycles at 2 A g^{-1}	[102]
Carbon-coated hierarchical SnS nanotubes	440.3 mAh g^{-1} after 100 cycles at 0.2 A g^{-1}	[103]
MoS_2/graphene nanoflakes	441 mAh g^{-1} after 250 cycles at 0.3 A g^{-1}	[104]
Metallic MoS_2 sandwich grown on graphene tube	313 mAh g^{-1} after 200 cycles at 0.05 A g^{-1}	[105]

Layered transition metal sulfides are excellent electrode materials because the interlayer spacing can facilitate sodium ion insertion and removal [92–97]. Two-dimensional nanoflakes are frequently used as electrode materials in SIBs. The utilization of 2D nanoflakes as SIB electrode materials and their electrochemical activities are summarized in Table 7.1 [98–105].

Liu *et al.* [106] used red phosphorus as the starting material for the synthesis of nanoscale red phosphorus (NRP) by using a phosphorus-amine-based technique and then on reduced graphene oxide (rGO) to produce a composite of NRP/rGO. This composite exhibited impressive long-term cyclability with a specific performance of 2,057 mAh/g at a current density of 0.1 A/g. At current densities of 2 A/g and 5 A/g, the significant long-term cyclability without performance degradation has capacity declines of 14.5% and 34.4% cycles after 1,400 and 5,000 cycles.

A unique mesoporous $B_2O_3/BC_2O/C$ nanosheet was designated by Wang *et al.* [107]. It contains uniformly embedded ultrafine PbTe nano dots (Figure 7.3). The 2D PbTe/BC nanosheets had a flat surface and thickness of approximately 10–56 nm. The electrical conductivity of the carbon matrix was greatly improved by these currently existing boracic phases, which also offered an abundance of structural flaws and active reaction sites. This material as the anode demonstrated an excellent sodium storage capacity of more than 1,000 mAh/g at current densities of 0.1 A/g and minimal capacitance deterioration of 10% at 10 A g^{-1} even after 500 cycles.

The electrostatic interaction between positively charged 1D CNTs and negatively charged 2D MXene nanosheets produced a porous Ti_3C_2 MXene/CNT composite without restacking of the MXene nanosheets. This facilitated ion access to the electrode and electrolyte transport. Ti_3C_2

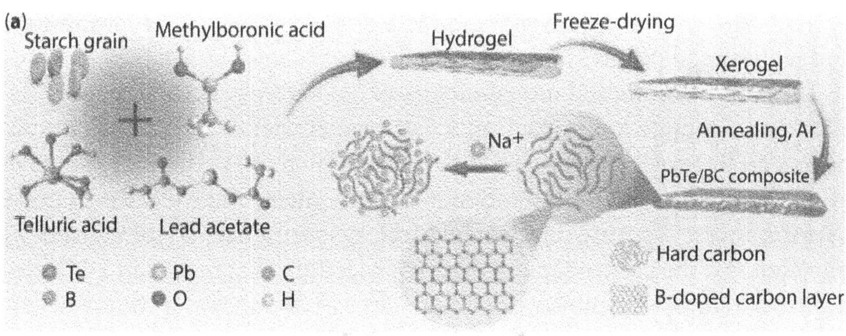

Figure 7.3 (a) Schematic depiction of the synthesis technique for PbTe/BC nanosheets [107] (reprinted with permission from copyright 2020, Elsevier).

MXene/CNT porous films demonstrated as an excellent SIB anode because of its structural and excellent cycling stability having performance of 421 mAh/cm^3 at a current density of 20 mA/g [108].

7.4 Application of 2D Nanomaterials in Potassium Ion Batteries

Potassium-ion batteries (PIBs) are among the leading candidates for enormous energy storage systems because of their affordability, suitable redox potential, and abundance in nature. Owing to the unique benefits of PIBs over lithium-ion batteries, significant work is currently being conducted on emerging stationary energy storage systems [109–111]. PIBs are more economical than LIBs because the potassium sources are plentiful and inexpensive [112]. In addition, the reduction potential of potassium is closer to that of LIBs, allowing for an operating voltage similar to that of LIBs, guaranteeing a high energy density [113, 114]. Thirdly, the rapid transport of K^+ in electrolytes is assisted by the small Stokes radius of K^+ ions [115–117]. However, the large size of K^+ significantly limits the performance of PIBs [118, 119]. Because of the large ionic radius of K^+, the K^+ reaction exhibits slow kinetics and poor rate performance, leading to lower ion diffusivity in the electrodes. In the meantime, the huge volume changes brought on by the insertion and extraction of K^+ result in pulverization of the electrode materials, which lowers the cyclability of PIBs [120, 121]. To achieve this, it is crucial to select host materials that can reversibly store K^+ and have sufficient structural stability to endure multiple potassization/depotassization processes. 2D materials are the best materials to address these challenges with pronounced capacity and an elongated lifespan, thus increasing the energy storage capability of PIBs. Numerous studies have been conducted on the development of 2D materials as electrodes in PIBs and widespread progress has been achieved.

The compositions and morphologies of 2D heterostructured nanomaterials have a significant impact on K^+ storage efficiency. The same components can be spatially arranged differently, which alters the morphologies of 2D-based nanomaterials and significantly alters their electrochemical performances. Because electrochemical performance is influenced by morphology, the controllable synthetic procedure that results in 2D-based heterostructures by minor deviations in the synthesis techniques will result in morphological changes in the 2D-heterostructures, influencing the electrochemical performance. Recent progress has been made in the use of 2D heterostructures such as graphene derivatives, TMDs, MXenes, and TMOs as PIB electrode materials. These 2D-based heterostructures details of synthetic methods and electrochemical properties are summarized in Table 7.2.

Guo and colleagues designed ultrathin VSe_2 ultrathin nanosheets as anode materials in KIBs with enhanced capacity, rate ability, and cyclability because of their structural stability, high surface area-to-volume ratio, rapid electron and K-ion transit and constrained self-aggregation [129]. Xie *et al.* [130] used a two-step solvothermal technique to yield a unique ultrathin rose-like MoS_2 nanoflake by chemically coupling MoS_2 and rGO. Owing to their increased MoS_2 interlayer spacing, the MoS_2@rGO composites demonstrated the best electrochemical performance. Layered metal dichalcogenides favor structural changes during potassium ion insertion/extraction by enlarging the interlayer gap [130–133].

Ma *et al.* designed a unique graphene co-doped with P/O (PODG) from raw materials triphenylphosphine and graphite oxide by thermal

Table 7.2 2D nanomaterials as electrodes in PIBs.

Electrode materials	Techniques used for synthesising materials	Types of electrode	Performance rate [mAh/g/mA/g]	Ref
$SnSb_2Te_4$/graphene	Multiple ball milling	Anode	368/100	[122]
N-doped graphene/$ReSe_2$/Ti_3C_2 MXene	Hydrothermal, freeze-drying and etching	Anode	138/10,000	[123]
$MoS_{2\text{-}0.1}$/MXene	Etching and hydrothermal	Anode	290.7/50; 168.2/500	[124]
$MoSe_2$@MoO_2 heterojunction	Hydrothermal	Anode	365/100; 172/2,000	[125]
Co_9S_8/NC@MoS_2	Hydrothermal, pyrolysis and annealing	Anode	122/1,000	[126]
Bi_2S_3/MoS_2@NC	Hydrothermal	Anode	300.2/3,000	[127]
VS4/SnS@C	Wet-chemical method	Anode	270.1/100; 162.1/5,000	[128]

annealing [134]. The carbon backbone of graphene is dual-doped with oxygen and phosphorus, causing a variety of vacancies and defects that may decrease the number of paths for K^+ transport. PODG demonstrated an excellent cycle life and an extraordinary flexible capacity of 474 mAh/g at current densities of 0.05 A/g with specific capacities sustained at current densities of 385, 235, and 160 mAh/g 0.5, 1, and 2 A/g for the 600th cycle.

Zhang *et al.* [135] utilized a two-step hydrothermal process to design a dispersed architecture with by orderly anchoring of Co_3Se_4 nanocrystallites on graphene sheets, which neatly fastened on graphene sheets and that could successfully delay structural deterioration. The capacity of the highly conductive graphene matrix and the dispersed Co_3Se_4 nanocrystallites to lower contact stress during K^+ intercalation led to an additional consistent and operative anode design compared to its two independent counterparts. After the 500th cycle, the optimised electrode shows outstanding cycling stability and rate capacity of 301.8 mAh/g at 1 A/g and 203.8 mAh/g at 5 A/g, respectively.

Jia *et al.* [136] designed inventive porous bamboo-like hollow tubes by composition of N-doped-C and MoS_2 layers that are joined vertically interchanging series to maintain their structural integrity. During the durable charge–discharge process, this porous bamboo-like shape significantly reduced the mechanical stress in the axial direction. It is possible to stack MoS_2 nanosheets with N-doped C layers to produce heterostructures that efficiently improve conductivity and preserve structural integrity. The attraction of the K atoms was enhanced by the N-doped-C/MoS_2 interface. This material is considered the best KIBs anode material because of its high specific capacity of 451 mAh/g at a current density of 50 mA/g during the first cycle. The steady performance of this material was maintained at 330 mAh/g even after 50 cycles. After 1,000 cycles, these materials demonstrated extended cyclability, with a reversible capacity of 151 mAh/g and a current density of 0.5 A/g.

MXenes are 2D nanomaterials that are widely employed as conductive substrates for the adsorption of other 2D materials to construct new types of MXene-based heterostructures. The mechanical properties of heterostructures can be strengthened by covalent bonds or weak van der Waals interactions, which enable them to respond to considerable changes in volume. Finally, MXene heterostructures were exploited for K storage applications. To further increase the electrocatalytic activity of MXene-based materials, several 2D–2D heterostructures have been designed using various strategic techniques. Li *et al.* reported layered heterostructures of $Ti_3C_2T_x$ MXene/MoS_2 as PIB electrode materials, which showed prolonged cyclability after 1,000 cycles, with a reversible capacity of 290.7 mAh/g at a current density of 0.05 A/g [137]. Recently, unique $MoSe_2$/MXene heterostructures with carbon coatings were designed using a multistep process by vertically attaching $MoSe_2$ nanosheets onto the MXene substrates [138]. Additionally, MXenes-based heterostructures provide more active sites for K^+ adsorption and transport on the surface of 2D materials. This heterostructure exhibited improved K^+ transfer kinetics and a potassium storage capacity of 355 mAh/g after 100 cycles. Furthermore, at a scan rate of 1.2 mV/s, the pseudocapacitive involvement of the $MoSe_2$/MXene@C heterostructures reached 90.2%.

SnS_2-NSs/MXene heterostructures were reported by Cao *et al.* [139] using a simple hydrothermal technique. Analysis by high-resolution transmission electron microscopy and scanning electron microscopy revealed the uniform growth of SnS_2 nanosheets on N,S co-doped MXene sheets and this heterostructure material has exceptional electrical conductivity and mechanical stability. Even after 800 cycles, this heterostructure still exhibited a high reversible capacity of 206.1 mAh/g and a current density of 0.5 A/g. Ultrathin $SnS_2/Ti_3C_2T_x$ heterostructures were designed by the congregation of the few-layer SnS_2 and $Ti_3C_2T_x$ [140]. This heterostructure was used as a PIB anode material, having a significant and sustained reversible capacities of 462.1 mAh/g at 0.1 A/g and 85.5 mAh/g at the higher current density of 2.0 A/g after 460 cycles. Zhao *et al.* [141] reported the layer porous heterostructures PDDA–NPCN/Ti_3C_2 by using N-rich porous carbon nanosheets/Ti_3C_2 and poly-diallyl dimethylammonium chloride (PDDA), which had more active sites and increased the layer gap between layers. High-performance potassium-ion batteries serves as PIB anodes.

7.5 Applications of 2D Nanomaterials in Supercapacitors

The production and storage of energy is a major concern of the twenty-first century. The development of alternative energy sources is crucial to avoid energy issues. Supercapacitors have been developed as an integral part of the energy industry as portable energy storage and supply devices. This differs from batteries and dielectric capacitors. Supercapacitors (SCs) play a significant role in energy conversion and storage devices because they provide an increased specific capacitance, a higher power density over an extended period of time, a faster charge-discharge rate, a better exchange feature, a low cost, and a harmless operating environment.

The high energy density compared with a dielectric capacitor, superior power density, and long cycling capabilities of these storage devices are currently attracting attention [142]. The performance of a supercapacitor is greatly influenced by the choice of the electrode material. The essential components of supercapacitors are electrodes and electrolytes, as well as current collectors, separators, and sealants, which have a significant impact on their electrochemical operation. The design and selection of electrode materials are among the most crucial aspects for enhancing the capacitance performance of supercapacitors. High specific surface area, corrosion resistance, conductivity, and strong thermal and chemical stability are required to enhance the performance of the supercapacitor. Owing to these properties, researchers are interested in 2D nanomaterials, such as chalcogenides, derivatives of graphene, TMOs, MXenes, and elemental 2D materials. Supercapacitors have numerous uses such as factory power backup, load cranes, and electric automobiles. For a supercapacitor to operate more effectively, these characteristics must be present [143]. Owing to their distinctive physicochemical features, two-dimensional layered nanomaterials have attracted considerable attention in the design of electrode materials.

Black phosphorus has extraordinary transporter mobility of up to 1,000 cm^2/Vs, a large predicted theoretical specific capacitance 2,596 mAh/g and an adjustable bandgap of 0.30 to 2.2 eV by controlling the number of layers and the layer thickness (phosphorene) [144]. Liquid-exfoliated black phosphorous nanoflakes are used as bendable electrode materials for supercapacitors with a high volumetric capacitance of 13.75 F/cm^3 which is greater than that of restacked graphene capacitance of ~1.0 F/cm^3 [145]. Yang *et al.* [146] designed a unique composite using CNT and black phosphorus at a ratio of 1:4. It is used as an electrode for a supercapacitor that has a mechanically induced high power density of 821.62 Wcm^3 and a high energy density of 5.71 $mWhcm^{-3}$ at 0.005 Vs^{-1}. Moreover, after 10,000 cycles, the initial capacitance lingered at 91.5%.

It is simple to modify graphene oxide nanomaterials, and perhaps its most notable quality is that it retains its excellent potential as an electrode material for supercapacitor applications. Because they have a large energy storage capacity and can be used to create nanomaterials with superior shapes, graphene-based supercapacitors have been identified as a promising study field. The synthesis of graphene sheets based on modified polymers was accomplished. This hybrid film was created *via* a network of interconnected carbon structures, and a successful supercapacitor electrode based on nanopores was fabricated. The computed capacity was 120 Fg^{-1} at a rapid scan rate of 1 V/s. The graphene/polyaniline nanocomposites and graphene were prepared using vacuum filtering,. These films exhibited excellent mechanical stability and flexibility. They have a strong conductivity and can be bent at large angles to create appropriate structures. The calculated electrochemical capacitance was 210 F/g with a discharge rate of 0.3 A/g [147, 148]. They are very conductive and have a large surface area, and extremely effective materials 2D materials such as graphene and its reduced oxide and metal chalcogenides are currently regarded as supercapacitor electrodes for high performance.

The TMD monolayers are held together by a weak van der Waals force of interaction and are composed of a sandwich of two sheets with a transition-metal layer of dichalcogenides in each layer. Several TMDs, such as $NbSe_2$, TiS_2, TaS_2, and MoS_2, have layered structures resembling those of graphite and are almost identical to those of graphene in terms of flexibility, thinness, and transparency. TMDs are attracting attention because of their ease of synthesis, outstanding conductivity, broad surface area, durability, and outstanding chemical and mechanical stability. Metal sulfides are electrochemically active materials that can be used as electrodes in supercapacitors. They are used as electrodes in tiny supercapacitors because they have applications in microsized devices. Spray-painting 2D MoS_2 on a Si/SiO_2 chip is a simple, low-cost method that results in microsized capacitors made of 2D MoS_2 films. The arranged microcapacitor had five electrodes per polarity, each measuring 820 μm in length and 5.5 mm in width. The thickness was 0.45 μm, and the distance between the two electrodes was 200 μm. For energy storage, the constructed supercapacitor exhibited substantial electrochemical efficiency. These materials have a greater capacitance than a microsupercapacitor of 8 mF/cm^2, which is more than the performance of graphene. A hydrothermal technique was used to develop a different type of $MoSe_2$ nanosheet-based supercapacitor. The calculated charge and discharge current density of the $MoSe_2$ nanosheet in supercapacitors was 5 A/g, giving a maximum capacity of 198.9 F/g. A high cycle capability is an important feature of supercapacitor devices [149, 150]. Vanadium disulfide (VS_2) nanosheets, have a broad surface area and exceptional conductivity, which make them excellent electrodes for supercapacitor applications. Owing to its metallic nature, VS_2 is an excellent material for various applications. The exceptional supercapacitor efficiency is due to the weak van der Waals forces of attraction between the interlayers of the conducting S-V-S thin films. A remarkably new metallic 2D material is produced by exfoliating the bulk type of VS_2 using an ammonia-assisted technique into ultrathin nanosheets that are stacked in five different S-V-S layers. These thin layers have been effectively used as electrode materials for in-plance supercapacitors. For the in-plane designs, the calculated capacitance of VS_2 was 4,760 F/cm^2. These materials retain their performance as supercapacitor electrodes even after 1,000 charging and discharging cycles [151]. Cobalt disulfide CoS_2 and a hybrid of CoS_2 and graphene have been shown to be potential electrode materials for energy storage in supercapacitors [152]. A 2D metal sulfide with good electrochemical characteristics is tungsten disulfide (WS_2). The S-W-S layers create a 2D covalent link with the formation, and van der Waals forces are used for separation. WS_2 is a direct-bandgap, 2D material. Thus, this material will draw attention to nanoelectronic applications such as energy storage devices in the future. According to previous reports, this substance can be used as a pseudocapacitive energy storage material. The combination of 2D reduced graphene oxide and WS_2 forms a hybrid material that is used as a high-performance supercapacitor electrode material. The increased surface area of these 2D composite materials improves the electrochemical performance of the supercapacitor. 2D nanoflakes have unique characteristics such as large surface area, outstanding electrical and thermal conductivity, and good mechanical stability, making them potential candidates for supercapacitor [153].

The hydrophilicity and metallic conductivity of MXenes make them an attractive alternative for storing energy using these materials as electrodes for supercapacitor applications [154]. Examples of MXenes that have been utilized in supercapacitors include electrode materials such as Ti_2CT_x, Ti_3C_2Tx, V_2CT_x, $Nb_4C_3T_x$, Nb_2CT_x, $Ta_4C_3T_x$, and Ti_2NT_x [155, 156]. These materials can store charge through pseudocapacitive mechanisms. Other methods were used by Qiu *et al.* to design Ti_3C_2 MXene, which has a conductance of 260 S cm^{-1}, which is two times the rGO conductivity of 1 S cm^{-1} [157]. Zhang *et al.* [158] designed a highly conductive and deformable unique MXene/MnO_2 composite on carbon cloth by combining of MXene and MnO_2. This composite was used as an electrode in capacitors, with an improved specific capacitance of 511.2 F/g as compared to MnO_2 nanorods on carbon cloth, which have capacitance of 302.3 F/g. The improved capacitance was due to the synergistic combination of MnO_2 and MXene. Zhao *et al.* [159] reported carbinization of Mn-based MOF and MnO_2 to get nanosheets of MnO_2/carbon composite (MnO_2/CNS). This composite was used as an electrode for supercapacitors, with an enhanced capacitance of 339 F/g at current densities of 0.5 A/g and using Na_2SO_4 as the electrolyte.

In recent years, many studies have been conducted on 2D nanoflake materials with various geometrical properties. 2D nanoflakes are anticipated to play a more important role in energy storage systems because of their distinctive features, such as more active sites, enhanced mechanical properties, rapid mass and electron transport kinetics, and larger specific surface areas. 2D nanoflake materials with large surface areas enable the intercalation of guest ions, rapid ion diffusion, and charge transfer, which are responsible for the electrochemical performance of supercapacitors. There are numerous reports on 2D nanoflake materials as electrochemically efficient electrode materials for supercacitor energy storage devices [160–166].

Conclusions

In this chapter, we discuss the preparation, design, and well-known applications of 2D-based materials in capacitors and batteries such as LIBS, NIBS, and PIBs, including graphene analogs, MXene-, TMD-, and TMO-based electrode materials. Novel molecular and morphological designs have triggered the development of a variety of 2D materials. Van der Waals interactions are typically present in various forms in 2D-based materials, which alter the capacity of intercalation reactions to occur quickly and at a high voltage, as well as the binding environment and storage locations for intercalated ions. A larger surface area, along with different types of functional groups and active sites, can be provided by the distinctive architecture of 2D materials, which facilitates reaction kinetics and leads to increased electrochemical performance. Therefore, these materials have been exploited in rechargeable batteries as electrodes, active components, structural supports, hosts for the active components, and separators. It is expecting that 2D-materials will eventually lead to significant developments and will increase the bounds of capacitors and rechargeable batteries through gradual advances.

References

1. Peng, L., Zhu, Y., Chen, D., Ruoff, R.S., Yu, G., Two-dimensional materials for beyond-lithium-ion batteries. *Adv. Energy Mater.*, 6, 1600025, 2016.
2. Pomerantseva, E. and Gogotsi, Y., Two-dimensional heterostructures for energy storage. *Nat. Energy*, 2, 17089, 2017.
3. Chowdhury, C. and Datta, A., Exotic physics and chemistry of two-dimensional phosphorus: Phosphorene. *J. Phys. Chem. Lett.*, 8, 2909–2916, 2017.
4. Kim, J.H., Jeong, J.H., Kim, N., Joshi, R., Lee, G.-H., Mechanical properties of two-dimensional materials and their applications. *J. Phys. D Appl. Phys.*, 52, 083001, 2019.

5. Tian, H., Seh, Z.W., Yan, K., Fu, Z., Tang, P., Lu, Y., Zhang, R., Legut, D., Cui, Y., Zhang, Q., Theoretical investigation of 2D layered materials as protective films for lithium and sodium metal anodes. *Adv. Energy Mater.*, 7, 1602528, 2017.
6. Geim, A.K. and Grigorieva, I.V., Van der Waals Heterostructures. *Nature*, 499, 419–425, 2013.
7. Anasori, B., Lukatskaya, M.R., Gogotsi, Y., 2D metal carbides and nitrides (mxenes) for energy storage. *Nat. Rev. Mater.*, 2, 16098, 2017.
8. Shi, L. and Zhao, T.S., Recent advances in inorganic 2D materials and their applications in lithium and sodium batteries. *J. Mater. Chem. A*, 5, 3735–3758, 2017.
9. Zeng, Z., Yin, Z., Huang, X., Li, H., He, Q., Lu, G., Boey, F., Zhang, H., Single-layer semiconducting nanosheets: High-yield preparation and device fabrication. *Angew. Chem. Int. Ed.*, 50, 11093–11097, 2011.
10. Wang, X., Weng, Q., Yang, Y., Bando, Y., Golberg, D., Hybrid two-dimensional materials in rechargeable battery applications and their microscopic mechanisms. *Chem. Soc. Rev.*, 45, 4042–4073, 2016.
11. Park, M., Zhang, X., Chung, M., Less, G.B., Sastry, A.M., A review of conduction phenomena in Li-ion batteries. *J. Power Sources*, 195, 7904–7929, 2010.
12. Casimir, A., Zhang, H., Ogoke, O., Amine, J.C., Lu, J., Wu, G., Silicon-based anodes for lithium-ion batteries: Effectiveness of materials synthesis and electrode preparation. *Nano Energy*, 27, 359–376, 2016.
13. Zhou, G., Wang, D.W., Li, F., Zhang, L., Li, N., Wu, Z.S., Wen, L., Lu, G.Q., Cheng, H.M., Graphene-wrapped Fe_3O_4 anode material with improved reversible capacity and cyclic stability for lithium ion batteries. *Chem. Mater.*, 22, 5306–5313, 2010.
14. Li, S., Wang, B., Li, B., Liu, J., Yu, M., Wu, X., Self-assembly of 2D sandwich-structured $MnFe_2O_4$/graphene composites for high-performance lithium storage. *Mater. Res. Bull.*, 61, 369–374, 2015.
15. Wang, B., Li, X., Zhang, X., Luo, B., Jin, M., Liang, M., Dayeh, S.A., Picraux, S.T., Zhi, L., Adaptable silicon-carbon nanocables sandwiched between reduced graphene oxide sheets as lithium ion battery anodes. *ACS Nano*, 7, 1437–1445, 2013.
16. Agyeman, D.A., Song, K., Lee, G.H., Park, M., Kang, Y.M., Carbon-coated Si nanoparticles anchored between reduced graphene oxides as an extremely reversible anode material for high energy-density Li-ion battery. *Adv. Energy Mater.*, 6, 1600904, 2016.
17. Lee, J.K., Smith, K.B., Hayner, C.M., Kung, H.H., Silicon nanoparticles-graphene paper composites for Li ion battery anodes. *Chem. Commun.*, 46, 2025–2027, 2010.
18. Zhou, H.-S., Kudo, T., Yoo, E.J., Hosono, E., Honma, I., Kim, J., Large reversible Li storage of graphene nanosheet families for use in rechargeable lithium ion batteries. *Nano Lett.*, 8, 2277–2282, 2008.

19. Wang, G., Shen, X., Yao, J., Park, J., Graphene nanosheets for enhanced lithium storage in lithium ion batteries. *Carbon*, 47, 2049–2053, 2009.
20. Bobnar, J., Lozinšek, M., Kapun, G., Njel, C., Dedryvère, R., Genorio, B., Dominko, R., Fluorinated reduced graphene oxide as a protective layer on the metallic lithium for application in the high energy batteries. *Sci. Rep.*, 8, 1–10, 2018.
21. Kim, S., Kim, S.K., Sun, P., Oh, N., Braun, P.V., Reduced graphene oxide/lii composite lithium ion battery cathodes. *Nano Lett.*, 17, 6893–6899, 2017.
22. Kim, S.D., Lee, J.G., Kim, T.G., Rana, K., Jeong, J.Y., Park, J.H., Yoon, S.S., Ahn, J.H., Additive-free electrode fabrication with reduced graphene oxide using supersonic kinetic spray for flexible lithium-ion batteries. *Carbon*, 139, 195–204, 2018.
23. Fu, C., Zhao, G., Zhang, H., Li, S., Evaluation and characterization of reduced graphene oxide nanosheets as anode materials for lithium-ion batteries. *Int. J. Electrochem. Sci.*, 8, 6269–6280, 2013.
24. Chen, Y., Fu, K., Zhu, S., Luo, W., Wang, Y., Li, Y., Hitz, E., Yao, Y., Dai, J., Wan, J., Danner, V.A., Li, T., Hu, L., Reduced graphene oxide films with ultrahigh conductivity as Li-ion battery current collectors. *Nano Lett.*, 16, 3616–3623, 2016.
25. Foroozan, T., Soto, F.A., Yurkiv, V., Sharifi-Asl, S., Deivanayagam, R., Huang, Z., Rojaee, R., Mashayek, F., Balbuena, P.B., Shahbazian-Yassar, R., Synergistic effect of graphene oxide for impeding the dendritic plating of Li. *Adv. Funct. Mater.*, 28, 1705917, 2018.
26. Fu, K., Wang, Y., Yan, C., Yao, Y., Chen, Y., Dai, J., Lacey, S., Wang, Y., Wan, J., Li, T., Wang, Z., Xu, Y., Hu, L., Graphene oxide-based electrode inks for 3D-printed lithium-ion batteries. *Adv. Mater.*, 28, 2587–2594, 2016.
27. Mukherjee, S., Ren, Z., Singh, G., Beyond graphene anode materials for emerging metal ion batteries and supercapacitors. *Nanomicro Lett.*, 10, 1–27, 2018.
28. Bolotin, K., II, Sikes, K.J., Jiang, Z., Klima, M., Fudenberg, G., Hone, J., Kim, P., Stormer, H.L., Ultrahigh electron mobility in suspended graphene. *Solid State Commun.*, 146, 351–355, 2008.
29. Wang, C., Li, D., Too, C.O., Wallace, G.G., Electrochemical properties of graphene paper electrodes used in lithium batteries. *Chem. Mater.*, 21, 2604–2606, 2009.
30. Fan, Y., Yang, Z., Hua, W., Liu, D., Tao, T., Rahman, M.M., Lei, W., Huang, S., Chen, Y., Functionalized boron nitride nanosheets/graphene interlayer for fast and long-life lithium–sulfur batteries. *Adv. Energy Mater.*, 7, 1602380, 2017.
31. Lei, W., Qin, S., Liu, D., Portehault, D., Liu, Z., Chen, Y., Large scale boron carbon nitride nanosheets with enhanced lithium storage capabilities. *Chem. Commun.*, 49, 352–354, 2013.

32. Mortazavi, B., Dianat, A., Cuniberti, G., Rabczuk, T., Application of silicene, germanene and stanene for na or Li ion storage: A theoretical investigation. *Electrochim. Acta*, 213, 865–870, 2016.
33. Li, L., Chen, L., Mukherjee, S., Gao, J., Sun, H., Liu, Z., Ma, X., Gupta, T., Singh, C.V., Ren, W., Cheng, H.-M., Koratkar, N., Phosphorene as a polysulfide immobilizer and catalyst in high-performance lithium-sulfur batteries. *Adv. Mater.*, 29, 1602734, 2017.
34. Makaremi, M., Mortazavi, B., Singh, C.V., 2D hydrogenated graphene-like borophene as a high capacity anode material for improved Li/Na ion batteries: A first principles study. *Mater. Today Energy*, 8, 22–28, 2018.
35. Mortazavi, B. *et al.*, Borophene as an anode material for Ca, Mg, Na or Li ion storage: A first-principle study. *J. Power sources*, 329, 456–461, 2016.
36. Li, W. *et al.*, Ultrafast and directional diffusion of lithium in phosphorene for highperformance lithium-ion battery. *Nano Lett.*, 15, 3, 1691–1697, 2015.
37. Guo, G.-C. *et al.*, Pristine and defect-containing phosphorene as promising anode materials for rechargeable Li batteries. *J. Mater. Chem. A*, 3, 21, 11246–11252, 2015.
38. Chen, L., Zhou, G., Liu, Z., Ma, X., Chen, J., Zhang, Z., Ma, X., Li, F., Cheng, H.M., Ren, W., Scalable clean exfoliation of high-quality few-layer black phosphorus for a flexible lithium ion battery. *Adv. Mater.*, 28, 510–517, 2016.
39. Zhuang, J., Xu, X., Peleckis, G., Hao, W., Dou, S.X., Du, Y., Silicene: A promising anode for lithium-ion batteries. *Adv. Mater.*, 29, 1606716, 2017.
40. Wang, X., Tang, C., Zhou, X., Zhu, W., Cheng, C., Theoretical investigating of graphene/antimonene heterostructure as a promising high cycle capability anodes for fast-charging lithium ion batteries. *Appl. Surf. Sci.*, 491, 451–459, 2019.
41. Leong, C.C., Pan, H., Ho, S.K., Two-dimensional transition-metal oxide monolayers as cathode materials for Li and Na ion batteries. *Phys. Chem. Chem. Phys.*, 18, 7527–7534, 2016.
42. Deng, S., Wang, L., Hou, T., Li, Y., Two-dimensional MnO2 as a better cathode material for lithium ion batteries. *J. Phys. Chem. C*, 119, 28783–28788, 2015.
43. Tan, H.T., Sun, W., Wang, L., Yan, Q., 2D transition metal oxides/hydroxides for energy-storage applications. *ChemNanoMat*, 2, 562–577, 2016.
44. Shen, L., Dong, Q., Zhu, G., Dai, Z., Zhang, Y., Wang, W., Dong, X., Versatile MnO_2/CNT Putty-Like Composites for High-Rate Lithium-ion Batteries. *Adv. Mater. Interfaces*, 5, 1800362, 2018.
45. Jia, R., Yue, J.L., Xia, Q.Y. *et al.*, Carbon shelled porous $SnO_{2-\delta}$ nanosheet arrays as advanced anodes for lithium-ion batteries. *Energy Storage Mater.*, 13, 303–311, 2018.
46. Rui, X., Lu, Z., Yu, H., Yang, D., Hng, H.H., Lim, T.M., Yan, Q., Ultrathin V_2O_5 nanosheet cathodes: Realizing ultrafast reversible lithium storage. *Nanoscale*, 5, 556–560, 2013.

47. Zhu, Y. and Cao, C., Remarkable electrochemical lithium storage behaviour of two-dimensional ultrathin α-Ni$(OH)_2$ nanosheets. *RSC Adv.*, 5, 83757–83763, 2015.
48. Mei, J., Liao, T., Ayoko, G.A., Sun, Z., Two-dimensional bismuth oxide heterostructured nanosheets for lithium-and sodium-ion storages. *ACS Appl. Mater. Interfaces*, 11, 28205–28212, 2019.
49. Ma, N., Jiang, X.Y., Zhang, L., Wang, X.S., Cao, Y.L., Zhang, X.Z., Novel 2D layered molybdenum ditelluride encapsulated in few-layer graphene as high-performance anode for lithium-ion batteries. *Small*, 14, 1703680, 2018.
50. Guo, S., Yu, H., Liu, P., Liu, X., Li, D., Chen, M., Ishida, M., Zhou, H., Surface coating of lithium-manganese-rich layered oxides with delaminated mno_2 nanosheets as cathode materials for Li-ion batteries. *J. Mater. Chem. A*, 2, 4422–4428, 2014.
51. Han, X., Gong, Y., Fu, K., He, X., Hitz, G.T., Dai, J., Pearse, A., Liu, B., Wang, H., Rubloff, G., Mo, Y., Thangadurai, V., Wachsman, E.D., Hu, L., Negating interfacial impedance in garnet-based solid-state Li metal batteries. *Nat. Mater.*, 16, 572–579, 2017.
52. Li, Y., Yao, J., Uchaker, E., Yang, J., Huang, Y., Zhang, M., Cao, G., Leaf-like V_2O_5 nanosheets fabricated by a facile green approach as high energy cathode material for lithium-ion batteries. *Adv. Energy Mater.*, 3, 1171–1175, 2013.
53. Patil, S.B., Kim, H.J., Lim, H.K., Oh, S.M., Kim, J., Shin, J., Kim, H., Choi, J.W., Hwang, S.J., Exfoliated 2D lepidocrocite titanium oxide nanosheets for high sulfur content cathodes with highly stable Li-S battery performance. *ACS Energy Lett.*, 3, 412–419, 2018.
54. Wang, L. and Sasaki, T., Titanium oxide nanosheets: Graphene analogues with versatile functionalities. *Chem. Rev.*, 114, 9455–9486, 2014.
55. Chen, D., Peng, L., Yuan, Y., Zhu, Y., Fang, Z., Yan, C., Chen, G., Shahbazian-Yassar, R., Lu, J., Amine, K., Yu, G., Two-dimensional holey Co3O4 nanosheets for high-rate alkali-ion batteries: From rational synthesis to *in situ* probing. *Nano Lett.*, 17, 3907–3913, 2017.
56. Sun, H., Hanlon, D., Dinh, D.A., Boland, J.B., Del Rio Castillo, A.E., Di Giovanni, C., Ansaldo, A., Pellegrini, V., Coleman, J.N., Bonaccorso, F., Carbon nanotubes-bridged molybdenum trioxide nanosheets as high performance anode for lithium ion batteries. *2D Mater.*, 5, 015024, 1–10, 2018.
57. Cen, Y., Li, S., Zhou, Y., Cai, X., Wang, X., Xiang, Q., Hu, B., Yu, D., Liu, Y., Chen, C., Ultrathin VO_2(B) nanosheets as cathode material for high-performance hybrid magnesium-lithium ion batteries. *J. Electrochem. Soc.*, 166, A1660–A1667, 2019.
58. Azam, A., Kim, J., Park, J., Novak, T.G., Tiwari, A.P., Song, S.H., Kim, B., Jeon, S., Two-dimensional WO_3 nanosheets chemically converted from layered WS_2 for high-performance electrochromic devices. *Nano Lett.*, 18, 5646–5651, 2018.

59. Zou, Y. and Wang, Y., NiO nanosheets grown on graphene nanosheets as superior anode materials for Li-ion batteries. *Nanoscale*, 3, 2615–2620, 2011.
60. Chhowalla, M., Shin, H.S., Eda, G., Li, L.-J., Loh, K.P., Zhang, H., The chemistry of two-dimensional layered transition metal dichalcogenide nanosheets. *Nat. Chem.*, 5, 263–275, 2013.
61. Yang, E., Ji, H., Jung, Y., Two-dimensional transition metal dichalcogenide monolayers as promising sodium ion battery anodes. *J. Phys. Chem. C*, 119, 26374–26380, 2015.
62. Li, Y., Enhanced Li adsorption and diffusion on MoS_2 zigzag nanoribbons by edgeneffects: A computational study. *J. Phys. Chem. Lett.*, 3, 16, 2221–2227, 2012.
63. Jing, Y., Metallic VS_2 monolayer: A promising 2D anode material for lithium ion batteries. *J. Phys. Chem. C*, 117, 48, 25409–25413, 2013.
64. Im, H.S., Lim, Y.R., Cho, Y.J., Park, J., Cha, E.H., Kang, H.S., Germanium and tin selenide nanocrystals for high-capacity lithium ion batteries: Comparative phase conversion of germanium and tin. *J. Phys. Chem. C*, 118, 21884–21888, 2014.
65. Sen, U.K. and Mitra, S., High-rate and high-energy-density lithium-ion battery anode containing 2D MoS_2 nanowall and cellulose binder. *ACS Appl. Mater. Interfaces*, 5, 1240–1247, 2013.
66. Chen, H.W., Liu, R.M., Wu, Y. *et al.*, Interface coupling 2D/2D $SnSe_2$/graphene heterostructure as long-cycle anode for all-climate lithium-ion battery. *J. Chem. Eng.*, 407, 126973, 2021.
67. Anasori, B., Lukatskaya, M.R., Gogotsi, Y., 2D metal carbides and nitrides (MXenes) for energy storage. *Nat. Rev. Mater.*, 2, 16098, 2017.
68. Eames, C. and Islam, M.S., Ion intercalation into two-dimensional transition-metal carbides: Global screening for new high-capacity battery materials. *J. Am. Chem. Soc.*, 136, 16270–16276, 2014.
69. Ahmed, B., Anjum, D.H., Hedhili, M.N., Gogotsi, Y., Alshareef, H.N., H_2O_2 assisted room temperature oxidation of Ti_2C MXene for Li-ion battery anodes. *Nanoscale*, 8, 7580–7587, 2016.
70. Mashtalir, O., Lukatskaya, M.R., Zhao, M.Q., Barsoum, M.W., Gogotsi, Y., Amine-assisted delamination of Nb_2C MXene for Li-ion energy storage devices. *Adv. Mater.*, 27, 3501–3506, 2015.
71. Yabuuchi, N., Kubota, K., Dahbi, M., Komaba, S., Research development on sodium-ion batteries. *Chem. Rev.*, 114, 11636–11682, 2014.
72. Wang, H., Hamanaka, S., Yokoyama, T., Yoshikawa, H., Awaga, K., *In-situ* XAFS studies of mn12 molecular-cluster batteries: Super-reduced Mn12 clusters in solid-state electrochemistry. *Chem. Asian J.*, 6, 1074–1079, 2011.
73. Gu, M., Belharouak, I., Zheng, J.M., Wu, H.M., Xiao, J., Genc, A., Amine, K., Thevuthasan, S., Baer, D.R., Zhang, J.G., Browning, N.D., Liu, J., Wang, C.M., Formation of the spinel phase in the layered composite cathode used in Li-ion batteries. *ACS Nano*, 7, 760–767, 2013.

74. Pramudita, J.C., Pontiroli, D., Magnani, G., Gaboardi, M., Ricco, M., Milanese, C., Brand, H.E.A., Sharma, N., Graphene and selected derivatives as negative electrodes in sodium- and lithium-ion batteries. *Chemelectrochem*, 2, 600–610, 2015.
75. Grosjean, C., Miranda, P.H., Perrin, M., Poggi, P., Assessment of world lithium resources and consequences of their geographic distribution on the expected development of the electric vehicle industry. *Renew. Sustain. Energy Rev.*, 16, 1735–1744, 2012.
76. Hasa, I., Mariyappan, S., Saurel, D., Challenges of today for Na-based batteries of the future: From materials to cell metrics. *J. Power Sources*, 482, 228872, 2021.
77. Kim, Y.J., Ha, K.H., Oh, S.M., High-capacity anode materials for sodium-ion batteries. *Chemistry*, 20, 11980–11992, 2014.
78. Shi, L. and Zhao, T., Recent advances in inorganic 2D materials and their applications in lithium and sodium batteries. *J. Mater. Chem.*, 5, 3735–3758, 2017.
79. Mao, J., Zhou, T., Zheng, Y., Gao, H., H.k. Liu, Z., Guo, Two-dimensional nanostructures for sodium-ion battery anodes. *J. Mater. Chem.*, 6, 3284–3303, 2018.
80. Wu, Y. and Yu, Y., 2D material as anode for sodium ion batteries: Recent progress and perspectives. *Energy Stor. Mater.*, 16, 323–343, 2019.
81. Liu, J., Borophene as a promising anode material for sodium-ion batteries with high capacity and high rate capability using DFT. *RSC Adv.*, 8, 32, 17773–17785, 2018.
82. Sun, J., Lee, H.W., Pasta, M., Yuan, H.T., Zheng, G.Y., Sun, Y.M., Li, Y.Z., Cui, Y., A phosphorene–graphene hybrid material as a high-capacity anode for sodium-ion batteries. *Nat. Nanotechnol.*, 10, 980, 2015.
83. Liu, Y.H., Zhang, A.Y., Shen, C.F., Liu, Q.Z., Cao, X., Ma, Y.Q., Chen, L., Lau, C., Chen, T.C., Wei, F., Zhou, C.W., Red phosphorus nanodots on reduced graphene oxide as a flexible and ultra-fast anode for sodium-ion batteries. *ACS Nano*, 11, 5530, 2017.
84. Xu, X.D., Zeng, H.L., Han, D.Z., Nitrogen and sulfur Co-doped graphene nanosheets to improve anode materials for sodium-ion batteries. *ACS Appl. Mater. Interfaces*, 10, 37172–37180, 2018.
85. Liu, Y.H., Liu, Q.Z., Zhang, A.Y., Room-temperature pressure synthesis of layered black phosphorus-graphene composite for sodium-ion battery anodes. *ACS Nano*, 12, 8323–8329, 2018.
86. Wang, Y.X., Chou, S.L., Liu, H.K., Dou, S.X., Reduced graphene oxide with superior cycling stability and rate capability for sodium storage. *Carbon*, 57, 202, 2013.
87. Yang, J.Q., Zhou, X.L., Wu, D.H., Zhao, X.D., Zhou, Z., S-doped N-rich carbon nanosheets with expanded interlayer distance as anode materials for sodium-ion batteries. *Adv. Mater.*, 29, 1604108, 2017.

88. Tang, K., Fu, L., White, R.J., Yu, L., Titirici, M.M., Antonietti, M., Maier, J., Hollow carbon nanospheres with superior rate capability for sodium-based batteries. *Adv. Energy Mater.*, 2, 873, 2012.
89. Wenzel, S., Hara, T., Janek, J., Adelhelm, P., Room-temperature sodium-ion batteries: Improving the rate capability of carbon anode materials by templating strategies. *Energy Environ. Sci.*, 4, 3342, 2011.
90. Guan, Z., Liu, H., Xu, B., Hao, X., Wang, Z.X., Chen, L.Q., Gelatin-pyrolyzed mesoporous carbon as a high-performance sodium-storage material. *J. Mater. Chem. A*, 3, 7849, 2015.
91. Liu, X.W., Gao, M., Yang, H., Zhong, X.W., Yu, Y., 2D sandwich-like nanosheets of ultrafine Sb nanoparticles anchored to graphene for high-efficiency sodium storage. *Nano Res.*, 10, 4360, 2017.
92. Nayak, P.K., Yang, L., Brehm, W., Adelhelm, P., From lithium-ion to sodium-ion batteries: Advantages, challenges, and surprises. *Angew. Chem. Int. Ed.*, 57, 1, 102, 2018. https://doi.org/ 10.1002/anie.201703772.
93. Mao, J.F., Zhou, T.F., Zheng, Y., Gao, H., Liu, H.K., Guo, Z.P., Two-dimensional nanostructures for sodium-ion battery anodes. *J. Mater. Chem. A*, 6, 8, 3284–3303, 2018. https://doi.org/10.1039/c7ta10500b.
94. Wang, P.F., You, Y., Yin, Y.X., Guo, Y.G., Layered oxide cathodes for sodium-ion batteries: Phase transition, air stability, and performance. *Adv. Energy Mater.*, 8, 8, 1701912, 2018. https://doi.org/10.1002/aenm.201701912.
95. Chao, D., Zhu, C., Yang, P., Xia, X., Liu, J., Array of nanosheets render ultrafast and high-capacity na-ion storage by tunable pseudocapacitance. *Nat. Commun.*, 7, 12122, 2016. https://doi.org/10.1038/ncomms12122.
96. Guo, Q., Ma, Y., Chen, T., Xia, Q., Yang, M., Xia, H., Yu, Y., Cobalt sulfide quantum dot embedded N/S-doped carbon nanosheets with superior reversibility and rate capability for sodium-ion batteries. *ACS Nano*, 11, 12, 12658–12667, 2017. https://doi.org/10.1021/acsnano.7b07132.
97. Xu, C., Xu, Y., Tang, C., Wei, Q., Meng, J., Carbon-coated hierarchical $NaTi_2(PO_4)_3$ mesoporous microflowers with superior sodium storage performance. *Nano Energy*, 28, 224–231, 2016. https://doi.org/10.1016/j.nanoen.2016.08.026.
98. Zhou, J., Wang, L., Yang, M., Wu, J., Chen, F., Hierarchical VS_2 nanosheet assemblies: A universal host material for the reversible storage of alkali metal ions. *Adv. Mater.*, 29, 35, 170206, 2017. https://doi.org/10.1002/adma.201702061.
99. Zhao, X., Cai, W., Yang, Y., Song, X., Neale, Z., Wang, H.-E., Sui, J., Cao, G., $MoSe_2$ nanosheets perpendicularly grown on grapheme with Mo–C bonding for sodium-ion capacitors. *Nano Energy*, 47, 224–234, 2018. https://doi.org/10.1016/j.nanoen. 2018.03.002.
100. Xie, X.Q., Makaryan, T., Zhao, M.Q., Van Aken, K.L., Gogotsi, Y., Wang, G.X., MoS_2 nanosheets vertically aligned on carbon paper: A freestanding electrode for highly reversible sodium-ion batteries. *Adv. Energy Mater.*, 6, 5, 1502161, 2016. https://doi. org/10.1002/aenm.201502161.

101. Yu, D., Pang, Q., Gao, Y., Wei, Y., Wang, C., Chen, G., Du, F., Hierarchical flower-like VS_2 nanosheets-a high rate-capacity and stable anode material for sodium-ion battery. *Energy Stor. Mater.*, 11, 1–7, 2018. https://doi.org/10.1016/j.ensm.2017. 09.002.
102. Xie, D., Xia, X.H., Zhong, Y., Wang, Y.D., Wang, D.H., Wang, X.L., Tu, J.P., Exploring advanced sandwiched arrays by vertical graphene and N-doped carbon for enhanced sodium storage. *Adv. Energy Mater.*, 7, 3, 1601804, 2017. https://doi.org/10.1002/aenm.201601804.
103. He, P., Fang, Y., Yu, X.Y., Lou, X.W.D., Hierarchical nanotubes constructed by carbon-coated ultrathin sns nanosheets for fast capacitive sodium storage. *Angew. Chem. Int. Ed.*, 56, 40, 12202–12205, 2017. https://doi.org/10.1002/anie.201706652.
104. Sun, D., Ye, D., Liu, P., Tang, Y., Guo, J., Wang, L., Wang, H., MoS_2/graphene nanosheets from commercial bulky MoS_2 and graphite as anode materials for high rate sodium-ion batteries. *Adv. Energy Mater.*, 8, 10, 1702383, 2018. https://doi.org/10.1002/aenm.201702383.
105. Geng, X.M., Jiao, Y.C., Han, Y., Mukhopadhyay, A., Yang, L., Zhu, H.L., Freestanding metallic 1T MoS2 with dual ion diffusion paths as high rate anode for sodium-ion batteries. *Adv. Funct. Mater.*, 27, 40, 1702998, 2017. https://doi.org/10.1002/adfm.201702998.
106. Liu, W.L., Ju, S.L., Yu, X.B., Phosphorus-amine-based synthesis of nanoscale red phosphorus for application to sodium-ion batteries. *ACS Nano*, 14, 974–984, 2020.
107. Wang, H.Q., An, D., Li, N. *et al.*, PbTe nanodots confined on ternary B2O3/BC2O/C nanosheets as electrode for efficient sodium storage. *J. Power Sources*, 461, 228110, 2020.
108. Xie, X.Q., Zhao, M.Q., Anasori, B.K., Maleski, K., Ren, C.E., Li, J.W., Byles, B.W., Pomerantseva, E., Wang, G.X., Gogotsi, Y., Porous heterostructured MXene/carbon nanotube composite paper with high volumetric capacity for sodium-based energy storage devices. *Nano Energy*, 26, 513, 2016.
109. Zhang, C., Zhao, H., Lei, Y., Recent research progress of anode materials for potassium-ion batteries. *Energy Environ. Mater.*, 3, 105, 2020.
110. Sha, M., Liu, L., Zhao, H., Lei, Y., Review on recent advances of cathode materials for potassium-ion batteries. *Energy Environ. Mater.*, 3, 56, 2020.
111. Hwang, J.-Y., Myung, S.-T., Sun, Y.-K., Recent progress in rechargeable potassium batteries. *Adv. Funct. Mater.*, 28, 1802938, 2018.
112. Xue, L., Li, Y., Gao, H., Zhou, W., Lü, X., Kaveevivitchai, W., Manthiram, A., Goodenough, J.B., Low-cost high-energy potassium cathode. *J. Am. Chem. Soc.*, 139, 2164, 2017.
113. Pramudita, J.C., Sehrawat, D., Goonetilleke, D., Sharma, N., An initial review of the status of electrode materials for potassium-ion batteries. *Adv. Energy Mater.*, 7, 1602911, 2017.

114. Han, J., Li, G.-N., Liu, F., Wang, M., Zhang, Y., Hu, L., Dai, C., Xu, M., Investigation of K3V2(PO4)3/C nanocomposites as high-potential cathode materials for potassium-ion batteries. *Chem. Commun.*, 53, 1805, 2017.
115. Li, L., Liu, L., Hu, Z., Lu, Y., Liu, Q., Jin, S., Zhang, Q., Zhao, S., Chou, S.-L., Understanding high-rate K^+-solvent co-intercalation in natural graphite for potassium-ion batteries. *Angew. Chem.*, 132, 13017, 2020.
116. Liu, Y., Gao, C., Dai, L., Deng, Q., Wang, L., Luo, J., Liu, S., Hu, N., The features and progress of electrolyte for potassium ion batteries. *Small*, 16, 2004096, 2020.
117. Zhou, M., Bai, P., Ji, X., Yang, J., Wang, C., Xu, Y., Electrolytes and interphases in potassium ion batteries. *Adv. Mater.*, 33, 2003741, 2021.
118. Zhang, C., Xu, Y., Zhou, M., Liang, L., Dong, H., Wu, M., Yang, Y., Lei, Y., Potassium prussian blue nanoparticles: A low-cost cathode material for potassium-ion batteries. *Adv. Funct. Mater.*, 27, 1604307, 2017.
119. Zhang, R., Huang, J., Deng, W., Bao, J., Pan, Y., Huang, S., Sun, C.-F., Safe, low-cost, fast-kinetics and low-strain inorganic-open-framework anode for potassium-ion batteries. *Angew. Chem.*, 131, 16626, 2019.
120. Zhang, L., Wang, W.A., Ma, X., Lu, S., Xiang, Y., Crystal, interfacial and morphological control of electrode materials for nonaqueous potassium-ion batteries. *Nano Today*, 37, 101074, 2021.
121. Wang, J., Liu, Z., Zhou, J., Han, K., Lu, B., Insights into metal/metalloid-based alloying anodes for potassium ion batteries. *ACS Mater. Lett.*, 3, 1572, 2021.
122. Wu, Z., Liang, G., Pang, W.K., Zhou, T., Cheng, Z., Zhang, W., Liu, Y., Johannessen, B., Guo, Z., Coupling topological insulator $SnSb_2Te_4$ nanodots with highly doped graphene for high-rate energy storage. *Adv. Mater.*, 32, 1905632, 2020.
123. Xia, Z., Chen, X., Ci, H., Fan, Z., Yi, Y., Yin, W., Wei, N., Cai, J., Zhang, Y., Sun, J., Designing N-doped graphene/$ReSe_2$/Ti_3C_2 MXene heterostructure frameworks as promising anodes for high-rate potassium-ion batteries. *J. Energy Chem.*, 53, 155, 2021.
124. Li, J., Rui, B., Wei, W., Nie, P., Chang, L., Le, Z., Liu, M., Wang, H., Wang, L., Zhang, X., Nanosheets assembled layered MoS_2/MXene as high performance anode materials for potassium ion batteries. *J. Power Sources*, 449, 227481, 2020.
125. Jiang, Q., Hu, S., Wang, L., Huang, Z., Yang, H., Han, X., Li, Y., Lv, C., He, Y., Zhou, T., Hu, J., Boosting potassium storage in nanosheet assembled $MoSe_2$ hollow sphere through surface decoration of MoO_2 nanoparticles. *Appl. Surf. Sci.*, 505, 144573, 2020.
126. Han, Y., Li, W., Zhou, K., Wu, X., Wu, H., Wu, X., Shi, Q., Diao, G., Chen, M., Bimetallic sulfide Co_9S_8/N-C@MoS_2 dodecahedral heterogeneous nanocages for boosted Li/K storage. *ChemNanoMat*, 6, 132, 2020.
127. Qin, Y., Zhang, Y., Wang, J., Zhang, J., Zhai, Y., Wang, H., Li, D., Heterogeneous structured Bi_2S_3/MoS_2@NC nanoclusters: Exploring the superior rate

performance in sodium/potassium ion batteries. *ACS Appl. Mater. Interfaces*, 12, 42902, 2020.
128. Cao, L., Luo, B., Xu, B., Zhang, J., Wang, C., Xiao, Z., Li, S., Li, Y., Zhang, B., Zou, G., Hou, H., Ou, X., Ji, X., Stabilizing intermediate phases via efficient entrapment effects of layered VS_4/SnS@C heterostructure for ultralong life-span potassium-ion batteries. *Adv. Funct. Mater.*, 31, 2103802, 2021.
129. Yang, C., Feng, J., Lv, F., Zhou, J., Lin, C. *et al.*, Metallic graphene-like VSe_2 ultrathin nanosheets: Superior potassium-ion storage and their working mechanism. *Adv. Mater.*, 30, 27, 1800036, 2018. https://doi.org/10.1002/adma.201800036.
130. Xie, K., Yuan, K., Li, X., Lu, W., Shen, C., Liang, C., Vajra, R., Ajayan, P., Wei, B., Superior potassium ion storage via vertical MoS_2 "nano-rose" with expanded interlayers on grapheme. *Small*, 13, 42, 1701471, 2017. https://doi.org/10.1002/smll.201701471.
131. Lian, P., Dong, Y., Wu, Z.-S., Zheng, S., Wang, X. *et al.*, Alkalized Ti_3C_2 mxene nanoribbons with expanded interlayer spacing for high-capacity sodium and potassium ion batteries. *Nano Energy*, 40, 1–8, 2017. https://doi.org/10.1016/j.nanoen.2017.08.002.
132. Jian, Z., Luo, W., Ji, X., Carbon electrodes for K-ion batteries. *J. Am. Chem. Soc.*, 137, 36, 11566–11569, 2015. https://doi.org/10.1021/jacs.5b06809.
133. Naguib, M., Adams, R.A., Zhao, Y., Zemlyanov, D., Varma, A., Nanda, J., Pol, V.G., Electrochemical performance of mxenes as K-ion battery anodes. *Chem. Commun.*, 53, 51, 6883–6886, 2017. https://doi.org/10.1039/c7cc02026k.
134. Ma, G.Y., Huang, K.S., Ma, J.S. *et al.*, Phosphorus and oxygen dual-doped graphene as superior anode material for room-temperature potassium-ion batteries. *J. Mater. Chem. A*, 5, 7854–7861, 2017.
135. Zhang, H., Cheng, Y., Zhang, Q., Fast and durable potassium storage enabled by constructing stress-dispersed Co_3Se_4 nanocrystallites anchored on graphene sheets. *ACS Nano*, 15, 10107–10118, 2021.
136. Jia, B., Yu, Q., Zhao, Y., Qin, M., Wang, W., Liu, Z., Lao, C., Liu, Y., Wu, H., Zhang, Z., Qu, X., Bamboo-like hollow tubes with MoS_2/N-doped-C interfaces boost potassium-ion storage. *Adv. Funct. Mater.*, 28, 1803409, 2018.
137. Li, J., Rui, B., Wei, W., Nie, P., Chang, L., Le, Z., Liu, M., Wang, H., Wang, L., Zhang, X., Nanosheets assembled layered MoS_2/MXene as high performance anode materials for potassium ion batteries. *J. Power Sources*, 449, 227481, 2020.
138. Huang, H., Cui, J., Liu, G., Bi, R., Zhang, L., Carbon-coated $MoSe_2$/MXene hybrid nanosheets for superior potassium storage. *ACS Nano*, 13, 3448, 2019.
139. Cao, Y., Chen, H., Shen, Y., Chen, M., Zhang, Y., Zhang, L., Wang, Q., Guo, S., Yang, H., SnS_2 nanosheets anchored on nitrogen and sulfur co-doped mxene sheets for high-performance potassium-ion batteries. *ACS Appl. Mater. Interfaces*, 13, 17668, 2021.
140. Liu, H., He, Y., Zhang, H., Wang, S., Cao, K., Jiang, Y., Liu, X., Jing, Q., Heterostructure engineering of ultrathin $SnS_2/Ti_3C_2T_x$ nanosheets for

high-performance potassium-ion batteries. *J. Colloid Interface Sci.*, 606, 167, 2022.
141. Zhao, R., Di, H., Hui, X., Zhao, D., Wang, R., Wang, C., Yin, L., Self-assembled Ti_3C_2 MXene and N-rich porous carbon hybrids as superior anodes for high-performance potassium-ion batteries. *Energy Environ. Sci.*, 13, 246, 2020.
142. Yu, A., Ultrathin, transparent, and flexible graphene films for supercapacitor application. *Appl. Phys. Lett.*, 96, 25, 253105, 2010.
143. Mukherjee, S., Ren, Z., Singh, G., Beyond graphene anode materials for emerging metal ion batteries and supercapacitors. *Nanomicro Lett.*, 10, 4, 70, 2018.
144. Liu, H., Phosphorene: An unexplored 2D semiconductor with a high hole mobility. *ACS Nano*, 8, 4, 4033–4041, 2014.
145. El-Kady, M.F. and Kaner, R.B., Scalable fabrication of high-power graphene micro-supercapacitors for flexible and on-chip energy storage. *Nat. Commun.*, 4, 1, 1–9, 2013.
146. Yang, B., Flexible black-phosphorus nanoflake/carbon nanotube composite paper for high-performance all-solid-state supercapacitors. *ACS Appl. Mater. Interfaces*, 9, 51, 44478–44484, 2017.
147. Chen, Y., High performance supercapacitors based on reduced graphene oxide in aqueous and ionic liquid electrolytes. *Carbon*, 49, 2, 573–580, 2011.
148. Wu, Q., Supercapacitors based on flexible graphene/polyaniline nanofiber composite films. *ACS Nano*, 4, 4, 1963–1970, 2010.
149. Balasingam, S.K., Lee, J.S., Jun, Y., Few-layered $MoSe_2$ nanosheets as an advanced electrode material for supercapacitors. *Dalton Trans.*, 44, 35, 15491–15498, 2015.
150. Cao, L., Direct laser-patterned micro-supercapacitors from paintable MoS_2 films. *Small*, 9(17), 905-2910, 2013.
151. Feng, J., Metallic few-layered VS_2 ultrathin nanosheets: High two-dimensional conductivity for in-plane supercapacitors. *J. Am. Chem. Soc.*, 133, 44, 17832–17838, 2011.
152. Zhang, L., Wu, H.B., Lou, X.W.D., Unusual CoS_2 ellipsoids with anisotropic tube-like cavities and their application in supercapacitors. *Chem. Commun.*, 48, 55, 6912–6914, 2012.
153. Zhang, Y., Controlled growth of high-quality monolayer WS_2 layers on sapphire and imaging its grain boundary. *ACS Nano*, 7, 10, 8963–8971, 2013.
154. Ghidiu, M., Lukatskaya, M.R., Zhao, M.Q., Gogotsi, Y., Barsoum, M.W., Conductive two-dimensional titanium carbide 'clay' with high volumetric capacitance. *Nature*, 516, 7529, 78–81, 2015.
155. Djire, A., Bos, A., Liu, J., Zhang, H., Miller, E.M., Neale, N.R., Pseudocapacitive storage in nanolayered Ti_2NT_x MXene using Mg-ion electrolyte. *ACS Appl. Nano Mater.*, 2, 5, 2785–2795, 2019.
156. Anasori, B., Two-dimensional, ordered, double transition metals carbides (MXenes). *ACS Nano*, 9, 10, 9507–9516, 2015.

157. Wu, X., Wang, Z., Yu, M., Xiu, L., Qiu, J., Stabilizing the MXenes by carbon nanoplating for developing hierarchical nanohybrids with efficient lithium storage and hydrogen evolution capability. *Adv. Mater.*, 29, 24, 1–8, 2017.
158. Zhou, H., MnO_2 nanorods/MXene/CC composite electrode for flexible supercapacitors with enhanced electrochemical performance. *J. Alloys Compd.*, 802, 259–268, 2019.
159. Zhao, K., Vertically aligned MnO_2 nanosheets coupled with carbon nanosheets derived from Mn-MOF nanosheets for supercapacitor electrodes. *J. Mater. Sci.*, 53, 18, 13111–13125, 2018.
160. Xia, H.C., Zhang, J.N., Yang, Z., Guo, S.Y., Guo, S.H., Xu, Q., 2D MOF nanoflake-assembled spherical microstructures for enhanced supercapacitor and electrocatalysis performances. *Nanomicro Lett.*, 9, 4, 43, 2017. https://doi.org/10.1007/s40820-017-0144-6.
161. Shang, P., Zhang, J.A., Tang, W.Y., Xu, Q., Guo, S.J., 2D Thin nanoflakes assembled on mesoporous carbon nanorods for enhancing electrocatalysis and for improving asymmetric supercapacitors. *Adv. Funct. Mater.*, 26, 43, 7766–7774, 2016. https://doi.org/10.1002/adfm.201603504.
162. Jabeen, N., Hussain, A., Xia, Q., Sun, S., Zhu, J., Xia, H., Highperformance 2.6 V aqueous asymmetric supercapacitors based on *in situ* formed $Na0.5MnO_2$ nanosheet assembled nanowall arrays. *Adv. Mater.*, 29, 32, 1700804, 2017. https://doi.org/10.1002/adma.201700804.
163. Li, P., Jin, Z., Peng, L., Zhao, F., Xiao, D., Jin, Y., Yu, G., Stretchable all-gel-state fiber-shaped supercapacitors enabled by macromolecularly interconnected 3D graphene/nanostructured conductive polymer hydrogels. *Adv. Mater.*, 30, 1800124, 2018. https://doi.org/10.1002/adma.201800124.
164. Qi, J.Q., Chang, Y., Sui, Y.W., He, Y.Z., Meng, Q.K., Wei, F.X., Ren, Y.J., Jin, Y.X., Facile synthesis of Ag-decorated Ni_3S_2 nanosheets with 3D bush structure grown on rGO and its application as positive electrode material in asymmetric supercapacitor. *Adv. Mater. Interfaces*, 5, 3, 1700985, 2018. https://doi.org/10.1002/admi.201700985.
165. Xu, J.S., Sun, Y.D., Lu, M.J., Wang, L., Zhang, J., Qian, J.H., Liu, X.Y., Fabrication of hierarchical $MnMoO_4$ center dot H_2-O@MnO_2 core-shell nanosheet arrays on nickel foam as an advanced electrode for asymmetric supercapacitors. *Chem. Eng. J.*, 334, 1466–1476, 2018. https://doi.org/10.1016/j.cej.2017.11.085.
166. Yang, X.J., Zhao, L.J., Lian, J.S., Arrays of hierarchical nickel sulfides/MoS2 nanosheets supported on carbon nanotubes backbone as advanced anode materials for asymmetric supercapacitor. *J. Power Sources*, 343, 373–382, 2017. https://doi. org/10.1016/j.jpowsour.2017.01.078.

8

Innovation in Photoinduced Antibacterial 2D Nanomaterials

Zubaid ul Khazir Rather[1*], Shabnam Kawoosa[2], Gulam Nabi Yatoo[1], Mohd Asif Hajam[1], Sajad Ahmed Bhat[1] and Javid Ahmed Banday[1†]

[1]*NPCL-Lab, Department of Chemistry, National Institute of Technology, Srinagar, J&K, India*

[2]*Department of Pharmaceutical Sciences, University of Kashmir, Srinagar, J&K, India*

Abstract

Bacterial disease epidemics continue to be a major escalating health concern worldwide. The advent of antibiotic overuse has since been linked to the continuous expansion of multi-drug resistant and "superbugs," which in turn has led to MDR diseases. The intriguing physicochemical characteristics of versatile nanomaterials hold the potential to evaluate and optimize their antibacterial properties. An advantageous aspect of nanomaterials is their dimensionality, which can alter their intrinsic properties and influence the antibacterial mechanisms of different materials. Furthermore, 2D nanomaterials have attracted considerable attention for overcoming the problems of resistance and eliminating other side effects owing to their comparatively simple synthesis and the potential to modify their inherent properties to render them sensitive to a wide variety of stimuli. Photoinduced bacterial inactivation has garnered much attention, because photoinduced antibacterial nanomaterials target a particular site, which makes it difficult for bacteria to acquire resistance when they are exposed. Nanomaterials exhibit significant photoconversion efficiency, thermalizing the photon energy that has been absorbed. In this approach, the maximum heat stress generated by the photoinduced mechanism can denature most proteins, causing functional impairment in the cell and eventually leading to microbial death. 2D nanomaterials have been found to exhibit a variety of indistinguishable antimicrobial mechanisms, such as the ability to physically penetrate membranes, extract phospholipids or produce oxygen reactive species. To combat bacterial pests, it is necessary to evaluate the effectiveness and safety of the readily available nanomaterials.

**Corresponding author*: rather.zubaid007@gmail.com

†*Corresponding author*: javidbanday@gmail.com

Subhendu Chakroborty and Kaushik Pal (eds.) 2D Nanomaterials: Synthesis, Properties and Applications, (155–184) © 2024 Scrivener Publishing LLC

***Keywords*:** 2D nanomaterials, anti-microbial agents, bacterial resistance, photo-induced nanosystems, graphene oxide nanomaterial

8.1 Introduction

The twenty-first century faced a substantial upsurge in multidrug-resistant pathogenic diseases that significantly endangered the healthcare sector. Although the existence of bacterial resistance to antibiotics has been recognized since the dawn of the antibiotic era, there has been a significant rise in the prevalence of resistant pathogenic strains in the last few decades [1]. Pathogenic bacterial and fungal infections can be prevented and cured using antimicrobial drugs. Although these medicines are generally effective, unbridled and excessive antibiotic use causes bacterial strains to mutate and develop antibiotic resistance, limiting the effectiveness of conventional antibiotic therapy or causing various side effects [2, 3]. Antimicrobial resistance decreases the efficiency of bactericidal activity. There is an ever-increasing demand for novel formulations that can specifically and effectively target microbial infections in tandem with resistant strains, thereby reducing the chances of recurrence of infections and alleviating the adverse effects of antimicrobial agents [4–6]. The WHO predicts that by 2050, pathogens that can now be treated will cause almost 10 million deaths annually unless new antimicrobial treatments are not developed [7]. Presently, the use of nanotechnology has rapidly impacted a diverse range of industries. Nanotechnology has been extensively exploited in the fields of healthcare and medicine, ranging from the development of pharmaceutically active materials to formulating medicines for diagnostics. Prevention of bacterial infections continues to be a major global therapeutic issue [8, 9]. This accelerating evolution of resistance emphasizes the immediate need for model and replaceable antibiotics. Considering these concerns, extensive attempts have been made to create multipurpose nanomaterials as novel antibacterial agents to treat different microbial diseases [10–12]. These nanomaterials have characteristic physicochemical and pharmaceutical properties, such as excellent membrane permeability and biocompatibility, along with prospective antibacterial activities, making them less likely to cause bacterial resistance than conventional antibiotics [13–15]. The ability of nanomaterials to modify their inherent properties is one of their main attributes. Various technologically advanced nanomaterials have been developed by altering their dimensional variances (i.e., zero-, one-, two-, and three-dimensional nanomaterials), with emphasis on the factors that affect their applicability, such as antibacterial efficacy, bioactivity for clinical use, and tunability towards other pathogenic microbes. The different biological activities of low-dimensional materials may be caused by dimensionality variance. Their antibacterial mechanisms vary owing to the dimensional differences between the materials. Physical contact with the bacterial surface is one of the most popular techniques for inhibiting bacterial growth by damaging the bacterial membranes. This causes expulsion of ions that inhibit enzymes, degrade nucleic acids, and produce ROS. According to research investigations, there is a surge of interest in developing stimulus-activated nanomaterials, which must be stimulated externally to act, providing "activatable" antibacterial materials. The two most frequently employed stimuli are magnetization and light (PDT/PTT) [16–18].

2Dimensional nanomaterials have layered or sheet-like structures, providing a substantial surface area, high photoreactivity, enhanced physical contact, and PDT/PTT efficacy. Furthermore, the dosage of pharmacologically active agents in antibiotics based on 2D nanomaterials is reasonably low compared to conventional antibiotics, which alleviates the issue of bacterial resistance and adverse effects associated with drugs [19–21]. An escalating range of 2D materials has been explored, which are claimed to possess better antibacterial efficiency than conventional antibiotics [22]. These nanomaterials act by different mechanisms to accomplish bactericidal activity; primarily, their enormous surface area facilitates interaction with bacterial membranes to permit functional modification; secondly, they may encompass some physicochemical catalytic properties that can directly inactivate or kill bacteria by physical interaction; and finally, some materials have exceptional potential to generate reactive oxygen species (ROS) in response to photo-irradiation, labeled as photothermal/photocatalytic/photodynamic capabilities, which can be used for sterilization [23–26].

More pertinently, a variety of 2D nanoparticles have been developed and purported to exhibit antibacterial properties. These include graphene-based nanomaterials (GBNMs), transition metal dichalcogenides (TMDCs), transition metal oxides, graphitic carbon nitride (g-C3N4), black phosphorus (BP), layered double hydroxides (LDHs), transition metal carbides, and nitrides (MXenes). Specifically, it has been demonstrated that graphene oxide (GO)-based 2D nanomaterials exhibit intrinsic antibacterial activity by stimulating the generation of ROS and physical interactions with bacteria, which destroy their cell membranes [27, 28]. A promising direction for numerous potential applications is light-induced prokaryotic cell death upon exposure to photoactive nanomaterials. Photoinduced treatments utilize light to trigger a cascade of events that lead to bacterial death. Various antibacterial phototherapy techniques have been developed based on the behavior of light-responsive nanomaterials. PDT uses photocatalysts or their intermediates, which can capture VL and generate ROS, leading to stress oxidation in the cells and are hazardous to them.

The photothermal effect has recently been found to have a high NIR-light-to-thermal conversion efficacy. The development of several photothermal nanomaterials that exhibit significant NIR light absorption and convert light into thermal energy is intriguing [29–31]. The most popular photothermal sterilization agents among the 2D nanomaterials are those derived from graphene. For example, reduced graphene oxide (rGO) nanosheets coated with multivalent sugar ligands demonstrated outstanding light-to-heat conversion efficiency during IR light irradiation, enabling powerful bacterial killing. Both biofilm formation and the evolution of antibiotic-resistant bacteria can be inhibited by the heat produced [32]. In this review, the key parameters determining antibacterial activity are addressed, along with the recent developments in 2D antimicrobial nanomaterials. Additionally, systematic demonstrations of the antibacterial mechanisms are summarized in detail. Figure 8.1 depicts the various types of 2D nanomaterials [33].

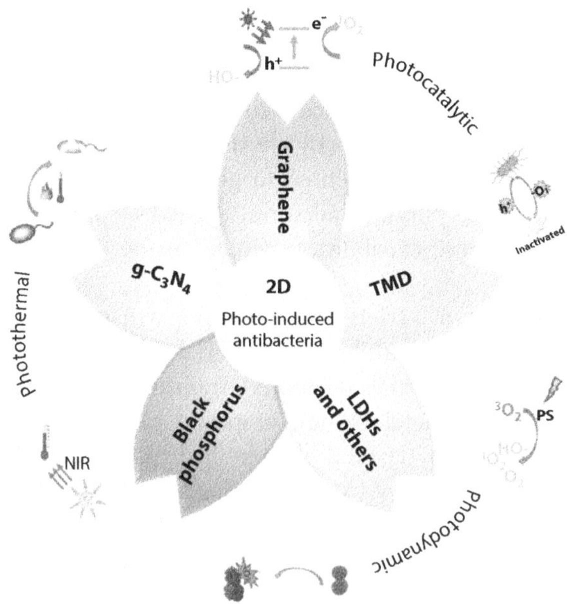

Figure 8.1 Schematic illustration depicting the types of photoinduced antibacterial nanomaterials [33].

8.2 Antibacterial Applications Based on Graphene-Induced Photostimulation

Carbon-derived nanosystems such as 3D graphene-based networks and 2D graphene nanosheets have been shown to have efficient antibacterial properties. Graphene-based nanosystems consist of 2D single-atomic-sheets of sp2-hybridized C atoms, which deliver a greater surface area, lower toxicity, strong water dispersibility, and enhanced electrical conductivity. Exfoliated graphite, graphene oxide (GO), and reduced graphene oxide (rGO) are the three most prevalent graphene-based nanomaterials depending on their chemical states. A sheet of oxidized graphene with various functional groups that incorporate oxygen is known as GO. GO transforms into rGO upon the reduction of functional oxygen-containing compounds. Hydrazine and dithiothreitol are usually used to decrease rGO nanosheets [34, 35]. It has been previously claimed that graphene and 2D materials have remarkable antibacterial properties. They can be sterilized directly through physical contact or indirectly by catalytic action because of their vast surface area and sharp edges. Furthermore, their surfaces have a variety of chemical properties and are conveniently modified or complexed with other compounds to form a nanocomposite, which increases antimicrobial action. GO is a highly dispersible graphene-derived product with polar oxygen groups on the planes of a two-dimensional structure. In particular, graphene-based materials possess the ability to retain biomolecules, and graphene and GO nanoparticles may have potent bactericidal effects on all variants of harmful microbes. This antibacterial mechanism has a complex structure and is entirely dependent on the inherent characteristics of the materials associated with graphene.

The major prerequisites to be considered for effective antibacterial action are the type of target microorganism, surface functionality and composition, and the environment of the cell and graphene interaction. Recent studies have shown that graphene oxide has polar oxygen groups, which makes it suitable for antibacterial applications as it may cause oxidative stress. Graphene, graphene oxide, and their hybrid materials have been used in several studies to be effective against both Gram (+ve) and Gram (−ve) bacteria. They may also interact chemically or physically with each other. Microorganisms may incur structural damage because of graphene, which is also capable of biologically separating cells from their surroundings, which would subsequently cause cell death. Additionally, the chemical reaction between microorganisms and graphene produces harmful compounds called reactive oxygen species (ROS). Moreover, the outer microbial surface acts as the source of electrons, from which they are progressively drained to the nearby environment for the electron transfer mechanism. This further increases the risk of ROS-independent oxidative stress, eventually resulting in biological death [36–40].

Additionally, the excellent photothermal, photocatalytic, and photodynamic performances of these nanomaterials when stimulated with light irradiation result from their high thermal and electrical conductivities. Because some GB materials exhibit strong NIR broad-spectrum absorption, they can quickly heat up to kill germs. This makes them the most favored 2D materials for photothermal sterilization. Because of their ability to absorb photons and convert them to heat, carbon-based compounds have a bacteriostatic effect. Most heating light sources fall between the 700 nm and 1,100 nm wavelength ranges. Reduced GO has an antibacterial effect because it absorbs radiation in this range and generates heat [41, 42].

Different elements, including carbon radical density, lateral dimension, and light, can directly or indirectly affect the antibacterial activity of GFNs. Through the stimulation of lipid peroxidation, carbon radical density plays an important role in the destruction of bacteria, and a GO system with a greater density of carbon radicals exhibits a stronger bactericidal effect. By accelerating electron transfer and phototransformation, sunlight might render bacteria more hazardous. High oxidative stress or the speed at which sunlight increases the flow of electrons from bacteria to GO damages bacterial antioxidant mechanisms and significantly disrupts bacterial membranes and is the primary reason for the improved antibacterial activity due to sunlight [43–45].

The antibacterial activity of GFNs depends on both the size and amount of oxygen in the sheets. Depending on the process employed for the oxidation of graphite, graphene oxides have a wide range of heterogeneous physical and chemical properties. The antibacterial efficacy of graphene oxide is size-dependent, as seen by the larger size of the same type of graphene oxide exhibiting better antibacterial activity. However, increasing size of GO can improve the capacity of bacteria to adhere to it, increasing the chances that bacteria will encounter it and render it inactive. More ROS can be produced because of the increasing oxidative content, which increases bactericidal potency. GO exhibited a higher level of oxidation than rGO. More oxidized groups (including C-OH, C = O, and C-O-C) of the same size can lead to a higher level of oxidation and increase the antibacterial power of GO. Conversely, the bactericidal properties of GO significantly decline following a reduction process (loss of oxygen-containing groups) [46–48].

In addition, the layout of the GFN sheet has a substantial effect on its antibacterial activity. The antibacterial effectiveness of GO sheets that were randomly, horizontally, and vertically aligned was reduced in that order [48]. Vertically aligned GO with sharp edges has a better capability to interact and penetrate the lipid bilayer of the bacterial cell wall. Therefore, an electron-transfer oxidation pathway is generated upon the insertion of vertical GO into the bacterial cell membrane, which significantly increases the amount of oxidative stress [49, 50].

In addition to the previously mentioned uses already mentioned, GBN coupled with metallic NPs is frequently employed in industrial and medical setups for purposes such as sterilization, water treatment and filtration, and as an anti-biofilm agent. The combination of metallic nanoparticles with graphene oxides (oxidized and reduced) upon exposure to wavelengths ranging from UV to NIR demonstrated a potent impact on bacterial cell death triggered by the generation of heat and ROS [51]. To increase their antibacterial effectiveness, GFN-based composites are functionalized with additional substances like Ag, metal nanoparticles (or metal oxides), photocatalysts, polymers, antibiotics, enzymes, and multicomponents. According to a recent study, under NIR laser light, Ag-NPs embedded in reduced GO (rGO/Ag) nanocomposites were successful in combating the multidrug-resistant bacterial infections. The Ag NPs and rGO enhanced

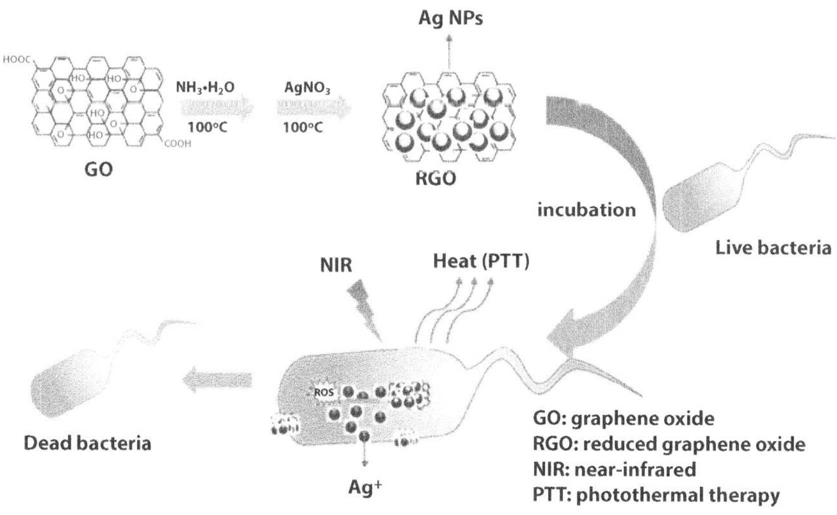

Figure 8.2 Schematic representation of photothermal activation of Graphene oxide and Ag nanoparticle complexes in the presence of NIR [54].

NIR absorption, which in turn improved the efficiency of photothermal conversion. rGO/Ag nanocomposites have been reported to have a higher photothermal efficiency than GO, rGO, and Ag NPs. A reduction in temperature was observed in conjunction with the poor density of NIR laser light and a limited concentration of rGO/Ag nanocomposites, finally leading to the irreversibly damage of the bacterial cell envelopes by NIR laser irradiation [52, 53]. Photothermal activation of GO and Ag nanoparticles has been depicted in Figure 8.2 [54].

In addition to the exhaustive exploration of graphene-based nanoparticles for their anti-microbrial properties, the focus of research has shifted towards Nanomaterials Beyond Graphene (NBG) in last the few decades. The remarkable antibacterial properties of 2D NBG have inspired the investigation of other non-graphene compounds, TMDC, C3N4, BP, MXenes, and other 2D NBG materials. Some of the recently formulated NBG's are summarized in Table 8.1.

Table 8.1 A tabulated summary of recently developed 2D NBG exhibiting anti-microbial effect.

Nanosystem	Mechanism	Type of bacteria	Reference
TMDC			
Chitosan/ Molybdium Sulfide (CS@ $MoS_{2)}$	PDT & PTT induced disintegration of cell membrane and leakage of cytoplasmic components	*E. coli, S. aureus*	[55, 56]
Magnetic MoS2	Improved bacterial and PTT conjugation	*E. coli, S. aureus*	[57]
PEG-MoS2	Synergistic antibacterial action of peroxidase-like catalysis and PTT	*E. coli, B. subtilis*	[58]
MoSe2	Peroxidase-like catalysis of H_2O_2 into •OH	*E. coli, B. subtilis*	[59]

(*Continued*)

Table 8.1 A tabulated summary of recently developed 2D NBG exhibiting antimicrobial effect. (*Continued*)

Nanosystem	Mechanism	Type of bacteria	Reference
MoO3	Disruption of the bacterial cell wall	*E. coli, S. aureus, E. faecalis, B. subtilis*	[60]
Metal-based			
Pd@Ag	Synergic effect of PTT & NIR-triggered Ag+ release	*E. coli, S. aureus*	[61]
TiO2	Photocatalytic antibacterial activity		[62]
Nitride-based			
g-C3N4	Photocatalytic	*E. coli*	[63]
Ag@g-C3N4	Photocatalytic generated ROS	*E. coli, S. aureus & P. aeruginosa*	[64]
Ag/PDA/g-C3N4	Photocatalytic & Ag NPs	*E. coli*	[65]
MXenes			
Ti3C2Tx	Physical interaction damages bacterial membrane	*E. coli, S. aureus*	[66]
MoS2/MXene	Sharp edges cause membrane damage		[67]
Ti3C2Tx MXene	Ti3C2Tx MXene	*E. coli, S. aureus*	[68]
Black phosphorus based			
BP	ROS-dependent oxidative stress & membrane damage	*E. coli, B. subtilis*	[69]

(*Continued*)

Table 8.1 A tabulated summary of recently developed 2D NBG exhibiting antimicrobial effect. (*Continued*)

Nanosystem	Mechanism	Type of bacteria	Reference
BP@silk fibroin	NIR-mediated PTT bactericida	*E. coli, B. subtilis*	[59]
AuNPs/BP	Catalytic synergistic Au	*E. coli*	[70]

8.2.1 Nanomaterials for Antibacterial Transition-Metal Dichalcogenides/Oxides

Owing to recent advances in key fields such as chemical and physical sciences, optoelectronics, semiconductors, and biomedicine, transition-metal dichalcogenide nanosheets (TMDC NSs), a novel class of inorganic 2D nanomaterials, have drawn increasing attention. The MX2 formula, which includes chalcogen and transition-metal atoms, is widely used to refer to 2D TMDC NSs. The initial M represents transition metal elements, including W, Mo, Ta, Nb, V, Hf, Zr, and Ti, whereas X represents chalcogen atoms (Te, Se, or S) [71–75]. 2D TMDC nanosystems have inspired a wide range of applications in the biomedical field, including biosensing, drug/gene delivery, bioimaging, and phototherapy. These materials retain a high surface-to-volume ratio and photoresponsive properties, which resemble those of graphene-based nano compounds. According to the latest reports, they may have an antibacterial effect owing to their distinct structure and physical/chemical characteristics. 2D TMDC NSs have shown the capacity to produce increased antibacterial efficacy, known as photoinduced antibacterial activity, particularly when combined with light irradiation [76–78]. Typically, 2D TMDC NSs can be photo-stimulated to produce numerous electrons and protons, which spontaneously react with H_2O and O_2 to produce hydrogen radicals and superoxide ions, respectively. These reactive species subsequently impede bacterial cell growth and lead to the death of microorganism [79, 80].

MoS2 nanosheets demonstrate a special quantum confinement effect that causes their bandgap to grow with decreasing domain size. Because of this feature, a successfully created type of vertically oriented few-layer MoS2 (FLV-MoS2) was able to produce ROS when exposed to sunlight, which effectively inactivated microorganisms in water. To compare the MoS2's inactivation efficiencies when incubated with *E. coli* (Gram-ve bacteria) under VL irradiation or in the dark, the photocatalytic antibacterial capabilities of this material were evaluated. The study confirmed that FLV-MoS2 exposed to VL had a higher level of disinfection efficacy than VL alone or FLV-MoS2 incubated in the dark for the same duration [81, 82].

However, the FLV-MoS2 film exhibited the highest disinfection rate compared to bulk MoS2 and horizontal nano-MoS2 made of MoS2 nanoflakes, whereas bulk MoS2 demonstrated the lowest antibacterial efficiency. This is due to the high number of active edge sites on the FLV-MoS2 film's surface and the higher intrinsic conductivity of e/h+ in MoS2. Additionally, a significant number of 2D TMDC NSs, including TiS2, WS2, MoSe2, and MoS2, have remarkable photothermal effects, which cause heat to be produced when NIR light is exposed [83, 84]. In addition to acting as photothermal agents to kill bacteria, 2D TMDC NSs with high NIR absorbance can expedite the production of hydroxide radicals (•OH) and the oxidation of GSH when used as intrinsic peroxidase-like catalysts for converting H2O2 to •OH. Furthermore, owing to their exceptional PD and photocatalytic capabilities, the majority of MoS2 nanosheets have been used as antibacterial nanoagents [85–88].

8.2.2 Antibacterial Nanomaterials Based on Carbon Nitride

Polymeric nitride-based nanosheets have a laminar structure and semiconductor properties with a narrow band gap. These materials have attracted widespread attention in the antibacterial domain because of their intrinsic qualities, such as surface and photocatalytic efficiency, adjustable electron band structure, and chemical stability [89, 90]. Therefore, 2D stacked g-C3N4 is considered a viable alternative to graphite. Generally, low-cost precursors, such as melamine, dicyandiamide, urea, and thiourea, can be thermally polymerized to produce bulk g-C3N4. As the thickness of the gC3N4 nanosheet is decreased, the resistance in the conductance of charge in the 2D conducting channels also decreased, resulting in an enhancement of the disinfection efficiency [91, 92].

Another cost-effective method for morphology-controlled g-C3N4 has been developed to improve light-driven photocatalytic disinfection [92]. The enhanced photocatalytic property due to the alteration of the g-C3N4 structure is attributable to better reactant/product drift along with multiple light scattering effects. Another method to enhance the efficacy of such materials is to develop a hybrid composite of g-C3N4 with other photoactive materials that primarily focuses on increasing the photogenerated carrier separation *via* the influence of the internal electrical field [93, 94]. In general, g-C3N4 nanosheets have attracted considerable attention because of their numerous properties. Bulk exfoliation is the most commonly used technique for preparing pristine g-C3N4. The most common technique used to address the shortcomings of pure g-C3N4 is the introduction of heterojunctions. g-C3N4 can inactivate MS2, bacteria, and other viruses. Additional research should be done [64, 95].

In addition to g-C3N4, compounds based on boron nitride (BN) can be employed as antibacterial agents. Three crystal shapes of BN—amorphous (a-BN), hexagonal (h-BN), and cubic—were identified (c-BN). Among these, hBN has undergone the most in-depth research. It is an isostructural 2D material that is atomically thick and has several features, including a wide bandgap and exceptional chemical and thermal stabilities. The h-BN nanomaterial's outstanding antibacterial activity was also demonstrated [96, 97].

8.2.3 Antibacterial Nanomaterials Based on Black Phosphorus

2D BP nanosheets have recently attracted considerable attention because of their remarkable thermal, optical, and electrical characteristics. Owing to weak van der Waals forces, these materials can be easily exfoliated into ultrathin 2D nanosheets. Owing to their thickness-dependent band gap and metal-free semiconductor characteristics, BP nanosheets exhibit extremely broad light absorption over the entire visible spectrum. As a result, 2D layered BP nanosheets can be generated under visible light and employed for PDT [98–102].

In 2015, PDT was applied in cancer treatment using BP as the source of singlet oxygen/ROS (under 660 nm) to kill cancerous cells. Subsequently, several researchers began to investigate whether BP might have an antimicrobial effect based on this characteristic. Based on BP, researchers have created a repeatable photodynamic bacterial inactivation platform that has also been applied to the treatment of wound infections. The hydrogel of CS hybridized with BP (CS-BP) exhibited better results in the eradication of bacteria (*E. coli* and *S. aureus*) than the CS hydrogel when exposed to VL irradiation for 10 min. CS-BP was effective in generating an abundant quantity of singlet oxygen in VL, which facilitated the inactivation of large colonies of *S. aureus*, even upon repetitive exposure [100, 103–106].

In addition to its exceptional photodynamic and photocatalytic properties, BP possesses remarkable photothermal attributes, making it an effective alternative for disinfection, wound healing, and photothermal therapy for cancer. Researchers have investigated BP nanosheets as NIR-light-controlled photonic nanomaterials for tumor photodynamic or photothermal therapy because they exhibit ideal photochemical characteristics under NIR irradiation. To increase the photothermal impact, Au was added to BP nanosheets to create BP/Au nanocomposites, which effectively rendered the pathogen *Enterococcus faecalis* and the antibiofilm susceptible to NIR radiation. Subsequent experiments revealed that the combined effects of physical contact damage, oxidative stress, and photothermic action provide strong sterilization performance for BP/Au nanocomposites. In conclusion, BP- and BP-related nanocomposites can be used as ideal antimicrobial agents [107–110].

8.2.4 Other 2D Antibacterial Nanomaterials

Recent studies have claimed that a new 2D material family of MXenes has exceptional antibacterial properties. Researchers have used delamination to create thin, clear, and hydrophilic Ti3C2Tx nanosheets (where T denotes surface termination, such as O, OH, or F). They discovered that Ti3C2Tx nanosheets had a larger surface area than multilayer Ti3C2Tx and Ti3AlC2. Bacterial growth is inhibited by the hydrogen bonds formed between the oxygenated groups of Ti3C2Tx MXene and cell membranes. Moreover, the high hydrophilicity and cation surface enhance bacterial interactions. The outcome was that Ti3C2Tx exhibited improved antibacterial activity compared to pure GO nanosheets. Owing to the synergistic interaction between the surface formed Ti3C2Tx and TiO2/C, the aged Ti3C2Tx MXene membrane exhibited better improved antibacterial activity than the newly created membrane. By comparing Ti2C and Ti3C2 MXenes, studies have shown that the atomic structure of MXenes significantly affects their antibacterial activity [66, 68, 111, 112].

Layered double hydroxides (LDHs) are a type of layered, inorganic, 2D material that resembles hydrotalcite materials. They have attracted a lot of interest and are useful in many applications, including drug delivery, adsorption, and photocatalysis. To improve the photothermal performance of LDH with a large surface area, a team of researchers developed GNR@ LDH nanomaterials using LDH nanosheets as the shell and gold nanorods (GNR) as the core. In a previous study, GNR@LDH was successfully used for bacterial inactivation of *E. coli* and *S. aureus* using photothermal excitation at 808 nm, which was subsequently used for photothermal therapy in wound repair [113–116].

In addition to the 2D NBG antibacterial nanomaterials already described, novel 2D nanomaterials, such as laponite (Lap), hexagonal boron nitride (BN), III2-VI3 compounds (In2Se3, Bi2Se3, Sb2Se3), and RuO2, have also been identified as intriguing antibacterial agents [117–120].

8.3 Antibacterial Mechanisms of Graphene-Based Family

To manipulate nanomaterials for prospective biomedical applications, graphene-based nanosystems have been extensively researched for their exceptional antibacterial activities owing to their astounding characteristics and predominant antibacterial mechanisms. The current antibacterial mechanisms of graphene-based nanomaterials primarily involve physical contact destruction, oxidative stress, disruption of protein–protein interactions, and photo-induced antibacterial effects [121–123] (illustrated in Figure 8.3). This mechanism is currently being explored, and scientific advances will unveil other intricate processes.

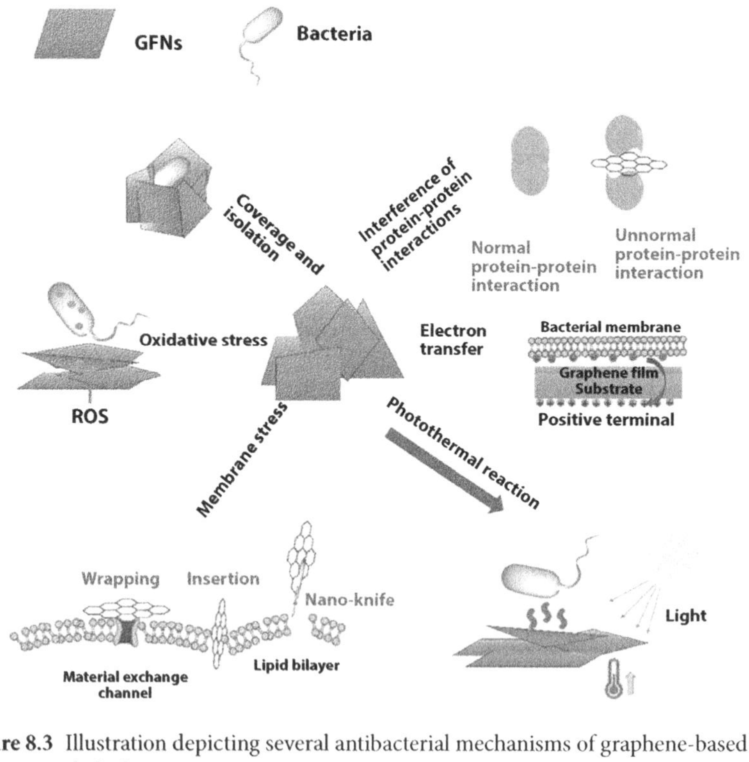

Figure 8.3 Illustration depicting several antibacterial mechanisms of graphene-based nanomaterials [36].

8.3.1 Physical Contact Destruction

The cell membranes of Gram-negative and Gram-positive bacteria differ. The former has two unique lipoprotein membranes separated by a thin layer of peptidoglycan. The inner layer of the dual membrane is covered with lipopolysaccharide, whereas the outer membrane is in the periplasmic space. Gram-positive bacteria, on the other hand, have a thick multilayer peptidoglycan lined outside the single lipid bilayer cell membrane. Bacterial membrane stress is associated with the contact sites, where the edges and corners have a greater impact than the planar surface of the nanosheets. It has been demonstrated that the sharp edges of nanosheets, sometimes referred as "nanoknives," were likely undermining the integrity of bacterial membranes upon physical contact, which could result in the outflow of intracellular constituents like nucleic acids (RNA/DNA), phospholipids, and proteins [50, 67, 123–125].

When bacterial membranes meet GFNs surfaces, the blade-like GFN nanosheets induce the membrane stress by inserting cuts into the surface of the membrane along with the formation of pores around the cell membrane, which results in the alteration of osmotic pressure of the bacterial cell and ultimately leads to swelling and death of the bacteria. Detrimental extraction of lipid molecules upon insertion is another indication of membrane stress. Since graphene has all sp2 hybridized carbons and a hydrophobic surface, it exhibits a high dispersion interaction with lipid molecules that contributes to strong lipid attraction. The basal plane of the cell membrane, composed of a phospholipid bilayer, serves as a barrier, trapping proteins, ions, and other chemicals, while preserving the cell from disruption. Destruction of the phospholipid bilayer in microbial cells is a major element of cell death. Membrane stress is composed of four actions: incision/cut, insertion, pore development, and lipid extraction [16, 126, 127].

8.3.2 Oxidative Stress

Oxidative stress can disrupt bacterial metabolism and permanently disrupt fundamental biological processes, rendering bacteria inactive and ultimately leading to cell death. Therefore, one of the most extensively used antibacterial techniques is induction of oxidative stress [128, 129]. Oxidative stress usually occurs *via* an ROS-dependent or ROS-independent pathway, and in both cases, it interferes with cellular processes, leading to cellular inactivation and death. Excessive accumulation of intracellular ROS such as hydroxyl radicals (OH•), hydrogen peroxide (H2O2), singlet molecular oxygen (IO2), and superoxide anions ($O2\bullet^-$) leads to ROS-dependent oxidative stress. The accumulation of intracellular ROS causes mitochondrial malfunction, cell necrosis, protein inactivation, cell membrane degradation, and lipid peroxidation. Graphene acts as an electron pump *via* an ROS-independent mechanism, where charge is transferred from the cell membrane to the GMs, leading to cell disruption and oxidation without the formation of reactive oxygen species [130, 131].

The antibacterial activity of GMs is mostly attributed to the generation of ROS in the bacteria, which may result in DNA fragmentation and bacterial cell lipid peroxidation. Moreover, ROS lower the activity of several bacterial enzymes, including glucosidase, xylosidase, glucosaminidase, and phosphatase. Additionally, oxidative stress has been shown to significantly increase the levels of mRNA, thioredoxin reductase, and intercellular reactive oxygen [72]. The number of oxygen-containing functional groups on GFNs affects ROS production. Therefore, because GO has many oxygen-containing functional groups on its surface, it may produce more ROS than rGO and graphene [132–134].

8.3.3 Disruption of Bacterial Protein Interactions

Protein–protein interactions play a crucial role in cellular metabolism and signal transmission in microorganisms. Biological dysfunction and even death result from blocking interactions between proteins. Intra-protein hydrophobic interactions in bacteria have been disrupted using large-scale all-atom molecular dynamic simulations; by positioning graphene into the hydrophobic interface of two interacting proteins [135].

8.3.4 Photo-Induced Mechanisms

Widely known photothermal techniques such as PTT, PCT, or PDT act as a source of photoactive components called fluorophores, which are activated by the sensitization of nanomaterials, involving the generation of heat or ROS in the presence of a specific wavelength of light. However, owing to their unique advantages, such as noninvasiveness, targeted selective therapy, and few adverse effects, these photo-induced antibacterial nanoagents been considered as a potential chemotherapeutic mechanism against bacterial growth in recent years [136–139].

An efficient antimicrobial strategy employing a photothermal activation mechanism involves the eradication of bacterial cells by the efficient production of heat when nanomaterials are irradiated under the ideal wavelength of light. The wavelength of the near-infrared region (NIR) is preferred in this paradigm because of its ability to penetrate multiple layers of biological tissue without disrupting normal cells. However, many 2D NBG could photo-thermally kill the bacterial cells when irradiated with wavelengths in the ultraviolet and visible range. Recent studies have shown that a variety of 2D NBG, including MoS2, metal-based nanomaterials, Sb2Se3, LDH-based compounds, and BP nanosheets, exhibit extraordinary antibacterial potential owing to their high photothermal conversion. Among them, MoS2 nanosheets have drawn significant attention as typical photothermal antibacterial agents because of their better light-to-heat conversion efficacy in the presence of NIR [116, 140–143]. With excellent NIR light absorbance/conversion abilities, photothermal nanomaterials have demonstrated that the heat produced not only can destroy or eradicate antibiotic-resistant bacteria, but also hinders the development of biofilms. To combat bacterial infections, 2D NBG based PTT is a reliable and effective method.

It is generally known that bacteria can experience distinctive oxidative stress due to photo-induced generation of ROS. The two main methods of photomediated ROS formation are photocatalytic and photodynamic processes. The photocatalytic approach to bacterial inhibition by photoirradiation is facilitated by the separation of holes and electrons in the valence band (VB) and conduction band (CB) of semiconductor nanomaterials. This process effectively produces highly reactive ROS. Different stages are usually included in photocatalytic disinfection: (i) the formation of light-induced distinct charged phases; (ii) the coming together of photo-induced charged species with the nearby H_2O and O_2 moieties; and (iii) the provision of bacterial binding to the large surface areas provided by the 2D NBG for the catalytic process. 2D semiconductors such as bismuth oxybromide (BiOBr) nanosheets, has gained popularity as bactericidal nanomaterials owing to their exceptional physicochemical properties and photocatalytic activity in the presence of VL [144–146].

PDT has also been reported as an alternative photoinduced ROS production method for minimally invasive antibacterial treatment along with photocatalytic-induced bacterial elimination. To efficiently destroy the bacterial membrane and intracellular components (nucleic acids, proteins, and DNA), external stimuli are primarily used (optimal excitation energy gained from a specific wavelength of light and free O_2 molecules in the vicinity) to selectively activate photosensitizers at the targeted site. PDT can also eliminate biofilms by reducing their excretory byproducts. Owing to its non-invasive nature, enormous loading capacity, and quick healing time, PDT is superior to other traditional antibacterial techniques [147–149].

8.4 Conclusion

In summary, nanomaterials provide a promising future for several industries including electronics, thermotics, and biological applications. 2D nanomaterials have drawn a significant amount of attention for the photoinduced inactivation of bacteria. They have been developed using different methods and materials, varying in size, surface area, morphology, and structure, whcih predominantly influence their antibacterial properties. The higher surface area of nanomaterials contributes to better surface characteristic properties and is therefore more favorable for microbial interactions. However, their smaller dimensions make it simpler for them to permeate the cell membrane. Furthermore, the change in the shape and structure of nanomaterials significantly improves the surface properties, such as the range of absorbed light, the rate of electron transformation, and hydrophobicity.

Most 2D materials exhibit exceptional photocatalytic or photothermal capabilities that are effective against an extensive variety of bacterial flora, making them a potential broad-spectrum antimicrobial nanosystem. Nano-based antibacterial formulations are more advantageous than conventional antibiotics owing to their sustainability, availability, ease of procurement, and low bacterial resistance. The primary mechanism of antibacterial action of photoinduced 2D nanomaterials includes photodynamic, photocatalytic, and photothermal-induced ROS/heat to destroy bacterial membranes, release components, or denature proteins under UV, VL, and NIR irradiation.

In conclusion, the various 2D materials explored are highly desirable for the photoinduced effect-mediated inactivation of microorganisms such as bacteria, protozoa, fungi, and viruses. To develop a "self-sterilization" system that can be extensively employed in gloves, public facility surfaces, films, gels, and coatings for food packaging, 2D materials can be incorporated. In this manner, it may be efficient to eliminate these unsettling, hazardous bacteria and viruses, such as COVID-19, that we are currently battling when exposed to light.

References

1. Appelbaum, P.C., 2012 and beyond: Potential for the start of a second pre-antibiotic era? *J. Antimicrob. Chemother.*, 67, 2062, 2001.
2. Interagency Coordination Group on Antimicrobial Resistance, *No Time To Wait: Securing The Future From Drug-Resistant Infections*, Report to

the secretary-general of the United Nations, World Health Organization, Geneva, Switzerland, 2019.

3. Boyce, K., Morrissey, O., Idnurm, A., Macreadie, I., Insights into the global emergence of antifungal drug resistance. *Microbiol. Aust.*, 40, 87, 2019.
4. Percival, S.L., Bowler, P.G., Russell, D., Bacterial resistance to silver in wound care. *J. Hosp. Infect.*, 60, 1, 1–7, 2005. DOI 10.1016/j.jhin.2004.11.014.
5. Helms, M., Vastrup, P., Gerner-Smidt, P., Mølbak, K., Excess mortality associated with antimicrobial drug-resistant salmonella typhimurium. *Emerg. Infect. Dis.*, 8, 5, 490–495, 2002. DOI 10.3201/eid0805.010267.
6. W. H. O., *Antibacterial Agents in Clinical Development: An Analysis of the Antibacterial Clinical Development Pipeline, Including Tuberculosis*, Geneva, 2017.
7. Humphreys, G. and Fleck, F., United Nations meeting on antimicrobial resistance. *Bull. World Health Organ.*, 94, 638, 2016.
8. Singer, A., Markoutsa, E., Limayem, A., Nanobiotechnology medical applications: Overcoming challenges through innovation. *Eurobiotech. J.*, 2, 146–160, 2018.
9. Willyard, C., The drug-resistant bacteria that pose the greatest health threats. *Nature*, 543, 15, 2017, doi: 10.1038/nature.2017.21550.
10. Miller, K.P., Wang, L., Benicewicz, B.C., Decho, A.W., Inorganic nanoparticles engineered to attack bacteria. *Chem. Soc. Rev.*, 44, 7787–807, 2015.
11. Chen, H., Gu, Z., An, H., Chen, C., Chen, J., Cui, R., Precise nanomedicine for intelligent therapy of cancer. *Sci. China Chem.*, 61, 1503–52, 2018.
12. Liu, Z., Lin, H., Zhao, M., Dai, C., Zhang, S., Peng, W., 2D superparamagnetic tantalum carbide composite mxenes for efficient breast-cancer theranostics. *Theranostics*, 8, 1648–64, 2012.
13. Li, Y., Liu, X., Tan, L., Cui, Z., Jing, D., Yang, X., Eradicating multidrug-resistant bacteria rapidly using a multi-functional g-c3n4@bi2s3 nanorod heterojunction with or without antibiotics. *Adv. Funct. Mater.*, 29, 1900946, 2019.
14. Wang, Y., Jin, Y., Chen, W., Wang, J., Chen, H., Sun, L., Construction of nanomaterials with targeting phototherapy properties to inhibit resistant bacteria and biofilm infections. *Chem. Eng. J.*, 358, 74–90, 2019.
15. Chitgupi, U., Qin, Y., Lovell, J.F., Targeted nanomaterials for phototherapy. *Nanotheranostics*, 1, 38–58, 2017.
16. Shi, L., Chen, J.G., Teng, L.J., Wang, L., Zhu, G.L., Liu, S., Luo, Z.T., Shi, X.T., Wang, Y.J., Ren, L., The antibacterial applications of graphene and its derivatives. *Small*, 12, 4165–4184, 2016.
17. Liu, T., Wang, C., Gu, X., Gong, H., Cheng, L., Shi, X.Z., Feng, L.Z., Sun, B.Q., Liu, Z., Drug delivery with PEGylated MoS2 nano-sheets for combined photothermal and chemotherapy of cancer. *Adv. Mater.*, 26, 3433–3440, 2014.
18. Han, J.Y., Xia, H.P., Wu, Y.F., Kong, S.N., Deivasigamani, A., Xu, R., Hui, K.M., Kang, Y.J., Single-layer MoS2 nanosheet grafted upconversion nanoparticles for nearinfrared fluorescence imaging-guided deep tissue cancer phototherapy. *Nanoscale*, 8, 7861–7865, 2016.

19. Wang, L.S., Gupta, A., Rotello, V.M., Nanomaterials for the treatment of bacterial biofilms. *ACS Infect. Dis.*, 2, 3–4, 2016.
20. Beyth, N., Houri-Haddad, Y., Domb, A., Khan, W., Hazan, R., Alternative antimicrobial approach: Nano-antimicrobial materials. *Evid. Based Complement. Alternat. Med.*, 2015, 246012, 2015.
21. Hemeg, H.A., Nanomaterials for alternative antibacterial therapy. *Int. J. Nanomedicine*, 12, 8211–25, 2017.
22. Mei, L., Zhu, S., Yin, W., Chen, C., Nie, G., Gu, Z., Zhao, Y., Two-dimensional nanomaterials beyond graphene for antibacterial applications: Current progress and future perspectives. *Theranostics*, 10, 2, 757–781, 2020.
23. Lambert, P.A., Bacterial resistance to antibiotics: Modified target sites. *Adv. Drug Delivery Rev.*, 57, 10, 1471–85, 2005.
24. Gu, Z., Zhu, S., Yan, L., Zhao, F., Zhao, Y., Graphene-based smart platforms for combined cancer therapy. *Adv. Mater.*, 31, 1800662, 2019.
25. Sun, Z., Li, Y., Guan, X., Chen, L., Jing, X., Xie, Z., Rational design and synthesis of covalent organic polymers with hollow structure and excellent antibacterial efficacy. *RSC Adv.*, 4, 40269–72, 2014.
26. Mangadlao, J.D., Santos, C.M., Felipe, M.J.L., de Leon, A.C.C., Rodrigues, D.F., Advincula, R.C., On the antibacterial mechanism of graphene oxide (GO) langmuir-blodgett films. *Chem. Commun.*, 51, 2886–9, 2015.
27. Kurapati, R., Kostarelos, K., Prato., M., Bianco, A., Biomedical uses for 2D materials beyond graphene: Current advances and challenges ahead. *Adv. Mater.*, 28, 29, 6052–74, 2016.
28. Sajjad, S., Leghari, S.A.K., Iqbal, A., Study of graphene oxide structural features for catalytic, antibacterial, gas sensing, and metals decontamination environmental applications. *ACS Appl. Mater. Interfaces*, 9, 43393–43414, 2017.
29. Chen, J., Fan, T., Xie, Z., Advances in nanomaterials for photodynamic therapy applications: Status and challenges. *Biomaterials.*, 237, 119827, 2020.
30. Kaur, N., Aditya, R.N., Singh, A. *et al.*, Biomedical applications for gold nanoclusters: Recent developments and future perspectives. *Nanoscale Res. Lett.*, 13, 1–12, 2018, doi: 10.1186/s11671-018-2725-9.
31. Wang, Y.W., Fu, Y.Y., Wu, L.J., Li, J., Yang, H.H., Chen, G.N., Targeted photothermal ablation of pathogenic bacterium, *staphylococcus aureus*, with nanoscale reduced graphene oxide. *J. Mater. Chem. B*, 1, 2496–2501, 2013.
32. Qi, Z., Bharate, P., Lai, C.H., Ziem, B., Boettcher, C., Schulz, A., Beckert, F., Hatting, B., Muelhaupt, R., Seeberger, P.H., Haag, R., Multivalency at interfaces: Supramolecular carbohydrate-functionalized graphene derivatives for bacterial capture, release, and disinfection. *Nano Lett.*, 15, 6051–6057, 2015.
33. Zhao, J., Huang, S., Ravisankar, P., Zhu, H., Two-dimensional nanomaterials for photoinduced antibacterial applications. *ACS. Appl. Bio Mater.*, 3, 12, 8188–8210, 2020.
34. Roy, V., Patra, S., Kumar, D., Madhuri, R., Sharma, P.K., *Biosens. Bioelectron.*, 68, 726, 2015.

35. Feng, L.L. and Liu, Z., Graphene in biomedicine: Oppurtunities and challenges. *Nanomedicine*, 6, 317, 2011.
36. Han, W., Wu, Z., Li, Y., Wang, Y., Graphene family nanomaterials (GFNs) —; promising materials for antimicrobial coating and film: A review. *Chem. Eng. J.*, 358, 1022–1037, 2019.
37. Ji, H., Sun, H., Qu, X., Antibacterial applications of graphenebased nanomaterials: Recent achievements and challenges. *Adv. Drug Deliv. Rev.*, 105, 2016, 176–189, 2016.
38. Zarafu, I., Turcu, I., Culită, D., Petrescu, S., Popa, M., Chifiriuc, M.C., Limban, C., Telehoiu, A., Ionită, P., Antimicrobial features of organic functionalized graphene-oxide with selected amines. *Materials*, 11, 1704, 2018.
39. Pulingam, T., Thong, K.L., Ali, M.E., Appaturi, J.N., Dinshaw, I.J., Ong, Z.Y., Leo, B.F., Graphene oxide exhibits differential mechanistic action towards gram-positive and gram-negative bacteria. *Colloids Surf. B Biointerfaces*, 181, 6–15, 2019.
40. Jia, X., Ahmad, I., Yang, R., Wang, C., Versatile graphene-based photothermal nanocomposites for effectively capturing and killing bacteria, and for destroying bacterial biofilms. *J. Mater. Chem. B*, 5, 2459–2467, 2017.
41. Wu, M.C., Deokar, A.R., Liao, J.H., Shih, P.Y., Ling, Y.C., Graphene-based photothermal agent for rapid and effective killing of bacteria. *ACS Nano*, 7, 2, 1281–90, 2013.
42. Shahnawaz Khan, M., Abdelhamid, H.N., Wu, H.F., Near infrared (NIR) laser mediated surface activation of graphene oxide nanoflakes for efficient antibacterial, antifungal and wound healing treatment. *Colloids Surf. B*, 127, 281–91, 2015.
43. Li, R., Mansukhani, N.D., Guiney, L.M., Ji, Z., Zhao, Y., Chang, C.H., French, C.T., Miller, J.F., Hersam, M.C., Nel, A.E., Xia, T., Identification and optimization of carbon radicals on hydrated graphene oxide for ubiquitous antibacterial coatings. *ACS Nano*, 10, 10966–10980, 2016.
44. Chong, Y., Ge, C., Fang, G., Wu, R., Zhang, H., Chai, Z., Chen, C., Yin, J.J., Light-enhanced antibacterial activity of graphene oxide, mainly via accelerated electron transfer. *Environ. Sci. Technol.*, 51, 10154, 2017.
45. Hou, W.C., Lee, P.L., Chou, Y.C., Wang, Y.S., Antibacterial property of graphene oxide: The role of phototransformation. *Environ. Sci. Nano*, 4, 647–657, 2017.
46. Gurunathan, S., Han, J.W., Dayem, A.A., Eppakayala, V., Park, M.R., Kwon, D.N., Kim, J.H., Antibacterial activity of dithiothreitol reduced graphene oxide. *J. Ind. Eng. Chem.*, 19, 1280–1288, 2013.
47. Liu, S., Zeng, T.H., Hofmann, M., Burcombe, E., Wei, J., Jiang, R., Kong, J., Chen, Y., Antibacterial activity of graphite, graphite oxide, graphene oxide, and reduced graphene oxide: Membrane and oxidative stress. *ACS Nano*, 5, 6971–6980, 2011.

48. Liu, S., Hu, M., Zeng, T.H., Wu, R., Jiang, R., Wei, J., Wang, L., Kong, V., Chen, Y., Lateral dimension-dependent antibacterial activity of graphene oxide sheets. *Langmuir*, 28, 12364–12372, 2012.
49. Faria, A.F., Perreault, F., Elimelech, M., Elucidating the role of oxidative debris in the antimicrobial properties of graphene oxide. *ACS. Appl. Nano Mater.*, 1, 1164, 2018.
50. Lu, X., Feng, X., Werber, J.R., Chu, C., Zucker, I., Kim, J.H., Enhanced antibacterial activity through the controlled alignment of graphene oxide nanosheets. *Proc. Natl. Acad. Sci. U. S. A.*, 114, E9793–E801, 2017.
51. Liu, R., Wang, X., Ye, J., Xue, X., Zhang, F., Zhang, H., Hou, X., Liu, X., Zhang, Y., Enhanced antibacterial activity of silverdecorated sandwich-like mesoporous silica/reduced graphene oxide nanosheets through photothermal effect. *Nanotechnology*, 29, 10, 105704, 2018.
52. Robinson, J.T., Tabakman, S.M., Liang, Y., Wang, H., Sanchez-Casalongue, H., Vinh, D., Dai, H., Ultrasmall reduced graphene oxide with high near-infrared absorbance for photothermal therapy. *J. Am. Chem. Soc.*, 133, 6825–6831, 2011.
53. Wang, S.G., Chen, Y.C., Chai, C.Y., Antibacterial gold nanoparticle-based photothermal killing of vancomycin-resistant bacteria. *Nanomedicine*, 13, 1405–1416, 2018, doi: 10.2217/nnm-2017-0380.
54. Tan, S., Wu, X., Xing, Y., Lilak, S., Wu, M., Zhao, J.X., Enhanced synergetic antibacterial activity by a reduce graphene oxide/ag nanocomposite through the photothermal effect. *Colloids Surf. B*, 185, 110616, 2020, doi: 10.1016/j.colsurfb.2019.110616.
55. Zhang, W., Mou, Z., Wang, Y., Chen, Y., Yang, E., Guo, F., Molybdenum disulfide nanosheets loaded with chitosan and silver nanoparticles effective antifungal activities: *In vitro* and *in vivo*. *Mater. Sci. Eng. C*, 97, 486–97, 2019.
56. Cao, W., Yue, L., Wang, Z., High antibacterial activity of chitosan-molybdenum disulfide nanocomposite. *Carbohydr Polym.*, 215, 226–34, 2019.
57. Zhang, W., Shi, S., Wang, Y., Yu, S., Zhu, W., Zhang, X., Versatile molybdenum disulfide based antibacterial composites for *in vitro* enhanced sterilization and *in vivo* focal infection therapy. *Nanoscale*, 8, 11642–8, 2016.
58. Yin, W., Yu, J., Lv, F., Yan, L., Zheng, L., Gu, Z., Functionalized nano-MoS_2 with peroxidase catalytic and near-infrared photothermal activities for safe and synergetic wound antibacterial applications. *ACS Nano*, 10, 11000–11, 2016.
59. Huang, X., Wei, J., Liu, T., Zhang, X., Bai, S., Yang, H., Silk fibroin-assisted exfoliation and functionalization of transition metal dichalcogenide nanosheets for antibacterial wound dressings. *Nanoscale*, 9, 17193–8, 2017.
60. Krishnamoorthy, K., Veerapandian, M., Yun, K., Kim, S.J., New function of molybdenum trioxide nanoplates: Toxicity towards pathogenic bacteria through membrane stress. *Colloids Surf. B*, 112, 521–4, 2013.

61. Mo, S., Chen, X., Chen, M., He, C., Lu, Y., Zheng, N., Two-dimensional antibacterial Pd@Ag nanosheets with a synergetic effect of plasmonic heating and Ag^+ release. *J. Mater. Chem. B*, 3, 6255–60, 2015.
62. Kaushik, R., Vineeth Daniel, P., Mondal, P., Halder, A., Transformation of 2-D TiO_2 to mesoporous hollow 3-D TiO_2 spheres-comparative studies on morphology-dependent photocatalytic and anti-bacterial activity. *Microporous Mesoporous Mater.*, 285, 32–42, 2019.
63. Li, R., Ren, Y., Zhao, P., Wang, J., Liu, J., Zhang, Y., Graphitic carbon nitride (g-C_3N_4) nanosheets functionalized composite membrane with self-cleaning and antibacterial performance. *J. Hazard. Mater.*, 365, 606–14, 2019.
64. Khan, M.E., Han, T.H., Khan, M.M., Karim, M.R., Cho, M.H., Environmentally sustainable fabrication of Ag@g-C_3N_4 nanostructures and their multifunctional efficacy as antibacterial agents and photocatalysts. *ACS Appl. Nano Mater.*, 1, 2912–2, 2018.
65. Wu, Y., Zhou, Y., Xu, H., Liu, Q., Li, Y., Zhang, L., Highly active, superstable, and biocompatible Ag/polydopamine/g-C_3N_4 bactericidal photocatalyst: Synthesis, characterization, and mechanism. *ACS Sustain. Chem. Eng.*, 6, 14082–94, 2018.
66. Rasool, K., Helal, M., Ali, A., Ren, C.E., Gogotsi, Y., Mahmoud, K.A., Antibacterial activity of $Ti_3C_2T_x$ MXene. *ACS Nano*, 10, 3674–84, 2016.
67. Alimohammadi, F., Gh, M.S., Attanayake, N.H., Thenuwara, A.C., Gogotsi, Y., Anasori, B., Antimicrobial properties of 2D MnO_2 and MoS_2 nanomaterials vertically aligned on graphene materials and Ti_3C_2 MXene. *Langmuir*, 34, 7192–200, 2018.
68. Rasool, K., Mahmoud, K.A., Johnson, D.J., Helal, M., Berdiyorov, G.R., Gogotsi, Y., Efficient antibacterial membrane based on two-dimensional Ti_3C_2Tx (MXene) nanosheets. *Sci. Rep.*, 7, 1598–609, 2017.
69. Xiong, Z., Zhang, X., Zhang, S., Lei, L., Ma, W., Li, D., Bacterial toxicity of exfoliated black phosphorus nanosheets. *Ecotoxicol. Environ. Saf.*, 161, 507–14, 2018.
70. Wu, Q., Liang, M., Zhang, S., Liu, X., Wang, F., Development of functional black phosphorus nanosheets with remarkable catalytic and antibacterial performance. *Nanoscale*, 10, 10428–35, 2018.
71. Wang, Q.H., Kalantar-Zadeh, K., Kis, A., Coleman, J.N., Strano, M.S., Electronics and optoelectronics of two-dimensional transition metal dichalcogenides. *Nat. Nanotechnol.*, 7, 11, 699–712, 2012.
72. Mak, K.F. and Shan, J., Photonics and optoelectronics of 2D semiconductor transition metal dichalcogenides. *Nat. Photonics*, 10, 4, 216–226, 2016.
73. Chhowalla, M., Shin, H.S., Eda, G., Li, L.J., Loh, K.P., Zhang, H., The chemistry of two-dimensional layered transition metal dichalcogenide nanosheets. *Nat. Chem.*, 5, 4, 263–75, 2013.
74. Kalantar-zadeh, K., Ou, J.Z., Daeneke, T., Strano, M.S., Pumera, M., Gras, S.L., Two-dimensional transition metal dichalcogenides in biosystems. *Adv. Funct. Mater.*, 25, 32, 5086–5099, 2015.

75. Tan, C. and Zhang, H., Two-dimensional transition metal dichalcogenide nanosheet-based composites. *Chem. Soc. Rev.*, 44, 9, 2713–31, 2015.
76. Zhu, H., Ni, N., Govindarajan, S., Ding, X., Leong, D.T., Phototherapy with layered materials derived quantum dots. *Nanoscale*, 12, 1, 43–57, 2020.
77. Zhu, H., Lai, Z., Fang, Y., Zhen, X., Tan, C., Qi, X., Ding, D., Chen, P., Zhang, H., Pu, K., Ternary Chalcogenide Nanosheets with Ultrahigh Photothermal Conversion Efficiency for Photoacoustic Theranostics. *Small.*, 13, 16, 1604139, 2017.
78. Ding, X., Peng, F., Zhou, J., Gong, W., Slaven, G., Loh, K.P., Lim, C.T., Leong, D.T., Defect engineered bioactive transition metals dichalcogenides quantum dots. *Nat. Commun.*, 10, 1, 41, 2019.
79. Wardman, P., Reduction potentials of one-electron couples involving free radicals in aqueous solution. *J. Phys. Chem. Ref. Data*, 18, 4, 1637–1755, 1989.
80. Liao, H. and Reitberger, T., Generation of Free OHaq Radicals by Black Light Illumination of Degussa (Evonik) P25 TiO2 Aqueous Suspensions. *Catalysts*, 3, 2, 418–443, 2013.
81. Liu, T., Wang, C., Cui, W., Gong, H., Liang, C., Shi, X., Li, Z., Sun, B., Liu, Z., Combined photothermal and photodynamic therapy delivered by PEGylated MoS2 nanosheets. *Nanoscale*, 6, 19, 11219–25, 2014.
82. Cheng, L., Liu, J., Gu, X., Gong, H., Shi, X., Liu, T., Wang, C., Wang, X., Liu, G., Xing, H., Bu, W., Sun, B., Liu, Z., PEGylated WS2 nanosheets as a multifunctional theranostic agent for *in vivo* dualmodal CT/photoacoustic imaging guided photothermal therapy. *Adv. Mater.*, 26, 12, 1886–93, 2014.
83. Lei, Z., Zhu, W., Xu, S., Ding, J., Wan, J., Wu, P., Hydrophilic MoSe2 nanosheets as effective photothermal therapy agents and their application in smart devices. *ACS Appl. Mater. Interfaces*, 8, 32, 20900–8, 2016.
84. Zhang, Y., Zheng, B., Zhu, C., Zhang, X., Tan, C., Li, H., Chen, B., Yang, J., Chen, J., Huang, Y., Wang, L., Zhang, H., Singlelayer transition metal dichalcogenide nanosheet-based nanosensors for rapid, sensitive, and multiplexed detection of DNA. *Adv. Mater.*, 27, 5, 935–9, 2015.
85. Yin, W., Yu, J., Lv, F., Yan, L., Zheng, L.R., Gu, Z., Zhao, Y., Functionalized nano-MoS2 with peroxidase catalytic and nearinfrared photothermal activities for safe and synergetic wound antibacterial applications. *ACS Nano*, 10, 12, 11000–11011, 2016.
86. Gao, Q., Zhang, X., Yin, W., Ma, D., Xie, C., Zheng, L., Dong, X., Mei, L., Yu, J., Wang, C., Gu, Z., Zhao, Y., Functionalized MoS2 nanovehicle with near-infrared laser-mediated nitric oxide release and photothermal activities for advanced bacteria-infected wound therapy. *Small*, 14, 45, e1802290, 2018.
87. Yuan, Z., Tao, B., He, Y., Liu, J., Lin, C., Shen, X., Ding, Y., Yu, Y., Mu, C., Liu, P., Cai, K., Biocompatible MoS2/PDA-RGD coating on titanium implant with antibacterial property via intrinsic ROS-independent oxidative stress and NIR irradiation. *Biomaterials*, 217, 119290, 2019.

88. Lin, Y., Han, D., Li, Y., Tan, L., Liu, X., Cui, Z., Yang, X., Li, Z., Liang, Y., Zhu, S., Wu, S., Ag2S@WS2 heterostructure for rapid bacteria-killing using near-infrared light. *ACS Sustain. Chem. Eng.*, 7, 17, 14982–14990, 2019.
89. Zhao, H., Chen, S., Quan, X., Yu, H., Zhao, H., Integration of microfiltration and visible-light-driven photocatalysis on g-C3N4 nanosheet/reduced graphene oxide membrane for enhanced water treatment. *Appl. Catal. B*, 194, 134–40, 2016.
90. Ong, W.J., Tan, L.L., Ng, Y.H., Yong, S.T., Chai, S.P., Graphitic carbon nitride (g-C3N4)-based photocatalysts for artificial photosynthesis and environmental remediation: Are we a step closer to achieving sustainability? *Chem. Rev.*, 116, 7159–329, 2016.
91. Jiang, J., Ou-yang, L., Zhu, L., Zheng, A., Zou, J., Yi, X., Tang, H., Dependence of electronic structure of g- C3N4 on the layer number of its nanosheets: A study by Raman spectroscopy coupled with first- principles calculations. *Carbon*, 80, 213–221, 2014.
92. Zhao, H., Yu, H., Quan, X., Chen, S., Zhang, Y., Zhao, H., Wang, H., Fabrication of atomic single layer graphitic-C3N4 and its high performance of photocatalytic disinfection under visible light irradiation. *Appl. Catal. B*, 152-153, 46–50, 2014.
93. Bing, W., Chen, Z., Sun, H., Shi, P., Gao, N., Ren, J., Qu, X., Visible-light-driven enhanced antibacterial and biofilm elimination activity of graphitic carbon nitride by embedded Ag nanoparticles. *Nano Res.*, 8, 5, 1648–1658, 2015.
94. Muñoz-Batista, M.J., Fontelles-Carceller, O., Ferrer, M., Fernandez-García, M., Kubacka, A., Disinfection capability of Ag/g-C3N4 composite photocatalysts under UV and visible light illumination. *Appl. Catal. B*, 183, 86–95, 2016.
95. Li, Y., Zhang, C., Shuai, D., Naraginti, S., Wang, D., Zhang, W., Visible-light-driven photocatalytic inactivation of MS2 by metal-free g-C3N4: Virucudal performance and mechanism. *Water Res.*, 106, 249–258, 2016.
96. Shi, Y., Hamsen, C., Jia, X.T., Kim, K.K., Reina, A., Hofmann, M., Hsu, A.L., Zhang, K., Li, H.N., Juang, Z.Y., Dresselhaus, M.S., Li, L.J., Kong, J., Synthesis of few-layer hexagonal boron nitride thin film by chemical vapor deposition. *Nano Lett.*, 10, 4134–4139, 2010.
97. Gao, G., Mathkar, A., Martins, E.P., Galvao, D.S., Gao, D., da Silva Autreto, P.A., Sun, C., Cai, L., Ajayan, P.M., Designing nanoscaled hybrids from atomic layered boron nitride with silver nanoparticle deposition. *J. Mater. Chem. A*, 2, 3148–3154, 2014.
98. Li, Z., Wu, L., Wang, H., Zhou, W., Liu, H., Cui, H., Synergistic antibacterial activity of black phosphorus nanosheets modified with titanium aminobenzenesulfanato complexes. *ACS Appl. Nano Mater.*, 2, 1202–9, 2019.
99. Lei, W., Liu, G., Zhang, J., Liu, M., Black phosphorus nanostructures: Recent advances in hybridization, doping and functionalization. *Chem. Soc. Rev.*, 46, 3492–509, 2017.

100. Luo, M., Fan, T., Zhou, Y., Zhang, H., Mei, L., 2D black phosphorus-based biomedical applications. *Adv. Funct. Mater.*, 29, 1808306, 2019.
101. Choi, J.R., Yong, K.W., Choi, J.Y., Nilghaz, A., Lin, Y., Xu, J., Black phosphorus and its biomedical applications. *Theranostics*, 8, 1005–26, 2018.
102. Ouyang, J., Liu, R., Chen, W., Liu, Z., Xu, Q., Zeng, K., A black phosphorus based synergistic antibacterial platform against drug resistant bacteria. *J. Mater. Chem. B*, 6, 6302–10, 2018.
103. Mao, C., Xiang, Y., Liu, X., Cui, Z., Yang, X., Li, Z., Repeatable photodynamic therapy with triggered signaling pathways of fibroblast cell proliferation and differentiation to promote bacteria-accompanied wound healing. *ACS Nano*, 12, 1747–59, 2018.
104. Huang, X., Wei, J., Zhang, M., Zhang, X., Yin, X., Lu, C., Water-based black phosphorus hybrid nanosheets as a moldable platform for wound healing applications. *ACS Appl. Mater. Interfaces*, 10, 35495–502, 2018.
105. Huang, X.W., Wei, J.J., Zhang, M.Y., Zhang, X.L., Yin, X.F., Lu, C.H., Song, J.B., Bai, S.M., Yang, H.H., Water-based black phosphorus hybrid nanosheets as a moldable platform for wound healing applications. *ACS Appl. Mater. Interfaces*, 10, 41, 35495–35502, 2018.
106. Lv, R., Yang, D., Yang, P., Xu, J., He, F., Gai, S., Li, C., Dai, Y., Yang, G., Lin, J., Integration of upconversion nanoparticles and ultrathin black phosphorus for efficient photodynamic theranostics under 808 nm near-infrared light irradiation. *Chem. Mater.*, 28, 13, 4724–4734, 2016.
107. Chen, W., Ouyang, J., Liu, H., Chen, M., Zeng, K., Sheng, J., Liu, Z., Han, Y., Wang, L., Li, J., Deng, L., Liu, Y.N., Guo, S., Black phosphorus nanosheet-based drug delivery system for synergistic photodynamic/photothermal/chemotherapy of cancer. *Adv. Mater.*, 29, 5, 1603864, 2017.
108. Shao, J., Ruan, C., Xie, H., Li, Z., Wang, H., Chu, P.K., Yu, X.F., Black-phosphorus-incorporated hydrogel as a sprayable and biodegradable photothermal platform for postsurgical treatment of cancer. *Adv. Sci. (Weinh)*, 5, 5, 1700848, 2018.
109. Shao, J., Xie, H., Huang, H., Li, Z., Sun, Z., Xu, Y., Xiao, Q., Yu, X.F., Zhao, Y., Zhang, H., Wang, H., Chu, P.K., Biodegradable black phosphorus-based nanospheres for *in vivo* photothermal cancer therapy. *Nat. Commun.*, 7, 12967, 2016.
110. Aksoy, I., Kucukkececi, H., Sevgi, F., Metin, O., Hatay Patir, I., Photothermal antibacterial and antibiofilm activity of black phosphorus/gold nanocomposites against pathogenic bacteria. *ACS Appl. Mater. Interfaces*, 12, 24, 26822–26831, 2020.
111. Ghidiu, M., Lukatskaya, M.R., Zhao, M.Q., Gogotsi, Y., Barsoum, M.W., Conductive two-dimensional titanium carbide 'clay' with high volumetric capacitance. *Nature.*, 516, 78–81, 2014.
112. Jastrzębska, A.M., Karwowska, E., Wojciechowski, T., Ziemkowska, W., Rozmysłowska, A., Chlubny, L., Olszyna, A., *J. Mater. Eng. Perform.*, 2018, Ahead of print.

113. Rives, V., Del Arco, M., Martin, C., Layered double hydroxides as drug carriers and for controlled release of non-steroidal antiinflammatory drugs (NSAIDs): A review. *J. Control. Release*, 169, 1-2, 28–39, 2013.
114. Ladewig, K., Xu, Z.P., Lu, G.Q., Layered double hydroxide nanoparticles in gene and drug delivery. *Exp. Opin. Drug Deliv.*, 6, 9, 907–22, 2009.
115. Zhao, Y., Wang, C.J., Gao, W., Li, B., Wang, Q., Zheng, L., Wei, M., Evans, D.G., Duan, X., O'Hare, D., Synthesis and antimicrobial activity of ZnTi-layered double hydroxide nanosheets. *J. Mater. Chem. B*, 1, 43, 5988–5994, 2013.
116. Ma, K., Li, Y., Wang, Z., Chen, Y., Zhang, X., Chen, C., Yu, H., Huang, J., Yang, Z., Wang, X., Wang, Z., Core-shell gold nanorod@layered double hydroxide nanomaterial with highly efficient photothermal conversion and its application in antibacterial and tumor therapy. *ACS Appl. Mater. Interfaces*, 11, 33, 29630–29640, 2019.
117. Leon Morales, C.F., Leis, A.P., Strathmann, M., Flemming, H.C., Interactions between laponite and microbial biofilms in porous media: Implications for colloid transport and biofilm stability. *Water Res.*, 38, 3614–26, 2004.
118. Das, S.S., Neelam, H.K., Singh, S., Hussain, A., Faruk, A., Laponite-based nanomaterials for biomedical applications: A review. *Curr. Pharm. Des.*, 25, 424–43, 2019.
119. Zhu, C., Shen, H., Liu, H., Lv, X., Li, Z., Yuan, Q., Solution-processable two-dimensional In2Se3 nanosheets as efficient photothermal agents for elimination of bacteria. *Chemistry*, 24, 19060–5, 2018.
120. Ding, W., Zhu, J., Wang, Z., Gao, Y., Xiao, D., Gu, Y., Prediction of intrinsic two-dimensional ferroelectrics in In2Se3 and other III2-VI3 van der Waals materials. *Nat. Commun.*, 8, 14956–64, 2017.
121. Sun, W. and Wu, F.G., Two-dimensional materials for antimicrobial applications: Graphene materials and beyond. *Chem. Asian J.*, 13, 3378–410, 2018.
122. Miao, H., Teng, Z., Wang, C., Chong, H., Wang, G., Recent progress in two-dimensional antimicrobial nanomaterials. *Chem. Asian J.*, 25, 929–44, 2019.
123. Zou, X., Zhang, L., Wang, Z., Luo, Y., Mechanisms of the antimicrobial activities of graphene materials. *J. Am. Chem. Soc.*, 138, 2064–77, 2016.
124. Schwechheimer, C. and Kuehn, M.J., Outer-membrane vesicles from Gram-negative bacteria: Biogenesis and functions. *Nat. Rev. Microbiol.*, 13, 605–19, 2015.
125. Brown, L., Wolf, J.M., Prados-Rosales, R., Casadevall, A., Through the wall: Extracellular vesicles in Gram-positive bacteria, mycobacteria and fungi. *Nat. Rev. Microbiol.*, 13, 620–30, 2015.
126. Pham, V.T.H., Vi Khanh, T., Quinn, M.D.J., Notley, S.M., Guo, Y., Baulin, V.A., Al Kobaisi, M., Crawford, R.J., Ivanova, E.P., Graphene induces formation of pores that kill spherical and rod-shaped bacteria. *ACS Nano*, 9, 8458–8467, 2015.

127. Tu, Y., Lv, M., Xiu, P., Huynh, T., Zhang, M., Castelli, M., Liu, Z., Huang, Q., Fan, C., Fang, H., Zhou, R., Destructive extraction of phospholipids from *Escherichia coli* membranes by graphene nanosheets. *Nat. Nanotechnol.*, 8, 594–601, 2013.
128. Zheng, H., Ma, R., Gao, M., Tian, X., Li, Y., Zeng, L., Antibacterial applications of graphene oxides: Structure-activity relationships, molecular initiating events and biosafety. *Sci. Bull.*, 63, 133–42, 2018.
129. Li, Y.F., Ouyang, S.H., Tu, L.F., Wang, X., Yuan, W.L., Wang, G.E., Caffeine protects skin from oxidative stress-induced senescence through the activation of autophagy. *Theranostics*, 8, 5713–30, 2018.
130. Li, J., Wang, G., Zhu, H., Zhang, M., Zheng, X., Di, Z., Antibacterial activity of large-area monolayer graphene film manipulated by charge transfer. *Sci. Rep.*, 4, 4359, 2015.
131. Gurunathan, S., Woong, H.J., Abdal Daye, A., Eppakayala, V., Kim, J., Oxidative stress-mediated antibacterial activity of graphene oxide and reduced graphene oxide in *Pseudomonas aeruginosa. Int. J. Nanomedicine*, 7, 5901–5914, 2012.
132. Hui, L., Huang, J., Chen, G., Zhu, Y., Yang, L., Antibacterial property of graphene quantum dots (both source material and bacterial shape matter). *ACS Appl. Mater. Interfaces*, 8, 20–25, 2016.
133. Chung, H., Kim, M.J., Ko, K., Kim, J.H., Kwon, H.A., Hong, I., Park, N., Lee, S.W., Kim, W., Effects of graphene oxides on soil enzyme activity and microbial biomass. *Sci. Total Environ.*, 514, 307–313, 2015.
134. Das, S., Singh, S., Singh, V., Joung, D., Dowding, J.M., Reid, D., Anderson, J., Zhai, L., Khondaker, S.I., Self, W.T., Seal, S., Oxygenated functional group density on graphene oxide: Its effect on cell toxicity. *Part. Part. Syst. Charact.*, 30, 148–157, 2013.
135. Luan, B., Tien, H., Zhao, L., Zhou, R., Potential toxicity of graphene to cell functions via disrupting protein-protein interactions. *ACS Nano*, 9, 663–669, 2015.
136. Xu, J., Yao, K., Xu, Z., Nanomaterials with a photothermal effect for antibacterial activities: An overview. *Nanoscale*, 11, 8680–91, 2019.
137. Wang, W., Li, G., Xia, D., An, T., Zhao, H., Wong, P.K., Photocatalytic nanomaterials for solar-driven bacterial inactivation: Recent progress and challenges. *Environ. Sci. Nano*, 4, 782–99, 2017.
138. Tavares, A., Carvalho, C.M.B., Faustino, M.A., Neves, M.G.P.M.S., Tomé, J.P.C., Tomé, A.C., Antimicrobial photodynamic therapy: Study of bacterial recovery viability and potential development of resistance after treatment. *Mar. Drugs*, 8, 91–105, 2010.
139. Wegener, M., Hansen, M.J., Driessen, A.J.M., Szymanski, W., Feringa, B.L., Photocontrol of antibacterial activity: Shifting from UV to red light activation. *J. Am. Chem. Soc.*, 139, 17979–86, 2017.

140. Chou, S.S., Kaehr, B., Kim, J., Foley, B.M., De, M., Hopkins, P.E., Chemically exfoliated MoS2 as near-infrared photothermal agents. *Angew. Chem. Int. Ed.*, 52, 4160–4, 2013.
141. Wang, X., Lv, F., Li, T., Han, Y., Yi, Z., Liu, M., Electrospun micropatterned nanocomposites incorporated with Cu2S nanoflowers for skin tumor therapy and wound healing. *ACS Nano*, 11, 11337–49, 2017.
142. Miao, Z., Fan, L., Xie, X., Ma, Y., Xue, J., He, T., Liquid exfoliation of atomically thin antimony selenide as an efficient two-dimensional antibacterial nanoagent. *ACS Appl. Mater. Interfaces*, 11, 26664–73, 2019.
143. Yu, J., Yin, W., Zheng, X., Tian, G., Zhang, X., Bao, T., Smart MoS2/Fe3O4 nanotheranostic for magnetically targeted photothermal therapy guided by magnetic resonance/photoacoustic imaging. *Theranostics*, 5, 931–45, 2015.
144. Liu, Y., Zeng, X., Hu, X., Hua, J., Zhang, X., Two-dimensional nanomaterials for photocatalytic water disinfection: Recent progress and future challenges. *J. Chem. Technol. Biotechnol.*, 94, 22–37, 2019.
145. Liu, C., Kong, D., Hsu, P.C., Yuan, H., Lee, H.W., Liu, Y., Rapid water disinfection using vertically aligned MoS2 nanofilms and visible light. *Nat. Nanotechnol.*, 11, 1098–105, 2016.
146. Wu, D., Wang, B., Wang, W., An, T., Li, G., Ng, T.W., Visible-light-driven BiOBr nanosheets for highly facet-dependent photocatalytic inactivation of *Escherichia coli*. *J. Mater. Chem. A*, 3, 15148–55, 2015.
147. De Melo, W.C., Avci, P., de Oliveira, M.N., Gupta, A., Vecchio, D., Sadasivam, M., Photodynamic inactivation of biofilm: taking a lightly colored approach to stubborn infection. *Expert Rev. Anti. Infect. Ther.*, 11, 669–93, 2013.
148. Fan, W., Huang, P., Chen, X., Overcoming the Achilles' heel of photodynamic therapy. *Chem. Soc. Rev.*, 45, 6488–519, 2016.
149. Biel, M.A., Photodynamic therapy of bacterial and fungal biofilm infections, in: *Photodynamic Therapy: Methods and Protocols*, pp. 175–194, Humana Press, Totowa, NJ, 2010.

9

2D Nanomaterials for Drug Delivery System

Syed Muzammil Munawar*, Dhandayuthabani Rajendiran and Kaleel Basha Sabjan

Department of Chemistry & Biochemistry, C. Abdul Hakeem College, (Autonomous) Melvisharam, Tamil Nadu, India

Abstract

Activation of drug structure at the nanometer level can prolong drug loading time, increase drug biotransport, and increase drug uniformity, thereby reducing drug toxicity and improving efficacy. Consequently, nanostep structures for drug delivery in the clinical discipline have attracted increasing interest, which has generated tremendous momentum just like ordinary pharmaceutical companies, biomedical and pharmaceutical companies have invested a great deal of money, energy and time in researching effective nanocarriers structure, 2D substances are of great interest because of their unique residence times in physical and chemical environments and their broad potential for applications in devices, biosensing, medicine, energy, and catalysis. The two multidimensional substances also have excellent drug delivery abilities compared to the conventional drug delivery structures. The main function of the 2D substance is the ultrafine nanogrid structure intentionally attached to the specific massive area of the ground thick structure for various external stimuli, 2D substances can exist as extremely green photodynamic and photothermal remedies for ailments. In addition, two-dimensional nanomaterial layer systems offer immense space for highly ecological medicines. Meanwhile, 2D substances with 2D terrain-changing material can have vivid orientation. Drugs can be properly stacked, and the effects of drugs can be reduced. However, NIR reaction, pH reaction with two-dimensional substances, two-dimensional SDDS can properly dispense tablets and effectively control blood consciousness to achieve great healing potential. These residences offer the possibility of multimodal healing of the disease. These special blessings are making 2D substance research more popular in the SDDS. However, it needs to be done

**Corresponding author*: syed_bio2004@yahoo.co.in

Subhendu Chakroborty and Kaushik Pal (eds.) 2D Nanomaterials: Synthesis, Properties and Applications, (185–208) © 2024 Scrivener Publishing LLC

much faster than can be accomplished in scientific exercises, but there are pressing issues that need to be addressed, including the transient long-term degradability and long-term toxicity of 2D materials. In addition, they study 2D substances *in vitro*, which are reserved for terrestrial animals, and *in vitro* experiments are also an important step towards their scientific implementation. Further intensive research and further *in vivo* experiments are expected to eventually implement SDDS-derived 2D coated compound packages as part of a scientific exercise. In this chapter we highlight the recent SDDS improvements, which are huge a wide range of 2D sensitive substance companies consisting of micelles, polymers, liposomes, hydrogels, nanofilms, nanoparticles, *etc.* Large organic packaging of 2D substances in 2D substances modified to be pharmaceutical companies function. 2D Stuff because pharmaceutical companies can improve disease drug treatment outcomes through managed launch, sustained launch, and stepped forward drug targeting. Therefore, 2D substances are perfectly suited to enhance the following SDDS technology.

Keywords: 2D nanomaterials, drug delivery system, biomedical applications, diagnostics

9.1 Introduction

Nanoscale drug transport facilities can increase the price of drug loading, increase the transport of organic drugs, provide a targeted drug, and avoid being overloaded with considerations, thereby quickly reducing drug toxicity, and increasing its effectiveness. As a result, nanoscale drug delivery structures are of concern in the field of medicine and have created strong momentum. The development of nanomaterials in steel represents a revolution in modern nanoscience archives. Before 1857, Faraday offered a triumph elementary biosynthesis of gold nanoparticles with advances in electron microscopy, which enabled the study of nanostructures, and extensive research has been conducted on the forbidden biosynthesis of colloidal gold nanomaterial steel nanostructures, their attractive home. To date, significant advances have been made in the biosynthesis of steel nanostructures using managed sizes, shapes, contents, and architecture. Crystal phase-directed biosynthesis has been proposed in different steel nanostructures. In essence, this failure recognizes the improvement in the range of 2D and near 2D steel nanostructures in research. Two-dimensional steel nanostructures with various thicknesses, shapes, sizes, crystalline phases, and compositions are available as nanowheels, nanoprisms, nanoribbons, nanoplates, nanodisks, nanorings, nanobelts, nanocones, and nanosheets. Similarly, herbal suppliers and biomedical and pharmaceutical organizations have invested a lot of money, power, and time in the analysis of high-performance nanocarrier structures. 2D substances are of great interest because of their different physicochemical foci and their vital importance in biosensors, catalysis, devices, drugs, force, etc. 2D substances explain the powerful function and delivery of drugs. Compared to traditional drug delivery structures, the most important distinguishing factor of 2D substances is the assembly of ultrafine nanoparticles, which is closely related to a good ground surface, thickness-dependent band gap, and response to multiple peripheral stimuli. 2D substances can be used for perfectly primed photodynamic and photothermal therapies for diseases, just as 2D nanosubstances are lamellar systems that offer ample space for drug overload. When the terrain of the 2D tissue changes, 2D substances can contain an active concentration. Drugs can be accurately added to the diagnosis, the side effects of drugs are reduced, with NIR reaction, pH reaction with two-dimensional substances, two-dimensional SDDS, and pills can be dropped accurately, and the consciousness of the blood can be effectively controlled to achieve a miraculous cure. This intentional specificity threatens the multimodality of the disease development. This major breakthrough makes substance discovery similar to 2D SDDS and extremely popular. Furthermore, two-dimensional substance studies are incomplete because *in vitro* animal experiments, such as *in vitro* studies, have critically advanced in their science packages. Given the specific studies and various *in vivo* experiments, it is not surprising that two-dimensional layered packages of derived substances in the form of SDDSs can be useful for scientific applications. As confirmation, we highlight the recent SDDS improvement for many suppliers of elegant 2D substances, such as hydrogels, nanoparticles, micelles, polymers, liposomes, and nanosheets. It is lonely and mostly organic 2D substance research that develops custom 2D substance packaging as a drug transport supplier. Using a 2D substance as a drug delivery agent leads to a better bottom line for treating diseases through prohibited release, sustained release, and increased drug awareness. Therefore, 2D substances have a unique opportunity in the latest SDDS cohorts. Elegant drug delivery nanostructures show remarkable physicochemical changes in response to less regulation and extra liberating pills attuned to the series of infections at the perfect goal and price. 2D substances have a higher theatrical status in many medical and technological disciplines owing to their vast, highly sought-after houses of elegant drug-carrying nanostructures. In this insolvency, we have summarized the improvement of emphasize several types of 2D substance drug supplier 2D substance types that can be extensively explored in elegant drug marketing frameworks, as well as the categorization of stimulus and reimbursement-responsive 2D substances and dangers in their packaging. Therefore, we rely on this insolvency. It is possible to stimulate a boom in new 2D substances with elegant drug-delivery structures and increase the affinity of scientists for elegant nanocarriers, mainly based on 2D substances. Smart Drug Transportation Structures (SDDS) are a type of drug delivery device that can transmit signals and respond to drug shipment and automatic rejection prevention. In concrete terms, all forms of intelligent drug delivery intend to deliver drugs in the right amount at the right time and in the right place. Control indicators of SDDS are not only incomplete to indoor nature, such as pH, redox, attention to particular biomolecules and enzymatic activity, but also incorporate outdoor nature, such as a mild range of wavelengths, magnetic fields, field strength, and ultrasound. Carriers are an idea of the SDDS. The stylish providers incorporate polymers, hydrogels, liposomes, micelles, nanoparticles, nano-sheets, etc.; 2D materials by-skip on to the reality that the dimensions of the cloth in a single measurement are decreased to the restriction of the thickness of the atomic layer, while in the different 2D the dimensions of the cloth are quite massive. In 2004, Novoselov *et al.* [1], suggested that graphene can be acquired by exfoliating graphite using a scotch tape. Hence has been proven to have sole and tremendous electric homes, because the 2D nanomaterials represented *via* way of graphene comprise hastily evolved, with novel 2D substances have emerged. In addition to 2D graphene, the circle of related substances also generally consists of transition steel dichalcogenides, MoS_2, WSe_2, TiS_2, SnS, single-element graphene, selenium, boron, tellurium, bismuth, and black phosphorus, etc., the main compounds of steel sulfide GaS, InSe, SnS, SnS_2, and various 2D substances [1]. These 2D substances also include a wide range of power-band systems, electric houses, photoelectronics, water separation, and force accumulation. Additionally, they present interesting focal points in medicine that can offer new opportunities for SDDS. SDDS illustrates a broad approach to this discipline. Pharmaceutical software, extensive research content, and various drug delivery structures can be quantitatively released with hard and fast factors over time. However, research on this type of drug has been the most advanced to date. However, transport structures are at the experimental stage, and there are many problems with their use in scientific practice. As cancer experts, researchers have found that conventional nanodrug delivery devices have glaring flaws. In short, the reduced drug burden severely limits multimodal drug software, and drugs, such as chemotherapy and immunotherapy, are limited to treating diseases. It strongly inhibits tumor accumulation and metastasis and achieves specific healing and cancer visualization. 2D substances have excellent photodynamic soft houses and green thermal conversion and retain a variety of clinical packaging benefits. This opens up great opportunities for them in clinical areas, including sensing, imaging, and healing. 2D substances exhibit unique drug delivery capabilities with multiple benefits. Two-dimensional nanomaterials have a specific critical layer system that provides a high surface area for medicine packaging. This is a key feature specific to 2D nanomaterial structures, leaving huge space for a large number of green drugs. In addition, graphene derivatives contain a single sp2 carbon bond with long-range soil contacts that can interact strongly with pills through hydrophobic interactions and supramolecular stacking. The 2D substances provide the corresponding capabilities for SDDS. This chapter highlights the contemporary improvement of SDDS for a wide range of fully intelligent hydrogels, liposomes, and nanocarriers, such as polymers, nanoparticles, nanosheets, and micelles in response to two-dimensional stimuli substances, such as 2D SDDS fabrics. Lastly, the blessings and downsides in 2D substances as SDDS are additionally represented in Figure 9.1. The diagrammatic representation of two-dimensional substances is smart drug release by multi-modal imaging.

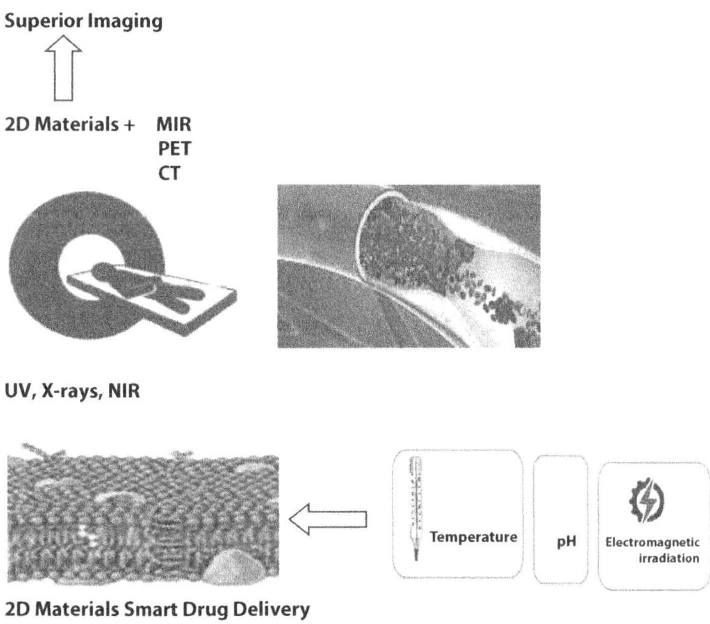

Figure 9.1 The diagrammatic representation of two-dimensional substances is smart drug release by multi-modal imaging.

9.2 2D Material Biosynthesis

Novoselo *et al.* [1] micromechanically exfoliated single-layer graphene. Many 2D material fabrication methods have been developed. Drug loading platforms can no longer use grounding methods. Micromechanical slough for· two-dimensional materials. The ablation of 2D micromechanical substances with long side lengths, low flaws, high purity, high surface area, and high crystallinity is excellent for precision packaging, photodetectors, and discipline transistors. Novoselov *et al.* [1]. Micromechanical fission creates gaseous monatomic graphene from cautious pyrolytic graphite by introducing 2D substances. The resistance converts the graphite support into a worn ribbon, creating cloth/graphite. Mechanical pressure breaks the brittle van der Waals pressure between the layers, strengthening the covalent connections. However, the efficiency was modest, and the small radial layers of graphene emotionally bonded into the ribbon were ultrasonically dissolved in acetone and sealed with the solution using a silicon plate. Preferably, the graph is much smaller than 10 nm, likely owing to van der Waals forces, and capillary forces can attach to the topmost silicon wafer and remove layers of maximum thickness by ultrasonically assisted acetone scrubbing to detect optical interference. The researcher can easily distinguish between thicker and much smaller graphs (1.5 nm), the former being visible under the light microscope and the latter not. Micromechanical peeling has revealed several new body locations and an unanticipated whole for delightful 2D substances. The yields and radii were low. Industrialization is limited by the length of the structured monolayer nanoparticles. Huang *et al.* [2] developed an innovative micromechanical peeling technology that is 20–60 times more efficient than the current process. Silicon wafers are used to divide the 2D texture from the dense volumetric texture generated by high-tech tape, unlike its initial passive application. To brighten the substrate and increase the van der Waals pressure between the silicon wafers and the outermost layer of the graphite film at the bottom of the strip, oxygen plasma is used to etch the substrate at 100 °C for 2–5 min. After removing a significant layer of appropriately aligned pyrolytic graphite to achieve the graphene diameter, the ribbon was best treated three or four times. Despite certain breakthroughs, some manual processes, and lack of relevant system resources, micromechanical peeling technology has always caused product randomization and limited efficacy when convenience exists. Thus, the liquid scratch method is the most effective approach for evaluating drug carriers made of 2D materials. Liquid peeling produces large-scale 2D materials. Micromechanical exfoliation produces tiny 2D materials that are suitable for biological applications, but not electrical ones. This method's 2D materials can be manipulated and duplicated by rotating the liquid waterfall, making it ideal for medical applications. Liquid peeling involves three steps. First, the force required to be pulled off was lowered. Many methods have worked better in creating this chart. Dispersing most of the material in a suitable solvent and breaking the van der Waals stresses of the distinct layers of the substance with the initial force of the solvent are the most typical approaches to creating this pattern. The remover shields the black phosphorous side of the 2D materials and provides a better peeling front end. Except for isopropyl alcohol and N-methyl-2-pyrrolidone, most solvents are hazardous and require more than two distilled water washes before they can be employed as drug delivery platforms. Fortunately, a few biocompatible solvents have been used to produce 2D materials in recent years. Tao *et al.* [3] used oxygen-free water to create a black phosphorus nanoparticle with an average hardness of 120 nm and a thickness of 1–2 nm, which has shown promising outcomes in cancer therapy. The active ingredients and a variety of properties confirm the best superiority of the following nanoparticles by Xue *et al.* [4], using ethanol as a solvent to radially exfoliate antimony at around 400 nm, around 3 nm wide, with incredible detectability of miRNAs. In addition to vegetable solvents, combined solvents or solvents for the production of 2D materials are also known. Surfactants are a common addition to beautifying scrubs overall performance of materials in aqueous solution. The problem of actually removing the surfactant from the substrate of another 2D substance limits the application of this generation in many traditional fields. However, the surfactant remains at the bottom of the 2D material, further enhancing the natural-looking results in the problem area. Wang *et al.* [5] accurately create 2D black phosphorus using human serum albumin as a surfactant, which increases nanoparticle dispersion, stability, and drug loading. Common solvent-based peel-strength reduction approach. Aided oxidation is better for graphene. A powerful oxidizing chemical creates additional oxygen-containing groups on each graphite layer substrate to increase the distance between the layers. Van der Waals notes that sonication weakens graphene oxide monolayers and produces large nanoparticles. Graphene oxide (GO) and reduced GO (rGO) substrates have various oxygen-containing groups, giving the drug a strong charge, modification, and hydrophilicity. Oxygen-containing groups increase the uncertainty of GO toxicity, which must be resolved before medicinal use. This borophene is closer to graphene in the periodic table owing to its cutting-edge file. A highly-oxidizable liquid cleanse can also be used. Borophen is the lightest 2D substance. However, with a much smaller surface layer, it is very difficult to cut thinner oxidized borophene using the classic peeling approach of Ji *et al.* [6] to generate a boron film thicker than 15 nm in the soil at a maximum temperature of 650 °C to convert it to B2O3. Dispersing the B2O3-coated boron film in water dissolved the B2O3 and exposed finer borophene with an average amplitude considerably below 100 nm and 5 nm. Selective etching removes liquid components to form 2D materials in addition to oxidation. Second, sufficient tensile stress must break the interlayer stress of the substance. Sonication and shear are the most frequent zoning procedures for separating fluid sections. Sonication takes less time, whereas shear requires a large manufacturing scale and a large amount of radial time. Liquid bubble cavitation, with crackling sounds and powerful surprise waves, can damage the middle layer of the 2D material chip collecting material. Contrary to popular belief, liquid sections, the path to their anisotropic chain form, and mild interchain van der Waals forces can be exfoliated to create 2D nanoparticles from unstratified materials, such as tellurium and selenium. Sonication can generate faults that are hazardous in electrically powered applications but advantageous in nanoscience emitter applications because of their increased photothermal performance. Because of their rapid speed, sonication-induced shear forces are unsuitable for use as drug transporters. Shear-assisted liquid peeling outperforms other special methods as an efficient and appealing manufacturing process. A rotor-stator mixer applies shear pressure to the material substrate, causing extensive delamination of the 2D material. This process produces several hundred liters of 2D material dispersions. Finally, the wiring is comparable. A major step toward durable 2D materials. Owing to its proximity to 2D assets and toxicity, the Cascading Liquid Vortex is ideal for natural benefits.

9.3 Encapsulation of 2D Materials

Nanotechnology in prescription medication delivery systems is of interest to researchers studying drug carrier-derived nanoscience. Nanoparticles (NPs) increase medication stability throughout blood transport and cell absorption when cells enter the process, improving drug biocompatibility. Smart NPs, which undergo substantial structural changes in their physical/chemical home in response to frequent environmental signals, also contribute to synergy and repression by evolving drug recognition in tissues or target cells and expanding. Active chemicals decrease natural tissue or cell release. 2D assets excel in the medication distribution because of their proximity to the ground. 2D materials can form conglomerates with molecules that are too large under physiological conditions, limiting their scientific usefulness. Scientists have used 2D tissue encapsulation to increase biocompatibility and dispersion. This section covers 2D asset-rendering techniques. 2D material encapsulation polymer SDDS breakthroughs have allowed 2D polymers to deliver medications at the proper time and location for function and structure, integrating drug delivery and rejuvenation demands. Researchers have employed 2D polymer homes to govern massive physical inversions or chemical variations, in response to tiny environmental changes, such as temperature, light, and pH. The sophisticated Hummers' method refines natural graphite into GO. GO was amino-functionalized using adipic acid dihydrazide. An amide bond between sodium alginate (SA) and GO collected the excess GO-SA. Polymers release large amounts of sub-physiological drugs into the tumor cell microenvironment. The decrease in pH increases the release rate. Li Tan *et al.* [7] also employed a PEGylated nanographene oxide-polyethyleneimine (PEI, 25 kDa) solution at 80 °C for 2 h. Thus, they successfully synthesized a nanographene oxide (nrGO) conjugate (PEGnrGO–PEI) with a gene delivery efficiency of 25 min irradiation (2 W/cm^2) with an 808 nm laser, increasing the transfection efficiency of the PEG-nGO/PEI gene but decreasing that of PEG–PEI by prolonging the transfection of the nrGO gene. PEGylated, water-soluble, and biocompatible black phosphorus (PB) nanoparticles were prepared using a solvent-free, one-pot mechanical milling technology with disproportionate energy because BP peak nanoparticles have the best graph-balance imaging and can effectively convert near-infrared (NIR) to heat, making them useful in thermal imaging studies for definitive cancer treatment and acoustic imaging. Yang *et al.* [8] observed that polyvinylpyrrolidone-coated MoS2 nanoparticles had excessive drug-loading capacity and a surprise synergistic impact in tumor chemotherapy and thermography. The porous MoS2 nanoparticles released the medication in a dual reaction regulated by NIR and pH, affecting the drug loading rate. Drug saturation of 7%–72% is adjustable. According to research, 2D textile polymers like SDDS may limit cancer development and reduce pill toxicity. The polymers of the 2D material may oppose absorption, assault tumor tissue with the most tumor cells, and swiftly release charges in response to stimulation.

9.4 Hydrogel Encapsulation—2D Materials

Hydrogels are three-dimensional hydrophilic polymer networks in water. Although they are water rich, they are insoluble. Their extracellular matrices are biocompatible. Precision 2D materials can be used to make hydrogels from physical, chemical, and natural shells. These adjustments can also cause chemical and physical changes or particular qualities of the materials by interacting with moderate or strong pH, temperature, and stimulation. Hydrogel machines and nanotubes are cold drug nanocarriers that can improve drug loading and delivery. A heat-sensitive and deformable hydrogel can regulate the release rate or enforce substrate-altering characteristics of polymer flakes to improve clinical treatment accuracy. NP-impregnated GO nanoparticles were produced *in situ* from hydroquinone-adsorbed Ag+ in a citrate buffer solution. The composite was found to be highly antibacterial. Exposure to intense sulfuric acid, dextrin, and cetyltrimethylammonium bromide (CTAB) produced graphene-carbon nanoplates. These leaves are oxidized to GO-like resources. Carbon nanoparticles did not affect normal cells but induced apoptosis in cancer cells. Li and Xianget *et al.* produced SP-MoS2 nanoparticles using a rotary evaporator. SP-MoS2 nanoparticles showed promise for photothermal breast cancer and live cell therapy *in vitro*. Smart 2D NIR photothermal materials can enter tissues, providing them miraculous biological advantages. These complex hydrogels use the photothermal conversion efficiency of polydopamine-modified black phosphorus to achieve molecular interaction, biocompatibility, and NIR drug responsiveness. Recently, 2D assets have been generated to fulfil Zhu's ultimate cancer therapy promise. Xiali *et al.* [9] prepared a thermo-touch and biodegradable hydrogel stream. Chitosan with β-glycerophosphate formed a hydrogel. Induction magnetotherapy with modified superparamagnetic PEI-GO minimally treated the maximum number of malignant lesions. All GO-based SDDS have a hydrogel-based Adriamycin (DOX) drug delivery mechanism. GO outperformed free DOX on MCF-7 cells *in vitro*, while Zhang *et al.* [10] created an advanced GO-based supramolecular nanohydrogel using non-covalent bonding to integrate GO as a cross-linking agent. The complex hydrogel model, doxorubicin hydrochloride, has been uploaded. Hydrogels inhibit molecular tumor proliferation and development *in vitro* and *in vivo*.

9.5 2D Material Encapsulation—Liposomes

Amphipathic phospholipid-based liposomes carry drugs. Molecular membranes depend on phospholipids. It has a phospholipid-based hydrophilic head and fatty acid-based hydrophobic tail. Phospholipids can spontaneously transfer mass in an aqueous environment, with the polar ends flowing through the water and non-polar ends, creating a bilayer. A double-layered box that dissolves in water or pulls medication into the water has long provided resources to the core. Liposomes are ideal medicinal carriers owing to their enormous capacity, low toxicity, biocompatibility, biodegradability, and immunogenicity. MoS2, WS2, and graphene interact with the liposomes. The 2D materials were then fully scanned. Recently, Sahu *et al.* [11] developed liposomes containing shielded graphene nano-oxide. NIR light activates liposome drug release through nrGO. Pulsed NIR lasers can intelligently regulate drug delivery in nrGO liposome suspensions. Thin-layer hydration also produced nanoliposomes. Graphene and Dox nanoparticles were introduced to liposomes at various lipid film hydrations. Also, Liu *et al.* [12] DOPC liposome-transition metal digenide hybrid materials. WS2 DOPC and DOX liposomes were adsorbed synergistically. After DOX adsorption onto WS2, the adsorption capability of the liposomes improved to 110%. DOPC alone adsorbed 1/4 of WS2. Hydrogen bonding helped the WS2-loaded liposomes bind to DOX.

9.6 2D Supply Encapsulation—Micelle

Polymeric micelles are amphiphilic drug carriers with hydrophilic and hydrophobic properties and accurate solvent self-assembly. It is two parts. Hydrophobic and hydrophilic coatings are options. Polymeric micelles are amphiphilic "intermediate structures." Hydrophobic agents transport drugs, whereas hydrophilic coatings improve the stability of drug delivery systems. Han *et al.* [13] claimed that prominent diaper firms have diaper liner resources. These movies show GO, branched 2D-shaped polyethyleneimine (bPEI) micelles, and amphiphilic polystyrene-b-poly(acrylic acid) block copolymers (BCMs). Every tissue is pH dependent; hence, this finding is significant. GO/BCM sheets released drugs quicker than (bPEI/BCM) sheets in PBS pH 2 and pH 7.4 buffers. bPEI/BCM multilayer GO overlay resources can also handle rapid drug release.

9.7 Stimuli Responsive 2D Material SDDSs—Classification

Owing to the unique quantum confinement effect and numerous energetic chemical domains on the surface, 2D materials are sensitive to certain external stimuli. Therefore, drug delivery devices based entirely on 2D materials can be used for image-guided reactive drug delivery. In this section, we present the responses of 2D materials to light, magnetic fields, and pH.

9.8 Light-Sensitive SDDSs

Small stimuli are often utilized in 2D stimuli-sensitive materials because they have realistic benefits and outcomes that are straightforward to evaluate *in vivo* and *in vitro*, and whether visible light and NIR are effective energy sources for SDDS applications. The NIR spectrum is safer and penetrates deeper tissues than the visible light. Biomedicine uses a 700 nm–1,000 nm NIR spectrum. Pharmacies that convert photons into heat using NIR-absorbing plasmonic materials may produce NIR-sensitive nanostructures. NIR radiation therapy penetrates deeper and causes less tissue damage than the conventional radiation therapy. A multifunctional anticancer drug delivery system based on GO has been developed for trimodal synergistic chemotherapy/PDT/PTT with poor NIR control. Under NIR laser radiation, the inexperienced Vedelolactone GO Indocyanine absorbs optical electricity and transforms it into heat, killing cancer cells with reactive oxygen species. Fan *et al.* [14] found that the unusual nanomedical sandwich medication with poor NIR sensitivity contains the NO donor BNN6 (N,N'-di-sec-butyl-N,N'-dinitroso-1,4-phenylenediamine). Nanomedicine has an unaltered NIR response, thermal stability, and drug loading using self-fulfilling GO Nanocard GO-BNN6. Polyurethane, as the thermosensitive fabric, and BP, as the near-infrared photothermal filler were used to create a state-of-the-art shape-memory polymer composite with the best shape memory, performance, and biodegradability. Far-NIR monitors tissue degradation following *in vivo* use. Nanodecorated sheets of BP-Ag nanoparticles can swiftly disrupt the bacterial membrane when exposed to NIR radiation. In mice, Ag@BP reduced bacterial load and tissue injury. In clinical applications, Ag@BP signifies optimum biocompatibility and bioprotection. PDLLA-PEG-PDLLA:PLEL was combined with BP nanoparticles. The BP@PLEL hydrogel rapidly transformed from a gel to an NIR-capable solution with the best-in-class NIR photothermal performance.

9.9 Magnetic Field-Responsive SDDSs

The high tissue penetration and spatial resolution of magnetic resonance imaging (MRI) make it popular in clinics. Excellent imaging requires MRI assessment. Traditional MRI readouts are not tumor-specific, making it difficult to differentiate between target and signal recordings to remedy this problem. Researchers have developed a fashionable MRI approach using 2D MnO2 nanoparticles. In 2014, Zhao *et al.* [15] reported tumor-specific MRI signals from MnO nanoplates with a period of 141 nm and width of 1.5 nm.

MnO_2 nanoplates can target more tumor cells using aptamers. After endocytosis of MnO_2 nanoplatelets in apical tumor cells, intracellular GSH lowered MnO_2 to Mn^{2+} and showed consistent MRI signals, whereas normal tissues showed no MRI signals. The reduced products had longitudinal (1/T1) and transverse (1/T2) relaxation factors 48 times higher than those of MnO_2 nanotubes. MRI is used in clinics to produce stunning images because of its high tissue penetration and spatial resolution. MRI evaluation requires support. Traditional MRI scanning lab signals are not tumor-specific, making them difficult to target and acquire. Scientists have developed a popular MRI approach using 2D MnO_2 nanoparticles to address this challenge. Zhao *et al.* [15] created tumor-specific MRI probes from MnO nanoplates with a period of 141 nm and a width of 1.5 nm in 2014. Aptamers help MnO_2 nanoplates to target most cancer cells. After tumor cells endocytose the MnO_2 nanoplates, intracellular GSH converts MnO_2 ions to Mn^{2+} and produces consistent MRI signals. In normal tissues, MnO_2 has no MRI controls. The condensed products had 48- and 120-fold higher longitudinal (1/T1) and transverse (1/T2) relaxation rates than MnO_2 nanotubes, giving MRI evaluation laboratories a huge nanosheet capacity.

9.10 Various Response Exhibits Diverse—Advantages/Disadvantages

Phototherapy works even at low tissue concentrations because laser light is directed. Laser penetration limits the application of clinical software in 2D tissue phototherapy. Magnetic fields penetrate tissues deeper than lasers. MRI, which is mostly based on the magnetic field response, has been thoroughly explored and has mature clinical applications; however, as much attention has been paid to the materials necessary, magnetic heat therapy has not been studied as much as photothermal therapy [16]. Drug delivery has focused on acidic pH in the tumor microenvironment. It can dramatically alter tumor drug release, tissue degradation, and drug targeting in acidic environments. According to various studies, cancer cell pH is independent and may only be alkaline. Remediation may disrupt pH-sensitive medication delivery pathways in cells.

9.11 2D Material SDDS Therapy—Cancer

Cancer is one of the major diseases that threatens human health and life. Surgery, chemotherapy, and radiotherapy have several disadvantages including extreme outcomes, medication resistance, and low efficacy.

PTT, PDT, and other maximal cancer treatment strategies have been developed to overcome these difficulties. Nanotechnology and nanoscience have advanced through decades of astounding advances and extraordinary promises for cancer treatment. Biomedicine research has focused on graphene, carbon nanotubes, magnetic iron oxides, quantum dots, scalable nanoparticles, and gold nanostructures. Recent nanostructures for SDDS have made 2D materials a hot topic owing to their chemical characteristics [17]. *In vivo* nanoscale cells that transport drugs are one of the finest anticancer techniques. Chemotherapy failure requires multidrug resistance in tumors. Thus, to maximize cancer cell-targeting, drug delivery, and multidrug resistance, pharmaceuticals must be biocompatible. The target nanohybrids no longer had the best anticancer effects, yet animals with B16 tumors had little systemic toxicity. Yu *et al.* [18] produced a reduced GO nanocomposite using a particular peptide-linked polydopamine (PDA). AAP10 was placed on a dopamine-modified rGO substrate. AAP10-pDA/rGO could remove breast cancers in mice using NIR irradiation and almost completely prevent tumor growth. Gold nanoparticles on GO nanoparticles were reduced to folate-modified dendrites and protected by mesoporous silica, forming a unique nanocomposite molecular sandwich imaging device and multifunctional smart nano-delivery agent for concentrated curcumin (Cur). This new drug delivery technology has large surface area, biodegradability, maximal biocompatibility, and moderate drug delivery. The nano-provider evaluated pH-sensitive homes and retarders. Today's medication delivery method can almost completely treat breast cancer with an effective anticancer agent. The inherent benefit of naked BP is its direct oxidation or breakdown and deposition in the tumor microenvironment, which can cause uneven photothermal impact and short-term scarring, according to Xing *et al.* Nanoscale BP and cellulose membranes based on fully injectable hydrogels exhibited the greatest PTT antitumor performance. High-quality nanoscale hydrogel platforms are biocompatible and non-toxic *in vivo* and *in vitro*.

9.12 Antibacterial

9.12.1 Central Nervous System

Since the invention of microbes in the 19th century, small pathogens have fought mankind. Antibiotic overuse and the growth of antibiotic-resistant bacteria for infectious disease therapy are important issues owing to the sluggish development of antibiotic indications. However, harmful microbial contamination poses a risk to public health. Therefore, the development of new antibacterial drugs is crucial. 2D biomaterials have gained considerable attention. Antifungal and antibacterial 2D materials were also present. Microwave radiation and seashell-inspired chemistry produced MoS2-PDA-Ag nanocomposites by Zeng *et al.* [19]. *In situ* microwave irradiation transparently transformed the MoS2-PDA-based PDA sheets for self-polymerization of the aqueous-alkaline dopamine-Ag nanoparticle solution. The complex is more bactericidal than PDA-MoS2 and plant MoS2. Arginine (Arg) in the GO/Ag nanostructure substrate produced electrospun polycaprolactone (PCL) nanocomposites. Nanocomposite (PCL)-GO/Ag/Arg has significant antibacterial activity against *Staphylococcus aureus* (*S. aureus*) and *Escherichia coli* (*E. coli*) and is biocompatible with L929 fibroblast cells; however, it is not practical. Athinarayanan *et al.* [20] produce Cu2O/graphene nanocomposites using a simpler and greener chemical method. Cu2O/G nanocomposites inhibited gram-positive and gram-negative bacteria at 300 μg/g, outperforming GO and Cu2O nanoparticles. Zhang *et al.* [21] presented a photon burst platform using BP quantum dots and liposome-encapsulated heat-sensitive pills to destroy drug-resistant bacteria and avoid modern bacterial resistance. The antimicrobial platform can deliver medications quickly, geographically, and dose-controlledly because BP quantum dots create heat to rupture liposomes upon NIR activation. These new SDDS minimize antibiotic doses, superbugs, and antibiotic overuse. Brain illness is a unique public health issue. Inflammatory thinking; stroke; and psychological, neurodegenerative, and neurodevelopmental brain issues have been treated. Therapeutics with high hydrophilic indices or molecular weights can be blocked by the BBB. Pills that address mental difficulties can cause BBB. GQDs' moderate hydrophobicity and mild toxicity made them carriers. Alzheimer's disease is associated with β-peptide aggregation (AD). Inhibition of peptide aggregation is one of the best treatments for AD. Small GQD can displace the BBB and inhibit β-peptide aggregation. GQDs compete with AD therapies. Go(120) had a higher effective contact mass and better inhibition than Go(60) [21]. GQDs inhibit Lewy neurites and body development and restore synapse loss and neuronal death. Thus, GQDs may worsen mitochondrial dysfunction. GQDs cross the BBB and suppress Parkinson's disease α-synuclein (Alphasyn) clustering. GQDs are also safe for treating Parkinson's disease *in vitro* and *in vivo*. Direct conversion into nerve cells without pluripotency is favored for the treatment of nervous system illnesses. Reset. In a mouse model, polyethylenimine oxide complexes delivered reprogramming factors to stimulate brain synthesis and alleviate Parkinson's disease symptoms.

9.13 Orthopedic

2D materials have unique physical, chemical, natural, bone repair, and drug delivery capabilities. Graphene promoted stem cell proliferation and osteogenic differentiation. Thus, it has an excellent orthopedic biomaterial sensitivity. You can examine and specialize in scaffolding components for bone tissue engineering, repair, substitute materials, and so on. Saravanan *et al.* [22] developed an injectable thermo-touch chitosan hydrogel containing GO to upregulate Runt-related transcription factor 2, collagen type 1, alkaline phosphatase, and osteocalcin, which facilitated mouse mesenchymal stem cell osteogenic differentiation under osteogenic conditions. Grapheme Nanosheet (GNS) was also made. The CaP-CS-based Micro-Arc AZ91D was extensively oxidized. Early GNS-CaP-CS/AZ91D extracts may require substantial calcium mineral deposits, alkaline phosphatase activity, and osteoblast-related gene expression in human bone marrow mesenchymal stem cells [23–26]. Zhao *et al.* lyophilized a GO/SF barrier film filled with SIM. Experiments on damaged Sprague–Dawley rat skulls showed that GO/SF/SIM has a higher potential for some of the four membranes (SF, GO/SF, GO/SF/SIM, and SF/SIM) and minor bone repair *in vitro* biocompatibility. Zhang *et al.* [27] extracted numerous layers of MoS2 nanoflakes from the substrate using a hydrothermal technique to create a nanostructured biointerface. The MoS2 interface with the nonporous form boosted MSC adhesion and proliferation without accelerating osteogenesis.

9.14 Diabetes Mellitus

Human islet amyloid aggregation polypeptide (hIAPP) is associated with insulin insufficiency and pancreatic beta cell death in type 2 diabetes. These new nanoparticles inhibited platelet aggregation. GO reduces and inhibits hIAPP aggregation. The strong GO-PEI combination also became a covalently bound PEI activator in GO. GO-PEI inhibited hIAPP fibril production better than GO. As validated by Zhou *et al.* [28], nanocomposites of nGO and PEG (EALYLV) modified insulin-derived peptides that inhibit hIAPP aggregation, protect INS-1 cells from toxicity, stabilize mitochondrial membrane potential, and minimize intracellular reactive oxygen species. Lee *et al.* [29] noted that gold particles and graphene are connected to a gold network to create the flexible, translucent patch. The patch had humidity, glucose, pH, and temperature sensors. When the patch regulates excessive sweat sugar levels, a heat-activated mechanism releases metformin to reduce the blood sugar levels.

9.15 2D Materials in Intelligent Drug Delivery System—Advantages

2D materials are more tunable than polymeric drug delivery platforms because they respond to environmental stimuli. The increased surface area of 2D materials allows fluorescent dyes to efficiently target regenerative molecules and medicines efficiently. Dual phototherapy is possible using several 2D materials. The BP-PDA-coated nanoparticles formed a nanopill, as reported by Gao *et al.* [30]. Then, the nanoparticle was primed with temercapto-PEG-folic acid and adriamycin. BP-based absolute nanoparticles exhibit increased photothermal conversion efficiency, biocompatibility, and maximum cancer cell counts. Under objective conditions, GQDs can load up to 50% cisplatin and reduce its systemic toxicity. It was slowly released in a pH-dependent manner. Cell tests revealed modest cytotoxicity, non-moderate selective absorption by tumor cells, and real tumor concentrations. In MDA-MB-231 cells, cisplatin-loaded nanocarriers are substantially more hazardous. The 2D GO nanoparticles with a thermosensitive matrix created hybrid nanogels with photothermal effects and good stability. GO nanoplates in nanogels increased the encapsulation of anticancer medication. The photothermal effects were considerably larger.

9.16 Disadvantages

2D nanomaterials can harm tissues, organs, cells, and biomolecules. Unlike other nanocarriers, ultra-weather resistant GNS medium is solid carbon. GNS benefits from non-moderate drug loading. However, the impressive place has a drawback. GNS controls blood component interactions and accelerates serum protein and scarlet blood cell production *in vivo*. GNS targets light cells for investigation. Graphene detonates intracellular ROS and disrupts mitochondrial membrane function for targeted cytotoxicity by activating β-boom-processing signaling pathways and activating mitogen-activated protein kinases. Graphene derivatives hemolyze dose-dependently. Unmodified graphene can induce apoptosis in lung epithelial cells *via* visual receptors. Graphics activate IP3 to release $Ca2+$. Progressive cystic $Ca2+$ increases $Ca2+$-mediated apoptosis by slow release; several layers of thick graphene, high dosages, and prolonged GO exposure can cause mitochondrial malfunction. GO and carboxygraphs govern mitochondrial shape and function according to Lamel *et al.* [31–40]. Respirable graphene can cause substantial lung damage. Graphene accumulates in the lungs of mice exposed to 10 mg/kg OB over 14 days, causing granulomatous alterations, pulmonary edema, inflammatory cell infiltration, and fibrosis [41–55].

9.17 Conclusion and Future Perspective

Improvements in 2D steel nanomaterials have increased dramatically over the past few decades, leveraging advances in many characterization tools and the growing name of target technology. In this chapter, we summarize the development of 2D steel nanomaterial research so far, from synthesis techniques to objects search for software exploration compared to 0D and 1D steel nanostructures; 2D steel nanomaterials are extremely difficult to obtain and excessive because of the inherently isotropic crystalline nature of metal floor strength in 2D morphology. However, many artificial techniques have been developed to provide reasonably stable two-dimensional metallic nanostructures with well-controlled parameters such as shape, length, composition, and crystal structure. For example, the shape and length of two-dimensional metal nanostructures can also be in the shape of a disc, triangle, square, hexagon, ring, cone, ribbon, ribbon, etc., at the same time, their thickness can vary from single atoms to masses of nanometers, and their transverse dimensions range from tens of nanometers to masses of micrometers. Many heterostructures, alloys, and 2D assemblies based entirely on steel nanomaterials have been fabricated efficiently. To observe the different mechanisms involved in creating a 2D morphology, we classify artificial techniques into bottom-up and top-down tactics and explain the evidence for the different techniques within each class. Although research on 2D physical and chemical houses of steel nanomaterials remains primitive, we rank the best-studied houses, including stability, crystal structure, plasmonic houses, magnetic houses, and electric houses, because one in each house of 2D steel nanomaterials with 2D properties has been well completed by comparing their normal performance in an entire group of applications. The excellent normal performance of 2D steel nanomaterials has proven to be a promising candidate for nanomaterials in new technologies.

Acknowledgements

We are thankful to the management and principal of C. Abdul Hakeem College (Autonomous) Melvisharam, Ranipet District, Tamil Nadu and to Dr. Subhendu Chakroborty, Associate Professor, IES University, Bhopal, India, for their encouragement, providing the necessary facilities and support in carrying out the work.

References

1. Novoselov, K.S., Geim, A.K., Morozov, S.V., Jiang, D., Zhang, Y., Dubonos, S.V., Grigorieva, I.V., Firsov, A.A., Electric field effect in atomically thin carbon films. *Science*, 306, 5696, 666–669, 2004.
2. Huang, Y., Sutter, E., Shi, N.N., Zheng, J., Yang, T., Englund, D., Gao, H.-J., Sutter, P., Reliable exfoliation of large-area high-quality flakes of graphene and other twodimensional materials. *ACS Nano*, 9, 11, 10612–10620, 2015.
3. Tao, W., Zhu, X., Yu, X., Zeng, X., Xiao, Q., Zhang, X., Ji, X., Wang, X., Shi, J., Zhang, H., Mei, L., Black phosphorus nanosheets as a robust delivery platform for cancer theranostics. *Adv. Mater.*, 29, 1, 1–9, 2017.
4. Xue, T., Liang, W., Li, Y., Sun, Y., Xiang, Y., Zhang, Y., Dai, Z., Duo, Y., Wu, L., Qi, K., Shivananju, B.N., Zhang, L., Cui, X., Zhang, H., Bao, Q., Ultrasensitive detection of miRNA with an antimonene-based surface plasmon resonance sensor. *Nat. Commun.*, 10, 1, 28, 2019.
5. Wang, S., Weng, J., Fu, X., Lin, J., Fan, W., Lu, N., Qu, J., Chen, S., Wang, T., Huang, P., Black phosphorus nanosheets for mild hyperthermia-enhanced chemotherapy and chemo-photothermal combination therapy. *Nanotheranostics*, 1, 2, 208–216, 2017.
6. Ji, X., Kong, N., Wang, J., Li, W., Xiao, Y., Gan, S.T., Zhang, Y., Li, Y., Song, X., Xiong, Q., Shi, S., Li, Z., Tao, W., Zhang, H., Mei, L., Shi, J., A novel top-down synthesis of ultrathin 2D boron nanosheets for multimodal imaging-guided cancer therapy. *Adv. Mater.*, 30, 36, e1803031, 1–11, 2018.
7. Li, T., Wu, L., Zhang, J., Xi, G., Pang, Y., Wang, X., Chen, T., Hydrothermal reduction of polyethylenimine and polyethylene glycol dual-functionalized nanographene oxide for high-efficiency gene delivery. *ACS Appl. Mater. Interfaces*, 8, 45, 31311–31320, 2016.
8. Yang, H., Zhao, J., Wu, C., Ye, C., Zou, D., Wang, S., Facile synthesis of colloidal stable MoS2 nanoparticles for combined tumor therapy. *Chem. Eng. J.*, 351, 548–558, 2018.
9. Zhu, X., Zhang, H., Huang, H., Zhang, Y., Hou, L., Zhang, Z., Functionalized grapheme oxide-based thermosensitive hydrogel for magnetic hyperthermia therapy on tumors. *Nanotechnology*, 26, 36, 1–13, 2015.

10. Zhang, Q., Deng, H., Li, H., Song, K., Zeng, C., Rong, L., Preparation of grapheme oxide-based supramolecular hybrid nanohydrogel through host-guest interaction and its application in drug delivery. *J. Biomed. Nanotechnol.*, 14, 12, 2056–2065, 2018.
11. Sahu, A., Kim, M., Ryu, J., Son, J.-G., Lee, E., Noh, D.Y., Tae, G., Nanographene oxide as a switch for CW/pulsed NIR laser triggered drug release from liposomes. *Mater. Sci. Eng. C Mater. Biol. Appl.*, 82, 19–24, 2018.
12. Liu, Y. and Liu, J., Hybrid nanomaterials of WS2 or MoS2 nanosheets with liposomes: Biointerfaces and multiplexed drug delivery. *Nanoscale*, 9, 35, 13187–13194, 2017.
13. Han, U., Seo, Y., Hong, J., Effect of pH on the structure and drug release profiles of layer-by-layer assembled films containing polyelectrolyte, micelles, and grapheme oxide. *Sci. Rep.*, 6, 1–10, 2016.
14. Fan, J., He, N., He, Q., Liu, Y., Ma, Y., Fu, X., Liu, Y., Huang, P., Chen, X., A novel selfassembled sandwich nanomedicine for NIR-responsive release of NO. *Nanoscale*, 7, 47, 20055–20062, 2015.
15. Zhao, Z., Fan, H., Zhou, G., Bai, H., Liang, H., Wang, R., Zhang, X., Tan, W., Activatable fluorescence/MRI bimodal platform for tumor cell imaging via MnO2 nanosheet– aptamer nanoprobe. *J. Am. Chem. Soc.*, 136, 32, 11220–11223, 2014.
16. Yang, H., Bremner, D.H., Tao, L., Li, H., Hu, J., Zhu, L., Carboxymethyl chitosan mediated synthesis of hyaluronic acid-targeted graphene oxide for cancer drug delivery. *Carbohydr. Polym.*, 135, 72–78, 2016.
17. Xie, B., Yi, J., Peng, J., Zhang, X., Lei, L., Zhao, D., Lei, Z., Nie, H., Characterization of synergistic anti-tumor effects of doxorubicin and p53 via graphene oxide-polyethyleneimine nanocarriers. *J. Mater. Sci. Technol.*, 33, 8, 807–814, 2017.
18. Yu, J., Lin, Y.-H., Yang, L., Huang, C.-C., Chen, L., Wang, W.-C., Chen, G.-W., Yan, J., Sawettanun, S., Lin, C.-H., Improved anticancer photothermal therapy using the bystander effect enhanced by antiarrhythmic peptide conjugated dopamine modified reduced graphene oxide nanocomposite. *Adv. Healthc. Mater.*, 6, 2, 1–9, 2017.
19. Zeng, G., Huang, L., Huang, Q., Liu, M., Xu, D., Huang, H., Yang, Z., Deng, F., Zhang, X., Wei, Y., Rapid synthesis of MoS2-PDA-Ag nanocomposites as heterogeneous catalysts and antimicrobial agents via microwave irradiation. *Appl. Surf. Sci.*, 459, 588–595, 2018.
20. Athinarayanan, J., Periasamy, V.S., Krishnamoorthy, R., Alshatwi, A.A., Evaluation of antibacterial and cytotoxic properties of green synthesized Cu2O/Graphene nanosheets. *Mater. Sci. Eng. C Mater. Biol. Appl.*, 93, 242–253, 2018.
21. Zhang, L., Wang, Y., Wang, J., Wang, Y., Chen, A., Wang, C., Mo, W., Li, Y., Yuan, Q., Zhang, Y., A photon-responsive antibacterial nanoplatform for synergistic photothermal-/pharmaco-therapy of skin infection. *ACS Appl. Mater. Interfaces*, 11, 1, 2018.

22. Saravanan, S., Vimalraj, S., Anuradha, D., Chitosan based thermoresponsive hydrogel containing graphene oxide for bone tissue repair. *Biomed. Pharmacother.*, 107, 908–917, 2018.
23. Khaleel Basha, S., Syed Muzammil, M., Dhandayuthabani, R., Sugantha Kumari, V., Development of nanoemulsion of alginate/aloe vera for oral delivery of insulin. *Mater. Today Proc.*, 36, 357–363, 2021.
24. Khaleel Basha, S., Syed Muzammil, M., Dhandayuthabani, R., Sugantha Kumari, V., Polysaccharides as excipient in drug delivery system. *Mater. Today Proc.*, 36, 280–289, 2021.
25. Khaleel Basha, S., Syed Muzammil, M., Dhandayuthabani, R., Sugantha Kumari, V., Nanoemulsion as oral drug delivery-a review. *Curr. Drug Res. Rev. Bentham Sci. Publishers*, 36, 12, 4–15, 2020.
26. Khaleel Basha, S., Dhandayuthabani, R., Syed Muzammil, M., Sugantha Kumari, V., Solid lipid nanoparticles for oral drug delivery. *Mater. Today Proc.*, 36, 313–324, 2021.
27. Zhang, X., Nie, J., Yang, X., Liu, Z., Guo, W., Qiu, J., Wang, S., Yu, X., Guan, Y., Liu, H., Li, L., Nanostructured molybdenum disulfide biointerface for adhesion and osteogenic differentiation of mesenchymal stem cells. *Appl. Mater. Today*, 10, 164–172, 2018.
28. Zhou, X., Cao, C., Chen, Q., Yu, Q., Liu, Y., Yin, T., Liu, J., PEG modified grapheme oxide loaded with EALYLV peptides for inhibiting the aggregation of hIAPP associated with type-2 diabetes. *J. Mater. Chem. B*, 3, 35, 7055–7067, 2015.
29. Lee, H., Choi, T.K., Lee, Y.B., Cho, H.R., Ghaffari, R., Wang, L., Choi, H.J., Chung, T.D., Lu, N., Hyeon, T., Choi, S.H., Kim, D.-H., A graphene-based electrochemical device with thermo responsive microneedles for diabetes monitoring and therapy. *Nat. Nanotechnol.*, 11, 6, 566–572, 2016.
30. Gao, N., Nie, J., Wang, H., Xing, C., Mei, L., Xiong, W., Zeng, X., Peng, Z., A versatile platform based on black phosphorus nanosheets with enhanced stability for cancer synergistic therapy. *J. Biomed. Nanotechnol.*, 14, 11, 1883–1897, 2018.
31. Yang, H., Bremner, D.H., Tao, L., Li, H., Hu, J., Zhu, L., Carboxymethyl chitosan mediated synthesis of hyaluronic acid-targeted graphene oxide for cancer drug delivery. *Carbohydr. Polym.*, 135, 72–78, 2016.
32. Xing, C., Chen, S., Liang, X., Liu, Q., Qu, M., Zou, Q., Li, J., Tan, H., Liu, L., Fan, D., Zhang, H., Two-dimensional MXene (Ti3C2)-integrated cellulose hydrogels: Toward smart three-dimensional network nanoplatforms exhibiting light-induced swelling and bimodal photothermal/chemotherapy anticancer activity. *ACS Appl. Mater. Interfaces*, 10, 33, 27631–27643, 2018.
33. Novoselov, K.S., Geim, A.K., Morozov, S.V., Jiang, D., Zhang, Y., Dubonos, S.V., Grigorieva, I.V., Firsov, A.A., Electric field effect in atomically thin carbon films. *Science*, 306, 666, 2004.

34. Huang, Y., Sutter, E., Shi, N.N., Zheng, J., Yang, T., Englund, D., Gao, H.-J., Sutter, P., Reliable exfoliation of large-area high-quality flakes of graphene and other two dimensional materials. *ACS Nano*, 9, 11, 10612–10620, 2015.
35. Coleman, J.N., Lotya, M., O'Neill, A., Bergin, S.D., King, P.J., Khan, U., Young, K., Gaucher, A., De, S., Smith, R.J., Shvets, I.V., Arora, S.K., Stanton, G., Kim, H.Y., Lee, K., Kim, G.T., Duesberg, G.S., Hallam, T., Boland, J.J., Wang, J.J., Donegan, J.F., Grunlan, J.C., Moriarty, G., Shmeliov, A., Nicholls, R.J., Perkins, J.M., Grieveson, E.M., Theuwissen, K., McComb, D.W., Nellist, P.D., Nicolosi, V., Two-dimensional nanosheets produced by liquid exfoliation of layered materials. *Science*, 331, 6017, 568–571, 2011.
36. Tao, W., Zhu, X., Yu, X., Zeng, X., Xiao, Q., Zhang, X., Ji, X., Wang, X., Shi, J., Zhang, H., Mei, L., Black phosphorus nanosheets as a robust delivery platform for cancer theranostics. *Adv. Mater.*, 29, 1, 1–9, 2017.
37. Xue, T., Liang, W., Li, Y., Sun, Y., Xiang, Y., Zhang, Y., Dai, Z., Duo, Y., Wu, L., Qi, K., Shivananju, B.N., Zhang, L., Cui, X., Zhang, H., Bao, Q., Ultrasensitive detection of miRNA with an antimonene-based surface plasmon resonance sensor. *Nat. Commun.*, 10, 1, 28, 2019.
38. Guo, Z., Chen, S., Wang, Z., Yang, Z., Liu, F., Xu, Y., Wang, J., Yi, Y., Zhang, H., Liao, L., Chu, P.K., Yu, X.F., Metal-ion-modified black phosphorus with enhanced stability and transistor performance. *Adv. Mater.*, 29, 42, 1–8, 2017.
39. Wang, S., Weng, J., Fu, X., Lin, J., Fan, W., Lu, N., Qu, J., Chen, S., Wang, T., Huang, P., Black phosphorus nanosheets for mild hyperthermia-enhanced chemotherapy and chemo-photothermal combination therapy. *Nanotheranostics*, 1, 2, 208–216, 2017.
40. Zhu, Y., Murali, S., Cai, W., Li, X., Suk, J.W., Potts, J.R., Ruoff, R.S., Graphene and graphene oxide: Synthesis, properties, and applications. *Adv. Mater.*, 22, 35, 3906–3924, 2010.
41. Naguib, M., Kurtoglu, M., Presser, V., Lu, J., Niu, J., Heon, M., Hultman, L., Gogotsi, Y., Barsoum, M.W., Two-dimensional nanocrystals produced by exfoliation of Ti3 AlC2. *Adv. Mater.*, 23, 37, 4248–4253, 2011.
42. Backes, C., Higgins, T.M., Kelly, A., Boland, C., Harvey, A., Hanlon, D., Coleman, J.N., Guidelines for exfoliation, characterization and processing of layered materials produced by liquid exfoliation. *Chem. Mater.*, 29, 1, 243–255, 2016.
43. Zeng, X., Luo, M., Liu, G., Wang, X., Tao, W., Lin, Y., Ji, X., Nie, L., Mei, L., Polydopamine-modified black phosphorous nanocapsule with enhanced stability and photothermal performance for tumor multimodal treatments. *Adv. Sci.*, 5, 10, 1–8, 2018.
44. Hu, Y., He, L., Ma, W., Chen, L., Reduced graphene oxide-based bortezomib delivery system for photothermal chemotherapy with enhanced therapeutic efficacy. *Polym. Int.*, 67, 12, 1648–1654, 2018.
45. Zhao, M., Dai, Y., Li, X., Li, Y., Zhang, Y., Wu, H., Wen, Z., Dai, C., Evaluation of long term biocompatibility and osteogenic differentiation of graphene

nanosheet doped calcium phosphate-chitosan AZ91D composites. *Mater. Sci. Eng. C Mater. Biol. Appl.*, 90, 365–378, 2018.
46. Zhao, B., Wu, F., Bai, Y.-Y., Fang, M., Wang, L., Preparation and biological properties of a graphene oxide/silk fibroin barrier membrane loaded with simvastatin. *New Carbon Mater.*, 33, 5, 460–468, 2018.
47. Zhang, X., Nie, J., Yang, X., Liu, Z., Guo, W., Qiu, J., Wang, S., Yu, X., Guan, Y., Liu, H., Li, L., Nanostructured molybdenum disulfide biointerface for adhesion and osteogenic differentiation of mesenchymal stem cells. *Appl. Mater. Today*, 10, 164–172, 2018.
48. Li, S., Lin, D., Hu, X., Yang, X., Directly probing the dissociation effects of grapheme oxide nanosheets on hIAPP fibrils. *Nanotechnology*, 29, 49, 2018.
49. Nedumpully-Govindan, P., Gurzov, E.N., Chen, P., Pilkington, E.H., Stanley, W.J., Litwak, S.A., Davis, T.P., Ke, P.C., Ding, F., Graphene oxide inhibits hIAPP amyloid fibrillation and toxicity in insulin-producing NIT-1 cells. *Phys. Chem. Chem. Phys.*, 18, 1, 94–100, 2016.
49. Lu, T., Wang, C., Zhu, Z., Jiang, W., Huo, M., Li, F., Polyethyleneimine functionalized graphene oxide against hIAPP amyloid aggregation. *Chem. J. Chin. Univ. Chin.*, 39, 6, 1274–1280, 2018.
50. Gao, N., Nie, J., Wang, H., Xing, C., Mei, L., Xiong, W., Zeng, X., Peng, Z., A versatile platform based on black phosphorus nanosheets with enhanced stability for cancer synergistic therapy. *J. Biomed. Nanotechnol.*, 14, 11, 1883–1897, 2018.
51. Soren, S., Chakroborty, S., Pradhan, L., Chandra, P., Sahu, J., Parhi, P., Hydrothermal synthesis of graphene modified sno nanocomposite for oxygen reduction reaction. *Mater. Today Proc.*, 57, 1, 72–76, 2022.
52. Soren, S., Chakroborty, S., Pal, K., Enhanced in tunning of photochemical and electrochemical responses of inorganic bi-metal oxide nanoparticles via rGO frameworks (MO/rGO): A comprehensive review. *Elsevier Mater. Sci. Eng. B*, 278, 115632, 2022.
53. Panda, P., Pal, K., Chakroborty, S., Smart advancements of key challenges in graphene-assembly glucose sensor technologies: A mini review. *Mater. Lett.*, 303, 130508, 2021.
54. Chakroborty, S., Bharadwaj, V., Sahoo, S.K., Sensing and biosensing with 2D nanosheets beyond grapheme, in: *Sensing and Biosensing with Optically Active Nanomaterials*, pp. 119–141, Elsevier, Amsterdam, 2021.
55. Chakroborty, S. and Panda, P., Nanovaccinology against infectious diseases, in: *Nanovaccinology Outbreak As Targeted Therapeutics*, Wiley-VCH, Verlag GmbH, 2022.

10

New Technology 2D Nanomaterials for Neural Tissue Engineering

Banti Baishya[1], Saurav Paul[1]*, Hillol Das[2], Utsab Singha[3] and Dipyaman Mohanta[4]

[1]*Department of Chemistry, Assam University, Silchar, Assam, India*
[2]*Department of Biotechnology, Assam University, Silchar, Assam, India*
[3]*Department of Zoology, The Assam Royal Global University, Guwahati, Assam, India*
[4]*Department of Chemistry, Cachar College, Silchar, Assam, India*

Abstract
The treatment of the nervous system in cases of injury is much more difficult compared to other body tissues because of the sheer complexity of the nervous system and its slow regeneration rate. Moreover, available therapeutic techniques have several drawbacks and cannot completely restore nervous system injuries. Recently, nanotechnology and tissue engineering approaches have attracted many researchers to effectively guide tissue regeneration. Owing to the exciting physicochemical properties such as optical, electronic, and biological properties of 2D nanomaterials, they have immense potential for robust applications in tissue engineering and regenerative medicine. The superior conductivity of 2D nanostructured materials augments their use in the development of novel scaffolds for neural tissue engineering applications. Moreover, the enhanced drug-loading capacity of 2D nanomaterials has attracted the attention of many researchers to utilize them as a drug/gene delivery method to treat various devastating nervous system disorders. This chapter summarizes the recent developments, current challenges, and potential prospects of 2D nanomaterials in the fields of neural tissue engineering and regenerative medicine.

***Keywords*:** 2D nanomaterials, tissue engineering, regeneration

**Corresponding author*: chemsaurav1@gmail.com

Subhendu Chakroborty and Kaushik Pal (eds.) 2D Nanomaterials: Synthesis, Properties and Applications, (209–240) © 2024 Scrivener Publishing LLC

10.1 Introduction

All organisms at regular intervals experience distressing injury or damage and must be repaired in time. The emergency to repair damaged tissue serves to ensure that the injury is not rapidly fatal and initiates the tissue remodeling process, wherein the damaged parts are repaired and replaced with a more stable organization that heals the wound. Regeneration to substitute or repair lost and damaged body parts is widespread and highly variable among animal kingdoms. The explanation for this variation remains a major challenge in biology [1]. In animals, the regeneration ability of lost body parts varies considerably. A variety of animals can regenerate whole individuals from small body fragments, such as cnidarians and flatworms, where some animal groups are principally incapable of regenerating structures, such as fish, nematodes, and birds [2].

One of the notable features of regeneration is the conformity with which the regenerating structures restore their original form. In some species, the repair and remodeling process persists until it refurbishes a minor version of the original structure, called morphallaxis, which then returns to its original structure. In some organisms, tissue regeneration is categorized as the construction of new growth and morphogenesis structures from a dedifferentiated primordium that later during developmental progression regenerates the original structure called epimorphosis [3, 4].

The tissue repair or regeneration process in the nervous system is a complex phenomenon that brings great confrontation to researchers. In recent years, tissue engineering has become an interdisciplinary area of modern research interest because it provides promising future challenges for the treatment and management of many diseases [5]. Tissue engineering is primarily meant to combat clinical conditions, such as neuronal diseases, skin wounds, liver diseases, cardiovascular diseases, and ligament disorders. Brisk progress in tissue engineering techniques has led to massive progress in damaged tissue regeneration. In recent years, various nanobiomaterials have been actively explored for use in tissue repair and engineering. Diverse arrays of nanobiomaterials have been explored in the biomedical field to fabricate complex arrangements of soft tissues, and they tend to imitate the degree of natural tissues or organs [6].

2D nanomaterials show potential for improving the functions of injured nervous system tissues owing to their high surface area. Recently, 2D nanomaterials have become of great importance for multivalent interfaces with biological systems, leading to new therapeutic tools. 2D nanomaterials, such as graphene, 2D chalcogenides, black phosphorus, and other classes, have many fascinating physicochemical properties that have generated a great deal of attention in the field of neural repair and regeneration [7, 8].

The regeneration potential of the human nervous system has some limitations; and sometimes due to this ability injuries to the nervous system often lead to the loss of sensory or motor function and neuropathic pain. Neural tissue engineering has emerged as an alternative approach to neural tissue regeneration. Several healing approaches have been used to assist nerve regeneration [9, 10]. In the central nervous system (CNS), the nerve regeneration process is still more challenging, and existing therapies are not sufficiently effective to restore normal functioning in the CNS. Therefore, tissue engineering may serve as an alternative strategy for repairing nervous tissue. In tissue engineering, many 2D nanomaterials have been used for the repair of injured neural tissues [10]. This chapter summarizes the recent developments, current challenges, and potential prospects of 2D nanomaterials in the field of neural tissue engineering and regenerative medicine.

10.2 Regeneration of Tissue and Organ Repair in Nature

10.2.1 The '*Curious Case*' *of Lizard*: A Nature's Classic

Lizards have evolved the capacity to regenerate a substantially large and complex organ and have a remarkable ability to regenerate amputated tails. The regenerated tissues of lizards have important and interesting implications, and repaired tails are known as flawed copies. The regenerated and original tails exhibit numerous morphological differences.

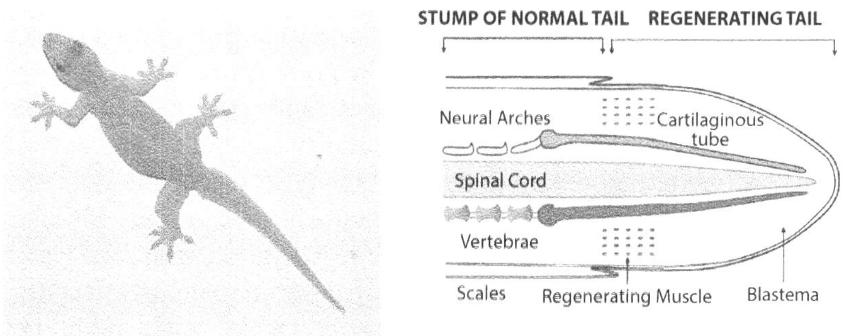

Figure 10.1 Schematic representation of lizard tail stump regeneration.

Regenerated lizard tail skeletons are cartilaginous [4, 11]. A schematic representation of lizard tail regeneration is shown in Figure 10.1.

The tail regeneration process in lizards is particularly appealing in terms of cartilage regeneration. In lizard tail regeneration, the generated tissue mass forms a blastema-like structure, and the regenerated frame consists of a cartilage cylinder adjoining the regenerated spinal cord. During the regeneration process, the reforming of the amputated tail engages blastema-based regeneration. The blastema organization confers an accumulation of cells that operate as progenitors for regenerated tissues. Lizards are the only amniotes that have the capability of blastema-based regeneration in adult forms, and the capability to form blastema is reduced considerably with age. After tail amputation, the tail stub ends in a terminal tail vertebra and engages a detached spinal cord. After some days, the injured epidermis migrates above the tail stub and forms a condensed apical cap. Cellular mass from the stub tissues assembles beneath the apical cap and propagates to form a blastema. After some time, the cartilage tube appeared in the region of the regenerated spinal cord and got in touch with the tail vertebra at its proximal end. During tail tissue regeneration in lizards, the vertebrae along with the spinal column of the original tail are regenerated and recognized as a single, unsegmented cartilage tube [12]. The proximal cartilage tube goes through hypertrophy and ossification, while the distal cartilage tube resists ossification for the lifetime of the lizard. In the animal kingdom, this type of regeneration process is unique. The redeveloped cartilage tube is a mature organ that is maintained for the duration of the lizard's life without transitioning to bone [11].

10.2.2 Regenerative Capabilities of Amphibians

The capability to reconstitute damaged or lost body parts widely varies among amphibians; some amphibians can regenerate their tails, limbs, and jaws. In amphibians, the regeneration process usually involves a sequence of morphogenetic activities that restore the normal structure of an organism following the loss of part or parts of it. Limb regeneration processes prevail most prominently in urodele amphibians, in both larval and adult stages. Among tetrapod vertebrates, the ability to regenerate limb structures after limb amputation is a unique feature of such groups. The anuran amphibians usually show evidence of limb regeneration in larval stages, but this ability is lost during metamorphosis. Regeneration of limb structure in urodele amphibians usually occurs through two major stages: (a) blastema structure development that bears a resemblance to the early embryonic limb bud and (b) blastema growth expansion and differentiation [13].

Limb regeneration in salamanders occurs through gradually more numerous populations of mesenchymatous cells that form a bud known as the blastema at the stub tip, which is similar in morphological form. Gradually, the blastema grows with proximodistal histogenesis and finally develops into the embryonic limb bud. The newly produced tissues of the limb stub represent the morphogenetic influence on the developing regeneration blastema. The developmental process of blastema reconstruction involves a cascade of cell dedifferentiation in damaged regions of tissues. During blastema development, cell proliferation mainly involved cellular interaction between the nerves and the epical epidermal cap. The skin and skeletal tissues of the newly formed limb elements derived from the blastema structures and differentiated with the existing parent tissues. The blood vessels and nerves redeveloped from the pre-existing blood vessels and axons, respectively [13, 14]. The developing blastema undergoes several morphological and histological changes to conquer a cellular mass of cells that progressively differentiate into proximal to distal and anterior to posterior directions, resulting in the formation or regeneration of distal divergence of the limbs. In urodeles, the capability to develop blastema structural features after limb amputation differentiates it from other anuran amphibians, reptiles, birds, and mammals.

10.2.3 Regeneration in Humans

Mammalian tissues regeneration strategies vary in response to injury. Mammalian tissues, such as the intestinal epithelium or liver, can mount broad regenerative capabilities to regain tissue function, whereas other tissues, such as the central nervous system (CNS), have limited or no regeneration potential [15].

The regeneration capability of the human liver after toxic injury and inflammation has some distinctive features. The damaged hepatic cells of the liver do not regenerate in the same way as a lizard's tail re-modified, but to a certain extent, it is a hyperplastic response such as increasing cell number in the residue liver, leading to an enlargement in liver size. The process involved in hepatic cell regeneration includes numerous cell types, intricate molecular pathways, and cellular communications. Liver regeneration is associated with the propagation of living cellular masses such as hepatocytes, endothelial cells, and Kupffer cells. The Proliferating hepatocytes produced the growth factors and were responsible for the cellular spur leading to the proliferation of other cells. The hepatocyte proliferation level is directly proportional to the rate of injury [16, 17].

The regenerative capabilities of neural tissue have been studied for well over a century. The nervous system mainly consists of the central nervous system (CNS) and the peripheral nervous system (PNS). The CNS mainly deals with the brain and spinal cord. The cranial and spinal nerves, along with their associated ganglia, formed the PNS. The role of the nervous system is to transmit information as a signaling pathway to the whole body. The signaling pathway of messaging takes place through a network of extensive neurons and supporting cells. The supporting cells involved in the neural communication network include glial cells, together with astrocytes, oligodendrocytes, and microglia [6]. Due to the limited endogenous regenerative capacity of the nervous system, injuries to the CNS are associated with numerous complications. During the neural injury, triggered glial cells such as microglia and astrocytes travel to the site and multiply to create a glial scar. The regeneration potential of nervous tissue has been studied for well over a century. The CNS has limited endogenous regenerative capacity. Peripheral nerve axons can regenerate under the proper circumstances. The nerve cell regeneration process in PNS, especially axonal rejuvenation, entails a composite interaction of supportive cells such as Schwann cells, growth factors, and extracellular matrix that forms a scaffold that governs the axon elongation [5, 18, 19]. When nerve cells in the PNS are damaged, especially the axon, the two ends of a nerve will draw back, resulting in the formation of a gap. The cell body of the neuron is attached to the proximal end, and the free-floating end contains the distal end. The injured nerve initiates the Wallerian degeneration in the distal stump, and the proximal stump initiates the axonal retrograde degeneration. The axons and myelin degenerate completely during Wallerian degeneration. The atomized neurons either die by necrosis or apoptosis or experience an attempt to repair [20].

10.3 Nanotechnology and Neural Tissue Engineering

Nanomaterials are of immense importance due to their wide range of optical, magnetic, chemical, and electronic properties. Nanostructured materials deliver enhanced physical and chemical properties due to their small size, which provides a large surface-to-volume ratio and makes them a better counterpart than bulk materials. These materials are used in many distinct fields of research, such as medicine, electronic devices, sunscreen, the chemical industry, fuel cells, sensors, etc.

10.3.1 Definition of Nanotechnology

Nanotechnology is the term given to those areas of science and engineering where phenomena take place at dimensions in the nanometer scale (1 nm–100 nm). Generally, nanotechnology is defined as a field concerned with creating machines along the scale of the nanometer, a billionth of a meter. For a material to become a nanomaterial, it is necessary to have at least one dimension in the nanometer range. However, there are several substances whose dimensions fall way beyond the nanoscale region but are still considered nanomaterials due to their unique physical and chemical properties.

Nanomaterials are classified into four major types depending on their particle dimensions, as illustrated in Figure 10.2.

a) Zero-dimensional (0-D)
b) One-dimensional (1-D)
c) Two-dimensional (2-D)
d) Three-dimensional (3-D).

a) Zero Dimensional (0-D)

Zero-dimensional nanomaterials are the materials where all the dimensions are within the nanometer range, i.e., 1 nm–100 nm. The most common examples of zero-dimensional nanomaterials are fullerenes and quantum dots. Some of the important characteristics of 0-D nanomaterials are that they can either be in an amorphous or crystalline state, they can be composed of single or multi-chemical elements, and there are different shapes and forms found in zero-dimensional nanomaterials. Zero-dimensional nanomaterials are mostly spherical.

b) One Dimensional (1-D)

1-D nanomaterials are those where, out of all the dimensions, at least one dimension is outside the conventional nanoscale region (>100 nm). Nanotubes, spirals, nanorods, and nanowires are the most common examples of this type of nanomaterial. One-dimensional nanoparticles could be amorphous or crystalline, and they may be obtained either in colloidal form or incorporated in a matrix.

c) Two Dimensional (2-D)

Two-dimensional nanomaterials are those in which at least two dimensions are not necessarily in the nanoscale region, i.e., greater than 100 nm in size.

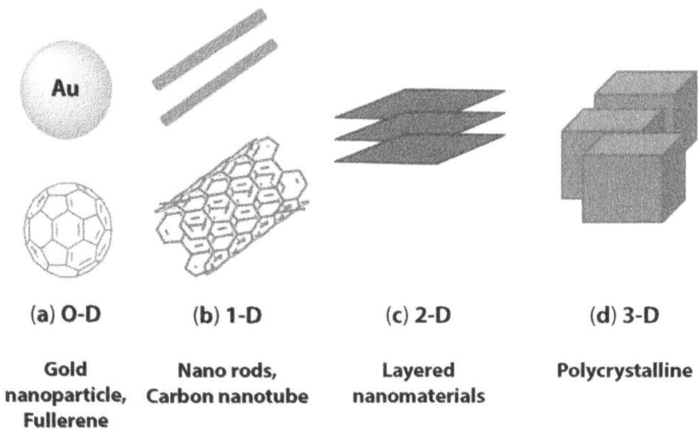

Figure 10.2 Classification of nanomaterials based on their dimensions (a) zero-dimensional (gold nanoparticles, fullerenes); (b) one-dimensional (nanorods, carbon nanotubes); (c) two-dimensional (layered nanomaterials); and (d) three-dimensional (polycrystalline, nanocubes).

Nanosheets, nanofilms and nanolayers are the most common examples of 2-D nanomaterials. 2-D nanomaterials could be amorphous or crystalline; these nanomaterials are utilized as mono- as well as multi-layered nanostructures.

d) Three-Dimensional (3-D)

When the materials are not restricted to the nanoscale in any of its dimensions, they are known as three-dimensional (3-D) or bulk nanomaterials. 3-D nanomaterials are comprised of three arbitrary dimensions greater than 100 nm. Bulk powders, bundles of nanowires and nanotubes, and dispersions of nanoparticles are some examples of three dimensions nanoparticles. The characteristics of these types of materials are that they could be amorphous or crystalline, and they can also be utilized as multi-layered nanostructures.

10.3.2 Synthesis of Nanomaterials or Nanoparticles

There are two different approaches through which nanoparticles are synthesized, which help to fabricate the structure of the nanoparticles. Using these two approaches, that is, top-down and bottom-up, we can reach the nanoscale. A schematic of both processes is shown in Figures 10.3 and 10.4.

a) Top-down approach
b) Bottom-up approach.

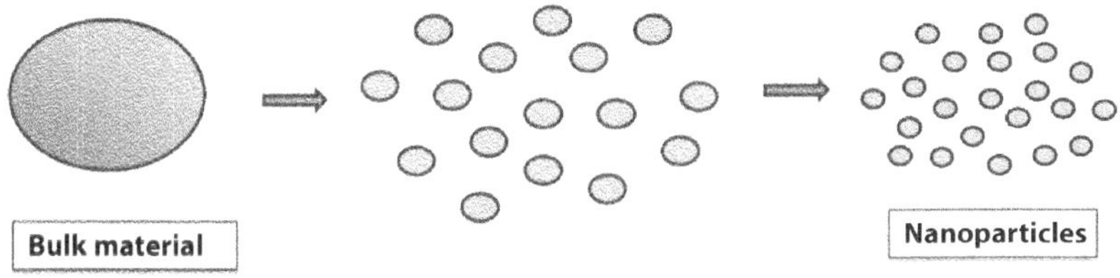

Figure 10.3 A schematic representation of the top-down approach.

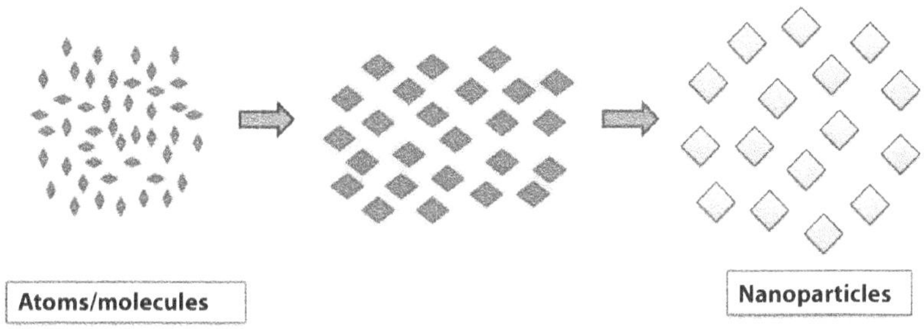

Figure 10.4 A schematic representation of the bottom-up approach.

Top-down approach: The top-down approach is used to synthesize nano-sized materials from bulk materials. This approach involves breaking up the bulk materials which are present in solids state by different physical processing method like mechanical method and lithographic techniques. In mechanical method, crushing, grinding, etching, and ball milling are used to cut the bulk materials into nonmaterials. This technique is not useful to synthesis nonmaterial as it is difficult to prepare nano size particles.

Bottom-up approach: The bottom-up approach is also called the self-assembly approach. In this synthesis process, atoms and molecules are combined to form a bulk material using physical and chemical processing methods. In physical techniques, different methods such as vaporization, sputtering, and plasma arcing are involved in the preparation of bulk materials starting from atom or molecules. The advantages of this route are

i) Ultra-fine nanoparticles, nanoshells, and nanotubes are prepared using this technique.
ii) The deposition parameters were controlled using the bottom-up approach.
iii) A narrow size distribution is possible.

10.4 2D Nanomaterials for Tissue Engineering Application

As discussed earlier in this chapter, nanomaterials refer to materials whose properties lie between those of the molecular and bulk materials. Nanoscale may be achieved in one-dimensional (1D), two-dimensional (2D), or three-dimensional (3D) directions [21]. 2D nanomaterials are known for their unique properties, as they possess and provide large nanoplanes. Graphene is the first 2D nanomaterial that has been experimentally demonstrated [22]. Many studies have been done conducted on the development of different 2D nanomaterials, such as black phosphorus (BPs) nanosheets, metal-organic frameworks (MOFs), and transition metal dichalcogenides (TMDs). Some of the 2D nanomaterials are shown in Figure 10.5. All these 2D nanomaterials exhibit excellent physiochemical properties, owing to which they have wide prospects in almost all aspects of tissue engineering. In addition to 2D nanomaterials, they also possess good biocompatibility, biodegradability, and surface functionality, which, when combined with cell biology, helps restore and regenerate damaged tissues or organs [23].

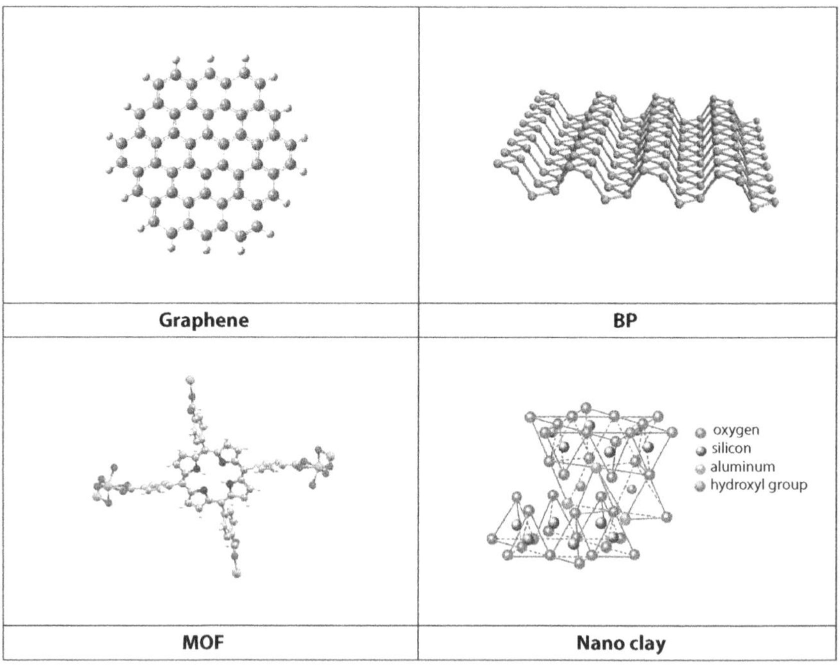

Figure 10.5 A schematic representation of some of the 2D nanomaterials.

Moreover, these 2D nanomaterials exhibit a high loading capacity, which makes them promising candidates for drug/gene delivery methods to treat several devastating neural diseases [24]. Recently, studies on the regulation and restoration of various nervous disorders by 2D nanomaterials have increased enormously worldwide and have found their efficacy in regulating several neural cell behaviors [25, 26]. Consequently, 2D nanomaterials have been extensively studied by scientists to exploit their functional properties in the treatment of neurological disorders. Several 2D nanomaterials are interesting for researchers in tissue engineering, some of which are discussed here.

10.4.1 Graphene-Based Nanomaterials in Tissue Engineering

The discovery of graphene created a sensation across the world, as it is the first 2D material with a wide range of applications in various fields [27, 28]. Graphene is a 2D carbon nanomaterial composed of a fused benzene ring-like structure arranged in a layered fashion [29]. Graphene has broad applications in display screens, sensors, new energy materials, photosensitive elements, and so on. Moreover, its large surface area, excellent mechanical strength, low cytotoxicity, and, most importantly, its biocompatibility and surface functionality make it a suitable candidate for use in biomedicine, exclusively in the field of tissue engineering. Some applications of graphene in tissue engineering are discussed in this section.

Graphene oxide (GO) is the oxidized from of graphene and have found applications in the regeneration of bone tissues. Reduced graphene oxide (rGO) is another form of GO that has excellent stability and biocompatibility and can be produced by chemical, thermal, and other methods to reduce oxygen content. Lee *et al.* [30] reported rGO/HAp NCs stimulates osteogenesis of MC3T3-E1 cells without osteoinductive agents. They found that the rGO/HAp composite significantly increased new bone formation in the calvarial defects. These findings have helped researchers across the globe to understand and exploit the role of rGO/HAp NCs in the development of bone regeneration techniques, as these NCs have a wide range of potential in stimulating osteogenesis. Serrano *et al.* [31] reported novel and flexible scaffold based on 3D GO and found potential utility in neural tissue regeneration. A highly practical interconnected network of neurons was formed on these scaffolds, which were rich in dendrites, axons, and synaptic connections. The versatility of these substrates promotes and encourages the potential application of these materials in the field of neural damage repair. Myocardial infarction (MI) causes death of heart cells due to disruption of the blood supply to the heart, resulting in heart failure.

The regeneration ability of heart muscles is poor, resulting in scar formation and abolishing normal electrical conduction, which impairs the heart by delaying or blocking proper signal propagation. Saravanan *et al.* [32] exhibited this failure in heart due to lack of conduction can be cured by using composites of GO and Au functionalized by natural polymer Chitosan (CS). Their work showed that the incorporation of A GO–Au composite in the polymer chitosan enhanced the electrical conduction with reduced cytotoxicity, and the scaffolds also improved myocardial contractility and reinforced ventricular function. In the last few decades, tissue engineering and repair have found prospective applications in myocardial infarction management. Cardiac patches are polymeric substrates that support myocardium proliferation. Ghasedmi *et al.* [33] reported a new type of composite material based on GO and polyethylene terephthalate (PET) nanofibers as electro-conductive cardiac patches [33]. Because cardiac cells are electroactive, they can respond to electro-conductive materials, and their work showed that the synthesized nanocomposites of GO/PET are promising electro-conductive cardiac patches for improving the regeneration of cardiac cell attachment. Skin wound healing is a challenging task for scientists. Graphene provides an excellent scaffold for regenerating skin wounds by overcoming hurdles, such as poor engrafting and scar formation. Li *et al.* [34] reported that graphene foam (GF) loaded with mesenchymal stem cells (MSCs) to facilitate the healing of skin wounds. The composites exhibited excellent biocompatibility with efficient regeneration of damaged cells. Their work demonstrated the efficient healing of skin wounds by GF-loaded MSCs while minimizing the scarring, and this work provides a new and better pathway for healing skin wounds, which will be of great clinical significance.

In summary GO and functionalized GO has numerous applications in tissue engineering such as repairing and regeneration of neural tissue, cardiac, bone tissue, etc. Some of them are summarized in Table 10.1. The pre-clinical applications of GO has been studied extensively but there are several fields where their use is yet to be explored and more efforts are required to exploit more of such potential applications of GO and functionalized GO.

Like GO black phosphorus is another class of 2D nanomaterials which have been studied quite comprehensively in the field of tissue engineering. Some of them are discussed and summarized as below.

10.4.2 Black-Phosphorus (BP)-Based Nanosheets in Tissue Engineering

Black phosphorus is similar to graphene in terms of appearance and structure, as shown in Figure 10.5. The excellent properties of BP in tissue engineering have attracted the attention of researchers over the past few years. Black phosphorus is a folded 2D sheet composed of phosphorus atoms. Apart from tissue engineering, this material has a wide range of applications in optoelectronics [35], such as a photothermal therapy [36] and photodynamic therapy [37].

The application of black phosphorus in fields such as bone tissue engineering and wound healing has gained much attention over the past few years. Wang *et al.* [38] reported that a composite of BP and $SrCl_2$ in PLGA when triggered with near infrared (NIR) light facilitated bone regeneration. The BP/$SrCl_2$/PLGA composites showed excellent cell viability and biodegradability, and tremendous bone regeneration in rats upon triggering the composites by NIR irradiation. Similarly, Raucci *et al.* [39] demonstrated that 2D BP can be used as an alternative to chemotherapy in osteosarcoma. They demonstrated that 2D BP could inhibit the metabolic activity of osteosarcoma cells without NIR irradiation. Furthermore, they showed that 2D BP enhanced anti-inflammatory cytokine generation and reduced the levels of pro-inflammatory mediators. Therefore, these materials are promising candidates for both bone regeneration and anticancer activity. However, the wound-healing ability of BP has also been reported by Shao *et al.* [40]. They found that BP nanosheets incorporated into a hydrogel provided excellent NIR photothermal performance and exhibited high biodegradability and biocompatibility. These BP hydrogels exhibited fast gelation, forming a gelled membrane on the wound. Compared with other reported photothermal therapeutic agents, these BP hydrogels exhibited high efficacy in eliminating residual tumor tissues. 2D black phosphorus has a wide range of applications in medical science and tissue engineering owing to its high biocompatibility and unique near-infrared photothermal responsive properties. Some of the other regeneration abilities of black phosphorus are summarized in Table 10.1.

10.4.3 Application of 2D Nanoclay in Tissue Engineering

Nanoclays are nanoparticles with a layered structure consisting of a 2D thick layer. One of such type is layered double hydroxide (LDH), which is generally known as hydrotalcite (HT). It is composed of hydroxides with layered crystalline arrangements of hydrotalcite containing more than two metal atoms. They have exceptional mechanical properties, good biocompatibility, and distinctive intercalating abilities, making them prospective candidates in the field of biomedical applications as well as in tissue engineering. Fayyazbakhsh *et al.* [41] showed that LDH improves bone regeneration in osteoporosis. The work shows that the pure LDH and LDH-hydroxyapatite nanocomposite (LDH-HAp) when blended with the gelatin vitamin D3 (VD3) complex, enhances the biological response and protects vitamin D3. The incorporation of VD3 reinforced the bioactivity and biodegradability. This work provides an insight into bone healing efficiency by simply encapsulating VD3 into LDH and LDH-HAp nanocomposites. Giriprasath *et al.* [42] reported a study in which composites of poly(3-Hydroxybutyric acid)-layered nanofibrous bone grafts were used for bone tissue engineering. The biocompatibility and activity of dual-layered bone grafts against MG63 cells have also been studied.

Table 10.1 Some applications of 2D nanomaterials in tissue engineering.

2D nanomaterial	Modification	Application	References
GO	nHAC/PLGA/GO	Improves and enhances the cell proliferation on osteoblast cells	[43]
	GO/PCL/CS/Collagen	Enhances the proliferation in osteogenesis	[44]
	Graphene films	Improves neural signals and enhances neural networks	[45]
	PET/GO	Recovers cardiac cell attachments and enhances the regeneration process in myocardial infarcted tissues	[46]
	Gel/GO	Enhances the maturation in cardiomyocytes	[47]
	Graphene foam	Promote wound healing ability and reduce scar formation	[48]

(*Continued*)

Table 10.1 Some applications of 2D nanomaterials in tissue engineering. (*Continued*)

2D nanomaterial	Modification	Application	References
Black Phosphorus	PAM/ChiMA/ BP-M/ AlgMA	Enhances the bone tissue regeneration process	[49]
	BP nanoscaffolds	Works in neural tissue engineering by showing axon regrowth and remyelination	[50]
	BP-PLEL hydrogels	Have prospective application in postoperative wound repairing	[51]
2D nanoclay (LDH)	Ca-LDH	Potential in bone tissue engineering	[52]
	NMF-LDH	Prospective in bone tissue engineering by enhancing the cell proliferation	[53]
	Mg-Fe-LDH	Hard tissue engineering by increasing proliferation and differentiation	[54]
	Hydrogel of PEGDMA-co-HEMA/ LDH	Prospective in tissue engineering and drug delivery	[55]

Osteoblasts, such as MG63 cells, exhibited first-rate proliferation in the top and bottom layers of nanofibrous bone grafts. Their study portrays promising bone implants for bone tissue engineering.

2D nanoclays have an exclusive intercalable structure and superior biocompatibility in tissue engineering, which makes them viable for the integration of modified release along with reinforcement in the regeneration and tissue repair ability. Table 10.1 presents some of the other applications of 2D LDH in tissue engineering.

10.5 2D Nanomaterials and Peripheral Nerve Engineering

Healing injuries in the peripheral nerve is one of the trickiest jobs in the clinical setting. Peripheral nerve injury (PNI) occurs in several types of cases including trauma and surgical complications. There are several medical treatments available for the cure of PNI, but complicated physiological structures and uncontrollable pathophysiological processes cause functional disability. The uses of nanotechnology in tissue engineering has been developed in the past few decades, providing a breakthrough and promising alternative in the treatment and repair processes of severe PNI. Recent developments in PNI regeneration processes using different types of nanomaterials are also discussed.

The nervous system comprises two main sections:

I) The central nervous system (CNS) and
II) The peripheral nervous system (PNS).

10.5.1 Peripheral Nerve

The PNS comprises the spinal nerves, cranial nerves, peripheral nerves (PNs), and neuromuscular connections. The CNS is well protected along the bones of the skull, and the spinal column produces a hard physical barrier that protects the CNS from abrasion. Three layers of connective tissue support the peripheral nerves: endoneurium, perineurium, and epineurium as shown in Figure 10.6. Each individual nerve fiber is enclosed in a protective sheath, known as the endoneurium. The fibers are held together into categories called fascicles, and each fascicle is encircled by a protective sheath known as the perineurium. Moreover, the epineurium, which comprises several fascicles, is amalgamated with a blood supply and fatty tissue within yet another sheath. Schwann cells and axons in the PNS are enclosed in a continuous basal lamina, unlike oligodendrocytes and their corresponding axons in the CNS. The PNS is not shielded by the blood–brain barrier or a strong bone layer; it is considerably more vulnerable to trauma and other injuries that affect the CNS [56]. From this perspective,

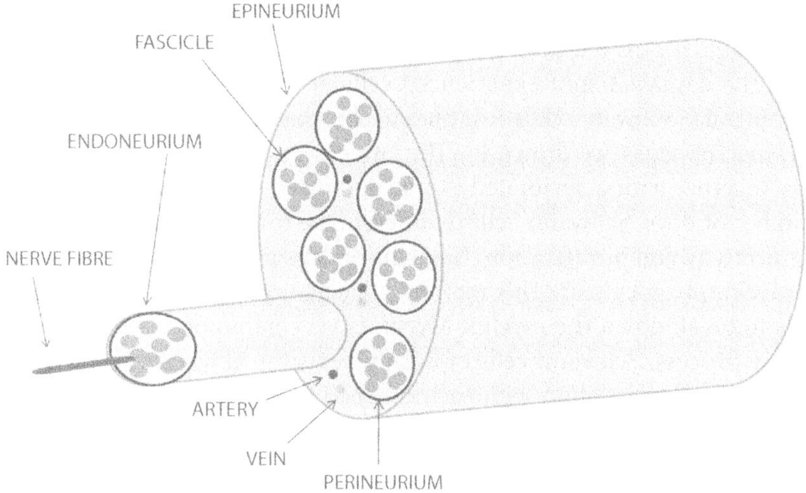

Figure 10.6 Layers of connective tissues in PNS.

nanotechnology and nano-based materials have attracted a great deal of engagement from scientific and medical sectors. Two-dimensional (2D) nanomaterials have experienced great curiosity because of their unprecedented electrochemical properties, established by the special role of atom-scale thickness, which allows for the free transfer of electrons on the material surface [57]. 2D nanomaterials possess sensational electrical properties, making them desirable candidates for improving the consequences of peripheral nerve injuries (PNI) [58]. These substances are regarded as reliable solutions for addressing the principal difficulties in Peripheral Nerve Regeneration (PNR). Peripheral Nerve Injury (PNIs) is significantly more common; however, PNs have a greater capacity for regeneration than CNS nerves. An instantaneous information transmitter tethering the human brain and rest of the body is the peripheral nerve [59].

10.5.2 Damage and Regeneration in Peripheral Nerve

As mentioned previously, peripheral nerve healing and regeneration are the biggest challenges in regenerative therapy and tissue engineering. Even with the greatest number of microsurgical procedures, peripheral nerve damage (PNI) is only marginally recoverable. Owing to the patients' loss of motor activities and sensory abilities, PNI is extremely devastating. Neurosurgery has long cast a shadow over the circumstances of peripheral nerve injury (PNI) and peripheral nerve regeneration (PNR) [60].

The Wallerian degeneration process begin immediately after some injuries. Although the cells themselves were not initially harmed, nerve stumps far from the damaged spot experienced cellular alterations. Axons begin to disintegrate, Schwann cells lose the myelin that has been covering them, and macrophages are drawn in. The latter is utilized to remove Schwann cells together with deteriorated axonal and myelin detritus. After a certain number of days, Schwann cells dedifferentiate owing to the loss of axonal connectivity and proliferation. Galectin-3 plays a key role in the activation of myelin phagocytosis. This method triggers macrophages and Schwann cells to break down the myelin. Myelin is very paramount for the degenerative process. Schwann cells of both preexisting and recently generated Schwann cell types align with each other to form the Bungner ligaments, highly aligned fibers formed by the Schwann cell basement membrane. After PNI, Schwann cells (myelinating cells of the PNS) de-differentiate and change to an improved phenotype, which plays an important role in the nerve regeneration process. Altogether, Schwann cells affect the PNR in three ways: proliferation, development of Bungner bands, and secretion of appropriate growth factors (GFs). To provide a favorable milieu for nerve regeneration and to fill the spaces between severed peripheral nerve stumps, a variety of potential natural and synthetic biomaterials with various molecular designs have been used over the past ten years. Additionally, the mechanical characteristics of nerve conduction should attest to the fact that stress at the injury site is avoided, and that nerve propagation does not rupture when the patient moves while at the same time having enough elasticity [56]. For broken axons to grow to the distal nerve trunk and its built-in target effectors, a natural bridge must form between the two nerve stumps during the regeneration process. A proregenerative region for axonal outgrowth over the injury site is provided by the Bunner's ligament, which is commonly referred to as the distal trunk. Endothelial cells, which contribute to the growth of new blood vessels and aid in the remodeling of the hostile environment, are essential in the regenerative process and play an inevitable role. Even so, the ambience inside the neural bridge is anomalously hostile for Schwann cells and axons. Tissue-engineered tabularization has been comprehensively studied as a possible means for filling neural gaps, and various types of nerve guiding channels (NGCs) have been proposed over the years, some of which have already been approved by the Food and Drug Administration (FDA) approved. NGCs are structured as hollow tubes that repair only short nerve gaps. The bulk of FDA-certified nerve guides are made of polyglycolic acid (PGA), polycaprolactone (PCL), and polylactic-co-glycolic acid (PLGA). Polyhydroxybutyrate (PHB) has neuroprotective properties and supports axonal regrowth. Polysaccharides such as chitosan (Cht) and alginate have also been considered as promising materials owing to their valuable results in PNR [61]. Therefore, the complex biological processes caused by injury lead to the exploitation of the biochemical and mechanical features of the microenvironment of neuronal tissue, reducing the recovery of neuronal function. Therefore, improving the existing blue plan to repair peripheral nerves more effectively is of great biomedical interest, especially in prototypes with neurodeficiencies [62].

The neural regeneration and degeneration processes are schematically represented in Figure 10.7. Degenerative alterations, including axon and myelin degradation, begin at the proximal and distal ends of the damaged site in peripheral nerve injuries, resulting in axonal disruption or nerve transection, as shown in Figure 10.7a. The distal end shows Wallerian (anterograde) degeneration and is cut off from the neuronal body. The components of the axonal cytoskeleton were disassembled and the axons were broken apart (Figure 10.7b). The myelin sheath breaks down because of these factors. Schwann cells and invading macrophages remove axons, myelins, and tissue waste. Axonal sprouts that grow from the cut proximal end of an axon are guided by bands of Büngner, which are formed by Schwann cells that multiply and orient along the outermost lamina (Figure 10.7c). When newly formed axonal sprouts are myelinated,

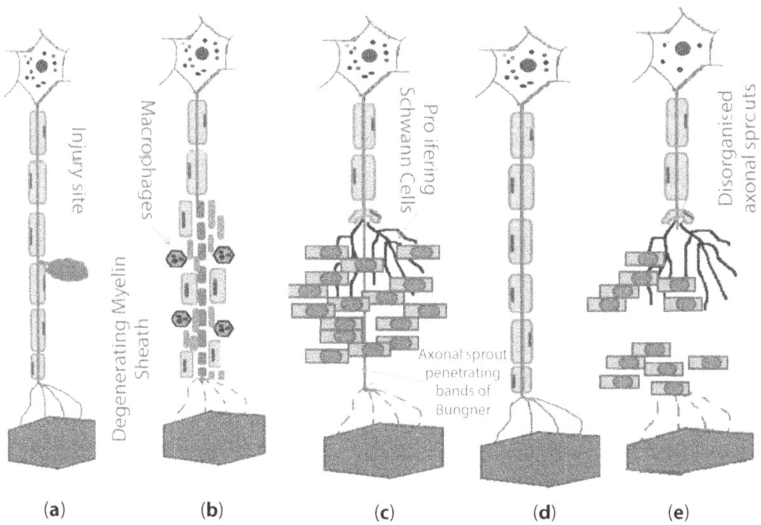

Figure 10.7 A schematic representation of neural regeneration and degeneration (a) normal neuron, (b) Wallerian degeneration, (c) initiating regeneration, (d) successful regeneration, and (e) unsuccessful regeneration.

they reach their destination and restore function (Figure 10.7d), whereas disorganized axonal sprouts lead to unsuccessful nerve regeneration, as shown in Figure 10.7e [63].

10.5.3 Key Features of Nanomaterials in Neural Tissue Engineering

Nerve damage can be repaired using various methods. Nanotechnology is a science that uses engineered materials on a nanoscale (nm) scale. Nanotubes (NT) and nanofibers (NF) can replicate the nervous system and the extracellular matrix (ECM) structure. Scaffolds can be combined by employing directed NTs, NFs, and other materials that have been functionalized to serve as cell-binding regions using nanotechnology. These scaffolds also served as topographical zeitgebers and lubricated the flow of nutrients and oxygen. Recent advances in nanotechnology, biomaterials, and tissue engineering have increased their tenability and have made it possible to use them with nanoscale control, leading to accelerated and targeted nerve regeneration. Nanostructured 2D scaffolds are of great importance for promoting increased cellular activity for nerve regeneration. Therefore, grafting is used to fabricate these materials to attain the optimal fundamental strength and surface characteristics for axonal regeneration. Multi-biomaterial nanostructures have been proposed to support axonal outgrowths. A variety of techniques have been used to fabricate nanostructured frameworks, such as electrospinning, self-assembly, and phase separation [60]. The nanofiber diameter should be superintended by tuning parameters such as polymer concentration, polymer solution flow rate, amino acid sequence, solvent conductivity and temperature in electrospinning, self-assembly and phase separation [59].

A technique called electrospinning is used to fabricate micro/nanoscale nanofibers (NFs) from natural and synthetic polymers or through polymer conglomeration. The natural, synthetic, or semi-synthetic materials to be electrospun are first dissolved in a suitable solvent to obtain a viscous solution. Gelatin has been used to counterfeit nanofibers, which can form potential scaffolds for tissue engineering. Self-assembly (SA) has recently emerged as a novel perspective for designing artificial frameworks that structurally and functionally emulate natural ECM. In contrast to electrospinning, this technique not only involves the specific biological components of ECM but also mimics the bottom-up process of ECM assembly [64]. SA is viewed as a method for altering and managing the effects of nerve-healing scaffolds. Polymer-rich and solvent-rich domains are produced during the phase separation of polymer solutionsand their morphologies can be fixed by low-temperature quenching. Phase separation is justified by the physical misfit between the solvent and polymer. To encourage gelation of the polymer state, the polymer was first dissolved in a solvent, and the resultant solution was kept at the gelation temperature [59].

A key feature of the promising nanobiomaterials is their porosity. Nanogels (NG) and nanotubes (NT) are the most important nanomaterials that can mimic the tubular organization found in cells and tissues. Hydrogels are highly very permeable and moisturized. Self-assembling peptides are particularly useful for nerve regeneration because they can be mixed with cells and/or growth agents that have liquid-phase amino acid sequences. Following self-assembly, cells and growth factors are immediately combined into nanomaterials, where cells connect and proliferate through pores, emitting growth factors with specific medication release rates. With dimensions ranging from 1 nm to 100 nm, nanotubes can penetrate cell membranes and mimic several intracellular structures, including microtubules and axons. They possess very high tensile strength. Because they have better mechanical, electrical, elastic, and rheological qualities than other materials, carbon nanotubes (CNTs) have been investigated and proposed for use in NGCs. Comparing graphene, the primary component of CNTs, with polystyrene substrates used in tissue culture, it has been found that graphene promotes neurite sprouting and proliferation. Moreover, CNTs have been shown to encourage synaptic contact formation and the regulation of cellular plasticity [65]. Despite the progress so far, further investigation of scaffold design and *in vivo* experiments is needed, which will be a challenging endeavor [64].

10.5.4 Mechanism of 2D Nanomaterial-Based Neural Regeneration

Advances in neural tissue engineering have offered hope for the treatment of neurological disorders. Two-dimensional nanomaterials help scaffold nerves, concentrate on the mechanism by which 2D nanomaterials promote new blood vessel formation and scrutinize their pro-regenerative effects on PNR [66]. In this chapter, the roles of different 2D nanomaterials in several tissue engineering processes are discussed. Here, we discuss the mechanism of regeneration and repair of various tissues by 2D nanomaterials such as graphene and black phosphorus, and a schematic representation of regeneration of neural tissues by these 2D nanomaterials is shown in Figure 10.8.

10.5.4.1 Graphene

Graphene is the most classical 2D nanomaterial suitable for enormous biological applications due to their high surface area and biocompatibility as already discussed. It also assists the scheme in which monolayer graphene can modulate neuronal behavior by affecting membrane function [67, 68] assembled graphene-based nerve guides to repair long peripheral nerve junctions. The expression of VEGF protein in graphene-based scaffolds was more than 18-month post-autograft of PNI, which is the outcome of many *in vivo* measurements. Interestingly, vigorously-shaped capillaries provide a nutrient-rich microenvironment for regenerating axons which results in the upregulation of nestin protein expression. It is uniquely present in neural stem cells and is crucial for the development of neural stem cells. These results show that the presence of graphene induces an angiogenesis-promoting milieu and increases neuronal fragility, which in turn promotes axonal regrowth [62].

10.5.4.2 Graphene Oxide

Graphene does not contain functional groups on its exterior, and it is unable to dissolve in solvents and tends to aggregate. Therefore, graphene oxide (GO) has been used in many studies because its surface is rich in oxygen-containing functional groups. The administration of graphene oxide, a highly oxidized derivative of graphene, increased the electrical conductivity of the polymer wire. Interestingly, graphene oxide-embedded nerve guides promote good regenerative capacity compared to their polymer analogs. Remarkably, the regeneration results of graphene oxide-containing scaffolds were considered of "gold standard" clinical autografts. Furthermore, the transmembrane phosphoglycoprotein CD34 used as a hematopoietic progenitor cell biomarker, was notably increased by the controlled release of graphene oxide [69]. The AKT-eNOS-VEGF signaling cascade implicated in the increase in endothelial cell proliferation may offer new insights into the physiological underpinnings of graphene activity in nerve regeneration.

10.5.4.3 Black Phosphorus (BP)

Similar to graphene, 2D black phosphorus has attracted great interest in the scientific community because of its extraordinary biodegradability, biocompatibility, and mechanical and optical properties [66]. Because of its thickness-dependent bandgap, in-plane anisotropic structure, high

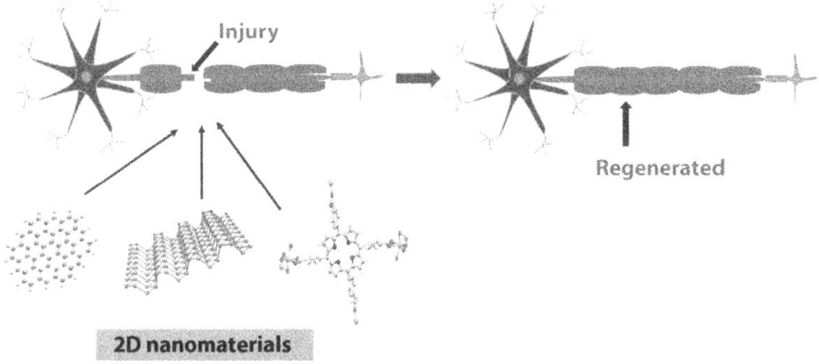

Figure 10.8 A schematic representation of regeneration of neural tissue by 2D nanomaterials.

carrier mobility, and biodegradable properties makes the potential to support peripheral nerve regrowth [70]. Black phosphorus showed tremendous promise for promoting the regrowth of peripheral nerves. For instance, adding 0.5% black phosphorus increases the electrical permeability of polymers to 9.8×10^{-3} S cm^{-1}. As a result, a nerve scaffold containing black phosphorus restores the bioelectrical continuity of injured nerves and encourages the expression of VEGF, helping to successfully construct new capillaries [71]. Additionally, growing data indicates that 2D nanomaterials may help PNR by producing electrical currents. These 2D materials have been linked to angiogenesis. Two interesting results with enormous potential for the development of nerve regeneration techniques are electrical currents that promote axonal growth, along with the production of new vessels. Electrical stimulation has long been demonstrated to be an effective therapeutic method for accelerating neuron growth.

10.6 Application of 2D Nanomaterials in Spinal Cord Repair

Injury to the spinal cord is a CNS injury characterized by sensory, motor, and autonomic dysfunction due to trauma, disease, or degeneration (non-traumatic), which can be wholly or partially. 2D materials like graphene exhibit excellent chemical and physical properties as well as physical properties with applications in numerous fields such as optoelectronics, energy storage, and biomedicine. The oxidized form of graphene, known as graphene oxide (GO), has many operating groups and catalytic activity. Another class of graphene derivatives is reduced GO (rGO), which exhibits excellent stability and biocompatibility. A conductive and anti-inflammatory GO-based hydrogel was developed along with a reverse GO-based scaffold for the improvement of spinal cord injury (SCI) [66]. It reduces inflammatory activity and promotes neuronal growth and axonal re-myelination in spinal cord injury repair. Domínguez-Bajo *et al.* [72] introduced reverse GO-modified microfibers (rGO-MF) to enhance regenerative functions, including axonal and vascular growth, in the damaged spinal cord. Some *in vivo* studies have shown that the presence of rGO-MF did not exacerbate the inflammatory response. Recently, researchers have enthralled advancing repair strategies for SCI by defeating inhibitors of recovery after SCI. Important paradigms such as mimicking the microenvironment, structural stability, and cellular or biomolecular deliverability, should be considered when developing technical scaffolds for SCI repair. Hydrogels are advantageous scaffolds for SCI regeneration because they serve as a temporary niche for cells that momentarily mimic soft tissue. In addition, they maintain high water concentrations and macromolecules that can mimic the moist milieu of the ECM and nearby tissues. Hydrogels can be directly applied to the site of inflammation as a viscous fluid or molded into various molds to generate a solid scaffold. Intriguingly, various researchers have enhanced the physicochemical properties and biological activity of hydrogels using 2D nanomaterials [73]. A novel injectable supramolecular hydrogel was created to repair spinal cord injuries. Graphene oxide was cross-linked with polyethylene glycol (PEG) capped with the anti-inflammatory medication diacerein to create hydrogels. Transition metal dichalcogenides (TMDs), including molybdenum disulfide (MoS2) are typical 2D layered TMDs that have attained much surveillance because of their unique optical, physicochemical, and biological significance. MoS2 has been used as a drug carrier for the treatment of SCI, and prolonged drug release was achieved by loading MoS2@PEG nanoflowers with etanercept (ET). The anti-inflammatory drug etanercept (ET)-MoS2@PEG has been proven to affect macrophagic behavior by stimulating M2 macrophages while inhibiting M1 macrophages in numerous *in vitro* experiments. The ability of ET-MoS2@PEG to provide neuroprotection and enhance musculoskeletal recovery was proven by *in vivo* tests, which are well established for healing and tissue repair [66].

As mentioned in this chapter, 2D nanomaterials such as layered double hydroxides (LDHs) possess high surface volumes and excellent electrical performance. An LDH, known as anionic clay, is composed of stable single-crystal units. LDH is a biodegradable 2D nanomaterial that can control immune cell types and limit inflammatory reactions in wounded tissues. By activating TGFBR2, LDH affects the polarization of the M2 phenotype in macrophages and microglia, enhancing the microenvironment of the immune system. LDH can be produced in SCI to re-establish the immunological microenvironment and can accumulate many neurotrophic factors, according to several studies. These investigations have also uncovered the material's receptor membrane and significant molecular pathways that promote neuronal regeneration to alleviate SCI and provide new methods for immune control [74].

10.7 2D Nanomaterials for Drug/Gene Delivery

Many researchers have focused on nanotechnology and tissue engineering approaches to effectively control tissue regeneration. In the context of tissue engineering, 2D nanomaterials play an immense role and have been extensively studied owing to their physiochemical and biological characteristics. The high electrical conductivity of 2D nanomaterials makes them interesting for the delopment of new scaffolds that can be used in neural tissue engineering. Nanomaterials, such as 2D, have a high loading capacity, which has led many investigators to use them as a gene delivery method to examine various serious diseases of the nervous system. Furthermore, their inflated aspect ratio makes them attractive materials for use in drug delivery systems. They allow them to adsorb a number of gene molecules and provide excellent control over drug release kinetics. TMDs are the first descendants of graphene that dispense similar structures to graphene, and they are derived from the quantum region, which has unique optical and electrical properties during the transition from indirect to direct bandgap, where they metamorphose the bulk material into a monolayer. TMD nanomaterials have also been used to regulate the neuronal differentiation of stem cells, which is pivotal for nerve tissue regeneration due to their high electrical conductivity. In contrast to graphene, TMDs exhibit stronger near-infrared (NIR) photothermal absorption, making them exceptional candidates for NIR-triggered drug delivery systems. The high surface area to volume ratio and enhanced drug/gene loading capacity of black phosphorus (BP) are due to its folded lattice structure, which makes them promising candidates for gene delivery. BPs have a wide range of biomedical applications including cancer imaging and therapy, biosensors, and drug delivery tools. Gene delivery systems in modern days are rooted in polyethylenimine (PEI)-functionalized 2D phosphorene nanosheets that have successfully delivered siRNA for synergistic photothermal and drug therapy. Fortunately, gene delivery systems based on 2D nanomaterials are grapheme-doped. Graphene-based materials (GBM) exhibit excellent properties that make them suitable candidates for gene/drug therapy applications. GBM toxicity rel0ies on several properties such as edgewise dimensions, chemical modulation, thickness, and colloidal dispersibility. Polyethylene glycol (PEG)-capped GBM adjuvants have also been used for gene delivery. When functionalized with PEG, GBMs show an excellent ability to facilitate the intracellular accumulation of materials. Overall, graphene-based materials (GBMs) are considered excellent delivery systems for gene delivery and clinical therapy because of their high propensity to take up genetic material and release it at determined target sites [75]. Layered double hydroxides (LDH) are 2D inorganic biodegradable materials that are used in various biomedical applications. LDH, which is an anionic hydrotalcite clay, has high charge density, which makes it a prominent carrier for drugs and biomolecules because it can guarantee a significant upstroke in drug loading by intercalating into the gap between the negatively charged and positively charged layers. Nanomaterials that respond to stimuli have been developed to deliver medication to tumours. LDHs react with H2O2 in the hypoxic environment of a tumor, which changes their structural makeup and causes drug release, increasing therapeutic efficacy. In the future, LDHs augmented with nanoparticles may be applied in a lower pH medium for more rapid and precise drug delivery to tumor targets [76]. Recently a potential 2D-MoS2 (molybdenum disulfide) a nanocarrier was used for loading etanercept (ET) for the treatment of spinal cord injury. Owing to their high biocompatibility, tunability, and intercalatable structures, two-dimensional TMD nanomaterials (such as MoS2) have become as an important segment for the blueprint of nanocarriers in drug delivery. Recent therapeutic strategies are insufficient for regenerating neural tissue from CNS and PNS injuries. Therefore, advanced tissue modification and drug delivery engineering are required. With the modernization and advancement of two-dimensional nanomaterials, the conciliated vasculature construction can be restored. Thus, two-dimensional nanomaterials have been proposed as an aptitude solution for long-range nerve defects [66].

10.8 Challenges and Prospects

Tissue engineering is the most attractive field of research in medical science. It has excellent prospects in the dealing with various devastating conditions. The nominal regenerative capacity of CNS is one of the most challenging aspects of tissue engineering. The PNS can regenerate after an injury. The regenerative power of the PNS is a slow process, and large injuries can be treated using neural tissue engineering tactics. The current strategies for the treatment of damaged nerve tissues are insufficient to treat or repair nerve tissues. Therefore, the requirement for new approaches in treating such damage is a great challenge for researchers worldwide. Various nanomaterial-based engineering approaches have been developed to prevent and treat nerve injuries. The 2D nanomaterials have the potential to improve injured tissues and offer various pathways for modulating tissue repair. 2D nanomaterial provides building blocks for the fabrication of 3D tissue scaffolds for various neural tissue engineering and other drug delivery systems to repair injured tissues. 2D nanomaterials have been gradually explored in tissue engineering scaffolds owing to their unique nanoscale features and biochemical properties, and they can be used as powerful tools in the development of tissue regeneration processes.

References

1. Khan, I., Neumann, C., Sinha, M., Tissue regeneration and reprogramming, in: *Wound Healing, Tissue Repair, and Regeneration in Diabetes*, pp. 515–534, 2020.
2. Baddour, J.A., Sousounis, K., Tsonis, P.A., Organ repair and regeneration: An overview. *Birth Defects Res. C*, 96, 1, 1–29, 2012.
3. Lozito, T.P. and Tuan, R.S., Lizard tail skeletal regeneration combines aspects of fracture healing and blastema-based regeneration. *Development*, 143, 16, 2946–2957, 2016.
4. Lozito, T.P. and Tuan, R.S., Lizard tail regeneration: Regulation of two distinct cartilage regions by Indian hedgehog. *Dev. Biol.*, 399, 2, 249–262, 2015.
5. Wang, X., So, K.F., Xu, X.M., Advances and challenges for neural regeneration research, in: *Neural Regeneration*, pp. 3–17, 2015.
6. Akter, F., Ibanez, J., Kotter, M., Neural tissue engineering, in: *Tissue Engineering Made Easy*, pp. 29–42, 2016.
7. Xue, J., Nanomaterial design and tests for neural tissue engineering, in: *Theranostic Bionanomaterials*, pp. 187–209, 2019.
8. Profire, L. and Constantin, S.M., Nanomaterials in tissue engineering, in: *Polymeric Nanomaterials in Nanotherapeutics*, pp. 421–436, 2019.
9. Binan, L. *et al.*, Approaches for neural tissue regeneration. *Stem Cell Rev. Rep.*, 10, 1, 44–59, 2014.
10. George, J. *et al.*, Neural tissue engineering with structured hydrogels in CNS models and therapies. *Biotechnol. Adv.*, 42, 107370, 2020.
11. Bely, A.E. and Nyberg, K.G., Evolution of animal regeneration: Re-emergence of a field. *Trends Ecol. Evol.*, 25, 3, 161–170, 2010.

12. Alibardi, L., The regenerating tail blastema of lizards as a model to study organ regeneration and tumor growth regulation in amniotes. *Anat. Rec.*, 302, 9, 1469–1490, 2019.
13. Stocum, D.L., Mechanisms of blastema formation and growth in regenerating urodele limbs, in: *Principles of Regenerative Medicine*, pp. 37–48, 2019.
14. Alibardi, L., Organ regeneration evolved in fish and amphibians in relation to metamorphosis: Speculations on a post-embryonic developmental process lost in amniotes after the water to land transition. *Ann. Anat.-Anat. Anz.*, 222, 114–119, 2019.
15. Delgado-Coello, B., Liver regeneration observed across the different classes of vertebrates from an evolutionary perspective. *Heliyon*, 7, 3, e06449, 2021.
16. Khan, A.Z. and Mudan, S.S., Liver regeneration: Mechanisms, mysteries and more. *ANZ J. Surg.*, 77, 1–2, 9–14, 2007.
17. Preziosi, M. and Monga, S., Update on the mechanisms of liver regeneration. *Semin. Liver Dis.*, 37, 02, 141–151, 2017.
18. Mo, X., Sun, B., Wu, T., EI-Hamshary, H., Nanofiber composites in neural tissue engineering, in: *Nanofiber Composites for Biomedical Applications*, pp. 395–410, 2017.
19. Papadimitriou, L. *et al.*, Biofabrication for neural tissue engineering applications. *Mater. Today Bio*, 6, 100043, 2020.
20. Sensharma, P. *et al.*, Biomaterials and cells for neural tissue engineering: Current choices. *Mater. Sci. Eng. C*, 77, 1302–1315, 2017.
21. Léonard, F. and Talin, A.A., Electrical contacts to one- and two-dimensional nanomaterials. *Nat. Nanotechnol.*, 6, 12, 773–783, 2011. https://doi.org/10.1038/nnano.2011.196.
22. Coleman, J.N. *et al.*, Two-dimensional nanosheets produced by liquid exfoliation of layered materials. *Science*, 331, 6017, 568–571, 2011.
23. Tu, Z. *et al.*, Multivalent interactions between 2D nanomaterials and biointerfaces. *Adv. Mater.*, 30, 33, 1706709, 2018.
24. Chimene, D., Alge, D.L., Gaharwar, A.K., Two-dimensional nanomaterials for biomedical applications: Emerging trends and future prospects. *Adv. Mater.*, 27, 45, 7261–7284, 2015.
25. Convertino, D. *et al.*, Effect of chemical vapor deposition WS2 on viability and differentiation of SH-SY5Y cells. *Front. Neurosci.*, 14, 592502, 2020.
26. Kang, Y. *et al.*, Graphene oxide and reduced graphene oxide induced neural pheochromocytoma-derived PC12 cell lines apoptosis and cell cycle alterations via the ERK signaling pathways. *Int. J. Nanomedicine*, 12, 5501–5510, 2017.
27. Zhu, Y. *et al.*, Graphene and graphene oxide: Synthesis, properties, and applications. *Adv. Mater.*, 22, 35, 3906–3924, 2010.
28. Rao, C.N.R. *et al.*, Graphene: The new two-dimensional nanomaterial. *Angew. Chem. Int. Ed.*, 48, 42, 7752–7777, 2009.
29. Lu, N. *et al.*, Graphene-based nanomaterials in biosystems. *Nano Res.*, 12, 2, 247–264, 2019.

30. Lee, J.H. *et al.*, Enhanced osteogenesis by reduced graphene oxide/hydroxyapatite nanocomposites. *Sci. Rep.*, 5, 1, 18833, 2015.
31. Serrano, M.C. *et al.*, 3D free-standing porous scaffolds made of graphene oxide as substrates for neural cell growth. *J. Mater. Chem. B*, 2, 34, 5698, 2014.
32. Saravanan, S. *et al.*, Graphene oxide-gold nanosheets containing chitosan scaffold improves ventricular contractility and function after implantation into infarcted heart. *Sci. Rep.*, 8, 1, 15069, 2018.
33. Ghasemi, A. *et al.*, Studying the potential application of electrospun polyethylene terephthalate/graphene oxide nanofibers as electroconductive cardiac patch. *Macromol. Mater. Eng.*, 304, 8, 1900187, 2019.
34. Li, Z. *et al.*, Three-dimensional graphene foams loaded with bone marrow derived mesenchymal stem cells promote skin wound healing with reduced scarring. *Mater. Sci. Eng. C*, 57, 181–188, 2015.
35. Song, Y. *et al.*, Vector soliton fiber laser passively mode locked by few layer black phosphorus-based optical saturable absorber. *Opt. Express*, 24, 23, 25933, 2016.
36. Qiu, M. *et al.*, Novel concept of the smart NIR-light–controlled drug release of black phosphorus nanostructure for cancer therapy. *Proc. Natl. Acad. Sci.*, 115, 3, 501–506, 2018.
37. Li, Y. *et al.*, Multifunctional nanoplatform based on black phosphorus quantum dots for bioimaging and photodynamic/photothermal synergistic cancer therapy. *ACS Appl. Mater. Interfaces*, 9, 30, 25098–25106, 2017.
38. Wang, X. *et al.*, Near-infrared light-triggered drug delivery system based on black phosphorus for *in vivo* bone regeneration. *Biomaterials*, 179, 164–174, 2018.
39. Raucci, M.G. *et al.*, Exfoliated black phosphorus promotes *in vitro* bone regeneration and suppresses osteosarcoma progression through cancer-related inflammation inhibition. *ACS Appl. Mater. Interfaces*, 11, 9, 9333–9342, 2019.
40. Shao, J. *et al.*, Black-phosphorus-incorporated hydrogel as a sprayable and biodegradable photothermal platform for postsurgical treatment of cancer. *Adv. Sci.*, 5, 5, 1700848, 2018.
41. Fayyazbakhsh, F. *et al.*, Release behavior and signaling effect of vitamin D3 in layered double hydroxides-hydroxyapatite/gelatin bone tissue engineering scaffold: An *in vitro* evaluation. *Colloids Surf. B Biointerfaces*, 158, 697–708, 2017.
42. Ramanathan, G. *et al.*, Fabrication of 3D dual-layered nanofibrous graft loaded with layered double hydroxides and their effects in osteoblastic behavior for bone tissue engineering. *Process Biochem.*, 64, 255–259, 2018.
43. Liang, C. *et al.*, Graphene oxide hybridized nhac/plga scaffolds facilitate the proliferation of MC3T3-E1 cells. *Nanoscale Res. Lett.*, 13, 1, 15, 2018.

44. Aidun, A. *et al.*, Graphene oxide incorporated polycaprolactone/chitosan/collagen electrospun scaffold: Enhanced osteogenic properties for bone tissue engineering. *Artif. Organs*, 43, 10, E264–E281, 2019.
45. Tang, M. *et al.*, Enhancement of electrical signaling in neural networks on graphene films. *Biomaterials*, 34, 27, 6402–6411, 2013.
46. Ghasemi, A. *et al.*, Studying the potential application of electrospun polyethylene terephthalate/graphene oxide nanofibers as electroconductive cardiac patch. *Macromol. Mater. Eng.*, 304, 8, 1900187, 2019.
47. Zhang, F. *et al.*, Easy applied gelatin-based hydrogel system for long-term functional cardiomyocyte culture and myocardium formation. *ACS Biomater. Sci. Eng.*, 5, 6, 3022–3031, 2019.
48. Li, Z. *et al.*, Three-dimensional graphene foams loaded with bone marrow derived mesenchymal stem cells promote skin wound healing with reduced scarring. *Mater. Sci. Eng. C*, 57, 181–188, 2015.
49. Wang, Z. *et al.*, Multifunctional nanoengineered hydrogels consisting of black phosphorus nanosheets upregulate bone formation. *Small*, 15, 41, 1901560, 2019.
50. Qian, Y. *et al.*, Concentrically integrative bioassembly of a three-dimensional black phosphorus nanoscaffold for restoring neurogenesis, angiogenesis, and immune homeostasis. *Nano Lett.*, 19, 12, 8990–9001, 2019.
51. Shao, J. *et al.*, Black-phosphorus-incorporated hydrogel as a sprayable and biodegradable photothermal platform for postsurgical treatment of cancer. *Adv. Sci.*, 5, 5, 1700848, 2018.
52. Fayyazbakhsh, F. *et al.*, Biological evaluation of a novel tissue engineering scaffold of layered double hydroxides (LDHs). *Key Eng. Mater.*, 493–494, 902–908, 2011.
53. Ramanathan, G. *et al.*, Fabrication of 3D dual-layered nanofibrous graft loaded with layered double hydroxides and their effects in osteoblastic behavior for bone tissue engineering. *Process Biochem.*, 64, 255–259, 2018.
54. Li, Q. *et al.*, Regulating the local pH level of titanium via Mg–Fe layered double hydroxides films for enhanced osteogenesis. *Biomater. Sci.*, 6, 5, 1227–1237, 2018.
55. Wang, P. *et al.*, Preparation and characterization of a novel hybrid copolymer hydrogel with poly (ethylene glycol) dimethacrylate, 2-hydroxyethyl methacrylate and layered double hydroxides. *J. Shanghai Jiaotong Univ. (Sci.)*, 17, 6, 712–716, 2012.
56. Carvalho, C.R. *et al.*, Nanotechnology in peripheral nerve repair and reconstruction. *Adv. Drug Deliv. Rev.*, 148, 308–343, 2019.
57. Rohaizad, N. *et al.*, Two-dimensional materials in biomedical, biosensing and sensing applications. *Chem. Soc. Rev.*, 50, 1, 619–657, 2021.
58. Qian, Y. *et al.*, Concentrically integrative bioassembly of a three-dimensional black phosphorus nanoscaffold for restoring neurogenesis, angiogenesis, and immune homeostasis. *Nano Lett.*, 19, 12, 8990–9001, 2019.

59. Cunha, C., Panseri, S., Antonini, S., Emerging nanotechnology approaches in tissue engineering for peripheral nerve regeneration. *Nanomed. Nanotechnol. Biol. Med.*, 7, 1, 50–59, 2011.
60. Carvalho, C.R., Oliveira, J.M., Reis, R.L., Modern trends for peripheral nerve repair and regeneration: Beyond the hollow nerve guidance conduit. *Front. Bioeng. Biotechnol.*, 7, 337, 2019.
61. Li, Y. *et al.*, Tissue engineering strategies for peripheral nerve regeneration. *Front. Neurol.*, 12, 768267, 2021.
62. Yan, Z. *et al.*, Two-dimensional nanomaterials for peripheral nerve engineering: Recent advances and potential mechanisms. *Front. Bioeng. Biotechnol.*, 9, 746074, 2021.
63. Hasirci, V. *et al.*, Peripheral nerve conduits: Technology update. *Med. Devices Evid. Res.*, 405, 7, 405–424, 2014.
64. Cunha, C., Panseri, S., Antonini, S., Emerging nanotechnology approaches in tissue engineering for peripheral nerve regeneration. *Nanomed. Nanotechnol. Biol. Med.*, 7, 1, 50–59, 2011.
65. Marti, M.E. *et al.*, Nanomaterials for neural tissue engineering, in: *Nanomaterials in Tissue Engineering*, pp. 275–301, 2013.
66. Cellot, G. *et al.*, Carbon nanotube scaffolds tune synaptic strength in cultured neural circuits: Novel frontiers in nanomaterial-tissue interactions. *J. Neurosci.*, 31, 36, 12945–12953, 2011.
67. Zheng, Y. *et al.*, 2D nanomaterials for tissue engineering and regenerative nanomedicines: Recent advances and future challenges. *Adv. Healthc. Mater.*, 10, 7, 2001743, 2021.
68. Pampaloni, N.P. *et al.*, Single-layer graphene modulates neuronal communication and augments membrane ion currents. *Nat. Nanotechnol.*, 13, 8, 755–764, 2018.
69. Qian, Y. *et al.*, Preclinical assessment on neuronal regeneration in the injury-related microenvironment of graphene-based scaffolds. *NPJ Regen. Med.*, 6, 1, 31, 2021.
70. Sidney, L.E. *et al.*, Concise review: Evidence for CD34 as a common marker for diverse progenitors. *Stem Cells*, 32, 6, 1380–1389, 2014.
71. Tao, J. *et al.*, Mechanical and electrical anisotropy of few-layer black phosphorus. *ACS Nano*, 9, 11, 11362–11370, 2015.
72. Qian, Y. *et al.*, Multilayered spraying and gradient dotting of nanodiamond–polycaprolactone guidance channels for restoration of immune homeostasis. *NPG Asia Mater.*, 11, 1, 36, 2019.
73. Domínguez-Bajo, A. *et al.*, Graphene oxide microfibers promote regenerative responses after chronic implantation in the cervical injured spinal cord. *ACS Biomater. Sci. Eng.*, 6, 4, 2401–2414, 2020.
74. Zheng, X. *et al.*, Applications of nanomaterials in tissue engineering. *RSC Adv.*, 11, 31, 19041–19058, 2021.

75. He, X. *et al.*, Bioactive 2D nanomaterials for neural repair and regeneration. *Adv. Drug Deliv. Rev.*, 187, 114379, 2022.
76. Izbudak, B. *et al.*, Layered double hydroxide-based nanocomposite scaffolds in tissue engineering applications. *RSC Adv.*, 11, 48, 30237–30252, 2021.

11

Theranostic Approach of 2D Nanomaterials in Breast Cancer

Pravati Panda[1], Subhendu Chakroborty[2*] and Kaushik Pal[3]

[1]Department of Basic Sciences, RIE, Bhubaneswar, Odisha, India
[2]Department of Basic Sciences, IITM, IES University, Bhopal, Madhya Pradesh, India
[3]Department of Physics, University Centre for Research and Development (UCRD), Chandigarh University, Mohali, Punjab, India

Abstract
Globally, breast cancer is an alarming disease and is considered one of the most common forms of malignant tumors existing among women, causing a massive number of new cases and mortality each year. Though breast cancer treatment is available, it is associated with high treatment costs, high side effects, and a long period of diagnosis and treatment. To tackle the urgency of developing a cost-effective process for treating breast cancer and aim to increase breast cancer patient survival rates, 2D nanomaterials were chosen as an alternative. The small particle size, large surface area, and exceptional physical and chemical characteristics make it more demandable. Literature reports suggest that 2D nanomaterials can cause apoptosis in cancer cells as well as being less harmful to normal cells due to their carbon structure. Therefore, these molecules can work as nanocarriers to incorporate miniature molecules like drugs, nucleic acids, or photosensitizers inside the living cells to achieve the treatment goal. This chapter includes the preparation of 2D nanomaterials as well as their current status and future perspectives toward the treatment of breast cancer.

***Keywords*:** Theranostic approach, 2D nanomaterials, breast cancer, nanomedicine, imaging, drug delivery, biomarkers, targeted therapy

**Corresponding author*: subhendu.cy@gmail.com

Subhendu Chakroborty and Kaushik Pal (eds.) 2D Nanomaterials: Synthesis, Properties and Applications, (241–262) © 2024 Scrivener Publishing LLC

11.1 Introduction

All over the world, the death rate due to cancer is very high and rising daily [1–3]. As per survey reports, it is the second most common life-threatening disease after cardiovascular disease. The uncontrolled growth of the cells generally causes it [4–8]. Among the different kinds of cancer, breast cancer is considered the most aggressive and difficult to treat. WHO reports that the morbidity and mortality rate in women due to cancer is generally higher than in men. It is first initiated by the angiogenesis of the cell [5–9]. Improper cell invasion will lead to the formation of a tumor, and finally, it will get converted into cancer. In the current scenario, photothermal therapy, chemotherapy, and surgery are the different treatments available for growing cancer. With respect to developing cancer treatment strategies, new cancer forms are becoming challenging for the scientific community [6–12]. Recently, one of the eight women who have triple-negative breast cancer. It caused the death of one woman of nine due to a lack of proper treatment. The traditional chemo- and photothermal therapies failed in this case. If a patient luckily gets rescued from it, there is still a fear of reoccurrence, leading to the patient's death [13–15]. Due to the high cost of cancer treatment, thousands of people fail to afford it and give up their lives. Many strategies have been invented to fight the battle against cancer by targeting different biomarkers like miRNA, mucin1, etc. Also, some drugs have been discovered to stop mitotic cell division, many of which are also in clinical trials [16–20]. Though the available cancer treatments are somehow effective, the major obstacles associated with the current therapies are that they still lack effective diagnosis in the early phase and lack target-specific drug delivery. Also, these cause significant side effects, including cancerous cells; healthy cells are also affected by the toxic effects of the drug and the radiation, as well as nausea, loss of appetite, vomiting, hair loss, etc. [21–25]. Therefore, the call of the day is the discovery of specific, target-oriented, and cost-effective anticancer drugs.

Recently, many reports elaborating on the synthesis and applications of various kinds of anticancer drugs have come to the forefront of the literature to encompass the existing drawbacks of conventional therapies and improve the lives of cancer patients. In this regard, nanomaterials come forward and act as a talented candidate to fight breast cancer more directly and effectively with fewer side effects. Biomedical applications of 2D nanomaterials are quite encouraging [26–30]. Enthusiastically, 2D nanomaterials have received a fascinating number of reports exploring their application for breast cancer therapy. Among the 2D nanomaterials, graphene and graphene oxide (GO) and its derivatives, Mxene, transition metal chalcogenides, etc., grabbed a lot of attention [31–35]. The surface of nanomaterials was functionalized with different kinds of functionalities like PEG, POx, PETx, etc., to enhance biocompatibility, chemical stability, surface area, and drug loading capacity. Highly specific and target-oriented anticancer drug-loaded, functionalized 2D nanomaterials were designed and studied for their anticancer abilities.

Additionally, the nanomaterials are used as a diagnostic tool for bioimaging to monitor the drug delivery at the specific site properly. To make nanomaterials, multidimensional people are also using combinational therapies [36–39].

This chapter summarizes the applications of different 2D nanomaterials as anti-breast cancer agents. The different forms of functionalized nanomaterials were designed by loading different anticancer drugs, and pleasingly, nanomaterials act as potential anticancer agents possessing high stability and biocompatibility with no adverse side effects.

11.2 Applications

Nowadays, to prevent the rapid spread of cancer and improve the versatility of cancer treatment, scientists are recommending combinational therapies, i.e., a combination of photothermal therapy with chemotherapeutics prepared by using nanomaterials. However, in some cases, it was observed that the polymer-coating over the nanomaterials during their synthesis hinders their solubility inside the living cells, hindering the drug uptake tendency of the cancerous cells. To address the challenge, dopamine-rGO (photothermal nano-agent) and resveratrol (RES; chemotherapeutic drug)-based hydrogels were prepared *in situ* by cross-linking chitosan and were applied for cancer chemo-PTT purposes. DOPA-rGO unveiled itself as an excellent photothermal nanoagent because of its outstanding cytocompatibility and greater photothermal capacity. Though Resveratrol possesses admirable anticancer properties, its applicability is limited due to its poor solubility. Hence, the authors planned to incorporate it inside the injectable hydrogel to improve its solubility. Experimental results revealed that the formulated hydrogels performed excellently well and displayed appreciable characteristics like *in situ* gelation, injectibility, fine cytocompatibility, and suitable physicochemical properties. Irradiation by NIR light increased the temperature of injectable hydrogels by about 11 °C. *In vitro* anticancer study data said that the hydrogels of DOPA-rGO@Gel

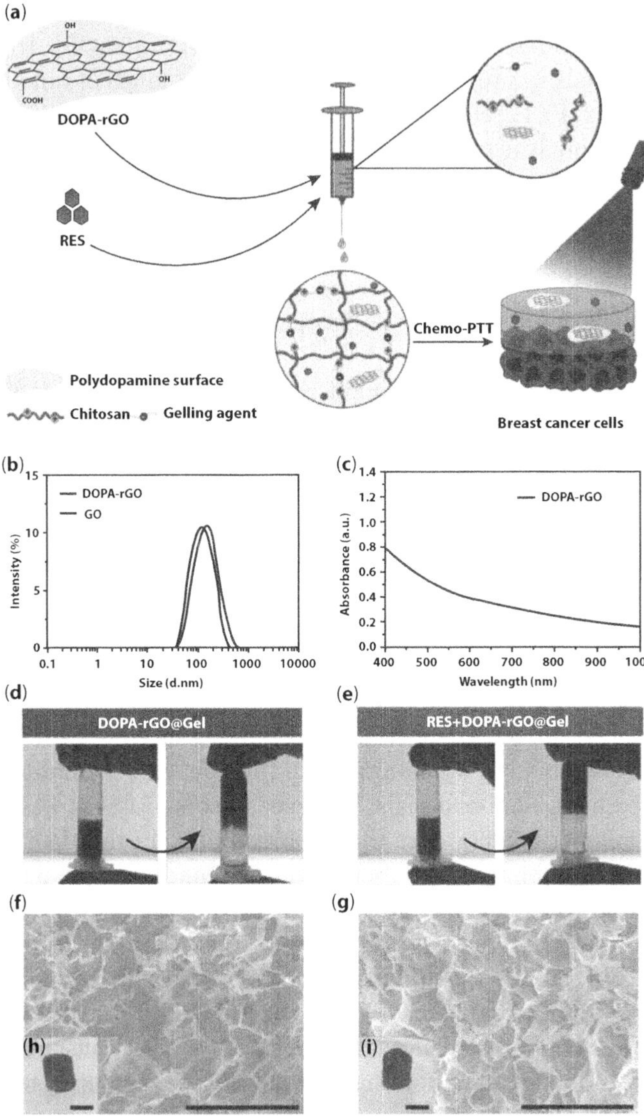

Figure 11.1 Analysis of the DOPA-rGO and the injectable in situ forming hydrogels physicochemical properties. Illustration of the RES+DOPA-rGO@Gel assembly and application in chemo-PTT of breast cancer cells (a). DLS size distribution of DOPA-rGO and GO (b). Visible-NIR absorption spectra of DOPA-rGO (at 25 μg/mL of DOPA-rGO) (c). Macroscopic images of the DOPA-rGO@Gel (d) and DOPA-rGO+RES@Gel (e) gelation. The dark arrow indicates an incubation period. SEM images of the cross-section of DOPA-rGO@Gel (f) and RES+DOPA-rGO@Gel (g). Scale bars represent 200 μm. Macroscopic images of the DOPA-rGO@Gel (h) and DOPA-rGO+RES@Gel (i) attained using the hollow cylindrical templates. Scale bars represent 1 cm. Reprint and copyright permission obtained from Elsevier publication [40].

prepared by excluding RES used for the photothermal therapy could diminish the viability of breast cancer cells by around 72%, whereas the RES+DOPA-rGO@Gel-based hydrogel's chemotherapy decreased the viability by around 75%. The combined chemo-photothermal therapy diminished the viability to 31% without hampering normal, healthy cells. Hence, the prepared RES+DOPA-rGO@Gel-based hydrogel can be applied for chemophotothermal therapy against breast cancer cells (Figure 11.1) [40].

Treatment of triple-negative breast cancer is quite hard because of the adverse effects of chemotherapy. To surmount this bottleneck, the authors designed and synthesized docetaxel (DTX) and GO-functionalized nanomaterials as advanced drug delivery agents. The stability and photothermal ability of GO were enhanced by the coating of polyethyleneimine (PEI). Then, the DTX was doped over the surface of GO through the π–π stacking effect. The nanomaterials were also combined with anti-miRNA21 by the electrostatic interaction between negatively charged anti-miRNA21 and PEI. The research group developed a multifunctional antitumor system with drug–gene and photothermal therapies. The constructed materials exhibited tremendous potential and inhibitory activity against triple-negative breast cancer cells (Figure 11.2) [41].

Polyvinylpyrrolidone (PVP) functionalized GO-based 2D nanomaterials were designed and developed and were subjected to study for their anticancer efficacy against MDA-MB-21 breast cancer cell lines. The PVP was loaded with *Juniperus squamata* root essential oil (JSEO); also, the synthesis and functionalization of the GO were carried out following advanced

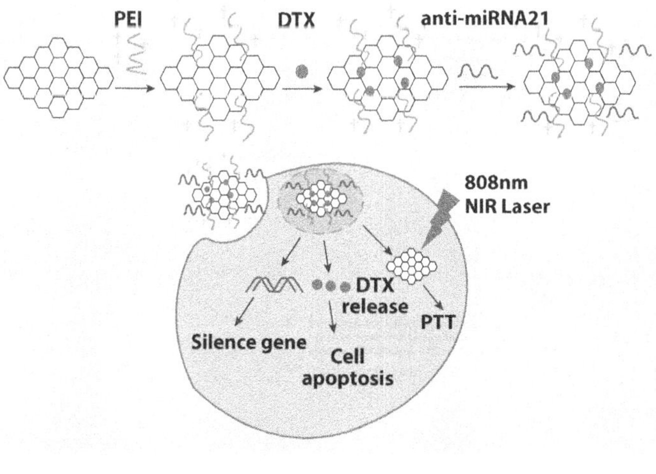

Figure 11.2 Schematic diagram indicating the synergistic antitumor therapy of the designed multifunctional drug delivery system. Reprint and copyright permission obtained from Elsevier publication [41].

Hummer's method. Pi-pi/hydrophobic interaction was used to load JSEO upon GO and GO-PVP. It was found that the JSEO loading capacity of GO-PVP was 17.4%, a little bit higher than the loading capacity of GO, which was 13.6%. Sixty-four biologically active compounds were analyzed using GC and GC–MS techniques. Also, the nanocomposites were characterized through FTIR, SEM, XRD, and Raman spectroscopy. It was observed that the cell survival rate upon treatment with GO-PVP-JSEO was 68% after 24 h, 32.74% after 48 h, and 21% after 72 h, respectively, at a 20 μg/mL concentration. The demonstrated results indicated the improved anticancer activity of the GO-PVP-JSEO nanocarrier over JSEO alone. Hence, after a little bit of study of the safety and efficacy of GO-PVP-JSEO, these nanomaterials can be subjected to breast cancer treatment applications (Figure 11.3) [42].

To enhance the activity and overcome the limitations (irreversible aggregation, poor dispersion, and potential cytotoxicity) associated with rGO sheets for breast cancer therapy, the group prepared PEGylated chitosan20k (PCS20k) functionalized rGO (PCS-rGO) nanosheets through one-pot reduction by using PCS as a stabilizing and reducing agent. The particle size and mass-to-charge ratio of the rGO nanosheets were effectively decreased using the functionalization strategy, allowing their effective dispersion. It was observed that, under near-IR radiation of 808 nm, the developed nanosheets displayed photothermal efficacy of h = 67.7%, superior to GO nanosheets with h = 55.0%. It effectively caused ablation in

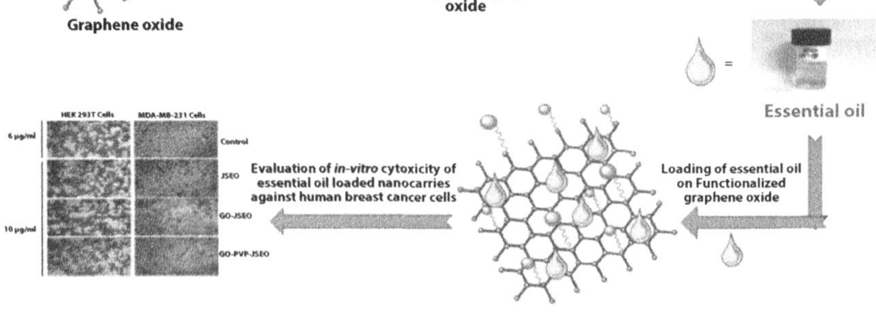

Figure 11.3 The flow chart of the procedure for loading of Juniperus squamata root essential oil onto functionalized GO. Reprint and copyright permission obtained from Elsevier publication [42].

cancer cells at 49 °C, suggesting its potential application for cancer photothermal therapy (Figure 11.4) [43].

Curcumin-functionalized rGO-based nanosheets were prepared and subjected to study for their anticancer ability against breast cancer cell lines. In this process, curcumin acts as a natural reducing and stabilizing agent. The pi-pi stacking interaction of curcumin with rGO was confirmed by

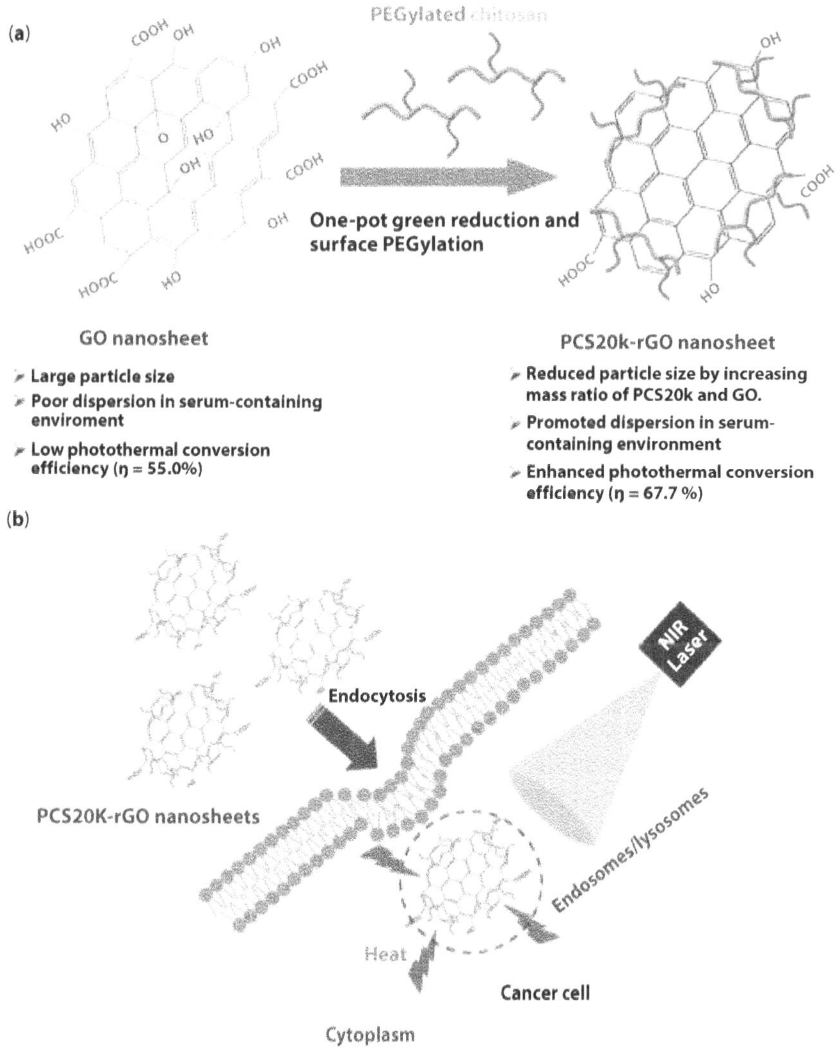

Figure 11.4 Illustrations of (a) one-pot green reduction and PEGylation of GO nanosheets with PCS20k and (b) anticancer effect of PCS20k-rGO nanosheets upon NIR-triggered hyperthermia. Reprint and copyright permission obtained from Elsevier publication [43].

FTIR and Raman spectroscopy. The zeta potential value of the nanosheets was decreased to −20 mV from −40 mV after successive reductions and functionalizations. The cytotoxicity was studied against SKBR3 and MDA-MB-231 cancer cell lines as well as against mouse fibroblast L929 normal cell lines. The nanosheets observed no morphological changes as well as no effective cytotoxicity for a concentration value of <70 μg/mL. In the meantime, the nanosheets were found to induce growth of the cancer cells by around 25% after 48 h of incubation. However, at >70 μg/mL concentration, cell morphological changes were initiated, and at a concentration of 100 μg/mL, apoptosis of the cells around 15%–25% occurred, which was confirmed by the MTT assay. Mechanistically, it was assumed that interaction between cells and rGO sheets resulted from apoptosis and morphological changes from elongated to roundup forms. Hence, it was concluded that a high concentration of rGO can be applied to treat breast cancer [44].

Angiogenesis, i.e., the formation of new blood cells, is considered the initiation mark of tumor cells, and then progression and metastasis of the cells occur. Treatment of tumors can be made by targeting the angiogenic markers present at the tumor site, which leads to more effective drug delivery as it does not require any extravasation. Monoclonal antibody (mAb)-conjugated nanographene oxide (GO)-based nanomaterials were constructed and applied against follicle-stimulating hormone receptors (FSHR). FSHR was found to be abundantly present both in primary and metastatic tumors and, thus, can be considered a highly selective tumor vasculature marker. Nanomaterials were found to have an average diameter of 120 nm, which was confirmed through TEM, AFM, and DLS. The authors doped the GO conjugates with radioactive 64Cu to enhance their visualization by PET inside the cells. Bioluminescence imaging (BLI) was used to monitor the tumor growth in breast cancer cells in female mice. *In vivo* and *in vitro* studies were done to analyze GO conjugates' targeting efficacy, stability, tissue distribution, and specificity. The targeting specificity of the developed nanosheets against cellular FSHR has been confirmed through fluorescence microscopy and flow cytometric analysis. The cbgLuc-MDA-MB-231 tumor cells exhibited a more potent and persistent uptake of 64Cu-NOTAGO-FSHR-mAb compared to 64Cu-NOTA-GO (non-targeted GO conjugates). The nanomaterials were found to possess an acceptable drug loading capacity (for DOX, 756 mg/g) and served as high-quality drug carriers. Fluorescence imaging signifies the effective drug delivery of DOX-loaded GO-FSHRmAb atcbgLuc-MDA-MB-231 at the metastatic site. Hence, this can be applied as a valuable tool for premature metastasis detection and targeted delivery of therapeutics [45].

The accurate and early-phase detection of breast cancer cells can enhance treatment, curability, and survival rates. To achieve early phase detection, the authors prepared six GO-based luminescent nanodots as fluorescent probes as a sensing material to effectively identify metastatic, cancerous, and healthy cells. The sensor system was found to be capable of giving fluorescent signals in the presence of cancer cells, thereby helping in the successful identification of breast cancer cells and easily distinguishing them as human epidermal growth factor receptor-2-positive, estrogen receptor-positive, and triple-negative phenotypes. High selectivity, along with ultra-high reproducibility, can be achieved by using this approach. Moreover, easy and cost-effective preparation, low cytotoxicity, excellent brightness, good aqueous solubility, and high photostability were significant characteristic features associated with the developed nanoprobe, making them unique and desirable. The sensor detected 49 out of 50 breast cancer samples accurately from the unknown cell sample with 98% detection accuracy. Hence, the developed nanoprobe-based luminescent material can be applied as a promising candidate for biomedical diagnostic purposes [46].

MiRNAs, i.e., microRNAs, are generally endogenous, small, and non-coding RNAs and can be considered a predictive biomarker due to their abnormal expression in many tumor cells, including breast cancer cells. Targeting miRNAs can help in the rapid and early diagnosis of cancer cells, hence receiving much attention. Here, the authors synthesized a label-free nano-genosensor for identifying breast cancer biomarkers, i.e., miRNA-21, made up of N-doped functionalized graphene (NFG), Ag NPs, and polyaniline (PANI). The developed nanocomposite was functionalized and immobilized by the specific sequence of an aminated complementary oligonucleotide of miRNA-21; detection was made through DPV (differential pulse voltammetry). The hybridization of miRNA-21 was monitored through oxidation peak current under optimal reaction conditions. The nanomaterial developed was found to be highly sensitive and specific, possessing a detection limit of 10 fM to 10 μM with 2.5 $\mu A\ cm^{-2}$ sensitivity and a LOD value of 0.2 fM. Simultaneously, it exhibited high reproducibility and recovery values of 94% and 107%, and thus, it can be applied for the identification of miRNA-21 in real clinical samples without doing RNA extraction, sample preparation, and amplification, etc. [47].

DOX-GPFH (doxorubicin-loaded graphene oxide functionalized with polyethyleneimine-folic acid and herapin) based on a dual-targeting nanobiomaterial was designed and synthesized. The nanobiomaterial was loaded with doxorubicin with an ultra-high loading capacity to enhance cellular uptake. GDC0941, a phosphatidylinositide 3-kinase acting as an Akt phosphorylation inhibitor, was selected to increase the anti-metastasis effect of DOX@GPFH by down-regulating the expression of matrix metalloproteinase. Integration by heparin increased the stability of the developed nanobiomaterial and enhanced the drug loading ratio to 125.1% from 64.4% by the presence of many carboxyl and sulfonic groups. *In vitro*, DOX@GPFH and GDC0941 exhibited inhibition rates of 61.2% ± 13.9% towards wound healing, 81.0% ± 3.6% towards cell migration, and 76.8% ± 5.2% towards cell invasion. Moreover, the authors also demonstrated the *in vivo* anti-metastasis efficacy of the developed nanomaterial, which was found to possess a 77.0% ± 7.6% rate towards tumor inhibition and a 73.7% ± 9.6% rate against pulmonary inhibition. Therefore, the developed nanobiomaterial can be significantly used for dual targeting purposes and can effectively be applied to the treatment of breast cancer metastasis (Figure 11.5) [48].

Globally, the mortality rates of women are high due to metastatic breast cancer, which is also difficult to treat with traditionally used chemo and photothermal therapy. Considering this hurdle, graphene-based nanoprobes were designed and developed as selective inhibitors of cancer cell motility. The authors successfully synthesized a highly stable and biocompatible PEG-GO-XN-based nanocomposite by coupling prenylated chalcone xanthohumol (considered a mitochondrial e-transport chain complex I inhibitor) with a PEGylated GO nanosheet. Experimentally, it was found that the nanoprobe selectively inhibits mitochondrial oxidative phosphorylation in MDA-MB-436 and MDA-MB-231 metastatic breast cancer cells. It inhibits cell growth by inhibiting the formation of ATP, disrupts the production of the F-actin cytoskeleton in the lamellipodia, and blocks the invasion and migration of breast cancer cells *in vitro* without disturbing the normal cells. Additionally, it inhibits the metastasis of MDA-MB-231 cells to the lung in a mouse cell model. It blocks the

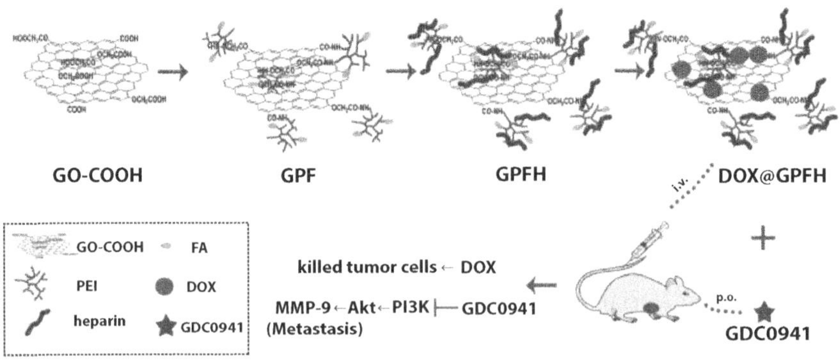

Figure 11.5 Schematic illustration of inhibiting the metastasis of breast cancer via PI3K/Akt/MMP-9 pathway by (DOX@GPFH) + GDC. Reprint and copyright permission obtained from Elsevier publication [48].

downregulation of E-cadherin and upregulation of N-cadherin, vimentin, Snail, and Twist induced by TGF-β1, thus maintaining the epithelial-like features and simultaneously diminishing the motility potential of breast cancer cells. Considering all these enormous properties of the nanomaterials, the developed nanoprobe can have promising applications for the treatment of metastatic cancer and, hence, be subject to a clinical trial [49].

Previously, it was reported that miRNA could be considered a bookmark for early cancer diagnosis, treatment, and prevention. Considering this idea, Hossein Naderi-Manesh and his group reported the synthesis of a novel electrochemical nanobiosensor for plasma miR-155 detection by merging the advantages associated with the electrochemical process with nanobiomaterials. The surface of the glassy carbon electrode was functionalized by GO and primarily decorated with thiolated probe-functionalized gold nanorods (GNRs). Differential pulse voltammetry was used to measure the signals produced by Oracet Blue (OB). The synthesis of nanobiomaterials and their successive dispersion over the electrode surface were confirmed through TEM, UV–Vis, CV, FE-SEM, and EDS spectroscopy. It was observed that the electrochemical signal possesses a linear relationship with the miRNA concentration, which ranges from 2.0 fM to 8.0 pM with a 0.6 fM limit of detection. The as-prepared electrochemical nanobiosensor exhibited high specificity and could easily distinguish between complementary target miRNAs and non-complementary miRNAs. Additionally, it was found to have superior selectivity and sensitivity, high storage capacity, and reproducibility and can be applied to real-sample analysis. Thus, the developed electrochemical nanobiosensor can be subjected to a clinical trial for direct detection of miRNA-155 in real samples for early breast cancer prevention [50].

Controlled drug release at the specific target site is also crucial for cancer treatment. To address this, GO-based multifunctional nanocarriers with controlled drug release capacity were prepared. The GO possesses a high surface area and many oxygen-rich functionalities, so it is chosen for controlled drug delivery. The surface area of the GO was functionalized with hyperbranched polyglycerol (HPG) *via* anionic ring-opening multi-branching polymerization (Anionic-ROMBP) using glycidol monomer. Then, after fabrication of the magnetic NPs, they were applied over the surface area of GO as well as between the branches following the co-precipitation method. Curcumin, a natural anticancer drug, was loaded in GO through π–π stacking and hydrophobic interaction. After analyzing the stability of the nanocarrier at different physiological pHs and its drug-releasing efficacy, it was subjected to a study for its cytotoxicity through an MTT assay. It was found that nanocarriers with a negative surface charge of −49.4 mV displayed a drug loading efficacy of 198%. Along with this, curcumin-loaded GO displayed good cytotoxicity and apoptotic ability inside the cancer cells as compared to curcumin-unloaded GO. Cost-effectiveness, biocompatibility, no toxicity, and no side effects were considered the advantages of the nanocarrier (Figure 11.6) [51].

Biocompatible carbon nanosheets integrated by multi-cyclodextrin were doped by doxorubicin (DOX), a known anticancer drug, as a nanocarrier designed and synthesized for successful drug delivery to breast cancer cells. TEM, SEM, XRD, EDS, etc. successively analyzed the morphology and structural features of the nanocarrier. To achieve the optimal formulation, the group also studied the drug-loading and encapsulation efficacy of the nanocarrier. The presence of cyclodextrin at the drug molecules' surface area enhances their stability and improves their drug loading and encapsulation efficiency. Its biocompatibility was analyzed through a hemolysis assay, and an MTT assay was performed to study the cytotoxic effect on breast cancer cells (MCF-7). The drug-releasing capacity of the nanocarrier was studied at different physiological pH conditions *in vitro*. As indicated by cytotoxicity analysis, the nanocarrier loaded with doxorubicin exhibited significant cytotoxicity against the MCF-7 cells. It diminishes the growth of the cancer cells by causing apoptosis, as confirmed by fluorescence microscopy observation. High stability with biocompatibility, target specificity, high drug loading, and uptake property with no toxicity to normal cells are the interesting characteristic properties associated with

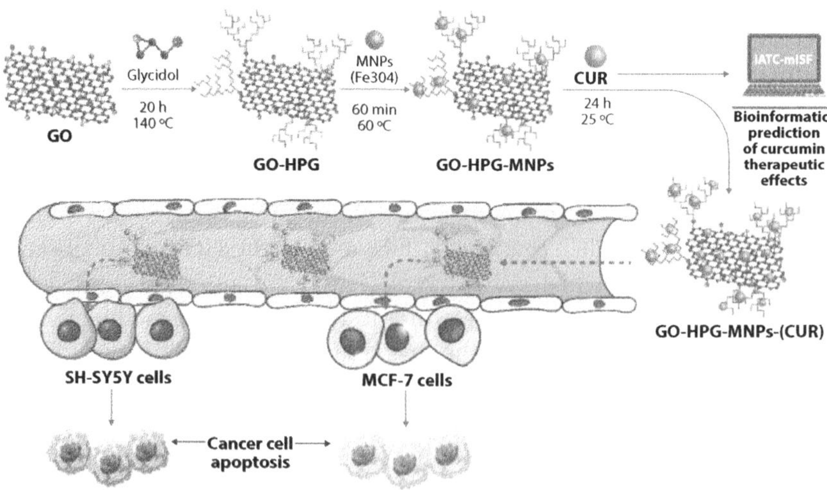

Figure 11.6 The schematic diagram of GO-HPG-MNPs nano-carrier synthesis and CUR loading. Reprint and copyright permission obtained from Elsevier publication [51].

the developed nanocarrier, making it a demandable material of choice for the delivery of other kinds of chemotherapeutic drugs to cancer cells [52].

Currently, triple-negative breast cancer (TNBC) is a major threat and is considered the most aggressive form of breast cancer. The available treatments are going in vain, and the death rate is growing high due to the limited availability of treatments and their high cost. If the patient somehow recovers from this, there is also the possibility of cancer recurrence. To tackle the present challenge, the authors prepared HA-GO-Met, i.e., hyaluronic acid-grafted metformin (a known anticancer drug)-loaded graphene oxide NPs. The developed nanobiomaterial was characterized through FTIR, TEM, SEM, DLS, UV–Vis spectroscopy, etc. The developed nanomaterial exhibited anticancer potential at lower dosages in TNBC *in vivo* and *in vitro* compared to the drug alone. It was found to cause apoptosis and cause cell migration in TNBC cells by targeting the miR-10b/PTEN axis via NFkB-p65. pAKT(473) expression was affected by the upregulation of PTEN, which induces cell apoptosis. On the other hand, it decreased the expression of pFAK/integrinβ1, and restrained cell migration. Hence, it inhibited the probability of cancer reoccurrence by inhibiting the epithelial-to-mesenchymal transition (EMT) and reducing cell stemness [53].

In this study, the authors studied the specific interaction of GO nanosheets with the metabolome present in breast cancer cells. Literature reveals many applications of oxidized GO nanosheets in the biomedical field. Taking this into account, the authors studied the exposure of GO to MCF-7 breast cancer cell lines by applying 1H-NMR profiling. They treated the viable cells with various concentrations of GO and demonstrated their effect on different metabolic activities. The nanomaterials not only influenced catabolism and anabolism but also significantly impacted proline and arginine metabolism and affected aminoacyl-t-RNA biosynthesis. At the same time, it increased the number of metabolic interruptions in cancer steroids, inhibiting cancer cell growth. Hence, these results suggest that GO can significantly influence the metabolism in MCF-7 breast cancer cells and thus can be applied for cancer treatment [54].

A superfast, sensitive electrochemical biosensor was developed by the deposition of phytohemagglutinin-L (PHA-L) and N-doped graphene quantum dots (NGQDs) over the screen-printed electrodes (SE) and was applied against the detection of breast cancer in the MCF-7 cell line in human serum. Nano QDs with 3 nm–12 nm particle diameter were fabricated upon the selected electrode using passion fruit juice through a microwave-assisted hydrothermal process and further conjugated with PHA-L *via* an amide linkage formed between the $-NH_2$ group of PHA-L and the COOH group of NGQDs. The electrical conductivity of the sensor is enhanced by the dual functionalization of NGQDs and displays superfast detection towards MCF-7. It displayed a linear range of detection and detected 20 to 10^6 cells mL^{-1} in human serum and 5 to 10^6 cells mL^{-1} in phosphate-buffered saline with LOD values of 2 and 1 cells mL^{-1}, respectively. Moreover, it displayed outstanding stability and hence acted as an excellent sensing probe to electrochemically detect very low concentrations of cancer cell growth, making early diagnosis and treatment possible [55].

The literature is enriched with many applications of PEGylated GO-based nanomaterials towards cancer treatment because of their superphotothermal capacity and excellent drug loading tendency. Intending to study the effect of functionalization, in this report, the author first functionalized the GO derivatives by poly(2-ethyl-2-oxazoline), an amphiphilic polymer, and then loaded them with the chemotherapeutic drug, DOX (doxorubicin) and D-α-Tocopherol succinate (TOS), generally used for breast cancer chemotherapy. Experimental results were found to be quite satisfactory, and it was investigated whether the 1:3 DOX:TOS molar ratio could exhibit therapeutic abilities toward breast cancer cells. It did not display any toxic effects on the normal cells. The drug delivery ability of DOX:TOS-loaded POxylated GO was outstanding and induced a great therapeutic effect against the studied cancer cells compared to the standard free drug alone. Overall, the POxylated GO loaded with DOX:TOS displayed higher cytotoxicity towards the breast cancer cells and can act as a superior phototherapeutic and drug delivery agent [56].

Transition metal dichalcogenides (TMDs) are one of the 2D nanomaterials. It possesses enormous versatility and, hence, has received many biomedical applications. Particularly, MoS2-based nanocomposites are highly interesting to study because of their advanced biological and physicochemical properties. β-CD functionalized MoS_2-based hybrid nanocomposites were synthesized using a liquid exfoliated hydrothermal process and examined for their morphological, structural, thermal, and optical properties. It exhibited remarkable anticancer properties and inhibited 70.45% of cell growth against the MCF-7 cell line. Additionally, it also displayed potential antibacterial properties. Thus, the developed nanocomposite can be further subjected to biomedical drug delivery and cancer therapy in the near future (Figure 11.7) [57].

Paclitaxel (PTX)-loaded fluorescent carbon nanogel (FCN) as a redox-responsive bioimaging probe was designed and prepared to achieve specific drug delivery to the target cancer cells. FCN was synthesized through the carbonization of disulfide and cross-linking it with HA (hyaluronic acid),

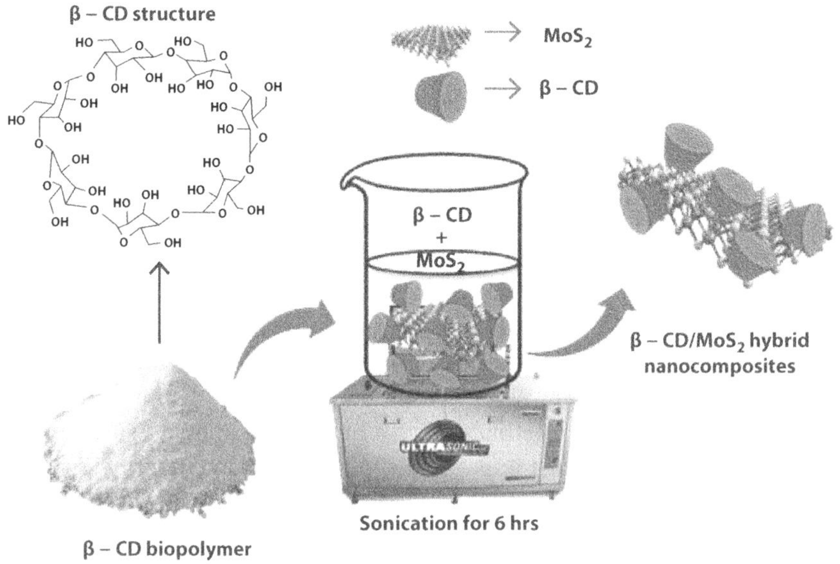

Figure 11.7 Schematic representation of the synthesis of β-CD/MoS2 hybrid NC prepared by liquid exfoliation process. Reprint and copyright permission obtained from Elsevier publication [57].

with subsequent encapsulation of MnO2 nanosheets for fluorescence on/off purposes. PTX-MnO_2/FCN displayed a sudden rise in fluorescence intensity in the presence of over-expression of glutathione (GSH) in cancer cells, leading to a reduction of Mn4+ to Mn2+ ruptures in the disulfide bond. As a result, PTX gets released from FCN. It displayed tremendous biocompatibility with healthy cells and high selectivity towards the target tumor cells, and thus can serve as a valuable material for biomedical applications [58].

A cDNA-ferrocene/MX-based nanoprobe was constructed as a competitive electrochemical aptasensor to detect Mucin1 (MUC1), considered a breast cancer biomarker. MXene (Ti_3C_2) nanosheets were selected as nanocarriers for the aptamer probes as they possess a large surface area and brilliant electrical conductivity. The cDNA-Fc/MXene probe was synthesized by the attachment of ferrocene-labeled complementary DNA, i.e., cDNA-Fc, to the surface of MXene. Then, through Au–S linkage, the MUC1 aptamer was attached to the electrode. The prepared sensing electrode was named Apt/Au/GCE. Merging the probe with the aptamer, the aptasensor was prepared and used to detect MUC1. The interaction of the aptasensor with the target resulted in the detachment of the probe from the sensing electrode, which was noticed by the diminished electrical signal. It offered a linear response for a wide range of concentrations ranging from 1.0 pM to 10 μM with a LOD value of 0.33 pM, which can have profound application for clinical diagnosis [59].

Conclusion

In conclusion, this chapter outlines the advantages of 2D nanomaterials over traditional methods for cancer treatment. Nanomaterials are quite easy to prepare and handle with low-cost synthesis and atom economics, making them more demandable. Traditional therapies are associated with some pitfalls, such as high side effects, high costs, being less target-specific, etc., which require the development of alternatives. Different nanomaterials were designed by targeting different biomarkers inside the cell; some target mitotic cell division and DNA replication, then functionalization was done by PEG or Pox and loaded with chemotherapy drugs like doxorubicin, paclitaxel, etc. The specificity of the nanomaterials was studied through bioimaging using fluorescent probes. Nanomaterials were found to exhibit strong anti-breast cancer potential by improving their pharmacokinetic properties as well as drug delivery efficacy. At the same time, they exhibited superior biocompatibility and high stability with no side effects. The presence of carbon in the nanomaterials makes them unique, desirable, and less toxic to healthy cells. This chapter will help the researcher gain an idea about 2D nanomaterial synthesis and application and will encourage them to explore their biomedical and clinical properties more in the future.

Acknowledgments

PP is thankful to the Department of Chemistry, Ravenshaw University, Cuttack, Odisha, India. SC is grateful to the School of Sciences, IES University, Bhopal, Madhya Pradesh, India.

References

1. Zhu, W., Li, H., Luo, P., Emerging 2D nanomaterials for multimodel theranostics of cancer. *Front. Bioeng. Biotechnol.*, 9, 769178, 2021.
2. Panda, P. and Chakroborty, S., Navigating the synthesis of quinoline hybrid molecules as promising anticancer agents. *ChemistrySelect*, 5, 10187–10199, 2020.

3. Panda, P., Nayak, S., Sahoo, S.K., Mohapatra, S., Nayak, D., Pradhan, R., Kundu, C.N., Diastereoselective synthesis of novel spiro indanone fused pyrano [3, 2-c] chromene derivatives following hetero-Diels–Alder reaction and *in vitro* anticancer studies. *RSC Adv.*, 8, 16802–16814, 2018.
4. Panda, P., Nayak, S., Bhakta, S., Mohapatra, S., Murthy, T.R., Design and synthesis of (Z/E)-2-phenyl/H-3-styryl-2 H-chromene derivatives as antimicrotubule agents. *J. Chem. Sci.*, 130, 127–142, 2018.
5. Raja, I.S., Kang, M.S., Kim, K.S., Jung, Y.J., Han, D.-W., Two-dimensional theranostic nanomaterials in cancer treatment: State of the art and perspectives. *Cancers (Basel)*, 12, 6, 1657, 2020.
6. Liu, J., Zhao, C., Chen, W.R., Zhou, B., Recent progress in two-dimensional nanomaterials for cancer theranostics. *Coord. Chem. Rev.*, 469, 214654, 2022.
7. Wang, J. and Yang, M., Two-dimensional nanomaterials in cancer theranostics, in: *Theranostic Bionanomaterials. Micro and Nano Technologies*, pp. 263–288, 2019.
8. Mohammadpour, Z. and Majidzadeh-A, K., Applications of two-dimensional nanomaterials in breast cancer theranostics. *ACS Biomater. Sci. Eng.*, 6, 1852–1873, 2020.
9. Raja, I.S., Kang, M.S., Kim, K.S., Jung, Y.J., Han, D.-W., *Cancers*, 12, 6, 1657, 2020.
10. Yang, W., Lyu, Q., Zhao, J., Cao, L., Hao, Y., Zhang, H., Recent advance in near-infrared/ultrasound-sensitive 2D-nanomaterials for cancer therapeutics. *Sci. China Mater.*, 63, 12, 2397–2428, 2020.
11. Wu, J., Hu, T., Zhao, G., Li, A., Liang, R., Two-dimensional transition metal chalcogenide nanomaterials for cancer diagnosis and treatment. *Chin. Chem. Lett.*, 33, 10, 4437–4448, 2022.
12. Gong, L., Yan, L., Zhou, R., Xie, J., Wu, W., Gu, Z., Two-dimensional transition metal dichalcogenide nanomaterials for combination cancer therapy. *J. Mater. Chem. B*, 5, 1873–1895, 2017.
13. Liu, L., Ma, Q., Cao, J., Gao, Y., Han, S., Liang, Y., Zhang, T., Song, Y., Sun, Y., *Cancer Nanotechnol.*, 12, 18, 2021.
14. Li, Z., Shao, J., Luo, Q., Yu, X.-F., Xie, H., Fu, H., Tang, S., Wang, H., Han, G., Chu, P.K., *Biomaterials*, 133, 37–48, 2017.
15. Hussein, E.A., Zagho, M.M., Nasrallah, G.K., Elzatahry, A.A., Recent advances in functional nanostructures as cancer photothermal therapy. *Int. J. Nanomed.*, 2897–2906, 2018.
16. Lee, W.-F., Functional nanomaterials and their 2D and 3D fabrications for drug delivery applications. *MATEC Web Conf.*, 353, 01023, 2021.
17. Navya, P.N., Kaphle, A., Srinivas, S.P., Bhargava, S.K., Rotello, V.M., Daima, H.K., Current trends and challenges in cancer management and therapy using designer nanomaterials. *Nano Converg.*, 6, 23, 2019.
18. Brindhadevi, K., AL Garalleh, H., Alalawi, A., Al-Sarayreh, E., Pugazhendhi, A., Carbon nanomaterials: Types, synthesis strategies and their application

as drug delivery system for cancer therapy. *Biochem. Eng. J.*, 192, 108828, 2023.
19. Li, X., Ma, F., Yang, M., Zhang, J., Nanomaterial based analytical methods for breast cancer biomarker detection. *Mater. Today Adv.*, 14, 100219, 2022.
20. Farheen, J., Hosmane, N.S., Zhao, R., Zhao, Q., Iqbal, M.Z., Kong, X., Nanomaterial-assisted CRISPR gene-engineering–A hallmark for triple-negative breast cancer therapeutics advancement. *Mater. Today Bio*, 16, 100450, 2022.
21. Wu, M., Niu, X., Zhang, R., Xu, Z.P., Nanomaterial-assisted CRISPR gene-engineering–A hallmark for triple-negative breast cancer therapeutics advancement. *Adv. Drug Deliv. Rev.*, 187, 114360, 2022.
22. Ashfaq, M., Talreja, N., Chauhan, D., Afreen, S., Sultana, A., Srituravanich, W., Two-dimensional (2D) hybrid nanomaterials for diagnosis and treatment of cancer. *J. Drug Deliv. Sci. Technol.*, 70, 103268, 2022.
23. Luo, G., Li, Z., Zheng, J., Yang, H., Li, X., Guishan, X., Duo, Y., Recent progresses on radiotherapeutics-based treatment of cancer with two-dimensional nanomaterials. *Appl. Mater. Today*, 29, 101584, 2022.
24. Panda, S., Hajra, S., Kaushik, A., Rubahn, H.G., Mishra, Y.K., Kim, H.J., Smart nanomaterials as the foundation of a combination approach for efficient cancer theranostics. *Mater. Today Chem.*, 26, 101182, 2022.
25. Xiang, Y., Hu, C., Wu, G., Xu, S., Li, Y., Nanomaterial-based microfluidic systems for cancer biomarker detection: Recent applications and future perspectives. *TrAC Trends Anal. Chem.*, 158, 116835, 2023.
26. Yang, Z., Wang, L., Zhang, J., Liu, J., Yu, X., *Nano Today*, 49, 101799, 2023.
27. Alamdari, S.G., Amini, M., Jalilzadeh, N., Baradara, B., Mohammadzadeh, R., Mokhtarzadeh, A., Oroojalian, F., *J. Control. Release*, 349, 269–303, 2022.
28. Wang, Y., Li, J., Li, X., Shi, J., Jiang, Z., Zhang, C.Y., Graphene-based nanomaterials for cancer therapy and anti-infections. *Bioact. Mater.*, 14, 335–349, 2022.
29. Ni, N., Zhang, X., Ma, Y., Yuan, J., Wang, D., Ma, G., Dong, J., Sun, X., Biodegradable two-dimensional nanomaterials for cancer theranostics. *Coord. Chem. Rev.*, 458, 214415, 2022.
30. Murali, A., Lokhande, G., Deo, K.A., Brokesh, A., Gaharwar, A.K., Emerging 2D nanomaterials for biomedical applications. *Mater. Today*, 50, 276–302, 2021.
31. Fan, H. and Guo, Z., Tumor microenvironment-responsive manganese-based nanomaterials for cancer treatment. *Coord. Chem. Rev.*, 480, 215027, 2023.
32. Murugan, C., Sharma, V., Murugan, R.K., Malaimegu, G., Sundaramurthy, A., Two-dimensional cancer theranostic nanomaterials: Synthesis, surface functionalization and applications in photothermal therapy. *J. Control. Release*, 299, 1–20, 2019.
33. Yoon, J., Shin, M., Lee, J.-Y., Lee, S.-N., Choi, J.-H., Choi, J.-W., RNA interference (RNAi)-based plasmonic nanomaterials for cancer diagnosis and therapy. *J. Control. Release*, 342, 228–240, 2022.

34. Jing, Z., Du, Q., Zhang, X., Zhang, Y., Nanomedicines and nanomaterials for cancer therapy: Progress, challenge and perspectives. *Chem. Eng. J.*, 446, 137147, 2022.
35. Zhao, Y., Wang, S.-B., Chen, A.-Z., Kankala, R.K., Nanoarchitectured assembly and surface of two-dimensional (2D) transition metal dichalcogenides (TMDCs) for cancer therapy. *Coord. Chem. Rev.*, 472, 214765, 2022.
36. Zheng, J.L., Meng, D.D., Zheng, X., Zhang, Y., Chen, H.F., Graphene-based materials: A new tool to fight against breast cancer. *Int. J. Pharm.*, 603, 120644, 2021.
37. Sachdeva, H., Khandelwal, A.R., Meena, R., Sharma, K., Khatik, N., Graphene-based nanomaterials for cancer therapy. *Mater. Today Proc.*, 43, 2954–2957, 2021.
38. Dolatkhah, M., Hashemzadeh, N., Barar, J., Adibkia, K., Aghanejad, A., Barzegar-Jalali, M. and Omidi, Y., Graphene-based multifunctional nanosystems for simultaneous detection and treatment of breast cancer. *Colloids Surf. B Biointerfaces*, 193, 111104, 2020.
39. Panda, P., Pal, K., Chakroborty, S., Smart advancements of key challenges in graphene-assembly glucose sensor technologies: A mini review. *Mater. Lett.*, 303, 130508, 2021.
40. Melo, B.L., Lima-Sousa, R., Alves, C.G., Moreira, A.F., Correia, I.J., de MeloDiogo, D., Chitosan-based injectable *in situ* forming hydrogels containing dopamine-reduced graphene oxide and resveratrol for breast cancer chemo-photothermal therapy. *Biochem. Eng. J.*, 185, 108529, 2022.
41. Chen, W., Li, S., Shen, Y., Cai, Y., Jin, J., Yang, Z., Polyethylenimine modified graphene oxide for effective chemo-gene-photothermal triples therapy of triple-negative breast cancer and inhibits metastasis. *J. Drug Deliv. Sci. Technol.*, 74, 103521, 2022.
42. Rana, A., Matiyani, M., Tewari, C., Negi, P.B., Arya, M.C., Das, V., Pal, M., Sahoo, N.G., Functionalized graphene oxide based nanocarrier for enhanced cytotoxicity of Juniperus squamata root essential oil against breast cancer cells. *J. Drug Deliv. Sci. Technol.*, 72, 103370, 2022.
43. Jhang, J.-W., Chou, Y.-H., Wang, T.-H., Hsieh, M.-H., Chiang, W.-H., One-pot green reduction and surface decoration of graphene oxide nanosheets with PEGylated chitosan for application in cancer photothermal therapy. *J. Taiwan Inst. Chem. Eng.*, 134, 104359, 2022.
44. Hatamie, S., Akhavan, O., Sadrnezhaad, S.K., Ahadian, M.M., Shirolkar, M.M., Wang, H.Q., Curcumin-reduced graphene oxide sheets and their effects on human breast cancer cells. *Mater. Sci. Eng. C*, 55, 482–489, 2015.
45. Yang, D., Feng, L., Dougherty, C.A., Luker, K.E., Chen, D., Cauble, M.A., Holl, M.M.B., Luker, G.D., Ross, B.D., Liu, Z., Hong, H., *In vivo* targeting of metastatic breast cancer via tumor vasculature-specific nano-graphene oxide. *Biomaterials*, 104, 361–371, 2016.

46. Tao, Y. and Auguste, D.T., Array-based identification of triple-negative breast cancer cells using fluorescent nanodot-graphene oxide complexes. *Biosens. Bioelectron.*, 81, 431–437, 2016.
47. Salahandish, R., Ghaffarinejad, A., Omidinia, E., Zargartale, H., Majidzadeh-A, K., Naghi, S.M., Sanati-Nezhad, A., Label-free ultrasensitive detection of breast cancer miRNA-21 biomarker employing electrochemical nano-genosensor based on sandwiched AgNPs in PANI and N-doped graphene. *Biosens. Bioelectron.*, 120, 129–136, 2018.
48. Li, J., Liang, X., Zhang, J., Yin, Y., Zuo, T., Wang, Y., Yang, X., Shen, Q., Inhibiting pulmonary metastasis of breast cancer based on dual-targeting graphene oxide with high stability and drug loading capacity. *Nanomed. Nanotechnol. Biol. Med.*, 14, 1237–1248, 2018.
49. Zhang, J., Yan, L., Wei, P., Zhou, R., Hua, C., Xiao, M., Tu, Y., Gu, Z., Wei, T., PEG-GO@XN nanocomposite suppresses breast cancer metastasis via inhibition of mitochondrial oxidative phosphorylation and blockade of epithelial-to-mesenchymal transition. *Eur. J. Pharmacol.*, 895, 173866, 2021.
50. Azimzadeh, M., Rahaiea, M., Nasirizadeh, N., Ashtari, K., Naderi-Manesh, H., An electrochemical nanobiosensor for plasma miRNA-155, based on graphene oxide and gold nanorod, for early detection of breast cancer. *Biosens. Bioelectron.*, 77, 99–106, 2016.
51. Paknia, F., Mohabatkar, H., Ahmadi-Zeidabadi, M., Zarrabi, A., The convergence of *in silico* approach and nanomedicine for efficient cancer treatment; *in vitro* investigations on curcumin loaded multifunctional graphene oxide nanocomposite structure. *J. Drug Deliv. Sci. Technol.*, 71, 103302, 2022.
52. Molaparast, M., Malekinejad, H., Rahimi, M., Shafiei-Iranneja, V., Biocompatible functionalized graphene nanosheet for delivery of doxorubicin to breast cancer cells. *J. Drug Deliv. Sci. Technol.*, 70, 103234, 2022.
53. Basu, A., Upadhyay, P., Ghosh, A., Bose, A., Gupta, P., Chattopadhyay, S., Chattopadhyay, D., Adhikary, A., Hyaluronic acid engrafted metformin loaded graphene oxide nanoparticle as CD44 targeted anti-cancer therapy for triple negative breast cancer. *Biochim. Biophys. Acta-Gen. Subj.*, 1865, 129841, 2021.
54. Raja, G., Selvaraj, V., Suk, M., Suk, K.T., Kim, T.-J., Metabolic phenotyping analysis of graphene oxide nanosheets exposures in breast cancer cells: Metabolomics profiling techniques. *Process Biochem.*, 104, 39–45, 2021.
55. Tran, H.L., Dang, V.D., Dega, N.K., Lu, S.-M., Huang, Y.-F., Doong, R., Ultrasensitive detection of breast cancer cells with a lectin-based electrochemical sensor using N-doped graphene quantum dots as the sensing probe. *Sens. Actuators B. Chem.*, 368, 132233, 2022.
56. de Melo-Diogoa, D., Costa, E.C., Alvesa, C.G., Lima-Sousa, R., Ferreir, P., Louro, R.O., Correi, I.J., *Eur. J. Pharm. Biopharm.*, 131, 162–169, 2018.
57. Kasinathan, K., Marimuthu, K., Murugesan, B., Pandiyan, N., Pandi, B., Mahalingam, S., Selvaraj, B., Cyclodextrin functionalized multi-layered MoS2 nanosheets and its biocidal activity against pathogenic bacteria and

MCF-7 breast cancer cells: Synthesis, characterization and *in-vitro* biomedical evaluation. *J. Mol. Liq.*, 323, 114631, 2021.
58. Choia, C.A., Ryplid, B., In, I., Park, S.Y., Selective redox-responsive theragnosis nanocarrier for breast tumor cells mediated by MnO2/fluorescent carbon nanogel. *Eur. J. Pharm. Sci.*, 134, 256–265, 2019.
59. Wang, H., Sun, J., Lu, L., Yang, X., Xia, J., Zhang, F., Wang, Z., Competitive electrochemical aptasensor based on a cDNA-ferrocene/MXene probe for detection of breast cancer marker Mucin1. *Anal. Chim. Acta*, 1094, 18–25, 2020.

12

2D Nanomaterials for Photocatalytic Hydrogen Production

Grandprix T. M. Kadja[1,2,3*], St Mardiana[1] and Arxhel S. F. Nanda[1]

[1]*Division of Inorganic and Physical Chemistry, Faculty of Mathematics and Natural Sciences, Institut Teknologi Bandung, Bandung, West Java, Indonesia*
[2]*Center for Catalysis and Reaction Engineering, Institut Teknologi Bandung, Bandung, West Java, Indonesia*
[3]*Research Center for Nanosciences and Nanotechnology, Institut Teknologi Bandung, Bandung, West Java, Indonesia*

Abstract

Photocatalytic hydrogen production has been in the spotlight owing to its ability to generate clean and sustainable energy sources. Two-dimensional (2D) nanomaterials have also emerged extensively in photocatalytic hydrogen production applications owing to their extraordinary optical and electronic properties. This chapter provides comprehensive information on the fundamentals of photocatalytic hydrogen production. Furthermore, an in-depth understanding of photocatalytic hydrogen production using 2D nanomaterials, including graphene-based materials, carbon nitrides, transition metal dichalcogenides, and MXene, is also elaborated in detail. In this context, the roles and performance of these nanomaterials are discussed. Furthermore, strategies for enhancing photocatalytic performance are also summarized. Finally, the future development and prospects of 2D nanomaterials for photocatalytic hydrogen production were analyzed.

***Keywords*:** Photocatalytic hydrogen production, 2D nanomaterials, graphene, carbon nitrides, transition metal dichalcogenides, MXene

**Corresponding author*: grandprix.thomryes@itb.ac.id

Subhendu Chakroborty and Kaushik Pal (eds.) 2D Nanomaterials: Synthesis, Properties and Applications, (263–288) © 2024 Scrivener Publishing LLC

12.1 Introduction

The rapid increase in the global population, which is growing in line with energy demand, has shifted the need for energy to prospect as a sustainable energy source [1–4]. In this sense, energy consumption has been threatened by restrictions on energy demand, climate issues, and accelerated fossil fuel depletion. Furthermore, public health issues are also threatened by pollutants. Therefore, alternative clean, renewable, harmless, and environmentally benign energy sources attracted considerable attention.

Hydrogen has become a promising alternative energy source because it is clean, abundant, sustainable, and environmentally friendly [5, 6]. In addition, hydrogen has the highest energy content per weight (122 kJ/mol) and generates only water as a product [2, 4]. Furthermore, the abundance of sunlight and water provided by nature allows hydrogen production to address the energy crisis. However, to date, appropriate strategies for generating hydrogen are still being developed. Recently, photocatalytic hydrogen production as a hydrogen production reaction pathway has been widely considered and studied as a potential breakthrough.

Photocatalytic hydrogen production process was pioneered in 1972 by Honda and Fujishima [2, 7]. Photocatalytic reactions are widely known to be the most challenging and attractive for hydrogen production processes, especially for industrial applications [1, 8]. Moreover, the need to produce sustainable energy with environmental protection has led to the use of raw materials as well as waste and greenhouse gas reduction [9]. Hence, semiconductor materials for photocatalysts in hydrogen production have become the critical basis for accomplishing energy demand. Thus, a stable and efficient photocatalyst is expected to increase hydrogen production.

In the last decade, 2D nanomaterials have become better photocatalyst materials owing to their exceptional surface-area properties [6]. 2D nanomaterials are a new type of material with sheet-like structures and a few atoms or single thicknesses [10]. In addition, 2D nanomaterials have a unique ultrathin layer with a potential energy bandgap for photocatalytic production because of their size limitations and electron confinement [11]. In addition, 2D nanomaterial possess numerous surface-active sites, high surface area, superior electron mobility, and excellent electron transfer [11, 12]. Therefore, 2D nanomaterials are promising photocatalyst candidates in photocatalytic hydrogen production.

Herein, we describe the principles of photocatalytic hydrogen production. Furthermore, several 2D nanomaterials, *i.e.*, graphene-based materials, carbon nitrides, transition metal dichalcogenides, and MXene, have been elaborated in detail according to their role and performance in photocatalytic hydrogen production reactions. In this sense, we also provide several approaches for enhancing their performance in photocatalytic hydrogen production reactions. Finally, we conclude this chapter with insights into and future prospects for photocatalytic hydrogen production.

12.2 Basics of Photocatalytic Hydrogen Production

Photocatalytic hydrogen production can be defined as the conversion of solar energy to chemical energy using a compatible photocatalyst. Since its initiation by Honda and Fujishima in 1792, they introduced a photoelectrochemical (PEC) cell for water photoelectrocatalytic cleavage (Figure 12.1a); numerous approaches have been developed by researchers for hydrogen generation.

There are two primary mechanisms for photocatalytic hydrogen production, *i.e.*:

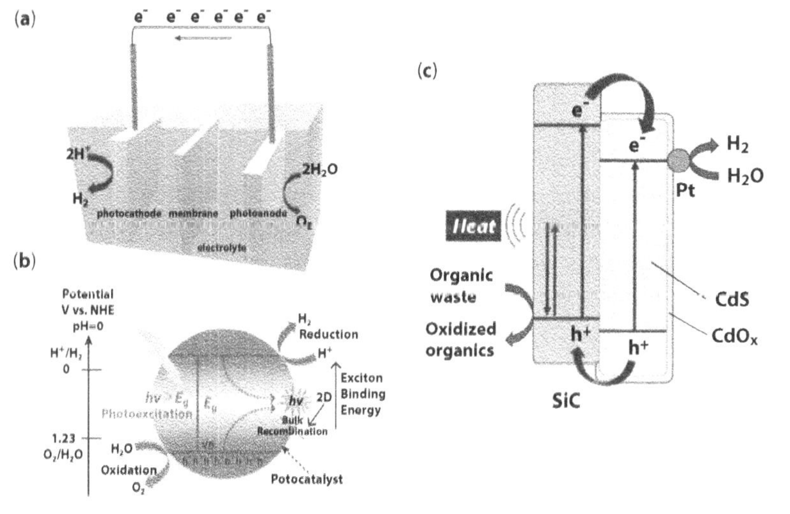

Figure 12.1 (a) Schematic diagram of the PEC. Reproduced with permission from Ref. [2], Copyright *2019* American Chemical Society. (b) Schematic illustration of the basic principles of photocatalytic water splitting. Reproduced with permission from Ref. [10], Copyright 2018 American Chemical Society. (c) Illustration of photoreforming organic of waste into hydrogen. Reproduced with permission from Ref. [13], Copyright 2021 American Chemical Society.

1. Photocatalytic water splitting.
2. Photoreforming of organic compounds.

The photocatalytic water splitting process produces hydrogen from water *via* two primary reactions: the oxygen evolution reaction (EOR) and the hydrogen evolution reaction (HER) (Figure 12.1b). This process produces hydrogen under light irradiation compatible with a semiconductor bandgap. Electron holes were continuously created by electrons excited from the valence band to the conduction band. Overall water splitting reaction:

$$H_2O \rightarrow H_2 + \frac{1}{2}O_2 \tag{12.1}$$

Furthermore, in the water splitting process, the lower part of the conduction band must be more negative than the H^+/H_2 reduction potential. In contrast, the upper part of the valence band must be more positive than the H_2O/O_2 potential oxidation. Moreover, the semiconductor bandgap must pass through the free energy, *i.e.*, 1.23 eV. Notably, electron hole pair recombination in the bulk and on their enroute to the bulk surface can occur during this reaction [10]. Furthermore, the desired photocatalyst must possess a much greater exposed surface and a shorter charge transfer distance. Hence, research on water splitting reactions has focused on the development of efficient photocatalyst materials.

Another alternative route for photocatalytic hydrogen production is photocatalytic reforming of organics. In contrast to the water-splitting reaction, this process yields different products, *i.e.*, H_2 and CO_2. The photoreforming process reaction:

$$C_xH_yO_z + (2x - z)H_2O \rightarrow (2x - z + \frac{y}{2})H_2 + xCO_2 \tag{12.2}$$

In this route, by consuming holes more quickly than the OH^- ions, the organic compound serves as a hole scavenger and gradually oxidizes into CO_2 and H_2O. In addition, for complex organic molecules, a variety of intermediates could be expected. Moreover, this process can also be exploited to convert organic compounds for water depuration and purification. In this respect, the organic compounds photoreforming can be regarded as a virtuous process of yielding hydrogen and, at the same time, depurating water from waste. Figure 12.1c exhibited the process of hydrogen production using the photoreforming of organic waste, which was investigated by Nakagawa and Nagata using a CdO_x/CdS/SiC photocatalyst. This process showed that the organic wastes could produce hydrogen using the photoreforming method.

Photoreforming of organic compounds usually uses methanol, ethanol, and sugar as organic compounds and TiO_2 and CdS as catalysts to generate hydrogen. This is because the organic compound acts as the proton source, while TiO_2 has a high overpotential for hydrogen production. Furthermore, it should be noted that the amount of metal in this process needs to be optimized because it can lead to the recombination center for electrons and holes.

12.3 2D Nanomaterials for Photocatalytic Hydrogen Production

12.3.1 Graphene-Based

Graphene (Figure 12.2a), as a 2D nanomaterial with a hexagonal structure consisting of a sp^2 hybridized carbon atom network and a 0 band gap semiconductor, has attracted enormous attention as a material for photocatalytic hydrogen production [14]. Besides, graphene has remarkable conductivity, exceptional optical transmittance, fast charge carrier mobility at room temperature, and a large surface area of 10^6 S cm^{-1}, ~97.7%, 200,000 cm^2 V^{-1} s^{-1}, and 2,630 m^2 g^{-1}, respectively [6, 14]. Hence, it turns out to be a potential candidate for photocatalytic hydrogen production. In this regard, the possibility of producing graphene from biomass also contributed to graphene's prominence. A schematic illustration of graphene's role in the photocatalytic process is exhibited in Figure 12.2b [15].

Generally, graphene can be generated by both bottom-up and top-down methods. Synthesis methods for graphene may have an effect on their intrinsic properties. Moreover, the thickness and flexibility of graphene layers might correspond to their electronic properties, such as chemical doping (p-doped) and electron-donating nitrogen functionalities (n-doped). Nevertheless, graphene oxide (GO) and reduced graphene oxide (rGO) as functionalized graphene, which allows chemical modification, could make the synthesis process of graphene-based materials more efficient [14]. The graphene preparation procedures are exhibited in Figure 12.2c.

Quiroz-Cardoso *et al.* reported the application of a graphene oxide–TiO_2 composite containing nickel nanoparticles (Ni/GO–TiO_2)

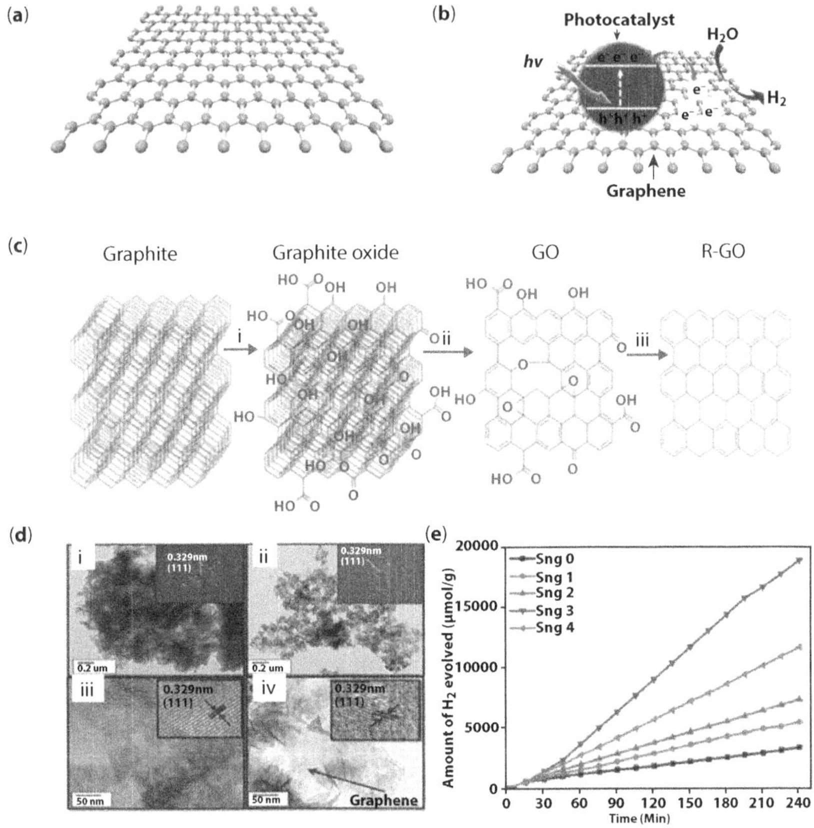

Figure 12.2 (a) The graphene structure illustration. (b) Graphene's role in the photocatalytic hydrogen production process. Reproduced with permission from Ref. [15]. Copyright 2016, Elsevier. (c) The synthesis procedures for graphene consist of oxidation, exfoliation, and final reduction. Reproduced with permission from Ref. [17]. Copyright 2019 MDPI. (d) FETEM images of Sn_3O_4 (i & iii) and Sn_3O_4/graphene (ii & iv). (e) The hydrogen production from the photocatalytic process using Sng0 (pristine Sn_3O_4) and 1%–4% graphene/Sn_3O_4 (coded as Sng1, Sng2, Sng3, and Sng4, respectively). Reproduced with permission from Ref. [18]. Copyright 2021 RSC.

for photocatalytic hydrogen production. The modification of TiO_2 with GO was expected to efficiently improve electron transfer. On the other hand, the presence of nickel nanoparticles was supposed to act as a co-catalyst, *i.e.*, act as an electron sink. The result showed that the hydrogen evolution rate of TiO_2 was successfully increased 1.3 times and 2.7 times with the addition of GO (GO–TiO_2 composite) and nickel nanoparticles (Ni/GO-TiO_2), respectively [4]. Previously, Gurbani *et al.* investigated nickel phosphate-based graphene (GO:NPO) nanocomposites to produce hydrogen using the photocatalytic method. In this sense, phosphate ions will act as bridging groups for metal precursors and graphene oxide [16]. The result exhibits hydrogen production activity up to 15.37 μmol H_2 h^{-1}.

Considering the modification of GO for hydrogen production, a novel cobalt ion implanted in 5,15-diphenyl-10,20-di(4-pyridyl)porphyrin-pillared graphene oxide (GO-Co^{2+}-DPyP) was reported by Luo *et al.* The electron transfer and the strong interaction of DPyP and graphene oxide increase photocatalytic activity. Moreover, the Co^{2+} ion serves as the linker of interfacials and is able to modify DPyP and graphene oxide morphology and structure [19].

Sethi *et al.* used Sn_3O_4 as an attractive semiconductor with graphene to form the heterostructure, which is expected to increase hydrogen production. This is regarding the ability of that combination to diminish the electron-hole pair recombination rate and enhance the active sites for the photocatalytic process [18]. The result shows a hydrogen production rate of 4,687 μmol h^{-1} g^{-1}. This value shows that Sn_3O_4 with graphene has the hydrogen production rate that is 5.7-fold higher than that of Sn_3O_4 alone (Figure 12.2e). It is worth noting that Sn_3O_4 nanoflakes were grown parallel and uniformly on the graphene surface (Figure 12.2b). Thus, the diffusion distance is short, and the electron transport is accelerated.

Recently, Singh *et al.* reported plasmonic Au–rGO–TiO_2 for hydrogen production *via* the photocatalytic process. Plasmonic is defined as a metal nanoparticle-sensitized heterostructured photocatalyst system [20]. The photocatalyst was prepared *via* the deposition-precipitation method. The result exhibited a rate of hydrogen evolution of 12.19 mmol g^{-1} h^{-1}. This is due to the presence of Au nanoparticles, which increase the local electric fields and light harvesting. Moreover, the combination of Au–rGO–TiO_2 exhibits the chance to modify the materials with different properties to increase hydrogen production.

12.3.2 Carbon Nitrides

Carbon nitrides are defined as polymeric materials that consist of carbon and nitrogen. The crystalline structure of carbon nitrides (C_3N_4) involves β-, α-, g-, cubic-, and defect zincblende-C_3N_4. Graphitic carbon nitride (g-C_3N_4) is the carbon nitride allotrope material, which is stable at room temperature and consists of a metal-free conjugated polymer in two-dimensional structures [21–23]. In addition, triazine (C_3N_3) and heptazine (C_6N_7) are the primary tectonic units for establishing the g-C_3N_4 allotropes (Figures 12.3a–b) [24]. Besides, g-C_3N_4 comprises a small amount of hydrogen and an amine group as the primary and/or secondary terminating agents (Figure 12.3c). The term graphitic originates from the similarity with graphite, which consists of sp^2-hybridized [22]. The difference lies in the existence of an intrinsic vacancy bordered by nitrogen atoms.

Moreover, the graphitic carbon nitrides have a band gap up to 2.56-2.70 eV, allowing these materials to be highly efficient photocatalyst materials for hydrogen production. Nevertheless, to gain more efficient performance, modification with metal and morphological alteration needs to be investigated. Further, the photocatalytic process still faces serious charge-carrier recombination.

Hu *et al.* reported the modification of graphitic carbon nitrides with Pt (Pt SAs/C_3N_4) for hydrogen production (Figure 12.3d). In this work, hydrogen production was also tested in pristine C_3N_4, C_3N_4 with the presence of NH_3 (C_3N_4-NH_3), and graphitic carbon nitrides with Pt nanoparticles (Pt NPs/C_3N_4). However, the Pt-N bridge bond can increase the hydrogen production rate of Pt SAs/C_3N_4 by 573.6 μmol h^{-1} due to the enhanced separation of charges and shortening the distance of electronic transfer. In this case, the Pt SAs/C_3N_4 catalyst denotes more exceptional activity than others (Figure 12.3e). Furthermore, the result exhibits the robust stability of Pt SAs/C_3N_4 after four cycles with a constant hydrogen production rate, as depicted in Figure 12.3f [25].

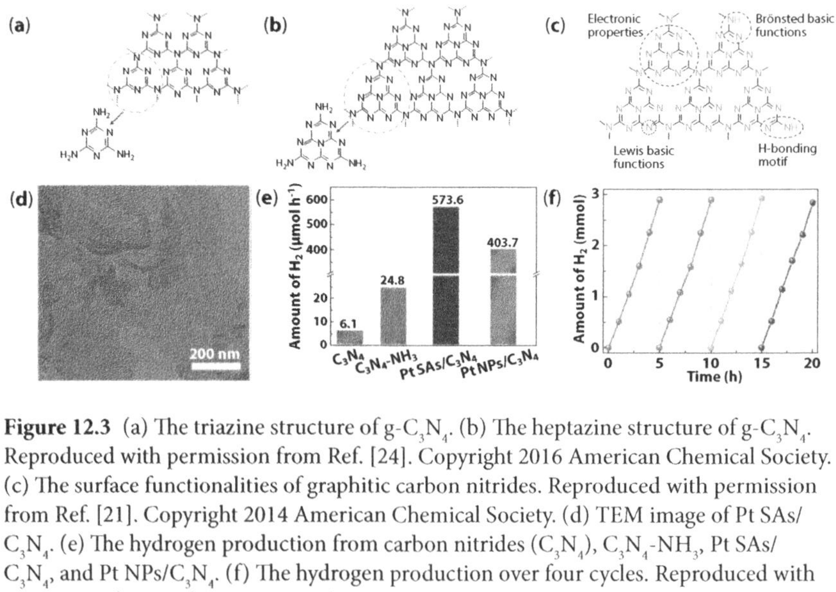

Figure 12.3 (a) The triazine structure of g-C_3N_4. (b) The heptazine structure of g-C_3N_4. Reproduced with permission from Ref. [24]. Copyright 2016 American Chemical Society. (c) The surface functionalities of graphitic carbon nitrides. Reproduced with permission from Ref. [21]. Copyright 2014 American Chemical Society. (d) TEM image of Pt SAs/C_3N_4. (e) The hydrogen production from carbon nitrides (C_3N_4), C_3N_4-NH_3, Pt SAs/C_3N_4, and Pt NPs/C_3N_4. (f) The hydrogen production over four cycles. Reproduced with permission from Ref. [25]. Copyright 2021 Elsevier.

Kheradmand *et al.* reported the iron oxides anchoring to the carbon nitrides nanotube to increase hydrogen production. This photocatalyst material is expected to provide a combination of non-toxic, low-cost, and readily available properties. The result showed that Fe_3O_4/C_3N_4 NT nanocomposite exhibits an excellent hydrogen production rate of 1.9 times than of the solely graphitic carbon nitrides. In this regard, this modification could decrease the recombination rate [23].

12.3.3 Transition Metal Dichalcogenides

Recently, 2D transition metal dichalcogenides (TMDs) have gained significant interest due to their readiness to exploit catalytic processes. In this sense, TMDs have peculiar electronic and optical structures required for photocatalytic hydrogen production [26–28]. TMDs are composed of two chalcogen atoms (X) sandwiched among two hexagonal layers of metal atoms (M), having the stoichiometry MX_2. Moreover, covalent bonding holds the sheets together, and van der Waals interactions hold the sheets next to each other [2]. Considering the TMD's outstanding properties, the synthesis strategies to produce the 2D molybdenum disulfide (MoS_2), the most well-known and used of the TMD materials, have been widely investigated for development. For instance, Sim *et al.* have prepared the semiconductor MoS_2 *via* chemical exfoliation and solvent thermal treatment methods (Figure 12.4a). The result reveals that those combination methods could enhance the stability of this material dispersion for up to 10 months [29].

Yi *et al.* have improved the photocatalytic performance of hydrogen evolution by engineering the noble metal-free molybdenum diselenide ($MoSe_2$). In this case, the $MoSe_2$, which acts as a co-catalyst on 2D-C_3N_4, was prepared by the solvothermal method to produce the interlayered-expanded $MoSe_2$ with an octahedral coordination phase (1T). The result showed a hydrogen evolution rate of up to 1,672.6 $mg^{-1}h^{-1}$ over a 420 nm wavelength. This excellent result is due to the improvement of charge extraction and structure optimization. Moreover, it is worth bearing in mind that although the fixed ante from interlayer engineering and the synergistic phase in this work is still elusive, those strategies successfully increase the photocatalytic performance of the TMD materials [30].

Ma *et al.* have prepared the 2D hybrid CdS nanosheets with MoS_2, which acts as a co-catalyst to enhance the photocatalytic activity. The result exhibits that the presence of MoS_2 as a co-catalyst could enhance the hydrogen production of CdS nanosheets by 2.03 folds, *i.e.*, 1.75 mmol $g^{-1}h^{-1}$. Moreover, the photocatalytic performance of CdS with MoS_2 is also

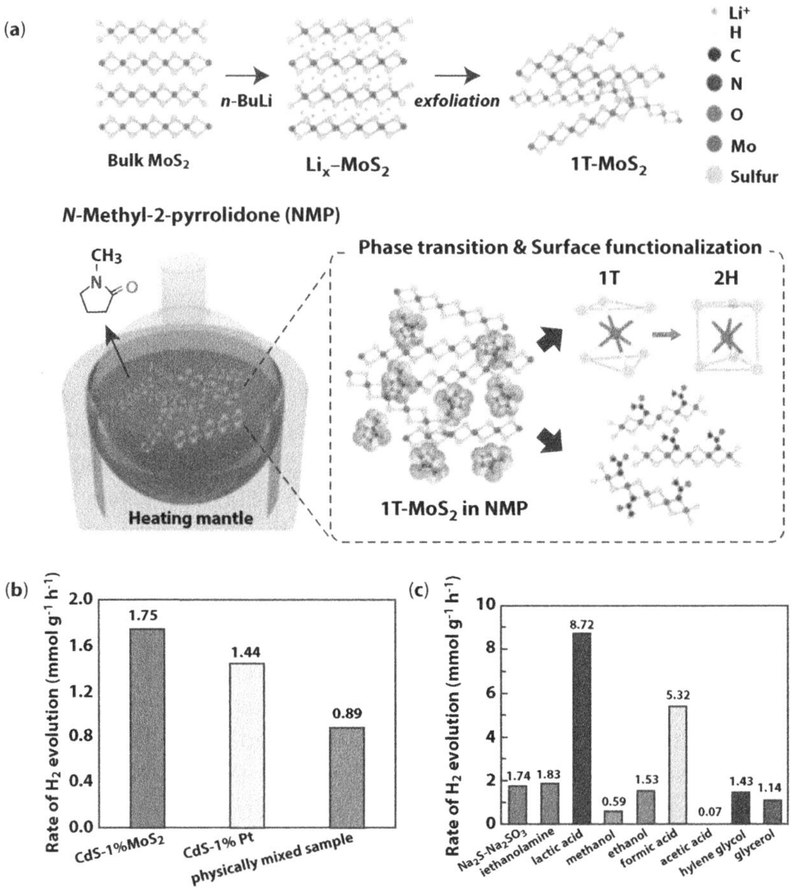

Figure 12.4 (a) The schematic process for preparing the dispersion of 2H-MoS_2 using chemical exfoliation and solvent thermal treatment. Notably, the chemical exfoliate and solvent thermal treatment, respectively, were performed using n-butyl lithium and N-methyl-2-pyrrolidone (NMP) as solvents. Reproduced with permission from Ref. [29]. Copyright 2017, American Chemical Society. (b) The hydrogen evolution rate of CdS-1% MoS_2 compared with CdS-1% Pt and CdS-1% MoS_2 with physical mixed. (c) The CdS-1% MoS_2 hydrogen evolution rate utilizes a variety of sacrificial agents. Reproduced with permission from Ref. [27]. Copyright 2017, Elsevier.

better than that of 1% Pt and the mixed sample method (Figure 12.4b). Furthermore, this work also reveals that formic and lactic acids serve as the best sacrificial agents, with hydrogen evolution rate of 8.72 mmol $g^{-1}h^{-1}$ and 5.32 mmol $g^{-1}h^{-1}$, respectively, as depicted in Figure 12.4c [27].

Ye and colleagues have investigated the crystallinity effect of the MoS_2 monolayer with CdS regarding the hydrogen evolution performance.

This work was conducted using thermal decomposition and liquid exfoliation at various temperatures, rendering the MoS_2 monolayers crystallinity-dependent. The result exhibits that the low temperature and low crystallinity of the MoS_2 monolayer yielded the highest HER rate due to the structural disorder. Moreover, the highest catalytic performances are obtained concurrently with a high quantum efficiency of up to 71.6%. However, it should be noted that this value will decrease gradually as the temperature increases [30].

Zeng *et al.* reported $ZnIn_2S_4$-$MoSe_2$ hybrid system preparation, a ternary chalcogenide ($ZnIn_2S_4$) with the presence of $MoSe_2$ as a co-catalyst, to efficiently improve hydrogen production. The result showed that the hydrogen production rate of $ZnIn_2S_4$-$MoSe_2$ is 2,228 $mmol^{-1}$. In this case, this value exhibited that the presence of $MoSe_2$ at a concentration of as much as 2% has successfully enhanced the photocatalytic activity of $ZnIn_2S_4$ by 2.2 times. Furthermore, from this result, we can conclude that $MoSe_{2,}$ a co-catalyst for the ternary chalcogenide, is comparable to MoS_2 regarding photocatalytic performance [31].

Li *et al.* reported the use of a nanohybrid of MoS_2-ZnO to enhance photocatalytic hydrogen production. In this sense, this work was conducted due to its ability to improve electron and hole production as well as recombination reduction. It can be seen from the hydrogen production enhancement of up to five times that of solely ZnO, i.e., 27.69 $mmol^{-1}$. Furthermore, it is worth noting that the dispersion method can substantially decrease the activity of the photocatalyst due to the absence of a complete interface [32].

12.3.4 MXene

A new class of transition metal-based 2D materials called MXene has demonstrated a possible co-catalyst in photocatalysis. Excellent metallic conductivity, a wealth of surface hydrophilic functionalities, and string redox reactivity arising from the metal sites exposed at the terminal are just a few of the features that MXene exhibits [33, 34]. The MXene family is auspicious for creating effective composite photocatalysts made of MXene because of the aforementioned unique properties [35]. Many attempts have been made to promote titanium carbide ($Ti_3C_2T_x$) as the most promising co-catalyst candidate among MXenes [36]. Hence, the research about $Ti_3C_2T_x$ preparation was continuously developed. One of the synthesis strategy procedures to generate $Ti_3C_2T_x/TiO_2$ can be seen in Figure 12.5a [37].

Su *et al.* prepared a variety of monolayer and multilayer $Ti_3C_2T_x/TiO_2$ composite photocatalysts using the co-catalyst (as shown in Figure 12.5b).

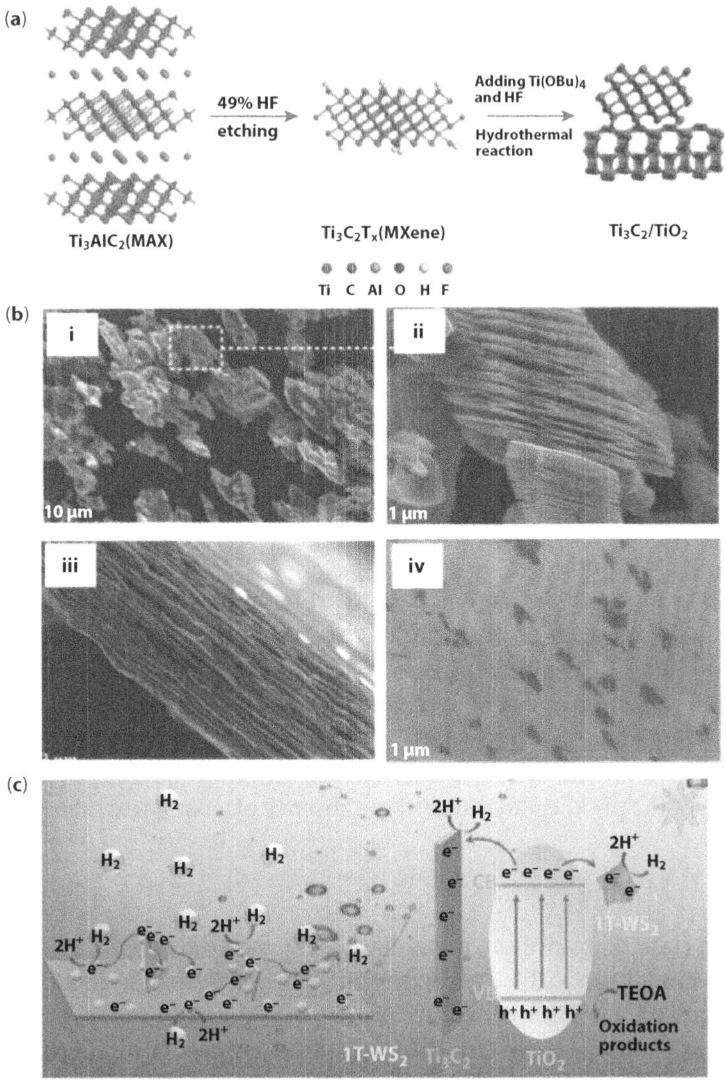

Figure 12.5 (a) Synthesis of the Ti_3C_2/TiO_2 composite. Reproduced with permission from Ref. [37] Copyright 2020, Elsevier. (b) Monolayer $Ti_3C_2T_x$ (i & ii) and multilayer $Ti_3C_2T_x$ (iii & iv) as the cocatalysts. Reproduced with permission from Ref. [38] Copyright 2019, ACS (c) Schematic photocatalytic mechanism of $1TWS_2@TiO_2@Ti_3C_2$ composites. Reproduced with permission from Ref. [39] Copyright 2020 Nano-Micro Letter.

The results showed that, compared to the pure TiO_2 and its multilayer counterpart, which exhibit hydrogen generation rates up to 290 μmol·g_{cat}^{-1}·h^{-1} and 920 μmol$g_{cat}^{-1}h^{-1}$, respectively, a monolayer $Ti_3C_2T_x/TiO_2$ had the highest hydrogen generation rate up to 2,650 μmol$g_{cat}^{-1}h^{-1}$. In addition, performance improvement is obtained due to the monolayer $Ti_3C_2T_x$ electrical conductivity enhancement and the $Ti_3C_2T_x/TiO_2$ interface's efficient charge-carrier separation [38].

On the other hand, Li and his research team have succeeded in preparing a nanohybrid of Ti_3C_2 and TiO_2, in which the former has a 2D structure and the latter exhibits a truncated octahedron bipyramid structure with exposed (001) and (101) faces. It has been demonstrated that heterojunction between facets (101) and (001) of TiO_2 precludes the more rapid recombination of light-generated carriers. Additionally, the leftover Ti_3C_2 will act as a co-catalyst to speed up the process due to its high electronic conductivity. Here, an aqueous solution containing 10% glycerin was used as a sacrificial agent for the H_2 production test, carried out at ambient pressure and temperature. The TF (F Ti_3C_2) sample outperforms the TOH (OH Ti_3C_2) sample in the catalytic reaction involving two synthesized samples because it has a higher photocurrent and lowers HER resistance [40].

Ding *et al.* reported a CdS/$Ti_3C_2T_x$ composite for hydrogen production, which was prepared *via* a sequence of electrostatic self-assembly and solvothermal methods. The result shows the hydrogen production rate of 3,226 μmol $h^{-1}g^{-1}$, which is five folds higher compared to the pure CdS, *i.e.*, 615 μmol$g^{-1}h^{-1}$. Furthermore, the nanosheet interface between CdS and $Ti_3C_2T_x$ could enhance the photocatalytic results due to the convenience of charge separation as well as hinder the recombination reaction. Further, it can be concluded that the CdS/$Ti_3C_2T_x$ composite increases photocatalytic activity and stability [35].

Yin *et al.* have successfully designed a heterostructure material of Ti3C2 MXene@Au@CdS for hydrogen production improvement using the photocatalytic method. This work obtained a hydrogen production rate of up to 17,070.43 (μmol g^{-1} h^{-1}). This result showed an improvement in hydrogen production compared to solely CdS up to 85 times. In addition, the fast transfer among the charges in the interface and the excellent stability were accomplished due to this material's ability to circumvent the drawback of the rich surface function [41]. Moreover, Yujie *et al.* reported that the composite 1T-WS_2@TiO_2@Ti_3C_2 could be used for photocatalytic hydrogen evolution with the mechanism shown in Figure 12.5c. The results exhibit the photocatalyst activity of those materials to be 50 times greater than that of pure TiO_2, with a value of 3,409.8 mol g^{-1} h^{-1} [39].

Wang *et al.* reported the modified CdS photocatalyst in the presence of a co-catalyst in the form of Ti_3C_2 MXene. The nanoflower form of the combination of 2D Ti_3C_2 MXene and 3D CdS can improve the performance of CdS by increasing carrier separation and transfer. The HER value of the Ti_3C_2@CdS composite with the existence of Ti_3C_2 as much as 15% was the highest photocatalytic value, i.*e.*, 88,162 μmol g^{-1} h^{-1}. These outcomes can surpass the performance of pure CdS by up to 91.57 folds. Hence, it should be noted that the Ti_3C_2 MXene addition could enhance electron transfer as well as photocatalytic performance [42].

12.4 Enhancing the Photocatalytic Performance

Several efforts to develop hydrogen production to tackle the energy crisis have been carried out by researchers. Moreover, the demand to provide excellent photocatalytic performance is still under expansion for commercial applications in photocatalytic hydrogen production. Therefore, many attempts have been made to increase the photocatalytic activity, especially strategies to enhance the catalytic performance of 2D nanomaterials, such as electronic structure manipulation, defect engineering, and co-catalyst modification.

Deng *et al.* have analyzed graphene's electronic and structural properties and those of other 2D materials. The result exhibits that the disruption in the graphene structure results in the modification of density as well as the improvement of catalytic properties. Furthermore, proficiency in controlling the electronic state renders unprecedented control over the chemical reactivity [43]. In 2018, Sun *et al.* successfully employed the surface charge manipulation of carbon nitrides (Figure 12.6a). There are two steps to manipulating the surface charge [44]. The first step is exfoliating the bulk precursor materials to generate the 2D nanosheet charge. The latter is obtaining the positive charge species because of 2D materials heteroatom site interaction with metal clusters due to their disparate electronegativities. This work is expected to also increase the photocatalytic performance due to the high densities and strong durability of urea electrolysis.

Lately, defect engineering has been defined as expectant strategies for modifying the surface's active sites and the physicochemical properties of photocatalysts [46]. Defect engineering is also widely known and has been exploited in 2D nanomaterials to increase the active site due to the

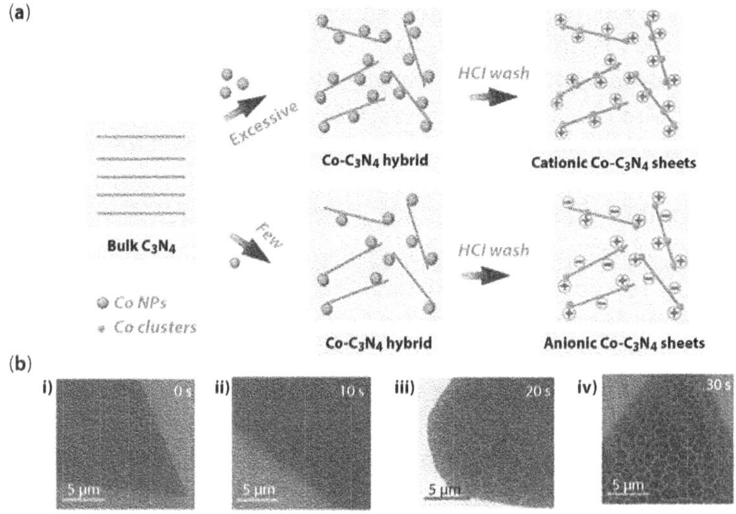

Figure 12.6 (a) The surface charge modification of two-dimensional Co-C_3N_4 nanosheets. Reproduced with permission from Ref. [44]. Copyright 2018 American Chemical Society. (b) SEM images of MoS_2 with oxygen plasma explosure of 0 (i), 10 (ii), 20 (iii), and 30 (iv) s, respectively. Reproduced with permission from Ref. [45]. Copyright 2016 American Chemical Society.

propensity of the basal plane to be thermodynamically stable but catalytically inert [47]. Thus, it could render excellent properties, *i.e.*, shortened diffusion length and great surface area. Moreover, the ultrathin thickness of 2D nanomaterials renders a high chance of adjusting the desired properties, including energy band structure, electrical conductivity, microstruc ture, and carrier concentration [48]. In addition, oxygen defect engineering has the ability to result in electronic structure modulation [49].

Ajayan and coworkers have investigated the defect engineering for MoS_2 regarding hydrogen production. The pristine MoS_2 was treated with oxygen plasma exposure and subsequently treated with hydrogen treatment, rendering cracks and triangular holes (Figure 12.6b). Thus, numerous active sites could be generated. The result exhibited an enhancement of catalytic activity as well as an improvement of current density, onset potential, and Tafel slope [45].

Recently, Han *et al.* reported the mesoporous graphitic C_3N_4 prepared by solvothermal exfoliation and thermal polymerization to enhance photocatalytic activity as well as hydrogen production. In this regard, the generated mesoporous properties could increase the photocatalytic performance due to surface area enhancement. Hence, the catalyst's abilities in charge separation, electron transfer, and light absorption range may improve. Moreover, this work reveals the optimum hydrogen evaluation rate up to 8,510 μmol $h^{-1} g^{-1}$ [50].

Aside from defect engineering, which has been done before, the co-catalyst modification could also enhance the photocatalytic performance. This is due to the photochemical process acceleration and charge separation [2]. In addition, the co-catalyst modification also increases the photocatalyst durability [5]. Further, this co-catalyst modification could be done by incorporating metal nanoparticles and nonmetal heteroatoms. Finally, it is worth noting that the 2D nanomaterial is also used as a co-catalyst [5].

Sun *et al.* reported 2D MXenes as a co-catalyst for the photocatalytic method. At least there are four synthesis methods to obtain the 2D MXene as the co-catalyst, *i.e.*, mechanical mixing, *in situ* decoration, oxidation, and self-assembly. Furthermore, in this work, they exhibit the MXene mechanism as a co-catalyst, which involves the adsorption of visible light, electron excitation to the conduction band, and the valence band's holes. Subsequently, it generates hydrogen as the result of electron transfer to MXenes. Furthermore, those charge transfers improve the electron-hole pair separation without recombinant reactions [51]. Thus, it could improve the photocatalytic activity.

Hong *et al.* scrutinized the few-layered Cu-doped MoS_2 on CdS nanorods (FCM/CdS). The results show that the catalysts can efficiently increase hydrogen production. These material synthesis methods are depicted in Figure 12.7a. The pristine CdS has been widely known to possess beneficial properties for photocatalysts because of its band edge position and band gap energy. However, exploiting pure CdS without co-catalyst loading is challenging due to the fast recombination rate and the photocorrosion threat. In this sense, metal doping is conducted to increase the electronic properties and active sites of molybdenum sulfate.

Moreover, the few-layered thin layers that could be gained by liquid exfoliation could also increase the number of MoS_2 active sites. Thus, Cu-doped MoS_2 is expected to overcome the hurdles in CdS usage for photocatalytic hydrogen production. The result exhibited a tremendous hydrogen production rate of 194.18 μmol h^{-1}, 52 times higher compared to the solely CdS [52].

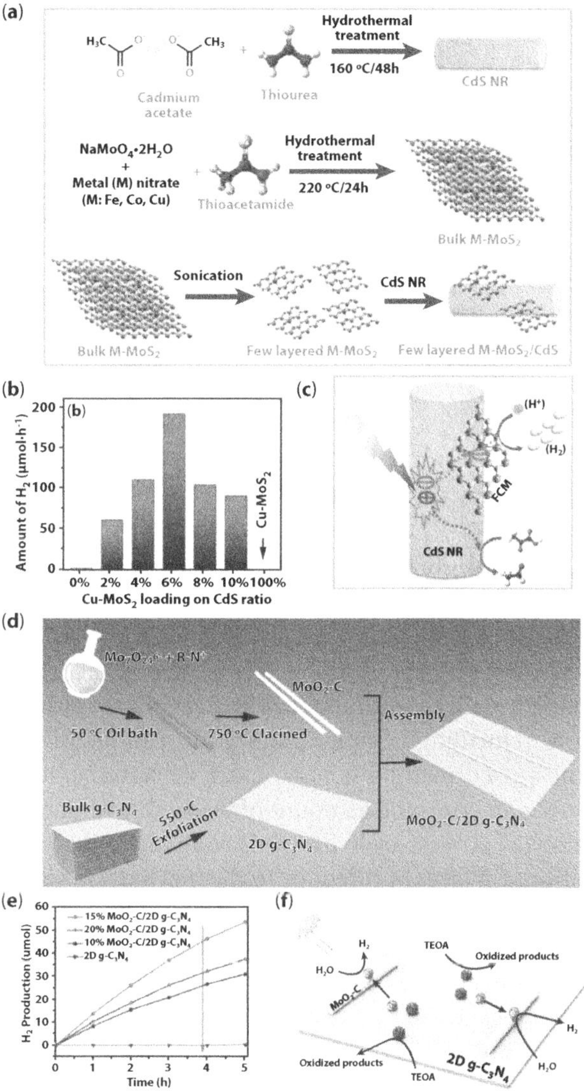

Figure 12.7 (a) Schematic synthesis procedures of a few layered Cu-MoS_2 on CdS nanorods. (b) The hydrogen production of FCM/CdS with different Cu-MoS_2 concentrations. (c) Schematic illustration of the proposed hydrogen production mechanism. Reproduced with permission from Ref. [52]. Copyright 2017, RSC. (d) Schematic illustration of MoO_2-C/2D g-C_3N_4 synthesis. (e) The hydrogen production of pure 2D g-C_3N_4 and MoO_2-C/2D g-C_3N_4 with MoO_2-C amount variation. (f) Schematic illustration of the hydrogen production mechanism. Reproduced with permission from Ref. [53]. Copyright 2018, Elsevier.

Chen *et al.* have analyzed the 1D metallic MoO_2-C, which acts as a co-catalyst over graphite carbon nitrides (MoO_2-C/2D g-C_3N_4). The synthesis mechanism of that catalyst is depicted in Figure 12.7d. In this work, the large 2D g-C_3N_4 surface area and the high electron-extracting ability of MoO_2 are expected to enhance hydrogen production. In addition, MoO_2 also has another outstanding property as a co-catalyst, *i.e.*, great active sites and density adsorption, as well as high conductivity. The result showed an increment of the hydrogen production of 157.5 folds higher than the pure 2D g-C_3N_4, *i.e.*, 1,071.0 μmol $g^{-1}h^{-1}$ Moreover, among the various amounts of MoO_2, *i.e.*, 10%–15%, 15% of MoO2 exhibited the optimum catalytic activity (Figure 12.7e). The photocatalytic mechanism is proposed as shown in Figure 12.7f [53].

12.5 Conclusion and Outlook

To date, hydrogen production from the photocatalytic process has attracted tremendous interest from researchers. In this sense, photocatalytic hydrogen production is defined as a promising pathway for realizing a sustainable energy alternative in the future. At least there are two ways of photocatalyzing hydrogen production: videlicet photocatalytic water splitting and photo-reforming organic compounds. Among those two mechanisms, the most used method is photocatalytic water splitting. However, although there is an enormous amount of research and publications about the photocatalytic process of hydrogen production, the development of industrialization still faces hurdles due to the mechanism's lack of complete understanding. Furthermore, the requirement of an effective photocatalyst material to elevate the catalytic activity also needs to be considered.

2D nanomaterials, which have outstanding properties, hold a notable promise for photocatalytic hydrogen production. A summary of some of them is provided in Table 12.1. Moreover, the 2D nanomaterials structural and chemical properties determine the functionalities of those materials. For instance, the thickness, defects, and surface charges present different architectures. Nevertheless, even though 2D nanomaterials have become promising candidates for photocatalytic processes, there remain several obstacles to overcome, especially for industrial realization.

With respect to the recent development of photocatalytic hydrogen production over 2D nanomaterials, several breakthroughs for future directions should be considered. In this regard, the preparation of photocatalyst materials, including 2D nanomaterials, should concern the desired properties, which can be achieved by tuning the structural properties.

Table 12.1 Summary of hydrogen production rates from various 2D nanomaterial photocatalysts.

Photocatalyst	Preparation method	Reaction condition	Light source	H_2 production	Ref.
Ni/GO-TiO_2	Photodeposition	Methanol–water solution (1:1)	Pen-Ray (λ = 254 nm, 2.2 mW cm^{-2})	2,171 µmol g^{-1} h^{-1}	[4]
GO:NPO	One-pot in-situ solar energy assisted	20% CH_3OH	300 W Xe lamp with a cutoff filter (λ ≥420 nm)	10.46 mmol H_2 g^{-1} h^{-1}	[16]
GO-Co^{2+}-DPyP	Pre-implanting by electrostatic and coordination interaction	10% vol Triethanolamine	300 W Xe lamp	1,677 µmol. g^{-1}	[19]
Sn_3O_4 nanoflake/graphene	Hydrothermal	20% CH_3OH	Under natural sunlight on sunny days between 10 am and 3 pm at Pune, Maharashtra State, India, in March to May	4,687 µmol g^{-1} h^{-1}	[18]
Au-rGO-TiO_2	Deposition–precipitation method	10% (v/v) methanol	450 W Hg lamp	60.99 mmol g^{-1}	[20]

(Continued)

Table 12.1 Summary of hydrogen production rates from various 2D nanomaterial photocatalysts. (*Continued*)

Photocatalyst	Preparation method	Reaction condition	Light source	H_2 production	Ref.
Pt SAs/C_3N_4	Ammonia-assisted immobilization	20% vol Triethanolamine	300 W xenon lamp	573.6 μmol h^{-1}	[25]
Fe_3O_4/C_3N_4 NT	Impregnation method	10% vol TEA solution	300 W xenon-arc lamp with a cutoff filter ($\lambda \geq 420$ nm)	1982.2 μmol g^{-1}	[23]
1T-$MoSe_2$/2D-C_3N_4	Solvothermal	Triethanolamine-water (1:9).	Xenon lamp (PLS-SXE 300C (BF) (λ = 420 nm, 300 W).	1,672.6 μmol$g^{-1}h^{-1}$	[30]
2D CdS NSs/MoS_2	Hydrothermal method and subsequent ultrasonic treatment	Na_2S-Na_2SO_3-water solution (4:1)	Xenon lamp (>400 nm, 300 W)	1.75 mmol $g^{-1}h^{-1}$	[27]
Crystalline monolayer MoS_2/CdS	Thermal decomposition	Lactic acid–water (3:7)	Xenon lamp (>420 nm, 300 W)	5.5 mmol h^{-1}	[26]
$ZnIn_2S_4$/$MoSe_2$	A one-pot polyol	0.35 M Na_2S-0.25 M Na_2SO_3	Xenon lamp (λ = 420 nm, W)	2,228 μmol $g^{-1}h^{-1}$	[31]

(*Continued*)

Table 12.1 Summary of hydrogen production rates from various 2D nanomaterial photocatalysts. (*Continued*)

Photocatalyst	Preparation method	Reaction condition	Light source	H_2 production	Ref.
MoS_2 nanosheet/ZnO	Dispersion	0.5 M Na2SO4	Xenon lamp (λ = 550 nm, 100 W)	27.69 mmol h^{-1}	[32]
Monolayer and multilayer $Ti_3C_2T_x/TiO_2$	Simple impregnation	Methanol–DI water (1:3)	Hg Lamp (285 nm–325 nm, 200 W)	–multilayer (920 $\mu mol g_{cat}^{-1} h^{-1}$) –monolayer (2,650 $\mu mol g_{cat}^{-1} h^{-1}$)	[38]
Ti_3C_2TF/TiO_2	Thermal oxidation	Glycerinum–water (1:10)	Xenon Lamp 350 W	125–130 $\mu mol g^{-1} h^{-1}$	[40]
$Ti_3C_2T_x/CdS$	Sequence of electrostatic self-assembly and solvothermal method	Lactic acid–water (1:1)	Xenon lamp (420 nm, 10.77 mW).	3,226 $\mu mol g^{-1} h^{-1}$	[35]
Ti_3C_2@Au@CdS	Self-reduction	Na_2S-Na_2SO_3-water solution (4:1)	Xenon lamp (420 nm, 300 W)	17,070.43 μmol $g^{-1} h^{-1}$	[41]
2D/3D Ti_3C_2 @CdS nanoflower	Hydrothermal	Sacrificial agent–DI water (1:4)	CHI 660E and Xenon lamp (405 nm, 350 W)	88.162 μmol $g^{-1} h^{-1}$	[42]

For instance, several efforts to enhance the photocatalytic process over 2D nanomaterials have been applied, *i.e.*, electronic structure manipulation, defect engineering, and co-catalyst modification. In addition, the crucial parts of the catalytic performance are catalytic activity and stability, which could be accomplished by controlling the morphology and defect states and manipulating the surface charge.

The computational studies using DFT simulation have an important effect on industrial realization due to the ability to predesign the desired 2D nanomaterials with prominent properties. In addition, the utilization of the machine learning method, whether supervised or unsupervised, is also capable of photocatalyst predesign, photoreactor technology preparation, and catalytic activity prediction. Hence, it will provide an efficient route for hydrogen production with low costs and highly effective results.

Lastly, we are convinced that preparing the 2D nanomaterial synthesis process with theoretical simulation using machine learning and DFT simulation, as well as developing the laboratory scale from an economic point of view, will embody the industrialization of hydrogen as an energy source. Thus, it will have beneficial impacts for humankind by mitigating climate change issues and achieving energy fulfillment worldwide.

Acknowledgments

This work was supported by Hibah PDUPT (first year) 2022 from the Ministry of Education, Culture, Research and Technology of the Republic of Indonesia.

References

1. Lakhera, S.K., Rajan, A., T.P., R., Bernaurdshaw, N., A review on particulate photocatalytic hydrogen production system: Progress made in achieving high energy conversion efficiency and key challenges ahead. *Renew. Sustain. Energy Rev.*, 152, 111694, 2021.
2. Ganguly, P., Harb, M., Cao, Z., Cavallo, L., Breen, A., Dervin, S., Dionysiou, D.D., Pillai, S.C., 2D nanomaterials for photocatalytic hydrogen production. *ACS Energy Lett.*, 4, 7, 1687–1709, 2019.
3. Zhang, J., Hu, W., Cao, S., Piao, L., Recent progress for hydrogen production by photocatalytic natural or simulated seawater splitting. *Nano Res.*, 13, 9, 2313–2322, 2020.

4. Quiroz-Cardoso, O., Suárez, V., Oros-Ruiz, S., Quintana, M., Ramírez-Rave, S., Suárez-Quezada, M., Gómez, R., Synthesis of Ni/GO-TiO2 composites for the photocatalytic hydrogen production and CO2 reduction to methanol. *Top. Catal.*, 65, 1015–1027, 2022.
5. Shanker, G.S., Biswas, A., Ogale, S., 2D materials and their heterostructures for photocatalytic water splitting and conversion of CO2 to value chemicals and fuels. *J. Phys. Energy*, 3, 2, 022003, 2021.
6. Saleem, Z., Pervaiz, E., Yousaf, M.U., Niazi, M.B.K., Two-dimensional materials and composites as potential water splitting photocatalysts: A review. *Catalysts*, 10, 4, 464, 2020.
7. Deng, F., Zou, J.P., Zhao, L.N., Zhou, G., Luo, X.B., Luo, S.L., Nanomaterial-based photocatalytic hydrogen production, In *Micro and Nano Technologies, Nanomaterials for the Removal of Pollutants and Resource Reutilization*, Elsevier Inc, Netherland, 2018.
8. Christoforidis, K.C. and Fornasiero, P., Photocatalytic hydrogen production: A rift into the future energy supply. *ChemCatChem*, 9, 9, 1523–1544, 2017.
9. Nawaz, A., Kuila, A., Rani, A., Mishra, N.S., Sim, L.C., Leong, K.H., Saravanan, P., Industrial application of light-driven nanomaterial, In *Micro and Nano Technologies, Industrial Applications of Nanomaterials,* Elsevier Inc., Netherland, 2019.
10. Su, T., Shao, Q., Qin, Z., Guo, Z., Wu, Z., Role of interfaces in two-dimensional photocatalyst for water splitting. *ACS Catal.*, 8, 3, 2253–2276, 2018.
11. Ioannidou, T., Anagnostopoulou, M., Christoforidis, K.C., Two-dimensional photocatalysts for energy and environmental applications. *Solar*, 2, 2, 305–320, 2022.
12. Natalya, S.A.C., Kadja, G.T.M., Azhari, N.J., Khalil, M., Fajar, A.T.N., Two-dimensional (2D) nanomaterials for enhanced oil recovery (EOR): A review. *FlatChem.*, 34, 100383, 2022.
13. Nagakawa, H. and Nagata, M., Photoreforming of organic waste into hydrogen using a thermally radiative CdOx/CdS/SiC photocatalyst. *ACS Appl. Mater. Interfaces*, 13, 40, 47511–47519, 2021.
14. Xie, G., Zhang, K., Guo, B., Liu, Q., Fang, L., Gong, J.R., Graphene-based materials for hydrogen generation from light-driven water splitting. *Adv. Mater.*, 25, 28, 3820–3839, 2013.
15. Cao, S. and Yu, J., Carbon-based H2-production photocatalytic materials. *J. Photochem. Photobiol. C Photochem. Rev.*, 27, 72–99, 2016.
16. Gurbani, N., Choudhary, R.J., Phase, D.M., Marumoto, K., Liu, R.S., Chouhan, N., Graphene oxide @ nickel phosphate nanocomposites for photocatalytic hydrogen production. *Chem. Eng. J. Adv.*, 6, 100105, 2021.
17. Albero, J., Mateo, D., García, H., Graphene-based materials as efficient photocatalysts for water splitting. *Molecules*, 24, 5, 906, 2019.
18. Sethi, Y.A., Kulkarni, A.K., Ambalkar, A.A., Panmand, R.P., Kulkarni, M.V., Gosavi, S.W., Kale, B.B., Efficient solar light-driven hydrogen generation

using an Sn_3O_4 nanoflake/graphene nanoheterostructure. *RSC Adv.*, 11, 48, 29877–29886, 2021.

19. Luo, Q., Zhu, K., Kang, S.Z., Qin, L., Han, S., Li, G., Li, X., A novel cobalt ion implanted pyridylporphyrin/graphene oxide assembly for enhanced photocatalytic hydrogen production. *J. Porphyr. Phthalocyanines*, 22, 9–10, 877–885, 2018.
20. Singh, N., Jana, S., Singh, G.P., Dey, R.K., Plasmonic Au–rGO–TiO_2 ternary photocatalyst for photocatalytic hydrogen production and dye degradation. *Nanotechnol. Environ. Eng.*, 7, 1, 119–137, 2022.
21. Zhu, J., Xiao, P., Li, H., Carabineiro, S.A.C., Graphitic carbon nitride: Synthesis, properties, and applications in catalysis. *ACS Appl. Mater. Interfaces*, 6, 19, 16449–16465, 2014.
22. Inagaki, M., Tsumura, T., Kinumoto, T., Toyoda, M., Graphitic carbon nitrides (g-C3N4) with comparative discussion to carbon materials. *Carbon N. Y.*, 141, 580–607, 2019.
23. Kheradmand, A., Wainwright, A., Wang, L., Jiang, Y., Anchoring iron oxides on carbon nitride nanotubes for improved photocatalytic hydrogen production. *Energy Fuels*, 35, 1, 868–876, 2021.
24. Ong, W.J., Tan, L.L., Ng, Y.H., Yong, S.T., Chai, S.P., Graphitic carbon nitride (g-C3N4)-based photocatalysts for artificial photosynthesis and environmental remediation: Are we a step closer to achieving sustainability? *Chem. Rev.*, 116, 12, 7159–7329, 2016.
25. Hu, Y., Qu, Y., Zhou, Y., Wang, Z., Wang, H., Yang, B., Yu, Z., Wu, Y., Single Pt atom-anchored C_3N_4: A bridging Pt–N bond boosted electron transfer for highly efficient photocatalytic H_2 generation. *Chem. Eng. J.*, 412, 128749, 2021.
26. Hai, X., Zhou, W., Chang, K., Pang, H., Liu, H., Shi, L., Ichihara, F., Ye, J., Engineering the crystallinity of MoS2 monolayers for highly efficient solar hydrogen production. *J. Mater. Chem. A*, 5, 18, 8591–8598, 2017.
27. Ma, S., Xie, J., Wen, J., He, K., Li, X., Liu, W., Zhang, X., Constructing 2D layered hybrid CdS nanosheets/MoS_2 heterojunctions for enhanced visible-light photocatalytic H_2 generation. *Appl. Surf. Sci.*, 391, 580–591, 2017.
28. Monga, D., Sharma, S., Shetti, N.P., Basu, S., Reddy, K.R., Aminabhavi, T.M., Advances in transition metal dichalcogenide-based two-dimensional nanomaterials. *Mater. Today Chem.*, 19, 100399, 2021.
29. Sim, D.M., Han, H.J., Yim, S., Choi, M.J., Jeon, J., Jung, Y.S., Long-term stable 2H-MoS2 dispersion: Critical role of solvent for simultaneous phase restoration and surface functionalization of liquid-exfoliated MoS2. *ACS Omega*, 2, 8, 4678–4687, 2017.
30. Yi, J., Li, H., Gong, Y., She, X., Song, Y., Xu, Y., Deng, J., Yuan, S., Xu, H., Li, H., Phase and interlayer effect of transition metal dichalcogenide cocatalyst toward photocatalytic hydrogen evolution: The case of $MoSe_2$. *Appl. Catal. B Environ.*, 243, 330–336, 2019.

31. Zeng, D., Xiao, L., Ong, W.J., Wu, P., Zheng, H., Chen, Y., Peng, D.L., Hierarchical $ZnIn_2S_4/MoSe_2$ nanoarchitectures for efficient noble-metal-free photocatalytic hydrogen evolution under visible light. *ChemSusChem*, 10, 22, 4624–4631, 2017.
32. Li, H., Dong, W., Zhang, J., Xi, J., Du, G., Ji, Z., MoS_2 nanosheet/ZnO nanowire hybrid nanostructures for photoelectrochemical water splitting. *J. Am. Ceram. Soc.*, 101, 9, 3989–3996, 2018.
33. Sagita, F., Radiman, C.L., Ledyastuti, M., Khalil, M., Kadja, G.T.M., Salt-modified MXene membrane for ultrafast and efficient cationic and anionic dyes removal. *J. Water Process Eng.*, 49, 10, 103133, 2022.
34. Khalil, M., Lesa, M., Juandito, A.G., Sanjaya, A.R., Ivandini, T.A., Kadja, G.T.M., Mahyuddin, M.H., Sookhakian, M., Alias, Y., A SBA-15-templated mesoporous $NiFe_2O_4$/MXene nanocomposite for the alkaline hydrogen evolution reaction. *Mater. Adv.*, 4 17, 3853–3862, 2023.
35. Ding, M., Xiao, R., Zhao, C., Bukhvalov, D., Chen, Z., Xu, H., Tang, H., Xu, J., Yang, X., Evidencing interfacial charge transfer in 2D CdS/2D MXene schottky heterojunctions toward high-efficiency photocatalytic hydrogen production, *Sol. RRL.*, 5, 2, 2000414, 2020.
36. Li, Y., Ding, L., Liang, Z., Xue, Y., Cui, H., Tian, J., Synergetic effect of defects rich MoS_2 and Ti_3C_2 MXene as cocatalysts for enhanced photocatalytic H_2 production activity of TiO_2. *Chem. Eng. J.*, 383, 123178, 2020.
37. Chen, L., Ye, X., Chen, S., Ma, L., Wang, Z., Wang, Q., Hua, N., Xiao, X., Cai, S., Liu, X., Ti_3C_2 MXene nanosheet/TiO_2 composites for efficient visible light photocatalytic activity. *Ceram. Int.*, 46, 16, 25895–25904, 2020.
38. Su, T., Hood, Z.D., Naguib, M., Bai, L., Luo, S., Rouleau, C.M., Ivanov, I.N., Ji, H., Qin, Z., Wu, Z., Monolayer Ti_3C_2 Tx as an effective Co-catalyst for enhanced photocatalytic hydrogen production over TiO_2. *ACS Appl. Energy Mater.*, 2, 7, 4640–4651, 2019.
39. Li, Y., Ding, L., Yin, S., Liang, Z., Xue, Y., Wang, X., Cui, H., Tian, J., Photocatalytic H_2 evolution on TiO_2 assembled with Ti_3C_2 MXene and metallic 1T-WS2 as co-catalysts. *Nanomicro Lett.*, 12, 1, 1–12, 2020.
40. Li, Y., Zhang, D., Feng, X., Liao, Y., Wen, Q., Xiang, Q., Truncated octahedron bipyramid TiO_2/MXene Ti_3C_2 hybrids with enhanced photocatalytic H_2 production activity. *Nanoscale Adv.*, 1, 1812–1818, 2019.
41. Yin, J., Zhan, F., Jiao, T., Wang, W., Zhang, G., Jiao, J., Jiang, G., Zhang, Q., Gu, J., Peng, Q., Facile preparation of self-assembled MXene@Au@CdS nanocomposite with enhanced photocatalytic hydrogen production activity. *Sci. China Mater.*, 63, 11, 2228–2238, 2020.
42. Wang, Y., Wang, X., Ji, Y., Bian, R., Li, J., Zhang, X., Tian, J., Yang, Q., Shi, F., Ti_3C_2 MXene coupled with CdS nanoflowers as 2D/3D heterostructures for enhanced photocatalytic hydrogen production activity. *Int. J. Hydrog. Energy*, 47, 52, 22045–22053, 2022.

43. Deng, D., Novoselov, K.S., Fu, Q., Zheng, N., Tian, Z., Bao, X., Catalysis with two-dimensional materials and their heterostructures. *Nat. Nanotechnol.*, 11, 3, 218–230, 2016.
44. Sun, Y., Duan, J., Zhu, J., Chen, S., Antonietti, M., Metal-cluster-directed surface charge manipulation of two-dimensional nanomaterials for efficient urea electrocatalytic conversion. *ACS Appl. Nano Mater.*, 1, 12, 6649–6655, 2018.
45. Ye, G., Gong, Y., Lin, J., Li, B., He, Y., Pantelides, S.T., Zhou, W., Vajtai, R., Ajayan, P.M., Defects engineered monolayer MoS2 for improved hydrogen evolution reaction. *Nano Lett.*, 16, 2, 1097–1103, 2016.
46. Zhang, Q., Chen, D., Song, Q., Zhou, C., Li, D., Tian, D., Jiang, D., Holey defected TiO_2 nanosheets with oxygen vacancies for efficient photocatalytic hydrogen production from water splitting. *Surf. Interfaces*, 23, 100979, 2021.
47. Xie, J., Yang, X., Xie, Y., Defect engineering in two-dimensional electrocatalysts for hydrogen evolution. *Nanoscale*, 12, 7, 4283–4294, 2020.
48. Xiong, J., Di, J., Xia, J., Zhu, W., Li, H., Surface defect engineering in 2D nanomaterials for photocatalysis. *Adv. Funct. Mater.*, 28, 39, 1–19, 2018.
49. Jia, T., Sun, C., Shi, N., Yu, D., Long, F., Hu, J., Wang, J., Dong, B., Li, J., Fu, F., Hu, S., Lee, J.H., Efficient oxygen vacancy defect engineering for enhancing visible-light photocatalytic performance over sno_2–x ultrafine nanocrystals. *Nanomaterials*, 12, 19, 3342, 2022.
50. Han, Q., Wang, B., Gao, J., Cheng, Z., Zhao, Y., Zhang, Z., Qu, L., Atomically thin mesoporous nanomesh of graphitic C3N4 for high-efficiency photocatalytic hydrogen evolution. *ACS Nano*, 10, 2, 2745–2751, 2016.
51. Sun, Y., Meng, X., Dall'Agnese, Y., Dall'Agnese, C., Duan, S., Gao, Y., Chen, G., Wang, X.F., 2D MXenes as Co-catalysts in photocatalysis: Synthetic methods. *Nanomicro Lett.*, 11, 1, 1–22, 2019.
52. Hong, S., Kumar, D.P., Kim, E.H., Park, H., Gopannagari, M., Reddy, D.A., Kim, T.K., Earth abundant transition metal-doped few-layered MoS_2 nanosheets on CdS nanorods for ultra-efficient photocatalytic hydrogen production. *J. Mater. Chem. A*, 5, 39, 20851–20859, 2017.
53. Chen, Z., Xia, K., She, X., Mo, Z., Zhao, S., Yi, J., Xu, Y., Chen, H., Xu, H., Li, H., 1D metallic MoO_2-C as co-catalyst on 2D g-C_3N_4 semiconductor to promote photocatlaytic hydrogen production. *Appl. Surf. Sci.*, 447, 732–739, 2018.

13

Supercapacitor Based on 2D Nanomaterials and Their Hybrid

Anupam Kumar* **and Arun Rathore**

Department of Electrical and Electronics Engineering, IES college of Technology, Bhopal, Madhya Pradesh, India

Abstract

In modern times, human lifestyle is heavily dependent on flexible electronic devices such as mobile phones, laptops, chargers, and electrical vehicles. One of the most pertinent requirements for these devices is an uninterruptible power supply. Currently, batteries and supercapacitors are the most widely used and researched electrochemical storage devices. Supercapacitors are preferred in applications that require rapid charge/discharge times. Other benefits of supercapacitors are their high-power density, extended durability, and environmentally friendly nature. As electronics devices used on a daily basis require lightweight and high-energy density energy storage systems, supercapacitors are preferred over batteries in these applications. Conventional supercapacitors suffer from the drawbacks of the requirement of an active electrode, the encapsulation requirement of the entire structure containing electrolyte and electrodes for prevention of leakage, and the rigid and bulky nature of the entire structure.

Supercapacitors with thinner and lighter technology incorporating flexible electrodes and electrolytic gel have been developed to overcome the above-mentioned drawbacks. High integrity improved electrochemical performance, and rational design are some of the considerations for forming modern flexible supercapacitors. In addition, high performing in terms of electrochemical performance and thin and flexible electrodes, which can be bent, rolled, or twisted depending on the application, are essential requirements for modern flexible supercapacitors. For rapid and stable ion kinetics, the electrolytic gels should be mechanically flexible and highly conductive for ion movement. Thus, the development of highly flexible and simple materials is required for energy storage systems.

**Corresponding author*: kanupam310@gmail.com

Subhendu Chakroborty and Kaushik Pal (eds.) 2D Nanomaterials: Synthesis, Properties and Applications, (289–314) © 2024 Scrivener Publishing LLC

Owing to the presence of large electrochemical sites, graphene, carbon nanotubes, and transition metal oxides have attracted significant interest from researchers. Owing to the presence of multiple atomic layers, 2D nanomaterials have attracted tremendous interest for the manufacture of supercapacitor electrodes. The higher active surface area, mechanical integrity owing to the stacked layer, and controllable electrical properties are other advantages of 2D nanomaterials. Owing to their layered structure, supercapacitors based on 2D nanomaterials exhibit a fast electrochemical performance and fast ion kinetics. In this chapter, the applications of supercapacitors in electrical technology (microgrids), their controllers, and developments in supercapacitor technology for the application of 2D nanomaterials in them are presented.

***Keywords*:** Supercapacitors, energy storage system, electrical vehicle, microgrids, graphene, carbon nanotubes, pseudocapacitors, nanosheets

13.1 Introduction

The storage and proper utilization of energy obtained from fossil fuels remains a challenging task. The energy supplied by storage elements should be uninterrupted and in an acceptable form. Energy-storing elements [1–3] are stored in different usable forms and at optimum locations for energy peak shaving [4–6]. Energy-storing elements are used to meet the energy gap between the energy-generating units and consumer utility units [7–10]. Typical applications of energy storage devices in electrical vehicles and uninterruptible power supplies are shown in Figures 13.1 and 13.2, respectively.

Electrochemical storage systems, inertia systems [11–13], and fuel cell systems [14] are among the storage systems available for modern power system applications. Batteries [15, 16] and supercapacitors generally facilitate the storage of electrical energy in the form of electrochemical energy. Owing to technical drawbacks, the presence of a large variety of contaminants and a shorter lifespan, conventional storage devices are finding themselves out of favor in modern day applications (electrical vehicles, microgrids, etc.).

The supercapacitor gives answers to all the above-mentioned problems. A major drawback of iron-based batteries is that they have a low energy density, which is covered by supercapacitors. Supercapacitors, when combined with other storage devices, can be used as energy storage devices in conventional and hybrid applications. In hybrid and electrical vehicles, supercapacitors are used as a primary source of power as well as to facilitate regenerative braking. Supercapacitors are preferred in applications

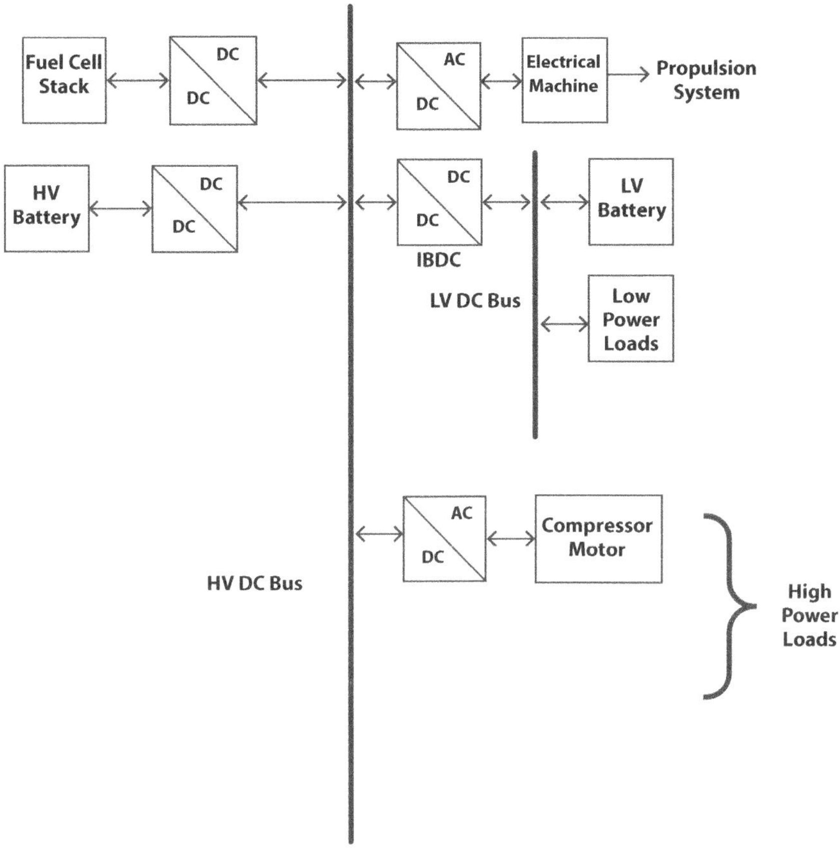

Figure 13.1 Application of energy storage device (battery) in an electric vehicle.

requiring quick charge/discharge times. Some other benefits of a supercapacitor are its high-power density, extended durability, and environmentally friendly nature. The cycle life and energy density of the energy storage system can be improved by changing the topologies of the energy storage systems (ESS) and incorporating hybrid topologies. For improvement of energy density, supercapacitors are directly placed in parallel with batteries and can store electrical energy, thus improving the response time and storage capacity of the conventional battery. Engineers from General Electric formed the first supercapacitors in 1957, which were further expanded by the SOHIO company and later patented as the electrolytic supercapacitor [17, 18]. Later in 1982, Pinnacle Research Institute (PRI) developed PRI ultracapacitors, which were subjected to military applications. In 1992, Maxwell laboratories introduced boost-caps which had increased charging

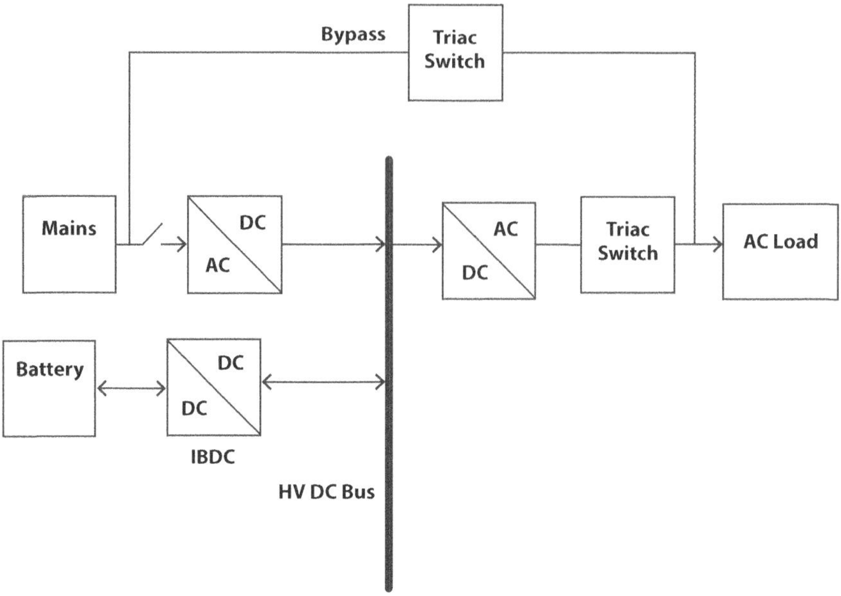

Figure 13.2 Application of energy storage device (battery) in an online Uninterruptible Power Supply (UPS).

and discharging current with higher capacitance values [19, 20]. These were specifically developed for higher power applications which were succeeded by tantalum electrolytic capacitors specially designed for specific application in the military [21, 22]. Electrostatic carbon electrodes are combined with electrochemical electrodes to form the hybrid capacitors having higher capacitance values [23]. Research is going on to enhance the supercapacitor's performance while decreasing the production cost [24].

A supercapacitor is structurally different from an electrolytic capacitor [25], as it contains a pair of solid electrodes, a separating membrane, and a liquid electrolyte. Ions of reverse polarity form an electrical double layer (EDL) across the electrodes, and this arrangement forms an ultracapacitor. Charges are separated in a Helmholtz double layer, and electrostatic charge storage is achieved. This process increases the capacitance of supercapacitors and forms the basis of EDLC (electrochemical double layer capacitance). The EDLC is shown in Figure 13.3.

Pseudocapacitance is another way of storing charge that is achieved by the redox faradaic reaction. Measurement techniques are used to separate the above-mentioned principles.

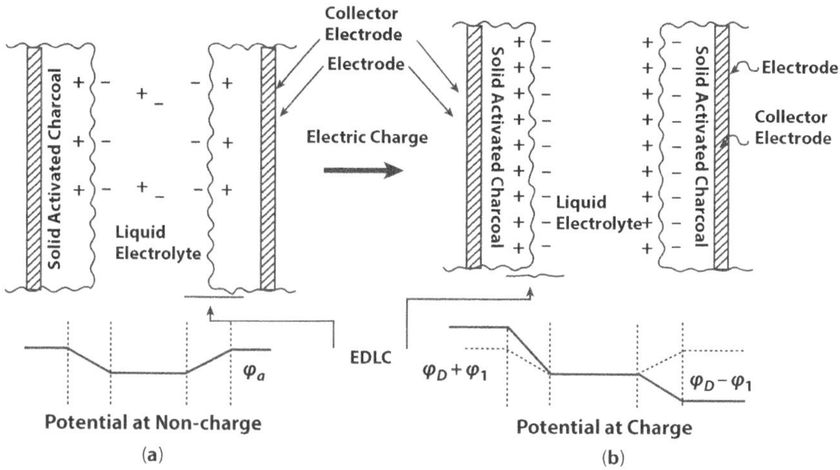

Figure 13.3 Mechanism of EDLC charge storage (reproduced with permission from [26]).

Supercapacitor operation is characterized by the interaction and mutual effects between electrode and electrolyte. Graphene, carbon fiber cloth, and carbon nanotubes (CNT) are some of the common 2D nanomaterials used for forming the electrode in an EDL supercapacitor. Polymers that can conduct electrons with low equivalent series resistance (ESR) are used for the formation of pseudocapacitors. Two electrodes, one having a higher pseudocapacitance and the other having a higher double layer capacitance value, are combined to yield a hybrid supercapacitor. The different types of supercapacitors are shown in Figure 13.4.

The supercapacitor provides so many advantages over conventional energy storage devices, but they still have some drawbacks, so the supercapacitor is used along with a conventional energy storage element in

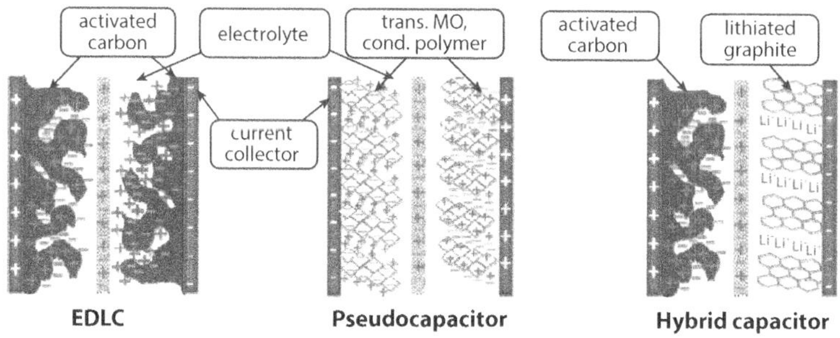

Figure 13.4 Schematic diagram of three types of supercapacitors (reproduced with permission from [27]).

many modern applications. The performance analysis of supercapacitors with different energy storage devices applied in various applications has been studied by many researchers [28–30]. One such application is the microgrid, where a supercapacitor and battery are used to supply the surplus power to the load. As the source is a solar PV array, irradiation is bound to vary. To lower irradiation, the supercapacitor and battery combined supply the load demand and maintain the load voltage at a constant value [31–33].

The supercapacitors offer low equivalent series resistance (ESR), high-energy density, and a long shelf life. The supercapacitor is applied in safe operating voltage zones and is subjected to low-voltage operation. Supercapacitors utilizing organic solvent electrolytes have a voltage range of 2.5 V–2.8 V, whereas the range of aqueous electrolyte-based supercapacitors lies between 2.0 V and 2.4 V. Supercapacitors are connected in series/parallel for higher voltage applications. The capacitor is better as compared to the supercapacitor in terms of power density, whereas the supercapacitor surpasses the capacitor in terms of energy density.

From Figure 13.5, it is easily understood that the supercapacitor acts as a bridge between the capacitor and the batteries. The power density of supercapacitors is superior when compared with batteries and fuel cells.

The characteristics of a supercapacitor are determined by the electrode material and the electrolyte used in the supercapacitor. The dependence of supercapacitor characteristics on the two parameters has been studied

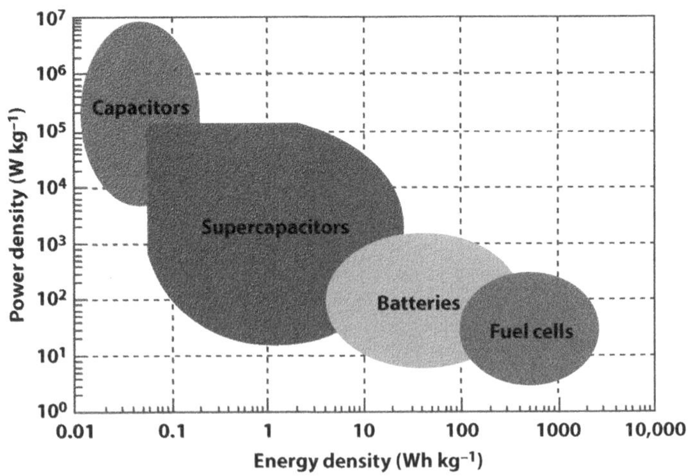

Figure 13.5 Ragone plot of various energy storage (reproduced with permission from [27]).

in detail by various researchers [34]. The supercapacitor's cost and performance efficiency directly depend on the materials used for supercapacitor development [35]. Investigations in the field of supercapacitor material are carried out regarding electrode materials, separators, collectors, and electrolytes.

Supercapacitor electrodes are amalgamated with current collectors in the form of thin sheets. The electrodes should be chemically stable, resist corrosion, be cost-effective, and be highly conductive. Carbon aerogel, carbon nanotubes (CNT), graphene, activated carbon, etc. are some of the carbon materials used for forming the electrodes in a supercapacitor. EDLC supercapacitors employ activated carbon as the electrode material, although it suffers from the serious drawback of lower electrical conductivity. Consolidated amorphous carbon (CAC) is one of the most frequently used electrode materials in supercapacitors. A derivative of activated carbon is activated carbon fiber (ACF), which has a diameter of 10 μm. Graphene is another carbon nanomaterial that is widely used for electrode formation in supercapacitors. Graphene is simply a one atom in dimension (thickness) layer of 3D graphite. Graphene is a hexagonal, 2-dimensional honeycomb lattice of graphite that can be rolled and modulated to form flexible electrodes. Various graphene layers are attracted

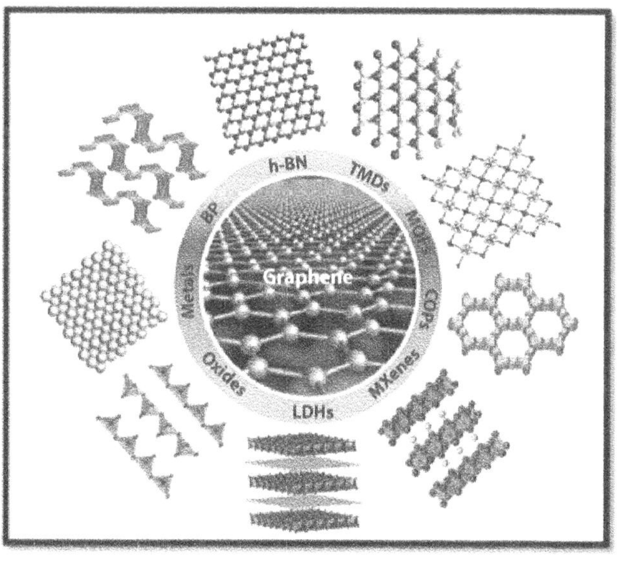

Figure 13.6 Various 2D nanomaterials for electrode application (reproduced with permission from [36]).

to each other with the help of van der Waal forces, which is equivalent to the text written by the pencil. The graphene layer is 0.345 nm thick. The graphene does not have an energy band gap between its valence and conduction bands and thus can be used for developing structures of different shapes. A diagram representing the 2-D materials required for the formation of electrodes is presented in Figure 13.6.

In pseudocapacitors, electrodes made up of Mno_2 and RuO_2 are used as they facilitate faradaic redox reactions and act as capacitive electrodes. Active electrode materials are created using faradaic redox reactions and lead to the formation of pseudocapacitance and an increase in specific energy.

The electrolyte of a supercapacitor consists of solvent and chemicals immersed in it, which results in the electrical conductivity of the entire electrolyte. The immersed chemicals are also responsible for ion concentration and their production in the electrolyte. The supercapacitor cell resistance and operational voltage window are greatly impacted by the electrolytes. Some of the variants of electrolytes are ionic, organic, and aqueous. Aqueous electrolytes and inorganic chemical materials are dissolved in water, thus forming an electrolyte. Low energy density and high-power density requirements are met by employing aqueous electrolytes in supercapacitors. The dissociation voltage of this supercapacitor is 1.15 V. For a higher separation voltage and a higher temperature range, an expensive class of organic electrolytes is used. Supercapacitors with voltages above 3.5 V comprise ionic electrolytes.

Separators play an important role in determining the performance efficiency of a supercapacitor. The ESR of the supercapacitor depends on the thickness and porosity of the separator. The separator market is led by polymer-based separators, which are highly durable, high in porosity, and highly cost-effective. As the active material on the electrode's surface is not highly efficient, a collector is required to facilitate the connection between capacitor electrodes. The performance of a supercapacitor is greatly increased by the application of a collector. Cell mounting is sealed to enhance the performance of supercapacitors in terms of charging and discharging currents. Galvanic cell housing corrosion is prevented in the collectors with the help of aluminum in the collectors.

A layer of slurry is brushed on the substrate for the fabrication of electrodes in the case of an EDLC supercapacitor. Binders, electrode materials, and additives for conductivities are mixed, and the mixture is cast on the substrate for the formation of electrodes. The subtract acts as a current collector. Conductor additives and electrodes are bound with the help of binders present in the electrode paste, as shown in Figure 13.7.

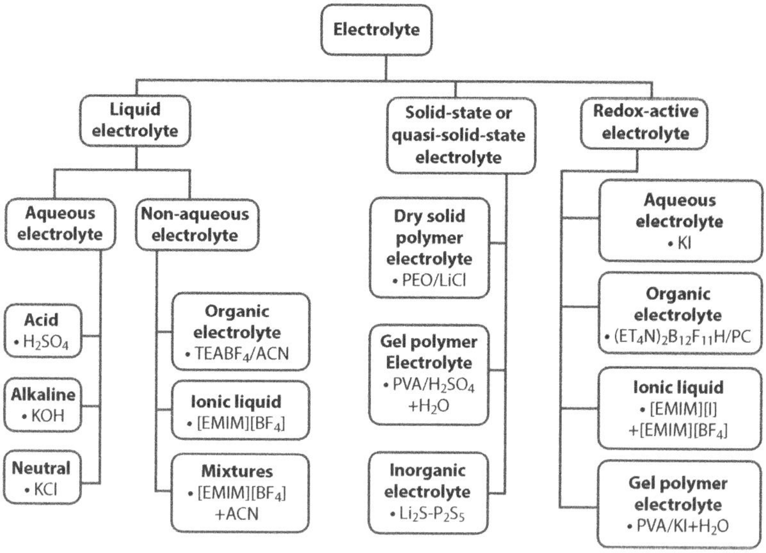

Figure 13.7 Classification of electrolytes (reproduced with permission from [37]).

In this chapter, the basic design structure, electrolytes, and 2D nanomaterials for supercapacitors are discussed.

13.2 Structure Design of 2D Nanomaterial-Based Supercapacitors

For improved electrochemical performance, the supercapacitors are designed by performing design trade-offs to achieve high energy density with proper safety for a large variety of flexible operations. The energy density of the storage supercapacitor for a given mass/volume is determined by paying appropriate attention to design considerations. The following materials are considered for designing an efficient energy storage device: a) conductive electrodes; b) electrolyte for ion conduction; c) separator for electron insulation; d) current collectors; and e) leakage prevention devices (encapsulation materials).

The following factors need to be considered for the design of supercapacitors:

a) Electrodes
b) Electrolyte

c) Cell design
d) Flexibility.

a) Electrode

The electrode material is prepared in the form of powder, and a coating of this material is carried out on the current collector. The electrode is grounded with binder and conductive carbon. Self-standing electrodes and supporting electrodes are the two variants of electrodes available in the field of supercapacitor electrodes. Thin films/fibers without any current collector form the self-standing electrode, which consequently reduces the device's weight and volume as they lack binders, additives, and current collectors. A graphite hybrid that is synthesized using chemical vapor disposition is used in synthesizing a supercapacitor device with a structure similar to that of a thin film. The freestanding electrode is very thin and can get deformed and, subsequently, reallocated away from the electrode construction.

This issue is overcome by employing active materials in supercapacitors. The active materials are coated and grown on lightweight substrates for this purpose. In the case of supporting electrodes, the electrode material is supported on a different substrate material. This substrate material acts as a current collector for the supercapacitor. Some examples of this substrate are flexible metal substrates (titanium, nickel, etc.), carbon substrates, and porous materials (textiles, plastics, and paper). High conductivity and the ability to be easily integrated with electronic devices are some of the requisites for the support of flexible supercapacitors.

b) Electrolyte

The electrolyte is responsible for the overall flexibility and conductivity of the supercapacitor. Gel or solid-state electrolytes have a wide operating temperature range, high reliability, easy handling, and easy fabrication and are preferred over liquid electrolytes. Quasi-solid-state or gel polymer electrolytes have a high conductivity of 10^{-4}–10^{-3} S cm^{-1} under normal operating conditions and are extensively used for flexible supercapacitors. The major components of a gel electrolyte are a conducting salt, a polymer host, and a solvent. The electrolyte ions are made to travel through the solvent, and the swelling of the polymer medium also takes place in the solvent. Polyethylene oxide (PEO), poly-vinylidene fluoride (PVDF), polyacrylonitrile (PAN), and polyvinyl alcohol (PVA) are some of the most commonly used polymer matrices.

Low dissociation energy is required for ions, and large anions should be constituents of polymer electrolytes containing electrolytic salt.

Some of the commonly used electrolytic salts are listed below:

a) Lithium perchlorate ($LiClO_4$)
b) Chloride salts of Lithium, Potassium, and Sodium (LiCl, KCl, and NaCl)
c) Sodium Sulfate (Na_2SO_4)
d) Phosphoric Acid (H_3PO_4)
e) Sulfuric Acid (H_2SO_4)

Four categories of gel polymers are listed below:

a) Organic gel electrolyte
b) Aqueous gel electrolyte
c) Redox active gel electrolyte
d) Ionic liquid gel electrolyte

Organic Gel Electrolyte

Improved ionic conductivity and a wide potential window are some of the features provided by the organic gel electrolyte. In addition to these two properties, the low interface formed by the supercapacitor devices is overcome by the organic gel electrolyte. Polymers having a high molecular weight constitute the organic gel electrolyte (ex. PMMA dissolved in organic solvent gelated with conducting salts). Dimethyl carbonate (DMC), ethyl methyl carbonate (EMC), ethylene carbonate (EC), and tetrahydrofuran (THF) are some of the commonly used organic solvents. These solvents provide high ionic conductivity, a lower viscosity, and a wider potential window by acting as a plasticizer for the involved polymer. The supercapacitor energy density is also improved by the application of organic gel electrolytes. Poly(2-hydroxy-3-phenoxypropyl acrylate) entrapping $LiClO_4$/PC is proposed to prepare a highly conductive and mechanically stable electrolyte [38]. The properties of a quasi-solid, dry, and liquid electrolyte are amalgamated in this newly researched electrolyte.

Aqueous Gel Electrolyte

In an aqueous gel electrolyte, surface tension traps the water molecules in a 3D polymer structure. The electrolyte-glass transition temperature depends on the degree of plasticization, which is determined by the water-to-polymer ratio in the electrolyte. In an aqueous gel electrolyte, for the host polymer matrix (PVA, PAA, or PEO), water is used for the purpose of plasticizer, as this results in superior ionic conductivity and reduced cost. A strong acid, a neutral, or a strong alkali can be used as an electrolytic salt. Due to its ease of preparation, better film formation ability, superior hydrophilicity, and lower cost, PVA is widely used for aqueous gel electrolytes. Relatively, a higher capacitance is displayed by PVA/H_3PO_4 due to the property of proton conduction.

The thermodynamic cycle of salt dissolution is shown in Figure 13.8.

Redox Active Gel Electrolyte

Redox entities are added to electrolytes to enhance the energy density of supercapacitors; the addition of these species also helps enhance the electrochemical properties of the supercapacitors [39–44]. Faradaic reactions are provided by the redox species and facilitate rapid electron transfer, thus providing the extra pseudocapacitance. The redox species are added to aqueous, ionic, and organic gel polymers to enhance the specific capacitance and electrochemical properties of the supercapacitor.

Some of the redox species are listed below:

a) Sodium Molybdate (Na_2MoO_4)
b) Potassium Ferricyanide ($K_3Fe(CN)_6$)
c) Methylene Blue (MB)
d) Ferrocene
e) Hydroquinone (HQ)
f) Indigo carmine (IC)
g) P-nitroaniline
h) Potassium Iodide.

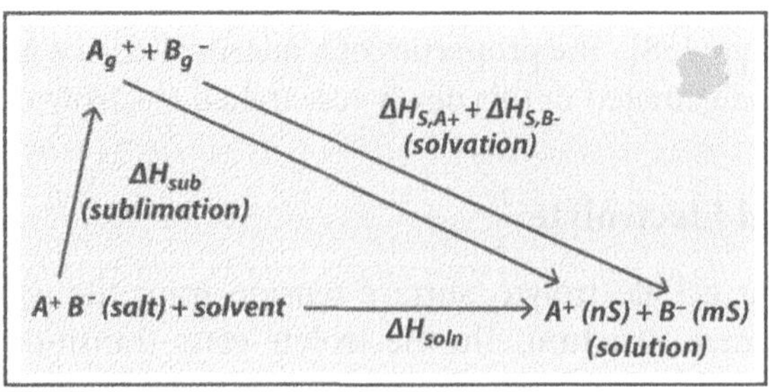

Figure 13.8 Thermodynamic cycle of salts dissolution (reproduced after permission from [38]).

An organic electrolyte of PMMA-PC-$LiClO_4$ is utilized for the formation of a micro-supercapacitor with the addition of HQ.

Ionic Liquid Gel Electrolyte

Nonvolatility, higher and wider potential, and high ionic conductivity are some of the advantages offered by the ionic gel electrolyte as compared with other electrolytes. Some polymer host matrix materials like PVDF, PAN, etc. use various solvents like 1-butyl-3-methylimidazolium chloride (BMIMCl), 1-butyl-3-methyl imidazolium bis(triflouromethylsulfonyl) imide (BMIM-TFSI) for the formation of several flexible supercapacitors. Superior mechanical stress performance is displayed by the ionic liquid gel electrolytes.

13.3 2D Nanomaterials for Supercapacitor Technology

The various 2D nanomaterials/nanosheets are a special set of crystals that are of the size of an atom. Due to their large specific surface area, electrical and chemical stability, excellent mechanical properties, and superior conductivity, 2D nanosheets are attracting the attention of global researchers. Graphene is one of the most important 2D nanomaterials and was invented in 2004 [45]. The chief benefits provided by the 2D are listed below:

a) Nanosheets structure having a thinness of an atom and a high exposure surface area, attributing to a large aspect ratio.
b) Superior capacitance values are attributed to a higher number of active electrochemical sites.
c) Extraordinary conductivity owing to restricted transport of charge to the 2D plane.
d) Excellent mechanical strength and flexibility offered to deformations are attributed to the large interlayer overlapping area.

All the above-mentioned benefits are instrumental in the formation of 2D nanomaterial-based supercapacitor electrodes and make 2D nanomaterials very apt for the development of flexible supercapacitors.

Some of the commonly used 2D nanomaterials are listed below:

a) Transition metal oxides (TMOs) and Transition metal hydroxides (TMHs)
b) Transition metal carbides/carbonitrides (MXenes)
c) Transition metal dichalcogenides (TMDs)
d) Black Phosphorous/Phosphorene (BP)

The above-mentioned 2D nanomaterials are discussed in the following section.

13.3.a Transition Metal Oxides (TMOs) and Transition Metal Hydroxides (TMHs)-Based Supercapacitor

TMOs and TMHs nanosheets have a large specific capacitance value, facilitate rapid redox reactions, have an abundance of active electrochemical sites, and offer high chemical stability. Due to these attributes, TMOs and TMHs are attracting a large variety of researchers from the field of supercapacitors. Due to their lower electron transport ability and different coordination styles, leading to restricted application in flexible supercapacitors, the TMOs and TMHs cannot be applied for the development of practical flexible supercapacitors [46]. Different hybrid composites are formed by various researchers such that they offer large surface areas, superior electronic conductivity, and a shorter diffusion path.

2D nanomaterials (TMOs and TMHs)-based electrodes and their recent developments are discussed in the next section.

13.3.a.1 Transition Metal Oxides

Transition metal oxides offer a large active surface area and high redox activity and are thus sought after by researchers for the fabrication of supercapacitor electrodes. The transition metal oxides offer high charge storage due to their large active surface area.

Some of the transition metal oxides are listed below:

a) RuO_2
b) Fe_3O_4
c) NiO
d) Co_3O_4

Due to its large and wide voltage window, superior reversible redox reactions, and higher proton conductivity, RuO_2 is the most sought-after and investigated 2D transition metal oxide for the fabrication of supercapacitor electrodes. RuO_2 also offers higher specific capacitances and three different oxidation states [47, 48]. Despite its above-mentioned advantages, RuO_2 suffers from a serious drawback of low availability, higher toxicity, and a much higher cost, resulting in its restricted application as a material for practical supercapacitor applications.

However, due to its low toxicity, superior theoretical capacitance (1,390 Fg^{-1}), higher abundancy, and lower cost, MnO_2 is extensively studied by various researchers for application in flexible supercapacitors. Higher crack resistance is exhibited by the electrodeposition of Birnessite-type MnO_2 on flexible carbon cloth. For example, a 2D nanosheet of MnO_2 (thickness of 2 nm) is developed on a porous gold substrate through the soft template method [49]. Furthermore, screen printing of MnO_2 nanosheets is carried out with high flexibility and adequate, controllable transparency [50].

The chemical bath deposition method is used for the preparation of porous MnO_2 (nanostructured). These nanostructured thin films are prepared on stainless steel [51, 52]. Similarly, PVP-$LiClO_4$ was used for developing a symmetric MnO_2 supercapacitor that can work at high voltages [53].

Due to its innately low electrical conductivity, the energy density of the MnO_2-based supercapacitor is quite low, even though the theoretical capacitance offered by the MnO_2 is quite high. The conductivity is increased by employing carbon-based additives like CNTs to form MnO_2 hybrids. Graphene is incorporated to form δ-MnO_2 and rolled onto PET substrate, resulting in the fabrication of a flexible supercapacitor. MnO_2 nanosheets are combined with carbon materials and metal oxides, resulting in the formation of hybrid composites and enhanced electrochemical supercapacitor performance [54]. Multichannel carbon nanofibers are used for growing MnO_2 nanosheets, whereas carbon nanofibers contain amorphous cobalt oxide.

Due to their high theoretical capacitance (greater than 2,700 Fg^{-1}), cobalt and nickel oxides are widely studied by various researchers. A hydrothermal approach was used for the preparation of 2D NiO nanosheets, which are supported by a Ni foam. In this material, even after 1,200 bending cycles, 91.3% of the capacitance is retained, thus, indicating very high mechanical flexibility for the fabricated material. A flexible supercapacitor offering only 5% variation in capacitance on 180° bending using a hybrid of ultrathin NiO nanosheet and carbon cloth.

13.3.a.2 Transition Metal Hydroxides

Transition metal hydroxides like $Co(OH)_2$ and $Ni(OH)_2$ are also employed for the formation of supercapacitor electrodes. Electron transport is enhanced, whereas the ion path is reduced by scaling bulk TMHs to 2D ultrathin nanosheets. This scaling further reduces cycling strain and increases the rate capability of supercapacitors. The synthesis of ultrathin 2D $Ni(OH)_2$ embedded in a CNT scaffold is carried out using a method assisted by CVD [54]. Further, a template-free hydrothermal approach is used for the development of a printable β-$Co(OH)_2$-based thin film supercapacitor.

Graphene was used to mitigate the low conductivity problem faced by the single metal oxide supercapacitor electrode.

13.3.b Transition Metal Carbide/Carbonitride (MXene)-Based Supercapacitor

MXenes are a few atoms thick transition metal nitrides, carbides, and carbonitrides, which are new variants of 2D nanomaterials employed in supercapacitor. MXenes have a general formula of $M_{n+1}X_nT_x$ derived from the MAX compound.

In $M_{n+1}X_nT_x$, M is the transition metal of the A group 13 and group 14 elements of the periodic table, where X is either nitrogen or carbon.

The surface termination group is represented by T_X. -OH, -O are some of the terminating groups. The number of terminating groups is represented by X. MXenes show a good hydrophilic nature as well as superior mechanical stability. Also, superior electronic conductivity and rich surface chemistry are displayed by the MXenes. The layer formed by MAX compounds forms a facile film and improves the storage ability of the storage device. This layer enhances the integration of different nanomaterials and facilitates rapid electron transfer.

As MXenes can be easily fabricated into freestanding films, they are used as current collectors and are apt for the development of flexible supercapacitors. The spray coating technique is used for the fabrication of all MXene microsupercapacitors by employing large and small flakes of $Ti_3C_2T_X$ as current collectors and electroactive layers, respectively [55]. 2D Ti_3AlC_2 layers are obtained by an intercalation of MXenes with cations (NH_4^+, Al^+, Li^+, etc.), which is filtered further to obtain MXene paper [56]. Ti_3C_2 MXene films were formed by etching Ti_3AlC_2 and roll milling the clay obtained. The electrochemical performances of MXenes are impeded by their irreversible aggravation and face-to-face stacking. The capacitive performance of the MXene is improved by incorporating other nanomaterials like graphene, conducting polymers, and carbon nanotubes. Electrostatic self-assembly is utilized between reduced graphene oxide (rGo) coated poly diallyl dimethyl ammonium chloride (PDDA) and MXene. The rGo coating is positively charged, whereas the MXene is negatively charged, thus giving us exfoliated MXene/rGo films.

3D MXene-graphene films facilitate higher mass transfer of electrolyte and thus increase the rate capability of the electrolyte as compared with 2D MXene–graphene. For applications involving smart wearable electronics, MXene–graphene composite aerogel is applied, which is fabricated by the conversion of a dispersion of liquid to aerogel through freeze-drying [57]. A wet spinning method having Ca^+ in the coagulation bath and utilizing electrostatic interaction for binding graphene framework and $Ti_3C_2T_X$ nanosheets is used for MXene–graphene hybrid nanosheets. Electrochemical stability is maintained by these yarn fibers under various structural transformations and bending angles. The issue of restacking is prevented by employing CNT as a composite with MXenes. Silk-derived carbon and activated carbon are utilized for the development of MXene composite-based supercapacitor electrodes.

Inorganic 2D hybrid film electrodes were formed from MXene composites, which were based on metal oxides. Vacuum filtration is used for assembling MnO_2 at the molecular level, which further led to the development of Ti_3C_2-MnO_2 hybrid films [58]. This Ti_3C_2-MnO_2-based supercapacitor showed great electrochemical stability even at bending angles in the range of 45°–180°.

The conductivity of MXenes depletes upon the oxidation of metal oxide used in the formation of composite MXenes. Thus, pseudocapacitance is included in the structure with the help of 2D double-layer hydroxides. Regarding this, a heterostructure supercapacitor electrode is developed by putting together NiCoAl-LDH and Ti_3C_2 MXene at the molecular level.

13.3.c Transition Metal Dichalcogenide (TMD)-Based Supercapacitor

Transition metal dichalcogenides (TMDs) represent a group of nanomaterials whose chemical formula is MX_2.

Where M represents the transition metal from groups 4, 5, and 6, X symbolizes a chalcogen [59]. Two layers of chalcogen will contain a layer of metal among them. The sandwiching will be held with the help of weak van der Waals forces [60, 61]. Exfoliation of this layered structure yields a layer of MX_2 nanosheet. This nanosheet will offer a large active surface area. This will further improve the interaction of electrolyte and electrode and enhance the capacitive capabilities of the supercapacitor. Storage of electrolyte ions is also improved through the exfoliation process [62]. Large redox currents will be obtained from the electrochemistry of TMDs. TMDs show large active edge sites and changeable oxidation states, thus resulting in superior electrochemical properties. These superior electrochemical properties will result in large energy storage capabilities.

MoS_2 displays a unique structure and resembles the generic MX_2 compound family. MoS_2 offers high values of capacitance and is abundantly available on earth, resulting in its wide application in flexible supercapacitors. MOS_2 displays low specific capacitance values as it contains 2H phases, which offer poor electrical conductivity [63, 64].

2D MoS_2 nanosheets are synthesized using the below-mentioned three techniques:

1) Exfoliation
2) CVD
3) Hydrothermal.

Hydrothermal and exfoliation methods are preferred over CVD methods due to their lower cost and ease of processing. The development of ultrathin MOS_2 paper is restricted by the lower aspect ratio and miniature size of MOS_2 [65]. Thus, various non-conductive materials or metals are used as framework/substrate for the fabrication of MOS_2-based supercapacitors [66].

A layer of MOS_2 is obtained by exfoliation in which Na^+, Li^+, and H^+ ions are intercalated, and further vacuum filtration is carried out to yield paper-thick MOS_2 films [67]. A Ti substrate and a TiO_2 substrate acting as a buffer layer are used for the synthesis of MOS_2-based supercapacitors that are shaped in the form of fiber [68]. The restacking poses a problem for 1-T MOS_2, as it reduces the specific capacitance of the supercapacitor, although the electrochemical performance of the synthesized supercapacitor improves.

The issue of restacking is solved using spacers, namely CNT and graphene. A vacuum filtration technique is applied to PVDF filter paper to yield a graphene/MOS_2 film of flexible nature. Graphene foam was grown on CVD and transferred to PDMS elastic to yield a 3D MOS_2/graphene-based supercapacitor device [69]. This device, when subjected to 180° bending, displayed capacitance retention of nearly 94%.

13.3.d Black Phosphorous-Based Supercapacitor

A single atomic layer of black phosphorous forms phosphorene, which is a 2D nanomaterial used for the fabrication of supercapacitors. The structure of phosphorene is puckered lamellae where the phosphorus layers are bonded with the help of van der Waal forces. The phosphorene offers superior electrical conductivity and thermodynamic stability.

The phosphorene offers quick ion diffusion and increases the capacitance conductivity.

BP nanoflakes are liquid exfoliated and used for developing a supercapacitor of high capacitance [70]. This supercapacitor offers improved conductivity with superior mechanical stability and high-stress handling. The fabricated BP nanoflakes result in bulkier structures that are unstable in nature. The electrochemical performance of supercapacitors is improved by incorporating CNTs, resulting in BP/CNT composite paper, which is used for supercapacitor fabrication with capacitance that will last for 12 months under various bending conditions. One-step masking of PET substrate is carried out to yield an interdigitated microsupercapacitor of phosphorene/graphene.

13.4 Conclusions

Energy storage devices play a very pertinent role in the overall functioning of the dc microgrid. Some of the applications of energy storage devices are electrical vehicles, UPS, etc. In the present chapter, types of supercapacitors are discussed, namely EDLC, pseudocapacitors, and hybrid supercapacitors.

The major components for the design of supercapacitors, namely electrodes, electrolyte, cell design, and flexibility, are discussed in detail in the present chapter. 2D nanomaterials used for electrode design are discussed. The various types of electrolytes, namely liquid, quasi-solid, and redox-active electrolytes. These electrolytes are discussed briefly in the chapter.

2D nanomaterials are apt for the formation of supercapacitor electrodes. Some of the commonly used 2D nanomaterials are listed as transition metal oxides (TMOs) and transition metal hydroxides (TMHs), transition metal carbides/carbonitrides (MXenes), transition metal dichalcogenides (TMDs), and black phosphorous/phosphorene (BP). All these above-mentioned 2D nanomaterials are discussed in detail in the present book chapter.

References

1. Novas, N., Alcayde, A., Robalo, I., Manzano-Agugliaro, F., Montoya, F.G., Energies and its worldwide research. *Energies*, 13, 6700, 2020.
2. Smith, S.C., Sen, P.K., Kroposki, B., Advancement of energy storage devices and applications in electrical power systems, in: *Proceedings of the 2008 IEEE Power and Energy Society General Meeting-Conversion and Delivery of Electrical Energy in the 21st Century*, Pittsburgh, PA, USA, July 20–24, 2008, pp. 1–8, 2008.
3. Sahin, M.E. and Blaabjerg, F., A hybrid PV battery/supercapacitor system and a basic active power control proposal in MAT-LAB/Simulink. *Electronics*, 9, 129, 2020.
4. Hou, N., Li, W.Y., Ding, L., Communicationless power management strategy for the multiple dab-based energy storage system islanded DC microgrid, in: *Proceedings of the 2020 IEEE Energy Conversion Congress and Exposition (ECCE)*, pp. 4656–4661, 2020, doi: 10.1109/ECCE44975.2020.9236420.
5. Kumar, A., Bhat, H.A., Agarwal, P., Comparative analysis of dual active bridge isolated dc to dc converter with buck-boost and flyback converters for bidirectional energy transfer. *J. Electr. Eng. Romania*, 20, 4, 139–151, 2020.
6. Kumar, A., Bhat, H.A., Agarwal, P., A single phase shift based isolated bidirectional DC-DC converter for bidirectional energy transfer in dc-microgrid with identification of optimum operating zone. *IJIED*, 5, 1, 26–40, 2019.
7. Kumar, A. and Bhat, H.A., Role of dual active bridge isolated bidirectional DC-DC converter in a DC microgrid, in: *Microgrids*, pp. 141–155, Academic Press, Massachusetts, USA, 2022.
8. Tiwari, K.S., Singh, B., Goel, K.P., Design and control of microgrid fed by renewable energy generating sources. *IEEE Trans. Ind. Appl.*, 54, 3, 2041–2050, 2018, doi: 10.1109/TIA.2018.2793213.
9. Thogaru, R.B. and Mitra, A., Regulation of DC microgrid voltage using optimized droop index control-based high gain converters in the presence of static and dynamic loads. *Electr. Eng.*, 104, 1649–1665, 2022. https://doi.org/10.1007/s00202-021-01403-5.
10. Kumar, M. and Hote, Y.V., Maximum sensitivity-constrained coefficient diagram method-based PIDA controller design: Application for load frequency control of an isolated microgrid. *Electr. Eng.*, 103, 2415–2429, 2021. https://doi.org/10.1007/s00202-021-01226-4.
11. Rawat, S.G. and Sathans, Modeling and performance analysis of renewable sources under islanded dc microgrid. *2018 International Conference on Inventive Research in Computing Applications (ICIRCA)*, pp. 498–503, 2018, doi: 10.1109/ICIRCA.2018.8597375.
12. Kim, J., Lee, H.S., Park, J., Inertia-free stand-alone microgrid—part II: Inertia control for stabilizing dc-link capacitor voltage of pmsg wind turbine system. *IEEE Trans. Ind. Appl.*, 54, 5, 4060–4068, 2018, doi: 10.1109/TIA.2018.2840083.

13. Kumar, A., Bhat, H., Agarwal, P., Real time performance investigation of extended phase shifted dual active bridge converter in dc-microgrid with optimum operating zone. *J. Electr. Eng. Romania*, 21, 02, 76–87, 2021.
14. Khalid, M., Akram, U., Shafiq, S., Optimal planning of multiple distributed generating units and storage in active distribution networks. *IEEE Access*, 6, 55234–55244, 2018, doi: 10.1109/ACCESS.2018.287278.
15. Zhao, L., Wang, X., Cui, X., Xu, P., Research on reliability evaluation of power generation system with solar thermal power, in: *Proceedings of the 5th Asia Conference on Power and Electrical Engineering (ACPEE)*, pp. 356–360, 2020, doi: 10.1109/ACPEE48638.2020.9136449.
16. Altun, H. and Sünter, S., Modeling, simulation and control of wind turbine driven doubly-fed induction generator with matrix converter on the rotor side. *Electr. Eng.*, 95, 157–170, 2020. https://doi.org/10.1007/s00202-012-0250-x.
17. Rathore, A. and Patidar, N.P., Reliability constrained socio-economic analysis of renewable generation based standalone hybrid power system with storage for off-grid communities. *IET Renew. Power Gener.*, 14, 2142–2153, 2020.
18. Rathore, A. and Patidar, N.P., Reliability assessment using probabilistic modelling of pumped storage hydro plant with PV-wind based standalone microgrid. *Int. J. Electr. Power Energy Syst.*, 106, 17–32, 2019.
19. Govindugari Venkata, S., Ganesh, V., Madichetty, S., Application of bacteria foraging algorithm for modular multilevel converter-based microgrid with effect of wind power. *Electr. Eng.*, 100, 2023–2036, 2018. https://doi.org/10.1007/s00202-018-0681-0.
20. Hirose, and Matsuo, H., Standalone hybrid wind-solar power generation system applying dump power control without dump load. *IEEE Trans. Ind. Electron.*, 59, 2, 988–997, 2011. doi: 10.1109/TIE.2011.2159692.
21. Shan, Y., Hu, J., Chan, W.K., Fu, Q., Guerrero, M.J., Model predictive control of bidirectional dc–dc converters and ac/dc interlinking converters—A new control method for pv-wind-battery microgrids. *IEEE Trans. Sustain. Energy*, 10, 4, 1823–1833, 2019, doi: 10.1109/TSTE.2018.2873390.
22. Shaterabadi, M. and Jirdehi, M.A., Smart scheduling of transmission line switching: Optimization of multi-objective microgrid's day-ahead energy scheduling with considering high penetration of green energies and INVELOX. *Electr. Eng.*, 103, 1753–1767, 2021. https://doi.org/10.1007/s00202-020-01193-2.
23. Peddeeti, S., Vuddanti, S., Babu, B.C., Effective power balancing and power quality enhancement in single-phase microgrid for remote areas with water pumping, battery storage and electronic load controller. *Electr. Eng.*, 104, 83–96, 2022. https://doi.org/10.1007/s00202-021-01377-4.
24. Eid, B.M., Guerrero, J.M., Abusorrah, A.M. *et al.*, A new voltage regulation strategy using developed power sharing techniques for solar photovoltaic generation-based microgrids. *Electr. Eng.*, 103, 3023–3031, 2021. https://doi.org/10.1007/s00202-021-01289-3.

25. Khalid, M., Akram, U., Shafiq, S., Optimal planning of multiple distributed generating units and storage in active distribution networks. *IEEE Access*, 6, 55234–55244, 2018, doi: 10.1109/ACCESS.2018.287278.
26. Sharma, P. and Bhatti, T.S., A review on electrochemical double-layer capacitors. *Energy Convers. Manage.*, 51, 12, 2901–2912, 20102010. https://doi.org/10.1016/ j.encon man.06.031.
27. Saikia, K.B., Benoy, M.S., Bora, M. *et al.*, A brief review on supercapacitor energy storage devices and utilization of natural carbon resources as their electrode materials. *Fuel*, 282, 118796, 2020. ISSN 0016-2361. https://doi.org/10.1016/j.fuel.2020.118796.
28. Sun, C., Negro, E., Vezzù, K., Pagot, G., Cavinato, G., Nale, A., Bang, Y.H., Di Noto, V., Hybrid inorganic-organic proton-conducting membranes based on SPEEK doped with WO3 nanoparticles for application in vanadium redox flow batteries. *Electrochim. Acta*, 309, 311–325, 2019.
29. Yüksek, Ö., Anılan, T., Saka, F., Problems and possible solutions to environmental impacts of small hydropower plants in Turkey. *Turkish J. Electromech. Energy*, 5, 37–47, 2020.
30. Sahin, M.E., A photovoltaic powered electrolysis converter system with maximum power point tracking control. *Int. J. Hydrog. Energy*, 45, 9293–9304, 2020.
31. Zhong, Y., Xia, X., Mai, W., Tu, J., Fan, H.J., Integration of energy harvesting and electrochemical storage devices. *Adv. Mater. Technol.*, 2, 1700182, 2017.
32. Zhang, H. and Sun, C., Cost-effective iron-based aqueous redox flow batteries for large-scale energy storage application: A review. *J. Power Sources*, 493, 229445, 2021.
33. Sahin, M.E., Blaabjerg, F., Sangwongwanich, A., A review on supercapacitor materials and developments. *Turk. J. Mater.*, 5, 10–24, 2020.
34. Liu, S., Wei, L., Wang, H., Review on the reliability of supercapacitors in energy storage applications. *Appl. Energy*, 278, 115436, 2020.
35. Jing, W., Lai, C.H., Wong, W.S.H., Wong, M.L.D., Battery-supercapacitor hybrid energy storage system in stand-alone DC microgrids: A review. *IET Renew. Power Gener.*, 11, 461–469, 2017.
36. Parnia, F. and Pillai, S.C., Two-dimensional (2D) electrode materials for supercapacitors. *Mater. Today Proc.*, 41, 3, 498–505, 2021. ISSN 22147853. https://doi.org/10.1016 /j.matpr. 2020.05.233.
37. Wuquan, Y., Haiyan, W., Jiqiang, N., Yijun, Z., Yong, H., New types of hybrid electrolytes for supercapacitors. *J. Energy Chem.*, 57, 219–232, 2021. ISSN 2095-4956. https://doi.org/10.1016/j.jechem.2020.09.016.
38. Frackowiak, E., Mellerr, M., Menzel, J., Gastol, D., Fic, K., Redox-active electrolyte for supercapacitor application. *Faraday Discuss.*, 172, 179–198, 2014.
39. Chodankar, R.N., Dubal, P.D., Lokhande, C.A., Patil, M.A. *et al.*, An innovative concept of use of redox-active electrolyte in asymmetric capacitor based on MWCNTs/MnO_2 and Fe_2O_3 thin films. *Sci. Rep.*, 6, 39205, 2016.

40. Pan, S., Deng, J., Guan, G., Zhang, Y., Chen, P., Ren, J. *et al.*, A redox-active gel electrolyte for fiber-shaped supercapacitor with high area specific capacitance. *J. Mater. Chem. A*, 3, 6286–6290, 2015.
41. Zhou, J., Cai, J., Cai, S., Zhou, X. *et al.*, Development of all-solid-state mediator-enhanced supercapacitors with polyvinylidene fluoride/lithium trifluoro-methanesulfonate separators. *J. Power Sources*, 196, 10479–10483, 2011.
42. Jang, S.H., Raj, J.C., Lee, G.W., Kim, C.B., Yu, H.K. *et al.*, Enhanced supercapacitive performances of functionalized activated carbon in novel gel polymer electrolytes with ionic liquid redox-mediated poly(vinyl alcohol)/phosphoric acid. *RSC Adv.*, 6, 75376–75383, 2016.
43. Sun, K., Feng, E., Peng, H., Ma, G. *et al.*, A simple and high-performance supercapacitor based on nitrogen-doped porous carbon in redox-mediated sodium molybdate electrolyte. *Electrochim. Acta*, 158, 361–367, 2015.
44. Ma, G., Sun, K., Peng, H., Mu, J., Lei, Z., High performance solid-state supercapacitor with PVA-KOH-K_3 $[Fe(CN)_6]$ gel polymer as electrolyte and separator. *J. Power Sources*, 256, 281–287, 2014.
45. Ma, G., Li, J., Sun, K., Peng, H., Mu, J., Lei, Z., A novel and high-effective redox-mediated gelpolymer electrolyte for supercapacitor. *Electrochim. Acta*, 135, 461–466, 2014.
46. Sun, J., Wu, C., Sun, X., Hu, H., Zhi, C., Hou, L. *et al.*, Recent progresses in high- energy-density all pseudocapacitive-electrode-materials-based asymmetric supercapacitors. *J. Mater. Chem. A*, 5, 9443–9464, 2017.
47. Kumar, S.K., Choudhary, N., Jung, Y., Thomas, J., Recent advances in two-dimensional nanomaterials for supercapacitor electrodes. *ACS Energy Lett.*, 3, 482–495, 2018.
48. Ferris, A., Garbarino, S., Guay, D., Pech, D., 3D RuO_2 microsupercapacitors with remarkable areal energy. *Adv. Mater.*, 27, 6625–6629, 2015.
49. Wei, W., Cui, X., Chen, W., Ivey, G.D., Manganese oxide-based materials as electro- chemical supercapacitor electrodes. *Chem. Soc. Rev.*, 40, 1697–1721, 2011.
50. Shi, S., Xu, C., Yang, C., Chen, Y., Liu, J., Kang, F., Flexible asymmetric supercapacitors based on ultrathin two-dimensional nanosheets with outstanding electrochemical performance and aesthetic property. *Sci. Rep.*, 3, 2598, 2013.
51. Guan, C., Zhao, C., Hu, Y. *et al.*, Cobalt oxide and N-doped carbon nanosheets derived from a single two-dimensional metal organic framework precursor and their application in flexible asymmetric supercapacitors. *Nanoscale Horiz.*, 2, 99–105, 2017.
52. Shi, P., Chen, R., Li, L., An, J., Hua, L., Zhou, J. *et al.*, Holey nickel hydroxide nanosheets for wearable solid-state fiber-supercapacitors. *Nanoscale*, 10, 5442–5448, 2018.
53. Wang, D., Zhang, Y., Yang, L., Fan, G., Lin, Y., Li, F., Direct *in situ* assembly of bimetallic Co Ni hydroxide/polyaniline-modified reduced graphene oxide

nanocomposite for asymmetric flexible supercapacitor electrode. *J. Mater. Sci. Mater. Electron.*, 31, 6467–6478, 2020.
54. Du, Y., Li, G., Zhao, L., Ye, L., Che, C., Liu, X. *et al.*, Core shell MnO2 nanotubes@nickel cobalt zinc hydroxide nanosheets for supercapacitive energy storage. *ACS Appl. Nano Mater.*, 3, 7462–7473, 2020.
55. Lei, C.J., Zhang, X., Zhou, Z., Recent advances in MXene: Preparation, properties, and applications. *Front. Phys-Beijing*, 10, 276–286, 2015.
56. Anasori, B., Xie, Y., Beidaghi, M., Lu, J., Hosler, C.B., Hultman, L. *et al.*, Two-dimensional, ordered, double transition metals carbides (MXenes). *ACS Nano*, 9, 9507–9516, 2015.
57. Yang, Q., Wang, Y., Li, X., Li, H., Wang, Z., Tang, Z. *et al.*, Recent progress of MXene-based nanomaterials in flexible energy storage and electronic devices. *Energy Environ. Mater.*, 1, 183–195, 2018.
58. Peng, Y.-Y., Akuzum, B., Kurra, N., Zhao, M.-Q., Alhabeb, M., Anasori, B. *et al.*, All-MXene (2D titanium carbide) solid-state microsupercapacitors for on-chip energy storage. *Energy Environ. Sci.*, 9, 2847–2854, 2016.
59. Han, Y., Ge, Y., Chao, Y., Wang, C., Wallace, G.G., Recent progress in 2D materials for flexible supercapacitors. *J. Energy Chem.*, 27, 57–72, 2018.
60. Mayorga-Martinez, C.C., Ambrosi, A., Eng, A.Y.S., Sofer, Z., Pumera, M., Transition metal chalcogenides (MoS2, MoSe2, WS2, WSe2) exfoliation technique has strong influence upon thier capacitance. *Electrochem. Commun.*, 56, 24–28, 2015.
61. Eng, A.S., Ambrosi, A., Sofer, Z., Simek, P., Pumera, M., Electrochemistry of transition metal dichalcogenides: Strong dependence on the metal-to-chalcogen composition and exfoliation method. *ACS Nano*, 8, 12185–12198, 2014.
62. Soon, J.M. and Loh, K.P., Electrochemical double-layer capacitance of MoS2 nanowall films. *Solid-State Lett.*, 10, A250–A254, 2007.
63. Yuan, S., Pang, S.-Y., Hao, J., 2D transition metal dichalcogenides, carbides, nitrides, and their applications in supercapacitors and electrocatalytic hydrogen evolution reaction. *Appl. Phys. Rev.*, 7, 021304, 2020.
64. Wang, S., Zhu, J., Shao, Y., Li, W., Wu, Y., Zhang, L. *et al.*, Three-dimensional MoS2@CNT/RGO network composites for high-performance flexible supercapacitors. *Chemistry*, 23, 3438–3446, 2017.
65. Liu, A., Lv, H., Liu, H., Li, Q., Zhao, H., Two dimensional MoS2/CNT hybrid ink for paper-based capacitive energy storage. *J. Mater. Sci. Mater. Electron.*, 28, 8452–8459, 2017.
66. Gao, L., Li, X., Cheng, J., Wang, B., Wang, Z. *et al.*, A coaxial yarn electrode based on hierarchical MoS2 nanosheets/carbon fiber tows for flexible solid-state supercapacitors. *RSC Adv.*, 6, 57190—57198, 2016.
67. Byun, S., Sim, M.D., Yu, J., Yoo, J.J., High-power supercapacitive properties of graphene oxide hybrid films with highly conductive molybdenum disulfide nanosheets. *ChemElectroChem*, 2, 1938–1946, 2015.

68. Acerce, M., Voiry, D., Chhowalla, M., Metallic 1T phase MoS2 nanosheets as supercapacitor electrode materials. *Nat. Nanotechnol.*, 10, 313–318, 2015.
69. Li, X., Cheng, J., Yuan, D., Ni, W., Guan, Q. *et al.*, Fiber-shaped solid-state supercapacitors based on molybdenum disulfide nanosheets for a self-powered photodetecting system. *Nano Energy*, 21, 228–237, 2016.
70. Gopalakrishnan, A. and Badhulika, S., Facile sonochemical assisted synthesis of a hybrid red black phosphorus/sulfonated porous carbon composite for high-performance supercapacitors. *Chem. Commun.*, 56, 7096–7099, 2020.

14

2D Nanomaterials Based for Electrocatalytic Application

Anchit Modi[1]*, D. K. Gupta[1], Jitendra Malviya[2] and N. K. Gaur[3]

[1]*Department of Basic Sciences, IITM, IES University, Bhopal (M. P.), India*
[2]*Department Life Sciences and Biological Sciences IES University, Bhopal (M. P.), India*
[3]*Department of Physics, Barkatullah University, Bhopal, (M. P.), India*

Abstract

This chapter comprehensively explores the significance of electrocatalysis and unique properties of 2D nanomaterials. The motivations for the incorporation of 2D nanomaterials into electrocatalytic systems have been thoroughly discussed. This chapter categorizes 2D nanomaterials into three main groups, with a detailed examination of graphene including its distinctive structure, properties, synthesis methods, and electrocatalytic applications. Transition metal dichalcogenides (TMDs) have also been extensively explored, including their structures, properties, synthesis methods, and electrocatalytic applications. Furthermore, this chapter presents a concise overview of other 2D nanomaterials such as black phosphorus and MXenes, showing their potential for electrocatalytic applications. Specific electrocatalytic reactions facilitated by 2D nanomaterials, such as the oxygen reduction reaction (ORR), hydrogen evolution reaction (HER), and carbon dioxide reduction reaction (CO2RR), have been thoroughly examined, emphasizing the role of 2D nanomaterials in enhancing the catalytic performance. Key findings and advancements in these areas have been highlighted. Synthesis and characterization techniques for 2D nanomaterials have been extensively discussed, emphasizing the intricate relationship between the synthesis, structure, and electrocatalytic performance. This chapter addresses the challenges of utilizing 2D nanomaterials for electrocatalysis and proposes strategies to overcome these obstacles. Additionally, future directions and emerging trends in the field of 2D nanomaterials for electrocatalysis are examined, summarizing the key points discussed throughout the chapter and underscoring the importance and potential

**Corresponding author*: anchitmodi87@gmail.com

Subhendu Chakroborty and Kaushik Pal (eds.) 2D Nanomaterials: Synthesis, Properties and Applications, (315–332) © 2024 Scrivener Publishing LLC

impact of 2D nanomaterials in enhancing the catalytic performance and driving advancements in energy conversion and storage technologies.

***Keywords*:** 2D nanomaterials, oxygen reduction reaction, hydrogen evolution reaction, carbon dioxide reduction reaction

14.1 Introduction

Electrocatalysis focuses on accelerating electrochemical reactions using catalysts. In electrochemical processes, reactions occur at the interface between an electrode and an electrolyte and involve the transfer of electrons and ions. Catalysts play a vital role in electrocatalysis by providing an alternative reaction pathway with lower energy barriers, facilitating electron transfer, and enabling faster and more efficient reaction kinetics. This acceleration of reactions has significant implications for various industries and scientific disciplines. Electrocatalysis has immense significance in energy conversion and storage [1–6]. They play a crucial role in fuel cells, where they enhance the conversion of chemical energy into electrical energy by catalyzing reactions such as oxygen reduction and hydrogen oxidation. Electrocatalysis is also crucial in electrolyzers, as it facilitates water splitting to produce hydrogen or oxygen, enabling energy storage through reversible electrochemical reactions. These applications have the potential to revolutionize clean and sustainable energy systems.

Furthermore, electrocatalysis has applications in electrosynthesis, enabling the synthesis of valuable compounds *via* electrochemical reactions. For example, converting carbon dioxide into fuels or fine chemicals, known as the carbon dioxide reduction reaction (CO2RR), offers a sustainable approach to chemical transformation. Electrocatalysis also plays a vital role in sensors and biosensors, wherein electrochemical detection methods rely on the catalytic properties of materials to achieve enhanced sensitivity, selectivity, and stability [7–9]. Moreover, electrocatalysis has been employed in environmental remediation for the electrochemical degradation of pollutants, contributing to the mitigation of environmental issues.

14.1.1 Introduction to 2D Nanomaterials and Their Unique Properties

In recent years, 2D nanomaterials have attracted significant attention for electrocatalysis. These materials possess a unique two-dimensional structure, typically with thicknesses ranging from a few atoms to a few nanometers and have much larger lateral dimensions. Among the various 2D nanomaterials, graphene is a well-known example of a monolayer of carbon atoms arranged in a hexagonal lattice. Another prominent class of 2D nanomaterials is transition metal dichalcogenides (TMDs), including molybdenum disulfide (MoS_2) and tungsten diselenide (WSe_2) [10–13].

What distinguishes 2D nanomaterials is their exceptional properties, resulting from their ultrathin nature and large surface area. Owing to their atomic-scale thickness, these nanomaterials have an extraordinarily high surface-to-volume ratio. This characteristic enables enhanced interactions between the surfaces and reactants or electrolytes, thereby facilitating improved electrocatalytic activity [10, 14]. In addition, the electronic structures of 2D nanomaterials can be finely tuned, allowing precise control over their catalytic properties. Furthermore, their mechanical flexibility, high thermal conductivity, and chemical stability contribute to their outstanding performance in electrocatalytic applications.

14.1.2 Motivation for Utilizing 2D Nanomaterials in Electrocatalytic Applications

The motivation for leveraging the potential of 2D nanomaterials in electrocatalytic applications stems from the need for highly efficient, stable, and cost-effective catalysts. Conventional catalysts often face limitations, such as low catalytic activity, inadequate selectivity, and susceptibility to degradation under harsh reaction conditions. The introducing of 2D nanomaterials as catalysts presents an opportunity to overcome these challenges and revolutionize the field of electrocatalysis [15, 16].

The exceptional properties of 2D nanomaterials make them attractive for electrocatalytic applications. Their substantial surface area enables the efficient utilization of active sites, thereby increasing catalytic activity and enhancing reaction rates. The tunability of their electronic structures allows for the precise engineering of their catalytic properties, tailoring them to meet specific reaction requirements. Moreover, the large surface area facilitates electron transfer between the reactants and electrode surface, enhancing the overall electrochemical performance [17, 18].

Electrocatalysis is vital for accelerating electrochemical reactions and is essential in various fields. The scalability and compatibility of 2D nanomaterials with diverse synthesis and modification techniques makes them promising candidates for electrocatalysis. Researchers can employ various methods to synthesize and functionalize these materials and customize their properties for specific applications [19]. This versatility opens new avenues for designing efficient and sustainable electrocatalytic systems.

14.2 Types of 2D Nanomaterials

14.2.1 Graphene

Graphene, a two-dimensional lattice of carbon atoms arranged in a honeycomb structure, possesses exceptional properties that make it highly attractive for electrocatalytic applications. Owing to high mechanical strength, excellent electrical and thermal conductivities, and large surface area, graphene enables efficient charge transport and intense reactant interactions, making it an ideal material for electrocatalytic applications. Synthesis methods such as mechanical exfoliation, chemical vapor deposition (CVD), epitaxial growth, and chemical synthesis offer versatile approaches for obtaining graphene with the desired properties. Its large surface area, high electrical conductivity, and active sites on its edges and defects contribute to its excellent electrocatalytic activity [20, 21]. Graphene-based electrocatalysts have demonstrated promising performance in oxygen reduction, hydrogen evolution, and carbon dioxide reduction reactions, making them highly valuable in applications such as fuel cells, water splitting, and electrochemical sensors owing to their remarkable catalytic activity, durability, and stability.

14.2.2 Dichalcogenides (TMDs)

Transition metal dichalcogenides (TMDs) are a class of 2D nanomaterials characterized by a layered structure consisting of a transition metal layer sandwiched between two chalcogen atoms such as sulfur, selenium, and tellurium. TMDs possess diverse properties, including adjustable bandgaps, high surface-to-volume ratios, and remarkable electrochemical activities [22, 23]. The specific electronic, optical, and catalytic properties of TMDs depend on the choice of transition metals and chalcogens, offering a wide range of possibilities. Several synthetic methods have been developed to fabricate TMDs, including physical vapor deposition (PVD), chemical vapor deposition (CVD), hydrothermal/solvothermal synthesis, and liquid-phase exfoliation. PVD and CVD techniques involve the controlled deposition of transition metal and chalcogen precursors onto substrates, enabling the controlled growth of TMD layers. Hydrothermal and solvothermal synthesis methods employ high-temperature and high-pressure conditions to facilitate TMD formation. Liquid-phase exfoliation allows bulk TMD crystals to be exfoliated into 2D nanosheets in a liquid medium. TMDs have demonstrated significant potential as electrocatalysts in various reactions. For instance, molybdenum disulfide (MoS2) exhibits outstanding catalytic activity for the hydrogen evolution reaction (HER), surpassing traditional platinum catalysts. TMDs, such as tungsten diselenide (WSe_2), molybdenum diselenide ($MoSe_2$), and tungsten disulfide (WS_2), have also displayed favorable catalytic properties for oxygen reduction reactions (ORR), making them suitable candidates for fuel cells and other energy conversion devices [24, 25]. The distinctive electronic and surface properties of TMDs make them appealing for diverse electrochemical reactions, including the CO2RR, water oxidation, and electrochemical sensing applications. Other 2D nanomaterials, such as black phosphorus and MXenes, also hold promise in various electrocatalytic applications, contributing to the advancement of this field [17, 26].

14.2.3 Brief Overview of Their Structures and Properties

In addition to graphene and transition metal dichalcogenides (TMDs), several other 2D nanomaterials have shown promise for electrocatalytic applications. Black phosphorus (phosphorene) is a layered material composed of phosphorus atoms. It possesses a direct bandgap and high charge-carrier mobility, making it suitable for applications in optoelectronics and energy storage. Another notable group is MXenes, which comprise 2D transition metal carbides, nitrides, and carbonitrides [17, 24]. MXenes exhibit excellent electrical conductivity, a high surface area, and tunable surface chemistry, making them appealing for applications in electrocatalysis, energy storage, and sensing.

Black phosphorus has demonstrated potential as an electrocatalyst for various reactions, including oxygen reduction, hydrogen evolution, and CO_2 reduction. Its distinctive band structure and active sites make it an attractive candidate for efficient energy conversion and storage [26]. MXenes, with their metallic conductivity, display excellent electrocatalytic properties for hydrogen evolution, oxygen evolution, and numerous other reactions [17]. Their high surface areas and unique surface terminations contribute to their catalytic activity, making them promising candidates for energy-related electrochemical applications.

In summary, 2D nanomaterials, such as graphene, TMDs, black phosphorus, and MXenes possess unique properties that render them suitable for diverse electrocatalytic applications. Understanding and utilizing their structures, synthesis methods, and electrochemical properties plays a crucial role in determining their efficacy as catalysts. By harnessing the potential of these materials, researchers have aimed to develop efficient and sustainable electrocatalytic systems for energy conversion, electrosynthesis, sensing, and environmental remediation.

14.3 Electrocatalytic Reactions Enabled by 2D Nanomaterials

14.3.1 Oxygen Reduction Reaction (ORR)

The oxygen reduction reaction (ORR) plays a crucial role in various electrochemical energy conversion devices, such as fuel cells and metal-air batteries. Efficient catalysts are essential for enhancing the ORR kinetics and improving the device performance. In recent years, 2D nanomaterials have emerged as promising candidates for ORR catalysis, owing to their unique properties. 2D nanomaterials, such as graphene, transition metal dichalcogenides (TMDs), and hexagonal boron nitride (h-BN), exhibit several advantages for ORR. First, their large specific surface areas provide abundant active sites for the catalytic reactions. The exposed edges and defects of these materials provide favorable sites for oxygen adsorption and subsequent reduction. Second, the inherent electronic properties of 2D materials can modulate the adsorption and reaction kinetics of the oxygen species, leading to enhanced ORR activity. For instance, the high electrical conductivity of graphene facilitates rapid charge transfer during the ORR. Lastly, the atomic thickness of 2D materials allows for efficient mass transport and accessibility to reactants, reducing the diffusion limitations commonly encountered in bulk catalysts [27–29].

Several key findings and advancements have been made regarding the use of 2D nanomaterials for ORR. A significant breakthrough is the discovery of highly active ORR catalysts based on graphene and its derivatives. Graphene-supported metal nanoparticles, such as platinum (Pt), have been extensively studied and have exhibited excellent ORR activity. Additionally, heteroatom-doped graphene, in which foreign atoms such as nitrogen or boron are incorporated into the graphene lattice, has shown remarkable ORR performance, rivaling Pt-based catalysts. Furthermore, TMDs such as molybdenum disulfide (MoS2) and tungsten diselenide (WSe2) have emerged as promising alternatives to graphene for the ORR [20, 23, 24]. These TMDs possess unique electronic band structures that enable efficient oxygen adsorption and charge transfer. Recent studies have demonstrated that TMD-based catalysts exhibit remarkable ORR activity and outperform commercial Pt/C catalysts.

To further enhance the catalytic performance of 2D nanomaterials, efforts have been directed toward the synthesis of hybrid structures and nanocomposites. For example, combining graphene with metal nanoparticles or coupling TMDs with carbon nanotubes has synergistic effects that enhance the ORR activity. Additionally, the controlled functionalization of 2D materials with catalytically active sites, such as transition metal ions or organic moieties, improves the ORR performance.

14.3.2 Hydrogen Evolution Reaction (HER)

The hydrogen evolution reaction (HER) is a crucial step in electrochemical water splitting and is important for renewable energy production. Efficient catalysts are essential for minimizing the overpotential required for the HER and enhancing the reaction kinetics. 2D nanomaterials have demonstrated outstanding potential for catalyzing the HER owing to their unique properties. Similar to the ORR, the large surface area and exposed edges of 2D nanomaterials provide numerous active sites for the HER. The intrinsic electronic properties of these materials can modulate the adsorption and desorption of hydrogen species, thereby facilitating the HER process [2, 6]. Moreover, the atomic thickness of 2D materials enables the efficient mass transport of reactants and products, thereby reducing diffusion limitations.

Significant progress has been made in utilizing 2D nanomaterials for the HER. Graphene has been extensively explored as a HER catalyst owing to its excellent electrical conductivity and high surface area. Various strategies, such as introducing defects, doping with foreign atoms, and functionalizing graphene with metal nanoparticles, have enhanced HER activity [21].

Moreover, TMDs have also shown promise as efficient catalysts for the HER. For instance, molybdenum diselenide (MoSe2) and tungsten disulfide (WS2) have demonstrated remarkable HER performance, comparable to or surpassing that of Pt-based catalysts [23–25]. The unique electronic properties of TMDs, combined with their large surface areas, allow for efficient hydrogen adsorption and charge transfer. Efforts have also been focused on developing hybrid structures and composite materials to further enhance the HER performance of 2D nanomaterials. Combining graphene or TMDs with other catalysts such as metal nanoparticles or metal oxides can achieve synergistic effects leading to improved HER activity. Additionally, engineering the morphology and surface chemistry of 2D nanomaterials through controlled synthesis techniques has proven effective in tailoring their catalytic properties for HER.

14.3.3 Carbon Dioxide Reduction Reaction (CO2RR)

The carbon dioxide reduction reaction (CO2RR) is a significant challenge for achieving sustainable carbon-based fuel production and mitigating greenhouse gas emissions. CO2RR converts carbon dioxide into valuable products, such as formate, methane, ethylene, or ethanol. 2D nanomaterials have shown promise in catalyzing CO2RR owing to their unique properties and tunable electronic structures [30].

Significant advancements have been made in harnessing the use of 2D nanomaterials for the CO2RR. 2D nanomaterials, including graphene, TMDs, and metal dichalcogenides (MX2, where M represents a transition metal) offer several advantages for the CO2RR. Their large surface area and unique edge structure provide abundant active sites for CO2 adsorption and subsequent reduction. The tunable band structures of these materials allow for efficient charge transfer and selectivity towards specific CO2RR products. Additionally, the atomic thickness of 2D materials facilitates the mass transport of reactants and products, thereby reducing diffusion limitations. Graphene-based catalysts functionalized with metal nanoparticles or organic moieties have shown good catalytic activity for the CO2RR [31–33]. These catalysts exhibit high selectivity towards specific products such as formate or ethylene, depending on the choice of metal nanoparticles or functional groups.

TMDs, particularly those containing single-transition metal atoms (e.g., single-atom catalysts), have also demonstrated remarkable CO2RR activity. The localized charge distribution around the single metal atoms enhances the adsorption of CO2 and facilitates selective reduction of the desired products. Moreover, combining TMDs with other catalysts, such as metal nanoparticles or metal oxides, improves CO2RR performance by creating synergistic effects [34].

Efforts have also focused on tailoring the structure and composition of 2D nanomaterials to enhance CO2RR. By controlling the synthesis parameters, such as the temperature, precursor concentration, and reaction time, the morphology, size, and defect density of 2D nanomaterials can be precisely tuned, leading to improved catalytic performance. Additionally, heteroatom doping and surface functionalization have been employed to modulate the electronic properties and active sites of 2D materials, further enhancing their CO2RR activities [30].

In conclusion, 2D nanomaterials have shown great potential for catalyzing various electrocatalytic reactions, including the ORR, HER, and CO2RR. Their large surface areas, unique electronic properties, and efficient mass-transport characteristics enable improved catalytic performance. Ongoing research and advancements in the synthesis and engineering of 2D nanomaterials hold promise for the development of efficient and sustainable electrocatalysts for energy conversion and environmental applications.

14.3.4 Synthesis and Characterization Techniques

14.3.4.1 Synthesis Methods for 2D Nanomaterials

The synthesis of 2D nanomaterials encompasses a range of techniques that enable the controlled growth and formation of materials with desired properties. Several synthesis methods have been developed, each offering distinct advantages and limitations regarding scalability, quality, and control over the material properties. One commonly used method is mechanical exfoliation, which involves mechanical cleavage of bulk materials, such as graphite, using adhesive tapes or other automated methods. This process results in the production of thin 2D flakes. Mechanical exfoliation is a straightforward and effective technique for obtaining high-quality 2D materials with exceptional properties [34, 35]. However, it is limited in terms of scalability and production of large materials.

Chemical vapor deposition (CVD) is another widely used method for synthesizing large-scale high-quality 2D nanomaterials. In CVD, precursor gases are introduced into a reaction chamber, decomposing and reacting on the substrate surface to form 2D layers [36]. This method allows precise control over growth parameters such as temperature, pressure, and precursor concentration, enabling tailoring of the size, morphology, and composition of the synthesized materials. CVD is advantageous owing to its scalability and ability to produce uniform and controllable 2D nanomaterials.

Liquid-phase exfoliation involves dispersing bulk materials in a liquid solvent and applying mechanical or ultrasonic forces to break them into thin 2D flakes. This approach enables large-scale production of 2D nanomaterials and is suitable for scalable synthesis. The choice of solvent in liquid-phase exfoliation can influence the exfoliation process and the properties of the resulting nanomaterials, offering further flexibility in tailoring their characteristics [37].

Templated growth techniques provide a means to achieve precise control over the structure and properties of the synthesized nanomaterials. Templated growth methods utilize templates or substrates with predefined patterns or designs to guide the growth of 2D materials. For example, epitaxial growth on a crystal substrate or nanostructured templates can be used to synthesize 2D materials with controlled dimensions and orientations [38].

The synthesis of 2D nanomaterials involves various methods, each with its own advantages and limitations. The choice of the synthesis method depends on the desired scale of production, material quality, and control over material properties, ultimately driving the development of advanced 2D nanomaterials for various applications. Mechanical exfoliation, chemical vapor deposition, liquid-phase exfoliation, and templated growth are essential techniques for the fabrication of 2D nanomaterials with tailored properties.

14.3.4.2 Characterization Techniques for 2D Nanomaterials

The accurate characterization of 2D nanomaterials is crucial for understanding their structure, morphology, composition, and properties. Various characterization techniques have been employed to study these materials at different length scales [39–45].

Atomic Force Microscopy (AFM): AFM is a widely used technique for imaging and characterizing the topography of 2D nanomaterials at the nanoscale. It uses a sharp tip that scans the surface of the materials and provides high-resolution height information.

Transmission Electron Microscopy (TEM): TEM allows imaging of 2D nanomaterials at the atomic scale. It provides valuable information about the morphology, crystal structure, defects, and lattice arrangement of materials. High-resolution TEM (HRTEM) can resolve individual atomic positions.

X-Ray Diffraction (XRD): XRD was used to analyze the crystal structure and orientation of the 2D nanomaterials. By analyzing the diffraction patterns produced by the X-rays interacting with the material, valuable information about the crystallographic phases and crystallinity can be obtained.

Raman Spectroscopy: Raman spectroscopy provides insights into the vibrational and optical properties of 2D nanomaterials. The interaction of laser light with the material produces characteristic Raman scattering spectra, which can reveal information about the chemical composition, defects, and layer thickness of the material.

14.3.4.3 Relationship Between Synthesis, Structure, and Electrocatalytic Performance

The synthesis parameters and methods employed to fabricate 2D nanomaterials directly influence their structural features and electrocatalytic performance. The relationship between the synthesis, structure, and electrocatalytic version is complex and requires a comprehensive understanding of the underlying mechanisms.

The morphology, size, thickness, and crystallinity of 2D nanomaterials significantly affect their surface areas, active sites, and mass transport properties. These structural characteristics, along with their intrinsic electronic properties, determine the catalytic activity and selectivity of materials toward specific electrocatalytic reactions [18, 25]. Furthermore, defects, dopants, or functional groups introduced during the synthesis process can influence the electrochemical behavior of 2D nanomaterials. These modifications can alter the charge transfer kinetics, adsorption/desorption properties, and reaction pathways, thereby affecting the overall electrocatalytic performance.

Understanding the structure–activity relationship requires combining advanced characterization techniques, electrochemical measurements, and theoretical modeling. Researchers can identify fundamental design principles and optimize synthesis strategies to develop highly efficient 2D nanomaterial-based electrocatalysts by systematically correlating synthesis parameters, structural features, and electrocatalytic performance [12].

In conclusion, the synthesis and characterization techniques employed for 2D nanomaterials are crucial for tailoring their properties and understanding their electrocatalytic performances. The relationship between the synthesis, structure, and electrocatalytic activity is multifaceted, and a thorough understanding of this interplay is essential for the rational design and development of advanced electrocatalysts for various applications.

14.4 Challenges and Future Perspectives

14.4.1 Current Challenges in Utilizing 2D Nanomaterials for Electrocatalytic Applications

Although 2D nanomaterials show great promise for electrocatalytic applications, several challenges must be addressed to fully harness their potential.

Scalable Synthesis: Many synthesis methods for 2D nanomaterials, such as mechanical and liquid-phase exfoliation, suffer from limited scalability. Developing scalable synthesis techniques that maintain the quality and properties of the materials is crucial for practical applications [30].

Catalyst Stability: Some 2D nanomaterials may exhibit poor stability under harsh reaction conditions. The electrochemical environment, such as high potentials and corrosive electrolytes, can lead to degradation or structural changes in catalysts, impacting their long-term performance. Ensuring the stability of 2D nanomaterials is a critical challenge [37].

Activity and Selectivity Enhancement: While 2D nanomaterials exhibit good electrocatalytic activity, further improvements are needed, particularly in achieving high catalytic activity and selectivity for specific reactions. Enhancing the intrinsic catalytic properties and tailoring the active sites for the desired responses pose ongoing challenges [41].

14.4.2 Potential Strategies to Overcome These Challenges

Several strategies can be pursued to address the challenges.

Rational Design: A deeper understanding of the structure–activity relationship is essential for rationalizing 2D nanomaterials with enhanced electrocatalytic properties. Researchers can tailor the material composition, morphology, and active sites through computational modeling and experimental investigations to optimize their performance [6].

Hybrid Structures and Composites: Integration of 2D nanomaterials with other catalysts such as metal nanoparticles or metal oxides can lead to synergistic effects and enhanced catalytic performance. The development of hybrid structures and composites allows the combination of different functionalities and can overcome the limitations of individual components [41].

Surface Engineering: Controlling the surface chemistry and functionalization of 2D nanomaterials can modulate their catalytic activity and selectivity. Introducing specific functional groups, doping with foreign atoms, or engineering defects can improve the catalytic properties and stability of materials [12].

14.4.3 Future Directions and Emerging Trends in the Field

The field of 2D nanomaterials for electrocatalysis is evolving rapidly, and several future directions and emerging trends are worth highlighting.

Integration With Other Energy Conversion Technologies: Integrating 2D nanomaterials with other energy conversion technologies, such as photovoltaics or energy storage systems, holds significant potential. Combining these technologies can enable efficient energy conversion, storage, and utilization, thereby contributing to the development of sustainable and scalable energy systems [1].

Multifunctional Catalysts: Researchers are exploring multifunctional catalysts that can simultaneously perform multiple electrocatalytic reactions. The development of catalysts that can perform the ORR, HER, and CO2RR simultaneously or sequentially can contribute to the efficient utilization of resources and energy [7].

***In Situ* and Operando Characterization:** Advancements in *in situ* and operando characterization techniques will provide real-time insights into the catalytic processes, reaction mechanisms, and structural changes of 2D nanomaterials during electrocatalysis. These techniques can offer valuable information for optimizing catalyst performance and understanding catalytic pathways [13].

Sustainability and Environmental Impact: As field progresses, it is crucial to consider the sustainability and environmental impact of 2D nanomaterials. Developing synthesis methods that minimize energy consumption, reduce waste generation, and employ environmentally friendly precursors is increasingly essential [16].

In summary, addressing the current challenges, such as scalable synthesis, catalyst stability, and activity enhancement, is essential for fully utilizing 2D nanomaterials for electrocatalytic applications. With ongoing efforts in rational design, hybrid structures, surface engineering, and emerging trends, this field holds promise for developing efficient and sustainable electrocatalysts, contributing to various energy conversion and environmental applications.

14.5 Conclusion

In this comprehensive chapter, we delve into the significant role of 2D nanomaterials in electrocatalytic reactions, focusing specifically on the oxygen reduction reaction (ORR), hydrogen evolution reaction (HER), and carbon dioxide reduction reaction (CO2RR). These reactions play vital roles in energy conversion and environmental applications, and 2D nanomaterials have emerged as promising candidates for enhancing their performance. One of the key advantages of 2D nanomaterials such as graphene and transition metal dichalcogenides (TMDs) is in their exceptional properties. Their large surface area provides abundant active sites for catalytic reactions, whereas their excellent electrical conductivity facilitates efficient charge transfer. Furthermore, their tunable electronic properties allow for precise control over catalytic activity and selectivity, making them highly desirable for electrocatalysis.

In the case of ORR, 2D nanomaterials have shown tremendous potential for improving catalytic activity. By leveraging their unique properties, researchers have enhanced electrocatalyst performance, surpassing the limitations of traditional catalysts. Similarly, in the HER, 2D nanomaterials exhibit remarkable performance comparable to, in some cases exceeding, Pt-based catalysts. Techniques such as hybrid structures and morphology engineering have further enhanced HER performance, making 2D nanomaterials attractive alternatives to expensive and scarce noble metal catalysts. Another exciting area where 2D nanomaterials have made significant strides is CO2RR. Owing to their large surface areas and tunable electronic structures, 2D nanomaterials offer advantages for catalyzing the conversion of CO2 into valuable products. Catalyst design, surface functionalization, and heteroatom doping have been explored to improve CO2RR selectivity and efficiency. By harnessing these advancements, efficient CO2RR catalysts based on 2D nanomaterials can mitigate greenhouse gas emissions and address environmental challenges.

The importance and potential impact of 2D nanomaterials on electrocatalysis cannot be overstated. Their unique properties enable the development of efficient and sustainable catalysts for energy-conversion reactions. Furthermore, their scalability and cost-effectiveness in synthesis methods make them practical for large-scale production, offering a more accessible alternative to expensive noble metals such as platinum. By utilizing 2D nanomaterials in electrocatalysis, researchers and industries can contribute to addressing environmental challenges by developing efficient CO2RR catalysts that can convert CO2 into valuable products. The versatility and multifunctionality of 2D nanomaterials are particularly noteworthy. These materials can be tailored and functionalized to exhibit multifunctional properties, thereby allowing them to perform multiple electrocatalytic reactions simultaneously or sequentially. This versatility opens opportunities for integrated energy conversion systems, where 2D nanomaterials can be employed to efficiently utilize resources and optimize energy conversion processes.

In conclusion, 2D nanomaterials have emerged as promising candidates for electrocatalytic reactions owing to their unique properties, scalability, and potential impact. Continued research and advancements in the field will further unlock their full potential, paving the way for development of efficient and sustainable catalysts for energy conversion and environmental applications. By harnessing the power of 2D nanomaterials, we can work towards a greener and more sustainable future.

References

1. Jaramillo, T.F., Jørgensen, K.P., Bonde, J., Nielsen, J.H., Horch, S., Chorkendorff, I., Identification of active edge sites for electrochemical H2 evolution from MoS2 nanocatalysts. *Science*, 317, 5834, 100–102, 2007. https://doi.org/10.1126/science.1141483.
2. Stamenkovic, V.R., Strmcnik, D., Lopes, P.P., Markovic, N.M., Energy and fuels from electrochemical interfaces. *Nat. Mater.*, 16, 1, 57–69, 2017. https://doi.org/10.1038/nmat4738.
3. Nørskov, J.K., Rossmeisl, J., Logadottir, A., Lindqvist, L., Kitchin, J.R., Bligaard, T., Jonsson, H., Origin of the overpotential for oxygen reduction at a fuel-cell cathode. *J. Phys. Chem. B*, 108, 46, 17886–17892, 20042004. https://doi.org/10.1021/jp047349j.
4. Heldt, F.S., Kupke, S.Y., Dorl, S., Reichl, U., Frensing, T., Single-cell analysis and stochastic modelling unveil large cell-to-cell variability in influenza A virus infection. *Nat. Commun.*, 6, 8938, 2015. https://doi.org/10.1038/ncomms9938.
5. Chen, Z., Higgins, D., Yu, A., Zhang, L., Zhangb, J., A review on non-precious metal electrocatalysts for PEM fuel cells. *Energy Environ. Sci.*, 4, 3167, 2011. https://doi.org/10.1039/c0ee00558d.
6. Li, L. and Xia, Y., Engineering the architecture of electrocatalysts for efficient water splitting. *Nano Today*, 11, 5, 511–532, 2016. https://doi.org/10.1016/j.nantod.2016.08.006.
7. Kim, B., Hillman, F., Ariyoshi, M., Fujikawa, S., Kenis, P.J.A., Effects of composition of the microporous layer and the substrate on performance in the electrochemical reduction of CO2 to CO. *J. Power Sources*, 312, 192–198, 2016. https://doi.org/10.1016/j.jpowsour.2016.02.043.
8. Kopljar, D., Wagner, N., Klemm, E., Transferring electrochemical CO2 reduction from semi-batch into continuous operation mode using gas diffusion electrodes. *Chem. Eng. Technol.*, 39, 2042–2050, 2016. https://doi.org/10.1002/ceat.201600198.
9. Kortlever, R., Tan, K.H., Kwon, Y., Koper, M.T.M., Electrochemical carbon dioxide and bicarbonate reduction on copper in weakly alkaline media. *J. Solid State Electrochem.*, 17, 1843–1849, 2013. https://doi.org/10.1007/s10008-013-2100-9.
10. Chhowalla, M., Shin, H.S., Eda, G., Li, L., Loh, K.P., Zhang, H., The chemistry of two-dimensional layered transition metal dichalcogenide nanosheets. *Nat. Chem.*, 5, 4, 263–275, 2013. https://doi.org/10.1038/nchem.1589.

11. Wang, Q.H., Kalantar-Zadeh, K., Kis, A., Coleman, J.N., Strano, M.S., Electronics and optoelectronics of two-dimensional transition metal dichalcogenides. *Nat. Nanotechnol.*, 7, 11, 699–712, 2012. https://doi.org/10.1038/nnano.2012.193.
12. Jaramillo, T.F., Jørgensen, K.P., Bonde, J., Nielsen, J.H., Horch, S., Chorkendorff, I., Identification of active edge sites for electrochemical H2 evolution from MoS2 nanocatalysts. *Science*, 317, 5834, 100–102, 2007. https://doi.org/10.1126/science.1141483.
13. Liu, Z., Wu, Z., Yao, Q., Cao, Y., Chaib, O.J.H., Xie, J., Correlations between the fundamentals and applications of ultrasmall metal nanoclusters: Recent advances in catalysis and biomedical applications. *Nano Today*, 36, 101053, 2020. https://doi.org/10.1016/j.nantod.2020.101053.
14. Voiry, D., Yang, J., Chhowalla, M., Recent strategies for improving the catalytic activity of 2D TMD nanosheets toward the hydrogen evolution reaction. *Adv. Mater.*, 28, 29, 6197–6206, 2016. https://doi.org/10.1002/adma.201600038.
15. Wang, Y., Zhang, Z., Mao, Y., Wang, X., Two-dimensional nonlayered materials for electrocatalysis. *Energy Environ. Sci.*, 13, 3993–4016, 2020. https://doi.org/10.1039/D0EE01714.
16. Duong, K.D.L., Yun, S.J., Lee, Y.H., Van der Waals layered materials: Opportunities and challenges. *ACS Nano*, 11, 12, 11803–11830, 2017. https://doi.org/10.1021/acsnano.7b07436.
17. Shinde, P.A., Patil, A.M., Lee, S., Junga, E., Jun, S.C., Two-dimensional MXenes for electrochemical energy storage applications. *J. Mater. Chem. A*, 10, 1105–1149, 2022. https://doi.org/10.1039/D1TA04642J.
18. Zhang, Z., Liu, P., Song, Y., Hou, Y., Xu, B., Liao, T., Zhang, H., Guo, J., Sun, Z., Heterostructure engineering of 2D superlattice materials for electrocatalysis. *Adv. Sci.*, 9, 2204297, 2022.
19. Patil, A.M., Wang, J., Li, S., Hao, X., Du, X., Wang, Z., Hao, X., Abudula, A., Guan, G., Bilateral growth of monoclinic WO3 and 2D Ti3C2Tx on 3D free-standing hollow graphene foam for all-solid-state supercapacitor. *Chem. Eng. J.*, 421, 2, 127883, 2021. https://doi.org/10.1016/j.cej.2020.127883.
20. Khan, M., Assal, M.E., Tahir, M.N., Khan, M., Ashraf, M., Hatshan, M.R., Khan, M., Varala, R., Badawi, N.M., Adil, S.F., Graphene/inorganic nanocomposites: Evolving photocatalysts for solar energy conversion for environmental remediation. *J. Saudi Chem. Soc.*, 26, 6, 101544, 2022. https://doi.org/10.1016/j.jscs.2022.101544.
21. Priyadharshini, S.D., Manikandan, S., Kiruthiga, R., Rednam, U., Babu, P.S., Subbaiya, R., Karmegam, N., Kim, W., Govarthanan, M., Graphene oxide-based nanomaterials for the treatment of pollutants in the aquatic environment: Recent trends and perspectives-a review. *Environ. Pollut.*, 306, 119377, 2022. https://doi.org/10.1016/j.envpol.2022.119377.
22. Arul, N.S. and Nithya, V.D., *Two Dimensional Transition Metal Dichalcogenides.* Springer, Singapore, 2019, https://link.springer.com/book/10.1007/978-981-13-9045-6.

23. Li, T. and Galli, G., Electronic properties of MoS2 nanoparticles. *J. Phys. Chem. C*, 111, 16192–16196, 2007. https://doi.org/10.1021/jp075424v.
24. Shi, Y., Li, H., Li, L., Recent advances in controlled synthesis of two-dimensional transition metal dichalcogenides via vapor deposition techniques. *Chem. Soc. Rev.*, 44, 2744–2756, 2015. https://doi.org/10.1039/C4CS00256C.
25. Huan, Y., Shi, J., Zou, X., Gong, Y., Xie, C., Yang, Z., Zhang, Z., Gao, Y., Shi, Y., Li, M., Yang, P., Jiang, S., Hong, M., Gu, L., Zhang, Q., Yan, X., Zhang, Y., Scalable production of two-dimensional metallic transition metal dichalcogenide nanosheet powders using NaCl templates toward electrocatalytic applications. *J. Am. Chem. Soc.*, 141, 47, 18694–18703, 2019. https://doi.org/10.1021/jacs.9b06044.
26. Elbanna, O., Zhu, M., Fujitsuka, M., Majima, T., Black phosphorus sensitized TiO2 mesocrystal photocatalyst for hydrogen evolution with visible and near-infrared light irradiation. *ACS Catal.*, 9, 4, 3618–3626, 2019. https://doi.org/10.1021/acscatal.8b05081.
27. Huan, Y., Shi, J., Zhao, G., Yan, X., Zhang, Y., 2D metallic transitional metal dichalcogenides for electrochemical hydrogen evolution. *Energy Technol.*, 7, 1980333, 2019.
28. Qazi, U.Y., Future of hydrogen as an alternative fuel for next-generation industrial applications; Challenges and expected opportunities. *Energies*, 15, 4741, 2022. https://doi.org/10.3390/en15134741.
29. Song, L., Li, H., Zhang, Y., Shi, J., Recent progress of two-dimensional metallic transition metal dichalcogenides: Syntheses, physical properties, and applications. *J. Appl. Phys.*, 131, 060902, 2022. https://doi.org/10.1063/5.0083929.
30. Lu, S., Lou, F., Yu, Z., Recent progress in two-dimensional materials for electrocatalytic CO2 reduction. *Catalysts*, 12, 2, 228, 2022. https://doi.org/10.3390/catal12020228.
31. Li, J.R., Ma, Y., McCarthy, M.C., Sculley, J., Yu, J., Jeong, H.K., Balbuena, P.B., Zhou, H.C., Carbon dioxide capture-related gas adsorption and separation in metal-organic frameworks. *Coord. Chem. Rev.*, 255, 1791–1823, 2011. https://doi.org/10.3390/catal12020228.
32. Keith, D.W., Holmes, G., St Angelo, D., Heidel, K., A process for capturing CO2 from the atmosphere. *Joule*, 2, 1573–1594, 2018. https://doi.org/10.1016/j.joule.2018.05.006.
33. Back, S., Lim, J., Kim, N., Kim, Y., Jung, Y., Single-atom catalysts for CO2 electroreduction with significant activity and selectivity improvements. *Chem. Sci.*, 8, 1090–1096, 2017. https://doi.org/10.1039/C6SC03911A.
34. Cui, H., Guo, Y., Guo, L., Wang, L., Zhou, Z., Peng, Z., Heteroatom-doped carbon materials and their composites as electrocatalysts for CO2 reduction. *J. Mater. Chem. A*, 6, 18782–18793, 2018. https://doi.org/10.1039/C8TA07430E.
35. Li, F., Sun, S., Chen, Y., Naka, T., Hashishin, T., Maruyama, J., Abe, H., Bottom-up synthesis of 2D layered high-entropy transition metal hydroxides. *Nanoscale Adv.*, 4, 2468–2478, 2022. https://doi.org/10.1039/D1NA00871D.

36. Ling, X., Lee, Y., Lin, Y., Fang, W., Yu, L., Dresselhaus, M.S., Kong, J., Role of the seeding promoter in MoS2 growth by chemical vapor deposition. *Nano Lett.*, 15, 2, 1402–1409, 2015. https://doi.org/10.1021/nl4033704.
37. Kang, T., Tang, T.W., Pan, B., Liu, H., Zhang, K., Luo, Z., Strategies for controlled growth of transition metal dichalcogenides by chemical vapor deposition for integrated electronics. *ACS Mater.*, 2, 6, 665–685, 2022. https://doi.org/1021/acsmaterialsau.2c00029.
38. Kang, T.W., Pan, B., Liu, H., Zhang, K., Luo, Z., Strategies for controlled growth of transition metal dichalcogenides by chemical vapor deposition for integrated electronics. *ACS Mater.*, 2, 6, 665–685, 2022.
39. Chhowalla, M., Shin, H.S., Eda, G., Li, L., Loh, K.P., Zhang, H., The chemistry of two-dimensional layered transition metal dichalcogenide nanosheets. *Nat. Chem.*, 5, 263–275, 2013. https://doi.org/10.1038/nchem.1589.
40. Zhang, Y., Tang, T., Girit, C., Hao, Z., Martin, M.C., Zettl, A., Crommie, M.F., Shen, Y.R., Wang, F., Direct observation of a widely tunable bandgap in bilayer grapheme. *Nature*, 459, 820–823, 2009. https://doi.org/10.1038/nature08105.
41. Laursen, L.B., Kegnæs, S., Dahla, S., Chorkendorff, I., Molybdenum sulfides efficient and viable materials for electro- and photoelectrocatalytic hydrogen evolution. *Energy Environ. Sci.*, 5, 5577–5591, 2012. https://doi.org/10.1039/c2ee02618j.
42. Dubey, K., Dubey, S., Sahu, V., Modi, A., Bamne, J., Haque, F.Z., Gaur, N.K., Defects and oxygen vacancies modified properties of transition metal doped Ce0.95X0.05O2 (X = Fe, Co, Ni) nanoparticles. *Mater. Sci. Eng. B*, 288, 116154, 2023. https://doi.org/10.1016/j.mseb.2022.116154.
43. Wang, T., Zhang, X., Mei, L., Ma, D., Liao, Y., Zu, Y., Xu, P., Yin, W., Gu, Z., A two-step gas/liquid strategy for the production of N-doped defect-rich transition metal dichalcogenide nanosheets and their antibacterial applications. *Nanoscale*, 12, 8415–8424, 2020. https://doi.org/10.1039/D0NR00192A.
44. Zaman, M.B., Poolla, R., Khandy, S.A., Modi, A., Tiwari, R.K., Thioglycolic acid assisted hydrothermal growth of SnS 2D nanosheets as catalysts for photodegradation of industrial dyes. *Nanotechnology*, 32, 245706, 2021. https://doi.org/10.1088/1361-6528/abec09.
45. Bano, A., Pandey, D.K., Modi, A., Gaur, N.K., MoB2 driven metallic behavior and interfacial charge transport mechanism in MoS2/MoB2 heterostructure: A first-principles study. *Sci. Rep.*, 8, 14444, 2018. https://doi.org/10.1038/s41598-018-32850-z.

15

Engineering 2D Nanomaterials for Biomedical Applications

Swaati Sharma*, **Hardeep Kaur, Mansi Thakur and Shinar Athwal**

Department of Biotechnology, University Institute of Biotechnology, Chandigarh University, Mohali, Punjab, India

Abstract

Two-dimensional nanomaterials are flattering a novel perception in the electrical, chemical, energy accumulator, and biomedical industries. Subsequently, with the innovation of graphene, scientists have endeavored to discover unique 2D nanomaterials. However, the investigation of 2D nanomaterials is still in its beginning, with most of the investigations emphasizing the interpretation of the description of exceptional material features with rare details aiming at the biomedical solicitation of 2D nanomaterials. Nanoscience and nanotechnology, and their extensive uses, have become widespread globally because nanomaterials have unusual and exceptional properties. The current progress in 2D nanomaterials has increased significantly, and breathtaking queries about their increased connections with biological moieties. A variety of 2D nanomaterials, including 2D metal carbides and nitrides (MXenes), transition metal dichalcogenides (TMDs), layered double hydroxides (LDHs), covalent organic frameworks (COFs), metal organic frameworks (MOFs), layered silicates (nanoclays), and polymer nanoparticles have been examined for their potential biomedical applications. In this chapter, we explain the distinctive physical, chemical, and biomedical properties, distinctive properties that mark 2D nanoparticles as precious or important, practices recently exposed to nanomaterial synthesis, biomedical uses of 2D nanomaterials, and their use in the management of diseases such as cancer. We also emphasize state-of-the-art biomedical approaches to 2D nanomaterials, together with modern progress, which are influencing the following evolving sector.

Keywords: 2D nanomaterials, modern sciences, biomedical, industries, nanosciences, nanotechnology, graphene, transition metal dichalcogenides

**Corresponding author*: swaati.e10067@cumail.in

Subhendu Chakroborty and Kaushik Pal (eds.) 2D Nanomaterials: Synthesis, Properties and Applications, (333–360) © 2024 Scrivener Publishing LLC

15.1 Introduction

Nanomaterials can be described as "materials that have external dimension along with internal structure or surface structure in the nanoscale," and to further understand the nanoscale, it is the "length ranging from approximately 1 nm to 100 nm" [1]. Nanomaterials bring tremendous enhancement in properties of bulk materials such as strength, electric and thermal conductivity, and lightness, with the addition of some new characteristics such as self-healing, self-cleaning, anti-freezing, anti-bacterial, and many more. One of the methods for classifying nanomaterials is dimensional classification, in which identical chemical compounds can demonstrate unexpected chemical assets when they are arranged in a zero-dimensional (0D) crystal structure. Here, altogether, the dimensions (x, y, and z) are measured within the nanoscale (no dimension is larger than 100 nm); one (1D) here, two dimensions (x and y) are at the nanometer scale and one dimension (z) is outside; two (2D) here, one dimension (x) is at the nanometer scale and two dimensions (y and z) are outside; and three (3D) they are hardly narrowed to the nanoscale in any dimension; these resources include three subjective dimensions above 100 nm [2, 3].

S. no.	Dimensions	Examples
1.	Fullerene 0-dimensional nanomaterial (0D)	Nanospheres, nanoclusters
2.	Carbon nano tube 1-dimensional nanomaterial (1D)	Nanotubes, nanorods, nanofibers
3.	2-dimensional nanomaterials (2D)	Nanofilms, nanolayers, nanocoatings
4.	Graphite 3-dimensional nanomaterials (3D)	Nanotubes, nanowires

Nanomaterials can be categorized into four different types based on their composition:

1. Carbon-centered nanomaterials: these nanomaterials mainly contain carbon in their structure and are mostly present in spherical shapes, allipsoid shapes, sheets, and tubes. Spherical and ellipsoid shapes are generally called

fullerenes (0D). Sheets are known as nanotubes (2D), and tubes are called nanotubes (1D).

2. Organic-based nanomaterials: these nanomaterials include organic molecules except for carbon structures inside their structure. These could be hydrophobic or have the various reacting groups on their side chains for example, dendrimers, cyclodextrin, liposomes, and micelles.
3. Inorganic-based nanomaterials: These nanomaterials include various types of metals and metal oxides. Metal-based nanomaterials are made up of mainly gold (Au), silver (Ag), cadmium (Cd), aluminum (Al), lead (Pb), iron (Fe), and their oxides.
4. Composite-based nanomaterials: These nanomaterials are a mixture of any of the carbon-based, organic-based, inorganic-based materials. These composite nanomaterials have a complex structure by combining two different nanostructures, like a metallic–organic framework.

In our biological system, mainly carbon-based and organic nanomaterials have been used to deliver vaccines, drugs, antibodies, detect diseases, and perform various other functions in an efficient manner, such as protein-based, nucleic acid-based, microorganism-based, etc.

In recent years, two-dimensional nanomaterials (2DNMs) have been acknowledged with exceptional response inside the scientific circles of new nanomaterials. 2DNMs are exceptionally favorable resources for biomedical usage because of their distinctive configuration and properties [4, 5]. 2DNMs are the finest nanomaterials because of their structure and dimensions; they possess an incrusted arrangement with convincing in-plane bonds and feeble van der Waals (vdW) among every layer. Even though the perfect state for 2DNMs is one layer, they are frequently constituted with a minimum of 10 layers [6].

A desperate breakthrough in the development of 2DNMs was the innovation of graphene, the initial 2D material, which is a uni-layer sheet of atoms of carbon that makes a hexagonal honeycomb framework from side-to-side sp^2 bonds [7]. While graphene and its related materials are not the key emphasis of this chapter, investigators worldwide in the field of science have significantly learned and testified on novel classes of 2DNMs that have unique graphine-like properties, e.g., hexagonal boron nitride (h-BN), layered double hydroxides (LDHs), transition metal dichalcogenides (TMDs), phosphorene, topological insulators (TIs), transition metal oxides (TMOs), black phosphorous (BP), antimonene (AM), bismuthene, nitrides and carbonitrides (MXenes), and metal–organic frameworks (MOFs) [8–11]. Over the past decade, these nanomaterials have attracted colossal interest for having adequate potential and extensive uses in various fields, viz., optoelectronics, electronics, sensors, catalysts, facilities for energy storage, lithium batteries, solar cells, amalgams, and many more [6].

Graphene is a marvelous material as it is light-weight and has a planar density of 0.77 mg/m^2. As shown in Figure 15.1, graphene has a unit structure of hexagonal rings of carbon with an area of 0.052 nm^2 [12]. In contrast, hexagonal boron nitride (hBN) is an insulated substance with layers in it and is isostructural to graphene, having a band gap of ~5.9 Ev. In Figure 15.1(b), in the arrangement of hBN, every atom is situated in the epicenter of the benzene ring on the superior and inferior layers, and in the construction of dichalcogenides (TMD), every metallic layer atom is crammed in the middle of two atomic layers of X [16].

Phosphorene, along with xenes, are some other examples of 2DNMs. Phosphorene is a mono-layered, steady allotrope of rudimentary phosphorus. It is a straight band-gap semiconductor in a packed honeycomb configuration. It is appropriate for optoelectronic models and transistors as it has good charge mobility (~1,000 cm^{2}V^{-1}s^{-1}). Depending on which material is dignified with significant and enormous properties. Monolayers of silicon (silicene), germanium (germanese), and tin (stanene) are together recognized as Xenes. All these have a hexagonal arrangement analogous to graphene, but are misshapen to various degrees. In contrast to graphene, they cannot be extracted from most substances and should be epitaxially produced on a substrate. They usually preserve a convincing communication with the substrate, whereas their immature, latent applications range from field-effect transistors to topologic insulators [6].

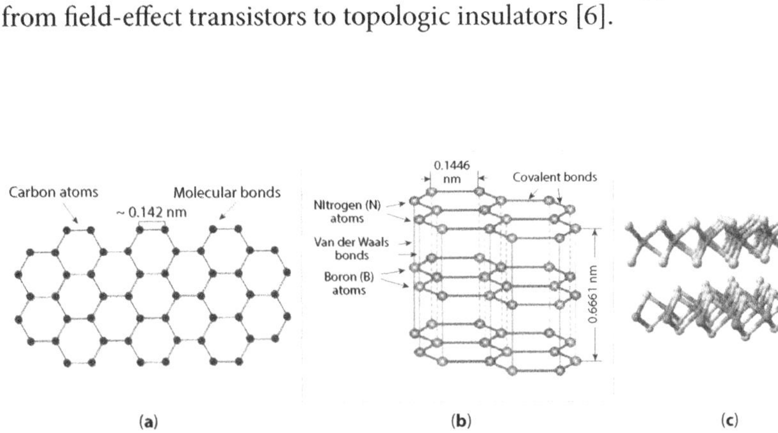

Figure 15.1 The structure of (a) single layer graphene, (b) hexagonal boron nitride (h-BN), and (c) transition metal dichalcogenide monolayers.

2DNMs can be grouped into three chief classes: organic, inorganic, and hybrid 2DNMs [13–15]. Metal-organic frameworks (MOFs), a third hybrid class of 2DNMs, also exist, as depicted by the existence of two groups: organic and inorganic [6]. 2DNMs offer a large surface area for interaction with units at the nanoscale, thus offering an exclusive association amid nano- and micro-scale schemes [17]. Excessive surface energy and elevated surface anchoring sites are stipulated due to the existence of a considerable number of surface atoms; this results in the enhanced interaction of 2DNMs with biological moieties, including biomolecules, cells, and cellular components. Some biomedical applications of 2DNMs include re-forming medicine, the production of additives, cancer therapeutic treatment, and biosensing [16]. Although advancement in biomedical applications has happened at a comparatively gradual pace among other materials. There have been contradictory descriptions concerning their toxicity because of the lack of systematic biocompatibility evaluation; as a result, applications with long-standing therapeutic approaches for tissue regeneration are approached with cautious optimism. In the last 10 years, traditional research on 2DNMs has expanded rapidly [18–21].

15.2 Synthesis of Nanomaterials

There are two diverse strategies' that have been established to produce nanomaterials, and both differ from one another based on quality, speed, and cost.

Top-down—This approach is essentially the break-down of bulk material Into nanoparticles with a size range of 1 nm–100 nm); this can be achieved by some advanced processes like precision engineering and lithography. Precision engineering is about designing the machine or instrument that provides the accuracy for designing such nanostructures.

Bottom-up—This approach will be referred to as building up of complex materials by using discrete units such as atoms–atoms or molecules–molecules by physical and chemical methods that lie inside a nanoscale range of 1 nm–100 nm using controlled manipulation of self-assembly of atoms and molecules.

Mechanical exfoliation—It was first used to create monolayer graphene and is also identified as 'scotch-tape method.' In this method, a portion of gummy tape is smeared on the surface of a layered material, and afterward it is unpeeled. The peeled tape contains some flakes that consist of a tiny number of layers. The flakes are then transferred by pressing the tape onto a substrate for further studies. This process has a low monolayer yield with no control over shape or size. The quality of the monolayer is excellent, though. This method is very popular for lab-based studies.

Liquid exfoliation—It is a mechanical method that involves exfoliation by means of an organic solvent as an intermediate to transmit automated force to the layered material suspended in the liquid. To improve the monolayer yield, variations are present; for example, we can introduce reactive ions that drive the sheets separately or promptly blend the solution to build surplus shear force on the layers. This method has some drawbacks, like the fact that the flakes are often less than 100 nm in size with a high density of defects, and the yield is also very low [6].

Chemical vapor deposition—In this process, one or more precursor gases are passed over a heated incinerator, where they countercompose and shape a slim layer of essential material. This procedure has proven to be very successful in producing graphene as well as TMDs. Several parameters like temperature, gas pressure, and reaction times need to be coordinated as they can disturb the width, properties, and conformation of the films. Additionally, this practice is more complicated and expensive than most top-down systems.

Solution-based chemical synthesis—This method includes high-temperature chemical reactions in solution, the function of nanoparticles into large nanosheets, interface-mediated growth, and many more. The flakes produced by this method are small (<100 nm). This technique is best for large-scale creation because of its properties like scalability, low cost, and versatility [6].

15.3 Nanomaterials for Cancer Treatment

In recent years, 2DNMs have been broadly utilized for cancer identification and management because of their distinctive benefits, which allow them to be required as biological medications. The viability of 2DNMs has proven that they can establish a favorable medically experimental nanoplatform for cancer treatment.

First, plentiful resources of 2DNMs are provided to meet various necessities, which include TMD, hBN, graphitic carbon nitride, group-VA semiconductors, and transition metal carbides. Second, desirable performances for diagnosis, imaging, or other remedies for cancer that can be practiced in the field of biomedical applications can be matched by 2DNMs since they have worthy physical, chemical, and biological properties, including optical, magnetic, electrical, or catalytic properties. Finally, good yields of 2DNMs can be prepared in laboratories.

In the cancer immunotherapy area, 2DNMs can be developed *via* encapsulating bioinspired cell membranes for cancer-directed treatment and can provide advancement in personalized immunotherapy. To understand therapeutic nanomedicine with outstanding efficiency of a persistence rate of 100% *in vivo* lacking the complicated nanoplatforms, a nontoxic and effective 2DNMs platform has been stated [22].

In the medical field, for future clinical applications of 2DNMs, the following six features should be focused on:

1. A more biocompatible and biosafe nanoplatform.
2. The imbalance between low clinical translation and increasing nanomedicines.
3. The in-progress concern for the degradability of the materials.
4. Functional optimization of materials.
5. Assortment of precise and efficient materials prior to certain types of cancer treatments.
6. Possible untargeted toxicity from the materials, which needs more organized therapeutic analysis.

15.4 Difference of 2D Materials from Bulk Materials

A bulk material that is layered has covalently bonded planes and is clutched along by weak van der Waals interactions. When some external force is applied to these materials, the van der Waals forces tend to be effortlessly astounded and break the material, producing its weak look. However, the covalent bonds that bind the atoms are truly strong, even though a monolayer will only have covalent bonds. By eliminating the weak associations from the material, it looks to become much stronger. Graphene has a ductile potency that is a thousand times superior to graphite, and a graphite pencil can easily be wrecked. Grapheme is over a hundred times stronger than steel. The surface area-to-volume ratio outlines how much a material is exposed to its environment. This is important for chemical reactions—the more the reactant is familiar with the material, the faster the response can appear, so 2D nanomaterials are more reactive than bulk materials. The electronic band structure of a material includes properties like electronic and optical; these properties designate how electrons shift over the material and are a consequence of the periodicity of its crystal structure. The periodicity is removed in a way perpendicular to the plane when a material goes from bulk to 2D, which can result in a change in the band structure [23].

This chapter provides a broad overview of the recent works on some 2DNMs, namely graphene, transition metal dichalcogenides (TMDs), and hexagonal boron nitride (hBN). Here one can discuss the current developments in the production, properties, applications, and some current problems in the development process of these nanomaterials.

15.4.1 Graphene

Nobel prize awardees, physicians Andre Geim and Konstantin Novoselov, first disclosed graphene in 2004 at the University of Manchester. Graphene is a crystal-like, transparent allotrope of carbon with an atomic scale and two dimensional, hexagonal patterns. In graphene, every carbon atom produces four bonds: three *s* bonds (sp^2 hybridized) with its three neighbors, and one *p* bond aligned out of plane. It is the elementary organizational element of all other allotropes like graphite, fullerene, nanocones, nanotubes, and many more; accordingly, it is called the mother of all carbon nanomaterials [6, 24].

Graphene is acknowledged for its excessive degree of rigidity, thermal conduction, strength, and impermeable nature towards liquids and gases. It is an applicant of choice in nano-electromechanical and contracted devices because of its outstanding chemical and physical properties and low cost

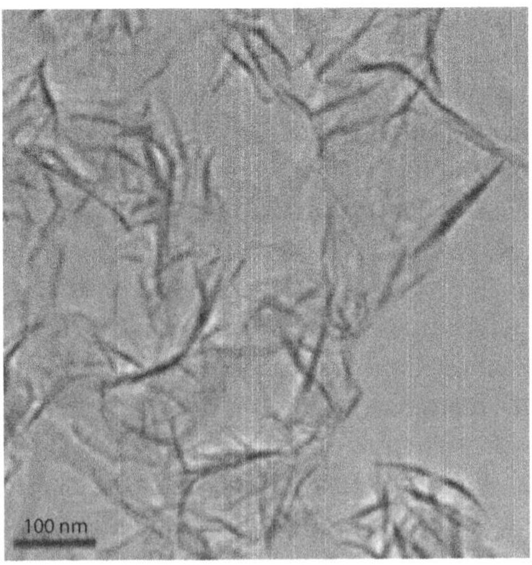

Figure 15.2 The transition electron microscope image of a crumpled graphene [27].

of production [25]. It possesses a feeble absorption coefficient of 2.3% of white light; consequently it is observed as white dust. The graphene surface area is 2,630 m^2/gr, which is double that of carbon nanotubes, which possess 1,315 m^2/gr (Figure 15.2) [26].

15.4.1.1 Synthesis of Graphene

There are several chemical and physical approaches that are projected to construct varieties of graphene, centered on top-down and bottom-up methods. Processes like Chemical vapor deposition (CVD) and multilayered growth [6, 28], plasma-enhanced chemical vapor deposition (PECVD) [29], mechanical cleavage [28, 30], scotch tape technique [31], chemical synthesis [31], liquid exfoliation [32, 33], and many more are extensively used to construct graphene.

Chemical vapor deposition (CVD) is a very frequent approach for graphene construction. CVD works on the principle of depositing and combining volatile gas molecules onto a substrate. In this process, the required substance is designed on the face of the substrate, and the undesirable gas is expelled out, which occurs in a reaction chamber. Graphene oxide is formed by the oxidation of graphite oxide, and the following oxidation procedure is very advantageous as it makes work the planes of graphene layers work along with numerous oxygenated functional group species [34].

There are three main categories for reducing graphene oxide (GO) into reduced graphene oxide (rGO): chemical reduction, thermal reduction, and electrochemical reduction. Chemical reduction utilizes highly toxic materials like hydrazine and often results in poor yields. rGO manufactured by this process commonly exhibits a small surface area. Thermal reduction includes a strong heating procedure that initiates excessive pressure, which in turn causes structural damage to the graphene layers. However, the rGO produced has a high surface area. Electrochemical decline demonstrates the finest outcomes in terms of construction and quality [35].

Micromechanical cleavage of bulk graphite is typically used to synthesize graphene nanoplatelets, which can only construct graphene flakes in inadequate amounts. The final GNP product on a large scale is often produced by mechanical cleavage followed by chemical reduction. Graphene oxide quantum dots (GQDs) also exhibit a large surface area; they can be amalgamated in several configurations, from single layers to tens of layers, but commonly below 30 nm. They are suitable materials for both electronic and optoelectronic applications. Construction *via* bottom-up approaches can generate GQDs with skillful measurements, but the making itself can be composite, which requires rigorous circumstances [36]. Graphene nanoribbons (GNRs) are quasi-one-dimensional materials with an extremely slim width. One of the most common techniques to manufacture GNRs comprises decompressing the walls of MWCTNs (multi-walled carbon nanotubes) together with the amalgamation of compounds like potassium as well as sodium, sonicating, and aerating under vacuum [37]. Aerogels are characteristically arranged by sol–gel procedures, a practice that alters molecular ancestor to extremely cross-linked organic/inorganic gels that can afterward be wiped by means of methods, viz., surplus drying, deep-freeze drying, and others, to reserve the insubstantial solid network [38].

Reduced graphene oxide (RGO) is the most used method to manufacture graphene. But the constructed graphene has many fundamental shortcomings in its structure, and a surplus amount of unconsumed oxygen is left over inside RGO, which renders its exceptional properties, including electrical and thermal conduction, and its mechanical properties as well [39, 40]. To reduce the defects inside the graphene caused by its oxidation, we start exfoliating the graphite inside the liquid solution to get a colloidal suspension of graphene layers. This process of exfoliating graphene inside the liquid suspension is termed liquid phase exfoliation (LPE) [41].

LPE mainly consists of graphene dispersion in solvent. Exfoliation by ultrasound or any other mechanical method and then purification [42, 43]. It is a four-step process, as illustrated by Shen and colleagues [44, 45].

i. Immersion of graphite inside a liquid medium—it diminishes the van der Waal forces of attraction amid the intact graphite layers.
ii. Interaction of liquid inside the graphite—the liquid particles intercalate between the layers of graphite and weaken their interactions (out of the plane interaction).
iii. Exfoliation of the layers—layers having weaker interactions when exposed to the mechanical force overcome the van der Waals interactions, and layers will tear apart.
iv. Stabilization of these graphene layers suspended inside the medium—exfoliated layers inside the liquid medium must be remain stable [44, 45].

Liquid phase exfoliation of -	Graphite oxides	Natural graphite	Expanded graphite	Graphite by organic solvent	Graphite by low boiling solvent
External driving forces	Sonication [48]	Ultrsonication/ shearing or electric field [46]	Ultra-sonication [47]	Sonication and centrifugation [42]	-
Descending van der Waals attraction	By oxidation and chemical interaction increases the distance between adjacent layers 3.4–7.0 A° [48, 49]	Overcome by strong interaction between solvent molecules and graphitic basal planes [46]	Alkali metals or acid molecules intercalates between graphite layers increase the distance between two adjacent layers by (3.4A°–5.3A°) [50]	When the solvent–graphite energy is equal to graphene–graphene interaction energy then the van der Waal force will overcome [51]	When Solvent–graphite energy will be equal to graphene–graphene energy will overcome the van der Waal energy [52]
Defects in graphene	Yes, due to oxidation	No defect	No defect	No defect	No defect

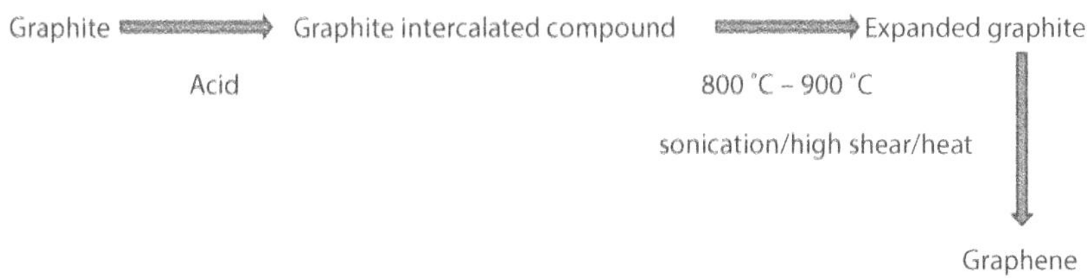

15.4.1.2 Graphene Properties

Graphene is basically found in a dispersion, on a substrate, or on a powder of graphene oxide. The material itself has some extraordinary properties, including electronic, optical, thermal, and mechanical, and many of these properties arise from the fact that the width of a layer of graphene is hardly one atom [53].

Graphene is a highly conductive material at ambient temperature, because of its high electron mobility (2 × 105 cm^2/Vs), together with 10^6 S/m of conductivity, and 31 Ω/sq of sheet resistance. Graphene is categorized both as a semimetal and a zero-bandgap semiconductor because it has an insignificant intersection among the valence and conduction bands.

A single layer of graphene is 10,000 times more electrically conductive than a few layers of graphene [6].

Graphene has a transparency of 97.7% and is highly transparent to visible light. Graphene has a unique optical property that produces elevated imperviousness for unilayers of atoms in a vacuum. Thermal transfer in graphene has attracted a lot of attention and is an area of expertise for its potential for thermal administrative uses. It has a thermal conductivity of 5,300 W/mK [55], in contrast to the heat conductivity of pyrolytic graphite of nearly 2,000 W/mK at room temperature [56]. The mechanical properties of open-standing monolayer graphene by utilizing nanoidentification were first measured by Hone and coworkers [57] and recognized graphene as "the strongest material ever measured." The fundamental potency of the monolayer membrane is measured to be 42 N/m, which is equal to the strength of 130 GPa [58].

15.4.1.3 Applications of Graphene

Many potential uses of graphene have been projected for development in areas including electronics, bioengineering, photovoltaics, energy accumulation, filtration, and strong composite materials [59, 60]. Graphene has many applications in various fields depending on its different types, monolayer or few layers.

Graphene can be used as a superlative material for sensors. Individual atoms present in graphene are displayed to their surroundings, which permits them to sense alterations in the atmosphere. The goal of graphene in chemical sensors is to identify just a single molecule of a potentially treacherous substance. In the agriculture industry, graphene sensors could advance the efficacy of examining very important crops, thus allowing the farmers to now monitor the existence of any injurious gases that could impact the crop field [61]. 'Smart' food packaging products can be created by using graphene oxide; this could help to prevent illness and rapidly cut down on unnecessary food waste. Graphene-coated packaging can identify atmospheric alterations by decomposing food [62].

To increase the effectiveness of the batteries and improve the discharge or charge cycle, graphene should be used in the anode and the cathode of several batteries. Graphene has also been utilized in batteries made of compounds like lithium-sulfur, lithium-ion, or many more, supercapacitors, and fuel cells because of its multifaceted abilities [63]. In electron emission, graphene is an ideal material as it demonstrates an elevated quality ratio, and suspension bonds at any end of the sheet permit proficient electron tunneling. Graphene can start an electric field at 0.1 V μm^{-2}, with a field augmentation feature of up to 3,700, which can upsurge up to 4,500 in screen-printed graphene films. To make the current material more solid and lighter, graphene is being integrated into many materials. For example, in the aviation industry, graphene is used as a compound material that is considerably lighter than steel, but will stipulate the essential strength and avoid the use of enormous wealth and fuel [64]. Like any additional material, graphene is also subject to fundamental shortcomings. Catalysts in metal ions can settle in these cavities and be sustained. Graphene, being inert in nature, does not interfere with the interactions between the catalyst and the substrate; it also delivers an even distribution of catalyst elements. In practical inks for electronic, thermal resistance, and anticorrosion purposes, graphene can be used. The conduction potentials connected with graphene encourage the ink as graphene is incorporated into ink construction, causing it to develop conductivity [65].

Graphene-centered spiral inductors manipulating kinetic induction at ambient temperature were initially verified at the University of California, Santa Barbara, led by Kaustav Banerjee, in January 2018 [6]. Graphene is very heat-conductive, strong, and light, which makes it a huge material for creating heat-dispersal solutions like thermal sinks or thermal dissipation films. This could be beneficial in both cases and additionally in various applications, viz., thermal foils for mobile devices [66].

15.4.2 Hexagonal Boron Nitride (hBN)

Hexagonal boron nitride has a layered structure like graphite. Boron and nitrogen are certain by tough covalent bonds within every layer, and these layers are composed by feeble van der Waals forces. hBN is an isoelectric correspondent of graphite and shares several physical properties along with very similar structural characteristics, so, it is also called white graphite. It is usually synthesized and is not present in nature [67]. This crystal structure provides exceptional lubricating qualities due to its short coefficient of friction of 0.15 to 0.70 [68]. hBN has varied industrialized uses in surface coating, amalgams, emollients, and insulators.

Boron nitride is composed of an identical number of nitrogen and boron atoms, organized in hexagonal accord, analogous to the carbon atom in graphene. Boron nitride, with its thermal and chemical resistance properties, is a refracted complex of boron and nitrogen, pertaining to the chemical formula BN. It occurs in several crystalline shapes that are isoelectronically like an analogously organized open framework of carbon. The hexagonal shapes consistent with graphite is very constant and subtle among BN polymorphs. This crystalline structure is hexagonal, and it is also known as h-BN, α-BN, g-BN, and graphitic boron nitride [6].

For nomenclature, the term "single-layer BN," i.e., monolayer of BN, is utilized, and in the case of multilayer, BNNSs. Displayed in Figure 15.1(b), monolayered BN includes a honeycomb arrangement comprising the benzene complex and iso-electric rings of borazine. B-N has a covalent class, other than that these bonds have an ionic potential of 1.45 Å length, due to the electronegativity alteration. The vastness among two borazine rings is 2.54 Å [69]. Figure 15.3 shows a transmission electron microscopy (TEM) micrograph of boron nitride plates, having a lateral dimension of 600 nm [70]. There is a difference in the organization of layers of atoms in BN and its nanomicrosheets together with graphene and graphite. The agreement in graphene is known as the AB stacking mode, whereas in the BN layers, AA is the stacking mode [71, 72]. In addition, the calculation shows that along the favorable energy, the layers of BN have comparative disarticulation from AA to AB stacking mode [73].

B-N bonds have ionic characteristics that are associated with C-C covalent bonds in graphene, and this is due to a difference in the electronegative charge, which can further lead to chemical connections or interactions as solders between the layers. This process helps to diminish the power and then alleviate the construction of several BNNS layers by diminishing the amount of suspension bonds at the edge [6, 74].

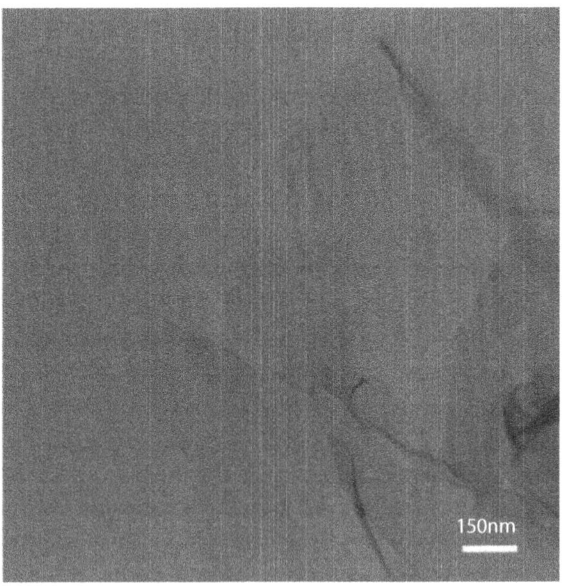

Figure 15.3 Image of the transmission electron microscope of two comparatively huge BNNSs, which have been centrally overlay [75].

15.4.2.1 *Hexagonal Boron Nitride (hBN) Synthesis*

Hexagonal boron nitride is produced by altering boric acid (H_3BO_3) or boron trioxide (B_2O_3) in the company of ammonia (NH_3) or urea ($CO(NH_2)_2$) in a nitrogen ambience [76]:

$$B_2O_3 + 2\ NH_3 \rightarrow 2\ BN + 3\ H_2O\ (T = 900\ °C)$$
$$B(OH)_3 + NH_3 \rightarrow BN + 3\ H_2O\ (T = 900\ °C)$$
$$B_2O_3 + CO(NH_2)_2 \rightarrow 2\ BN + CO_2 + 2\ H_2O\ (T > 1{,}000\ °C)$$
$$B_2O_3 + 3\ CaB_6 + 10\ N_2 \rightarrow 20\ BN + 3\ CaO\ (T > 1{,}500\ °C)$$

The consequential boron nitride is formless in nature and comprises 92%–95% BN and 5%–8% B_2O_3. To achieve BN concentrations >98%, the residual B_2O_3 can be vaporized in subsequent steps at temperatures >1,500 °C. Additionally, such strengthening can crystallize BN, and the dimension of the crystallite increases with a forging temperature [68, 77]. Boron nitride films can initially be produced by chemical vapor deposition (CVD) from nitrogen precursor and boron trichloride [68]. Ultrafine boron nitride utilized for emollients and astringents can be created by the combustion of boron powder in nitrogen plasma at 5,500 °C [78]. Tremendously subtle link explosive, NI_3 is produced in a very low yield at −30 °C by the effect of boron nitride with iodine fluoride in trichlorofluoromethane [79]. The oxidation of boron nitride along with nitrides of lithium, alkaline earth metals, and lanthanides is used to produce nitridoborate compounds [80].

15.4.2.2 *Properties of Hexagonal Boron Nitride (hBN)*

The partial ionic configuration in BN layers in hBN diminishes the electrical conduction and covalency; however, the interaction increases in between the layers, which results in the development of Hbn's hardness. Because of the lack of a large band gap and color in hBN electron, delocalization is also reduced. The solidity, electric, and heat conduction rise inside the plane rather than perpendicular to them [81]. hBN shows remarkable chemical and heat strengths. hBN is constant in disintegration at different temperatures in different spaces, like in air at 1,000 °C, in vacuum at 1,400 °C, and in an inert atmosphere at 2,800 °C. Boron nitride is unsolvable in standard acids, whereas it is solvable in alkaline igneous salts and nitrides like $NaOH\text{-}Na_2CO_3$, LiOH, $NaNO_2$, and KOH [82].

15.4.2.3 Applications of Hexagonal Boron Nitride (hBN)

Primarily, hexagonal BN was utilized in beauty products in the year 1940 in Japan, but it is presently restricted for this purpose because of its high price. It was later introduced in late 1990s with the optimization of the hBN construction process, and now it is consumed by all prominent manufacturers of cosmetic production [83]. Boron nitride porcelains are conventionally consumed as a high-temperature equipment because of their exceptional heating and chemical strength. hBN can be found in porcelains, alloys, rubbers, plastics, resins, and many more, giving them slippery properties [84].

Hexagonal BN is used as the current-outflow obstructive layer of the photo drum in xerographic processes and photo printers [85]. hBN, when assorted with binder (boron oxide), can be utilized to stick oxygen binders, which offers a response for regulating fuel movement. Here, the temperature firmness and protecting properties of hBN are used by the binder [86].

In precision target rifle, hBN has been utilized as a bullet and bore emollient as a substitute for molybdenum disulfide layer, normally mentioned as "moly." It is asserted that it decreases the aberration in point of effect connecting the uncontaminated bore initial shot and the subsequent shots, increases effective barrel life, and also helps to increase intervals along bore cleaning [87].

BN nanosheets (hBN) are also known as exceptional proton conductors; their elevated proton transfer rate, when combined with the electrical resistance, leads to their implementation in fuel cells and water electrolysis [88]. Another property of BN nanosheets is that at a temperature of ~1,100 °C, hBN can be accumulated by catalytic disintegration of borazine in a chemical vapor deposition system in approximately 10 cm^2 areas. They are utilized as substrates for graphene-dependent instruments because of their hexagonal atomic structure, tiny framework variance with graphene, and high regularity [86]. hBN is a worthy emollient at both low and high temperatures, this lubricant property of hBN is partially beneficial when the electrical conduction or chemical responsiveness of graphite would be difficult. hBN emollients can even be consumed in a vacuum, e.g., for cosmic space purposes. The grease potential of pulverized hBN can be used in paints, leads in pencils, cement used in dental surgeries, and many more [89].

15.4.3 Transition Metal Dichalcogenides (TMDs)

Encouraged by graphene, other graphenes like 2D nanomaterials, especially transition metal dichalcogenides (TMD) with only a small layer width came to light and have extended interesting attention in modern years, which involves MoS_2, WS_2, TaS_2, TiS_2, WSe_2, $MoSe_2$, etc. [90, 91]. They have MX_2 as their chemical formula, where M stands for a transition metal like molybdenum or tungsten and X is a chalcogen like sulfur or tellurium. Exfoliated layers of TMD are semiconductors and have a large energy band gap. Because of their similarities to graphene, they are very useful for discovering and constructing sensors, batteries made of lithium, optoelectronics, and energy accumulators. They also have the properties of fluorescence and photoluminescence [92]. TMD can have various crystalline structures. 2H-phase with trigonal symmetry is the most common crystalline structure for TMD, which results in semiconducting characteristics. These semiconducting materials have an unintended band gap when in the majority.

2D properties of layered compounds of TMD belonging to groups IV–VI and III–VI have a limited number of reports currently and, are largely unknown. After achieving a width of monolayer or multilayer, these materials are estimated to display unexpected properties. However, TMD illustrates a range of electric and optical potentialities that are appropriate for catalysts, nanotribology, optoelectronics, and lithium-ion batteries [93]. Layered structures of TMD are manufactured by convincing in-plane covalent bonding all together with feeble relations among the layers, especially for groups 4–7, while some of groups 8–10 are usually present in non-layered structures [93]. With the decrease in the number of layers, the properties of TMD materials change; thus, the band gap of single-layered MoS_2 (molybdenum disulfide) is 1.8 eV –1.9 eV, while that of bulk MoS_2 is approximately 1.3 eV [94], showing an indirect to direct band gap transition. After exfoliation into monolayers or a few layers, the surface area of TMD materials greatly increases.

15.4.3.1 Synthesis of Transition Metal Dichalcogenides (TMDs)

The mechanical exfoliation technique is classically accepted to formulate monolayer TMD samples, and the samples organized by this process have noble features and can be used for reviewing their fundamental properties [95–97] by scanning tunneling microscopy (STM), atomic force microscopy (AFM), transmission electron microscopy (TEM), and optical microscopy.

TMD material made by mechanical exfoliation is minute in size, roughly on the tens of micron scale.

Liquid phase exfoliation of sheets of clay materials was first reported near the beginning of the 1960s [98] and has outstanding analyses into processes of exfoliating sheets of TMD [99–101].

Large-scale growth is essential for their use in real devices, which can be attained by chemical vapor deposition (CVD). This technique can further be branched into two groups: the vapor phase reactions of metal oxides among chalcogen precursors, and the sulfurization (selenization) slim films of metals [102–105]. The vapor phase is a 2-step response; the one includes the configuration of MoO_{3-x}, which in addition to interacting with sulfur vapors, raises MoS_2 layers.

15.4.3.2 Transition Metal Dichalcogenides (TMDs) and its Properties

Mechanical Property:

The resilient abilities of a deferred share of MoS_2 nanomacrosheets enclosing 5 to 25 layers were measured by Castellanos-Gomez *et al.* [106]. The AFM (atomic force microscopy) tip was used to affect a weight in the middle of the deferred area of the MoS_2 nanosheets to determine its elastic mechanical effects. The flexible distortion of the nanomacro sheets (δ), the refraction of the AFM cantilever (Δz_c), and the dislocation of the examining piezotube of the AFM mechanism (Δz_{piezo}) are linked by the equation $\delta = \Delta Z_{piezo} - \Delta Z_{c.}$, when the tip and the sample come into contact. Here, Castellanos-Gomez and his team discovered that the Young's modulus of MoS_2 is very elevated, i.e., $E = E = 0.33 \pm 0.07$ TPa [107].

An alternative example of the mechanical properties of suspended MoS_2 is when Bertolazzi, along with his co-workers, stated the quantification of the rigidity and ultimate tensile strength of the mono-layer MoS_2. By using the AFM, MoS_2 nanosheets (single or bilayer) have been found to be deformed and eventually broken. The Young's modulus of monolayer MoS_2 is initiated at approximately 270 ± 100 GPa, together with the in-plane rigidity of 180 ± 60 Nm^{-1}. The superior hypothetical edge signifies that the product is extremely crystalline and nearly non-defective, as the force of a steady mono-layer membrane is 11% of its Young's modulus. Hence, the above proven consequences illustrate that the monolayer MoS_2 is appropriate for diverse purposes like strengthening rudiments in composites as well as to produce stretchy electronic devices [107].

Electrical Structure and Optical Property

TMD displays metallic properties when the orbitals are moderately engaged, whereas when they are fully unavailable, they display semiconductive ones. Chalcogen atoms have a minor influence on the electronic structure associated with metal atoms, though it is observed that the enlargement of the d bands decreases the band gap by accumulating the atomic number of the chalcogen [108].

Optical absorption is linked to the band structure of the TMD semiconducting materials with layers and is another important characteristic of the same. By measuring the contrast under broadband green illumination, Benameur and his group observed that they could differentiate single-, bi- and tri-layer MoS_2 and WSe_2 on a 90 nm and 270 nm SiO_2 substrate [109]. Li, along with his fellow workers, recently demonstrated a simple approach using the standard optical microscopy imaging technique and Image J software perhaps utilized to recognize single- to ti-layer MoS_2 nanosheets [110].

15.4.3.3 Applications of Transition Metal Dichalcogenides (TMDs)

To produce low-power electronic devices, monolayer MoS_2 nanosheets are mainly important supplementing materials to graphene because of their outstanding semiconducting properties along a straight band gap of 1.83 Ev [111]. Due to MoS_2's high work function, it acts as a hole injector. The broad-gap MoO_3 layers perform like an electron-restriction in the LED and depend on the MoO_3/MoS_2 hybrid arrangement, which improves the carrier balance in the tool and initiates improvements in light-emitting presentation and effectiveness [112].

The huge-scale manufacturing of mono- and multi-layer MoS_2 nanosheets can facilitate their broad applications for energy deposit devices, namely batteries. MoS_2 nanosheets were manufactured for use as electrodes for lithium-ion batteries (LIBs). The Li-storage ability of MoS_2 can be enhanced by including polymer molecules like poly (ethylene oxide) (PEO) in the Li intercalation solution [113], as there is a growth in the interlayer spacing of the MoS_2 nanosheets because of the presence of PEO. Recently, for the fabrication of electrodes, few-layer MoS_2 and decrease graphene oxide (rGo) hybrids were utilized that exhibit a unique capacity of approximately 1,100 mAh/g at a current of 100 mA/g along with exceptional cycling constancy. Chemical, biological, and gas sensors utilizing FET (field effect transistor) device structures have become very popular. Mono- and multi-layered MoS_2 nanosheets resting on devices are being engaged for NO observation [110].

15.5 2DNMS for Next-Generation Quantum and Electronic Devices

Single-layered 2DNMs exhibit some unusual properties that are of great use for a broad range of quantum and microelectronic systems. But one cannot ignore their defects; due to the defect at the atomic level, the material's microscopic function gets influenced and can lead to novel quantum behaviors. There are many kinds of defects that are yet to be understood by the researchers. Systematic study of these defects at relevant scales or with atomic resolution is one of the biggest challenges in the field.

Artificial intelligence suggests a way to map and identify defects in 2DNMs. An application of artificial intelligence, i.e., convolutional neutral networks, is used to quickly analyze data from autonomous experiments, which has recently become a powerful tool for imaging these exotic materials.

Defects can cause issues with the macroscopic function of the materials or can be used advantageously. The combination of AI and autonomous discovery can give us a nice way to screen for defects and measure them. 2DNMs have the possibility of being utilized in next-generation quantum and electronic devices, as the process could drastically reduce the time required to characterize them.

Discovery and understanding of the quantum properties of solids have enabled the use of technologies over the decades, such as transistors and lasers. As scientists discover other applications for quantum information science, such as quantum sensing and computing, it helps them better understand a phenomenon in solids called quantum coherence. Their goal is to improve scientists understanding and control of coherence in solids, which could lead to new devices and applications [114].

15.6 Functionalized Hybridization of 2D Nanomaterials

There have been a variety of hybridization methods by which we can use 2DNMs to manufacture a variety of 2D hybrids with advanced and upgraded properties and improved applications. Functional hybridization on TMD nanosheets can be performed through physical or chemical doping, linkage, and adsorption. Some more hybrids of 2DNMs are also announced based on their definite surface structure and necessities, such as hybridized graphene with carbon materials, metals and metal oxides, functionalized C_3N_4 nanosheets with protons, materials, and nanomaterials. Most of the discovered hybrids include the work of nanosheets in a few layers, except monolayer, which limits the discovery of some unique applications that fall in the form of monolayer nanosheets. It is an immense objective to manufacture monolayer nanosheets at an elevated production and bulky scale for most materials that are layered.

To hold back other important subjects, there is a huge requirement to either boost or advance controllability or maybe decrease struggles in hybridization of 2DNMs for making important achievements like unparalleled presentation and unidentified functional composites, for example. Existing tactics are unable to register a consistent allotment of preservative species and regulate their arranging density and size on nanosheets. Moreover, feeble non-covalent binding instead of strong covalent linkage is largely in use for attaining fewer coordinated hybridizations of two or several components.

It is also well recognized that oriented expansion or arrangement of anisotropic nanostructures like nanorods and nanobelts on 2DNMs is very important but has not been achieved yet; however, once achieved, many narrative properties may arise through independent consequences. Lately, an effort issued by Jarillo-Herrero and his co-worker established an unusual superconductivity in the twisted bilayer grapheme along a twist angle of $\approx 1.1°$ (i.e., a magic angle in the grapheme super lattice) [115].

Encouraged by the work, more work is swiftly experimented with and grown to a developing ground of twistronics that eventually unlocks plentiful opportunities to utilize the unique properties of 2D heterostructures synthesized in various arrangements.

References

1. *Nanotechnologies–Vocabulary–Part 1: Core Terms*, International Organization for Standardization, 2015. Retrieved January 8, 2018.
2. Golberg, D., Bando, Y., Stéphan, O., Kurashima, K., Octahedral boron nitride fullerenes formed by electron beam irradiation. *Appl. Phys. Lett.*, 73, 2441–2443, 1998.
3. Mohapatra, R.K. and Das, D., *Chemical Modification of Solid Surfaces by the Use of Additive*, Bentham Science, Singapore, 2020.
4. Gao, P., Xiao, Y., Li, L., Li, W., Tao, W., Biomedical applications of 2D monoelemental materials formed by group va and via: A concise review. *J. Nanobiotechnol.*, 19, 1–23, 2021.
5. Mei, X., Hu, T., Wang, Y., Weng, X., Liang, R., Wei, M., Recent advancements in two-dimensional nanomaterials for drug delivery. *Wiley Interdiscip. Rev. Nanomed. Nanobiotechnol.*, 12, e1596, 2020.

6. Rafiei-Sarmazdeh, Z., Zahedi-Dizaji, S. M., Kang, A. K., Chapter Metrics Overview, *Chapter-1 Two-Dimensional Nanomaterials, (Nanostructures)*, 807 December 15th, 2018 Reviewed: February 18th, 2019 Published: May 22nd, 2019.
7. Novoselov, K.S., Geim, A.K., Morozov, S.V., Jiang, D.E., Zhang, Y., Dubonos, S.V., Grigorieva, I.V., Firsov, A.A., Electric field effect in atomically thin carbon films. *Science*, 306, 666–669, 2004.
8. Carrow, J.K., Singh, K.A., Jaiswal, M.K., Ramirez, A., Lokhande, G., Yeh, A.T., Sarkar, T.R., Singh, I., Gaharwar, A.K., Photothermal modulation of human stem cells using light-responsive 2D nanomaterials. *Proc. Natl. Acad. Sci. U.S.A.*, 117, 13329–13338, 2020.
9. Hu, T., Mei, X., Wang, Y., Weng, X., Liang, R., Wei, M., Two-dimensional nanomaterials: Fascinating materials in biomedical field. *Sci. Bull.*, 64, 1707–172, 2019.
10. Bolotsky, A., Butler, D., Dong, C., Gerace, K., Glavin, N.R., Muratore, C., Robinson, J.A., Ebrahimi, A., Two-dimensional materials in biosensing and healthcare: From *in vitro* diagnostics to optogenetics and beyond. *ACS Nano*, 13, 9781–9810, 2019.
11. Tao, W., Kong, N., Ji, X., Zhang, Y., Sharma, A., Ouyang, J., Qi, B., Wang, J., Xie, N., Kang, C., Emerging two-dimensional monoelemental materials (Xenes) for biomedical applications. *Chem. Soc.*, 48, 2891–2912, 2019.
12. Zhen, Z. and Zhu, H., *Graphene: Fabrication, Characterizations, Properties and Applications*, p. 538, Academic Press, Version of Record 22 September 2017.
13. Lv, R., Robinson, J.A., Schaak, R.E., Sun, D., Sun, Y., Mallouk, T.E., Terrones, M., Transition metal dichalcogenides and beyond: Synthesis, properties, and applications of single- and few-layer nanosheets. *Acc. Chem. Res.*, 48, 1, 56–64, 2015.
14. Manzeli, S., Ovchinnikov, D., Pasquier, D., Yazyev, O.V., Kis, A., 2D transition metal dichalcogenides. *Nat. Rev. Mater.*, 2, 17033, Published: June 13, 2017.
15. Kalantar-Zadeh, K., Ou, J.Z., Daeneke, T., Mitchell, A., Sasaki, T., Fuhrer, M.S., Two dimensional and layered transition metal oxides. *Appl. Mater. Today*, 5, 73–89, December 2016.
16. Murali, A., Lokhande, G., Deo, K.A., Brokesh, A., Gaharwal, A.K., Emerging 2D nanomaterials for biomedical applications. *Mater. Today*, 50, 276–302, November 2021.
17. Lee, Y., Kim, J., Jung, H., Kim, M., Bae, H., Conductive polymer composites for soft tactile sensors. *Adv. Mater. Technol.*, 2, 9, 1700053, 2017.
18. Zhu, C., Du, D., Lin, Y., Graphene and graphene-like 2D materials for optical biosencing and bioimaging. *2D Materials*, pp. 1–17, IOP Publishing Ltd, 2, September 25, 2015.
19. Luo, Y., Li, Z., Ch. Zhu, X., Qu, L., Du, D., Lin, Y., Engineered nanomaterials for photoreduction. *Trends Biotechnol.*, Science Direct, 36, 11, 1145–1156, November 2018.

20. Yadav, V., Roy, S., Singh, P., Khan, Z., Jaiswal, A., 2D MoS2-based nanomaterials for the therapeutic, bioimaging and biosensing application. *National Library of Medicine*, 15, 1, November 22, 2018.
21. Bullock, C.J. and Bussy, C., Biocompatibility considerations in the design of graphene biomedical materials. *Adv. Mater. Interf.*, 6, 11, April 18, 2019.
22. Zhu, W., Li, H., Luo, P., Emerging 2D nanomaterials for multimodel theranostics of cancer. *Front. Bioeng. Biotechnol.*, November 19, 2021.
23. https://www.ossila.com/en-in/pages/introduction-2d-materials.
24. Mohapatra, R.K., *Engineering Chemistry with Laboratory Experiments*, PHI, Delhi, 2015.
25. Mohapatra, R.K., Das, D., Azam, M., *Chemical Modification of Solid Surfaces by the Use of Additive*, Bentham Science, Singapore, 2020.
26. Penkov, O., Kim, H.-J., Kim, H.-J., Kim, D.-E., Tribology of graphene: A review. *Int. J. Precis. Eng. Manuf.*, 15, 3, 577–585, 2014.
27. Pantelic, R.S., Meyer, J.C., Kaiser, U., Stahlberg, H., The application of graphene as a sample support in transmission electron microscopy. *Sol. State. Commun.*, 152, 15, 1375-1382, 2012. Available from: https://www/emsdiasum.com/microscopy/products/graphene/graphene_tem.aspx
28. Rao, C.N.R., Sood, A.K., Subrahmanyam, K.S., Govindaraj, A., Graphene: The new two-dimensional nanomaterial. *Angew. Chem. Int. Ed.*, 48, 7752–7777, 2009.
29. Shao, Y., Wang, J., Wu, H., Liu, J., Aksay, I.A., Lin, Y., Graphene based electrochemical sensors and biosensors: A review. *Electroanalysis*, 22, 1027–1036, 2010.
30. Geim, A.K. and Kim, P., Carbon wonderland. *Sci. Am.*, 298, 90–97, 2008.
31. Novoselov, K.S. *et al.*, Electric field effect in atomically thin carbon films. *Sci. Natl. Library Med.*, 306, 666–669, 2004.
32. Park, S. and Ruoff, R.S., Chemical methods for the production of graphenes. *Nat. Nanotechnol.*, 4, 217–224, 2009.
33. Sampath, S., Basuray, A.N., Hartlieb, K.J., Aytun, T., Stupp, S.I., Stoddart, J.F., Direct exfoliation of graphite to graphene in aqueous media with diazaperopyrenium dications. *Adv. Mater.*, 25, 2740–2745, 2013.
34. http://www.4spepro.org/pdf/004401/004401.pdf
35. Bak, S., Kim, D., Lee, H., Graphene quantum dots and their possible energy applications: A review. *Curr. Appl. Phys.*, 11, 1192–1201, 2016.
36. Peng, Z., Xiang, C., Yan, Z., Natelson, D., Graphene nanoribbon and nanostructured sno2 composite anodes for lithium ion batteries. *ACS Nano*, 7, 7, 6001-6006, 2013.
37. http://s3.amazonaws.com/academia.edu.documents/41175514/Advanced_carbon_aerogels_for_energy_appl20160114-15050-1liyorj.pdf?
38. Xia, Z.Y. *et al.*, The exfoliation of graphene in liquids by electrochemical, chemical, and sonication-assisted techniques: A nanoscale study. *Adv. Funct. Mater.*, 23, 4684–4693, 2013.

39. Chhowalla, M. *et al.*, The chemistry of two-dimensional layered transition metal dichalcogenide nanosheets. *Nat. Chem.*, 5, 263–275, 2013.
40. Nicolosi, V., Chhowalla, M., Kanatzidis, G., Strano, S., Coleman, N., Liquid exfoliation of layered materials. *Science*, 340, 6139, 2013.
41. Luckham, F. and Rossi, S., Colloidal and rheological properties of bentonite suspensions. *Adv. Colloid Interface Sci.*, 82, 1, 43–92, 1999.
42. Eda, G. and Chhowalla, M., Chemically derived graphene oxide: Towards large-area thin-film electronics and optoelectronics. *Adv. Mater.*, 22, 22, 2392–2415, 2010.
43. Nascimento, J., Serodre, T., Santos, J., Paulinelli, L., Santos, A. *et al.*, Molecular insights into the production of few-layer graphene in N-Cyclohexylpyrrolidone + water mixtures. *Carbon*, 171, 723–738, 2021.
44. Hernandez, Y., Nicolosi, V., Lotya, M., Blighe, F., Sun, Z. *et al.*, High-yield production of graphene by liquid-phase exfoliation of graphite. *Nat. Nanotechnol.*, 3, 9, 563–568, 2008.
45. Lotya, M., Hernandez, Y., King, P., Smith, R., Nicolosi, V. *et al.*, Liquid phase production of graphene by exfoliation of graphite in surfactant/water solutions. *J. Am. Chem. Soc.*, 131, 10, 3611–3620, 2009.
46. Zhou, M., Tian, T., Li, X. *et al.*, Production of graphene by liquid-phase exfoliation of intercalated graphite. *Int. J. Electrochem. Sci.*, 9, 2, 810–820, 2014.
47. Wang, M., Xu, X., Ge, Y., Dong, P., Baines, R. *et al.*, Surface tension components ratio: An efficient parameter for direct liquid phase exfoliation. *ACS Appl. Mater. Interfaces*, 9, 10, 9168–9175, 2017.
48. Narayan, R. and Kim, S., Surfactant mediated liquid phase exfoliation of graphene. *Nano Converg.*, 2, 1, 2015.
49. Shen, J., He, Y., Wu, J., Gao, C., Keyshar, K. *et al.*, Liquid phase exfoliation of two-dimensional materials by directly probing and matching surface tension components. *Nano Lett.*, 15, 8, 5449–5454, 2015.
50. Cui, X., Zhang, C., Hao, R., Hou, Y., Liquid-phase exfoliation, functionalization and applications of graphene. *Nanoscale*, 3, 5, 2118–2126, 2011.
51. Schniepp, H., Li, J., Mcallister, M., Sai, H., Herrera-Alonso, M. *et al.*, Functionalized single graphene sheets derived from splitting graphite oxide. *J. Phys. Chem. B*, 110, 17, 8535–8539, 2006.
52. Mcallister, M., Li, J., Adamson, D., Schniepp, H., Abdala, A. *et al.*, Single sheet functionalized graphene by oxidation and thermal expansion of graphite. *Chem. Mater.*, 19, 4, 4396–4404, 2007.
53. https://www.cheaptubes.com/graphene-synthesis-properties-and-applications/
54. Nair, R.R., Blake, P., Grigorenko, A.N., Novoselov, K.S., Booth, T.J., Stauber, T., Peres, N.M.R., Geim, A.K., Fine structure constant defines visual transparency of graphene. *Science*, 320, 5881, 1308, June 6, 2008.
55. Balandin, A.A., Ghosh, S., Bao, W., Calizo, I., Teweldebrhan, D., Miao, F., Lau, C.N., Superior thermal conductivity of single-layer graphene. *Nano Lett.*, 8, 3, 902–907, February 20, 2008.

56. Touloukian, Y.S., *Thermophysical Properties of Matter: Thermal Conductivity: Nonmetallic Solids*, vol. 7, IFI/Plenum, Plenum Publishing, 1970.
57. Lee, C., Wei, X.D., Kysar, J.W., Hone, J., Measurement of the elastic properties and intrinsic strength of monolayer grapheme. *Science*, 321, 385–388, 2008.
58. Briggs, B.D., Nagabhirava, B., Rao, G., Deer, R., Gao, H., Xu, Y., Yu, B., Electromechanical robustness of monolayer graphene with extreme bending. *Appl. Phys. Lett.*, 97, 22, 223102, 2010.
59. Shehzad, K., Xu, Y., Gao, C., Duan, X., Three-dimensional macro-structures of two-dimensional nanomaterials. *Chem. Soc. Rev.*, 45, 20, 5541–5588, 2016.
60. Al Faruque, M.A., Syduzzaman, M., Sarkar, J., Bilisik, K., Naebe, M., A review on the production methods and applications of graphene-based materials. *Nanomaterials MDPI*, 11, 2414, 1–34, 2021.
61. Hasan, T., Hu, G., Howe, R., Yang, Z., New graphene based inks for hugh-speed manufacturing of printed electronics, University of Cambridge, Oct. 19, 2015.
62. Lei, W., Li, C., Cole, M., Qu, K., Ding, S., Zhang, Y., Warner, J., Zhang, X., Wang, B., Milne, W., A graphene-based large area surface-conduction electron emission display. *Carbon*, 56, 255–263, 2013.
63. Oyefusi, A. *et al.*, Hydroxyapatite grafted carbon nanotubes and graphene nanosheets: Promising bone implant materials. *Spectrochim. Acta A Mol. Biomol. Spectrosc.*, 132, 410–416, 2014.
64. Cutting Edge Vision and Innovation in Graphene Application. Applied Graphene Materials PLC, Graphene dispersions. appliedgraphenematerials.com. Archived from the original on 27 May 2014. Retrieved 26 May 2014.
65. Kang, J., Matsumoto, Y., Li, X., Jiang, J., Xie, X., Kawamoto, K., Kenmoku, M., Chu, J.H., Liu, W., Mao, J., Ueno, K., Banerjee, K., On-chip intercalated-graphene inductors for next-generation radio frequency electronics. *Nat. Electron.*, 1, 46–51, 2018.
66. Paine, R.T. and Narula, C.K., Synthetic routes to boron nitride. *Chem. Rev.*, 90, 73–91, 1990.
67. Kawaguchi, M. *et al.*, Electronic structure and intercalation chemistry of graphite-like layered material with a composition of BC6N. *J. Phys. Chem. Solids*, 69, 5–6, 1171, 2008.
68. Lin, Y. and Connell, J.W., Advances in 2D boron nitride nanostructures: Nanosheets, nanoribbons, nanomeshes, and hybrids with graphene. *Nanoscale*, 4, 6908–6939, 2012.
69. Song, L. *et al.*, Large scale growth and characterization of atomic hexagonal boron nitride layers. *Nano Lett.*, 10, 3209–3215, 2010.
70. Shi, Y. *et al.*, Synthesis of few-layer hexagonal boron nitride thin film by chemical vapor deposition. *Nano Lett.*, 10, 4134–4139, 2010.
71. Marom, N. *et al.*, Stacking and registry effects in layered materials: The case of hexagonal boron nitride. *Phys. Rev. Lett.*, 105, 046801–046804, 2010.

72. Erickson, K.J. *et al.*, Longitudinal splitting of boron nitride nanotubes for the facile synthesis of high quality boron nitride nanoribbons. *Nano Lett.*, 11, 3221–3226, 2011.
73. Blasé, X., De Vita, A., Charlier, J.C., Car, R., Frustration effects and microscopic growth mechanisms for BN nanotubes. *Phys. Rev. Lett.*, 80, 1666–1669, 1998.
74. Rudolph, S., Boron nitride (BN). *Am. Ceram. Soc. Bull.*, 79, 50, 2007.
75. Engler, M., Hexagonal boron nitride (hbn)–applications from metallurgy to cosmetics. *Ceram. Forum Int. Ber. DKG*, 84, 2007.
76. Hubacek, M., *Synthesis of Boron Nitride from Oxide Precursors*, TMD Hosting, December 12, 2007.
77. Mirkarimi, P.B. *et al.*, Review of advances in cubic boron nitride film synthesis. *Mater. Sci. Eng. R Rep.*, 21, 2, 47–100, 1997.
78. Tornieporth-Oetting, I. and Klapötke, T., Nitrogen triiodide. *Angew. Chem. Int. Ed.*, 29, 6, 677–679, 1990.
79. Pan, Z. *et al.*, Harder than diamond: Superior indentation strength of wurtzite BN and lonsdaleite. *Phys. Rev. Lett.*, 102, 5, 055503, 2012.
80. Leichtfried, G. *et al.*, 13.5 Properties of diamond and cubic boron nitride, in: *Landolt-Börnstein–Group VIII Advanced Materials and Technologies: Powder Metallurgy Data. Refractory, Hard and Intermetallic Materials*, vol. 2A2, P. Beiss (Eds.), pp. 118–139, Springer, Berlin, 2002.
81. Housecroft, C.E. and Sharpe, A.G., *Inorganic Chemistry*, 2nd ed., vol. 318, Pearson Education, 2005.
82. Engler, M., Hexagonal boron nitride (hBN)–applications from metallurgy to cosmetics. *Ceram. Foeum Int. Ber. DKG*, 84, 0173–9913, 2007.
83. Greim, J. and Schwetz, K.A., Boron carbide, boron nitride, and metal borides, in: *Ullmann's Encyclopedia of Industrial Chemistry*, Wiley-VCH, Weinheim, 2005.
84. Schein, L.B., *Springer Series in Electrophysics, Electrophotography and Development Physics. Physics Today*, vol. 14, pp. 66–68, Springer-Verlag, Berlin, 1988.
85. Engler, M., Hexagonal boron nitride (hBN)–applications from metallurgy to cosmetics. *Ceram. Foeum Int. Ber. DKG*, 84, 0173–9913, 2007.
86. Kawaguchi, M. *et al.*, Electronic structure and intercalation chemistry of graphite-like layered material with a composition of BC6N. *J. Phys. Chem. Solids*, 69, 5–6, 1171, 2008.
87. Hu, S. *et al.*, Proton transport through one-atom-thick crystals. *Nature*, 516, 7530, 227–230, 2014.
88. Park, J.-H., Park, J.C., Yun, S.J., Kim, H., Luong, D.H., Kim, S.M., Choi, S.H., Yang, W., Kong, J., Kim, K.K., Lee, Y.H., Large-area monolayer hexagonal boron nitride on Pt foil. *ACS Nano*, 8, 8, 8520–8, 2014.
89. Chhowalla, M., Shin, H.S., Eda, G., Li, L.J., Loh, K.P., Zhang, H., The chemistry of 2D layered transition metal dichalcogenide nanosheets. *Nat. Chem.*, 5, 263–275, 2013.

90. Zhang, X., Lai, Z., Tan, C., Zhang, H., Solution-processed two-dimensional MoS2 nanosheets. *Angew. Chem. Int. Ed.*, 55, 8816–8838, 2016.
91. Mak, K.F., Lee, C., Hon, J., Shan, J., Heinz, T.F., Atomically thin MoS2: A new direct gap semiconductor. *Phys. Rev. Lett.*, 105, 136805–136808, 2010.
92. Xu, M., Liang, T., Shi, M., Chen, H., Graphene-like two-dimensional materials. *Chem. Rev.*, 113, 3766–3798, 2013.
93. Mak, K.F., Lee, C., Hone, J., Shan, J., Heinz, T.F., Atomically thin MoS2: A new direct-gap semiconductor. *Phys. Rev. Lett.*, 105, 13, 2010.
94. Yin, Z.Y., Li, H., Jiang, L., Shi, Y.M., Sun, Y.H., Lu, G., Zhang, Q., Chen, X.D., Zhang, H., Single-layer MoS2 phototransistors. *ACS Nano*, 6, 74, 2012.
95. Li, H., Yin, Z.Y., He, Q.Y., Huang, X., Lu, G., Fam, D.W.H., Tok, A.I.Y., Zhang, Q., Zhang, H., Fabrication of single- and multilayer MoS2 film-based field-effect transistors for sensing NO at room temperature. *Small*, 8, 63, 2012.
96. Ghatak, S., Pal, A.N., Ghosh, A., Nature of electronic states in atomically thin mos2 field-effect transistors. *ACS Nano*, 5, 7707, 2011.
97. Walker, G.F., Macroscopic swelling of vermiculite crystals in water. *Nature*, 187, 312, 1960.
98. Feng, J., Sun, X., Wu, C.Z., Peng, L.L., Lin, C.W., Hu, S.L., Yang, J.L., Xie, Y., Metallic few-layered VS2 ultrathin nanosheets: High two-dimensional conductivity for in-plane supercapacitors. *J. Am. Chem. Soc.*, 133, 17832, 2011.
99. Golberg, D., Exfoliating the inorganics. *Nat. Nanotechnol.*, 6, 200, 2011.
100. Feng, J., Peng, L.L., Wu, C.Z., Su, X., Hu, S.L., Lin, C.W., Dai, J., Yang, J.L., Xie, Y., Physical properties of inorganic two-dimensional nanomaterilas. *Chem. Soc. Rev.*, 24, 1969, 2012.
101. Dines, M.B., Intercalation in layered compounds. *J. Chem. Educ.*, 51, 211, 1974.
102. Han, S.A., Bhatia, R., Kim, S.-W., Synthesis, propperties and potential applications of two-dimensional transition metan dichalcogenides. *Nano Converg.*, 2, 17, 2015.
103. Lee, Y.-H., Zhang, X.-Q., Zhang, W., Chang, M.-T., Lin, C.-T., Chang, K.-D., Yu, Y.-C., Wang, J.T.-W., Chang, C.-S., Li, L.-J., Lin, T.-W., Synthesis of large-area MoS2 atomic layers with chemical vapor deposition. *Adv. Mater.*, 24, 2320, 2012.
104. Lee, Y.-H., Yu, L., Wang, H., Fang, W., Ling, X., Shi, Y., Lin, C.-T., Huang, J.-K., Chang, M.-T., Chang, C.-S., Dresselhaus, M., Palacios, T., Li, L.-J., Kong, J., Synthesis and transfer of single-layer transition metal disulfides on diverse surfaces. *Nano Lett.*, 13, 1852, 2013.
105. https://www.ossila.com/en-in/pages/introduction-2d-materials.
106. Castellanos-Gomez, A., Poot, M., Steele, G.A., van der Zant, H.S.J., Agraït, N., Rubio-Bollinger, G., Elastic properties of Freely Suspended MoS2 nanosheets. *Adv. Mater.*, 24, 772, 2012.
107. Bertolazzi, S., Brivio, J., Kis, A., Stretching and breaking of uitrathin MoS2. *ACS Nano*, 5, 9703, 2011.

108. Chhowalla, M., Shin, H.S., Eda, G., Li, L.J., Loh, K.P., Zhang, H., The chemistry of 2D dimensional layered transition metal dichalcogenide nanosheets. *Nat. Chem.*, 5, 263, 2013.
109. Benameur, M.M., Radisavljevic, B., Heron, J.S., Sahoo, S., Berger, H., Kis, A., Visibility of dichalcogenide nanolayers. *Nanotechnology*, 22, 125706, 2011.
110. Li, H., Lu, G., Yin, Z.Y., He, Q.Y., Zhang, Q., Zhang, H., Optical identification of single- and few-layered MoS2 sheets. *Small*, 8, 682, 2012.
111. Zhong, C., Duan, C., Huang, F., Wu, H., Cao, Y., *Chem. Mater.*, 23, 326, 2010.
112. Han, S.A., Bhatia, R., Kim, S.-W., Synthesis, propperties and potential applications of two-dimensional transition metan dichalcogenides. *Nano Converg.*, 2, 17, 1–14, 2015.
113. Xiao, J., Choi, D., Cosimbescu, L., Koech, P., Liu, J., Lemmon, J.P., A graphene-like MoS2/graphene nanocomposite as a high performance anode for lithium ion batteries. *Chem. Mater.*, 22, 4522, 2010.
114. Ornes, S., *Lawrence Berkeley National Laboratory*, Berkeley Lab. U.S. Department of Energy Office of Science, August 25, 2022.
115. Cao, Y., Fatemi, V., Fang, S., Watanabe, K., Taniguchi, T., Kaxiras, E., Jarillo-Herrero, P., Unconventional superconductivity in magic-angle graphene superlattice. *Nature*, 556, 43, 2018.

16

The Potential Applications of 2D Nanomaterials for Water Purification

Hardeep Kaur[1,2*], Swaati Sharma[2], Shinar Athwal[2], Mansi Thakur[2], Meenakshi Verma[1,3] and Vishal Mutreja[1]

[1]*Department of Chemistry, University Institute of Science, Chandigarh University, Mohali, Punjab, India*
[2]*Department of Biotechnology, University Institute of Biotechnology, Chandigarh University, Mohali, Punjab, India*
[3]*University Centre for Research and Development, Chandigarh University, Mohali, Punjab, India*

Abstract
In recent years, the scarcity of water has become a complicated challenge owing to the increase in global population, climate change, and economic expansion, thereby emphasizing the requirement for developing advanced technologies for water treatment that can offer clean purified water in an affordable, sustainable, scalable, and reliable manner. The present advancements in 2D-nanomaterials (2DM) provide a contemporary pathway for acknowledging the foremost challenges in treating water because of their distinctively superior structural properties. Surging 2-D nanostructures like MXene, g-C_3N_4, black phosphorus, graphene, and h-BN have considerably predicted an extraordinary surface-to-volume ratio, which ensures ultrafast processing time, ultrahigh treatment efficiency, and ultralow material use for the water purification process. In this review, we have discussed the progressive development of 2D nanomaterials to successfully carry out water purification involving adsorption, separation, or photocatalysis. Elementary strategies to design 2DM with a significant emphasis on physicochemical factors for efficient water purification have also been enlisted. The analysis of different 2DM-based functional membranes, such as filtration and solar desalination membranes, has also been conducted. Moreover, the vital role of 2DM in water purification was noted. 2DM exhibits considerable antiparasitic and antimicrobial activities against waterborne pathogens of main concern, such as *Entamoeba*

**Corresponding author*: hardeep.uibt@cumail.in

Subhendu Chakroborty and Kaushik Pal (eds.) 2D Nanomaterials: Synthesis, Properties and Applications, (361–396) © 2024 Scrivener Publishing LLC

histolytica, *Shigella dysenteriae*, and *Vibrio cholera*. In addition, 2DM can absorb noxious chemicals, dyes, and metals from water. The review also elaborates on the structure-processing-property relationship in 2DM, with the aim of fabricating next-generation 2DM units with intelligent, programmable, multifunctional, and selective water purification technologies. However, some bottlenecks such as adverse effects of 2DM on the environment and human health, seepage of 2DM in drinking water, and aggregation of nanoparticles need to be adequately addressed. The importance of purified clean water across the globe for future generations imparts light and inspiration for the development of affordable and efficient 2DM units to ensure a secure supply of clean water in the growing world.

Keywords: 2DM, nanomaterial, single-layer, water purification, nanosheets

16.1 Introduction

Water is an important aspect of all known life forms and has a positive impact on human civilizations and social enhancement. On Earth, water specifically occurs in salty seawater (1%–4%), while 3% of the Earth's water sources are fresh water (icecaps, glaciers, and groundwater). Due to the scarcity of freshwater resources on earth, the availability of purified water has decreased over the years. In addition, an increased global population has decreased the availability of sustainable and affordable clean water. The increase in water contamination has also increased the concern of purifying water resources to sustain ecosystem balance [1]. For the same reason, 2DM has gained significant acknowledgement for the sustainable purification of water resources. 2D nanomaterials (2DM) are nanomaterials with two dimensions above 100 nm and one dimension below 100 nm. The present advancements in 2D-nanomaterials (2DM) provide a contemporary pathway for acknowledging the foremost challenges in treating water because of their distinctively superior structural properties. Surging 2-D nanostructures like MXene, g-C_3N_4, black phosphorus, graphene (Figure 16.1), and h-BN have considerably predicted an extraordinary surface-to-volume ratio, which ensures ultrafast processing time, ultrahigh treatment efficiency, and ultralow material use for water purification [2]. In this review, we have discussed the progressive development of 2D nanomaterials to successfully carry out water purification involving adsorption, separation, or photocatalysis. Elementary strategies to design 2DM with a significant emphasis on physicochemical factors for efficient water purification have also been enlisted. The analysis of different 2DM-based functional membranes, such as filtration and solar desalination membranes, has also been conducted [3, 4]. Moreover, the Figure 16.1

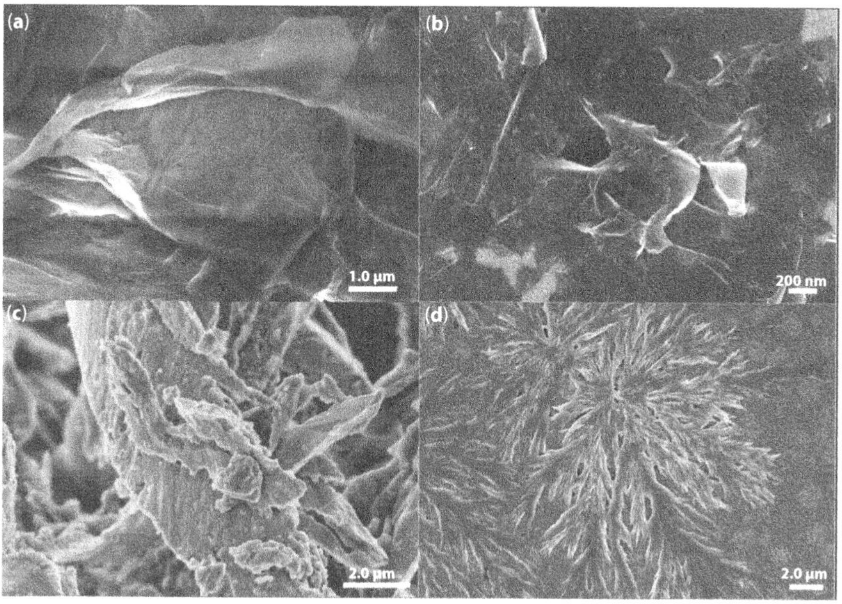

Figure 16.1 FE-SEM micrograph covering top view of (a) 1.0 mm magnification low-scale few-layered graphene sheets, (b) 200 magnification high-scale depicting nanofiber morphology on graphene-nanosheets, (d) Nano-tree designed morphology formed *via* homogeneous dispersion of CdS integrated HBLC nanowires (SA þ 9BAO) on graphene-nanosheets in the unidirectional form [5].

demonstrates a FE-SEM micrograph covering top view of 1.0 mm magnification low-scale few-layered graphene sheets, a 200 magnification high-scale depicting nanofiber morphology on graphene nanosheets, and the Nano-tree designed morphology formed *via* homogeneous dispersion of CdS integrated HBLC nanowires (SA þ 9BAO) on graphene-nanosheets in the unidirectional form [5]. 2DM exhibits considerable antiparasitic and antimicrobial activities against waterborne pathogens of main concern such as *Entamoeba histolytica*, *Shigella dysenteriae*, and *Vibrio cholera*. In addition, 2DM can absorb noxious chemicals, dyes, and metals from water. This review considers various purification processes based on the use of 2DM to purify water resources.

16.2 Contaminants Present in Water

Several contaminants owing to anthropogenic or natural activities are present as impurities within water, making it unfit or non-potable for consumption purposes. The consumption of such contaminated water can cause major health issues ranging from mild to acute, depending on the nature the contaminant. These contaminants in water solutions are of various kinds and include chemical contaminants (organic/inorganic), biological pollutants, and suspended solids [6].

(a) Chemical contaminants: Several chemical contaminants from various sources are present in water. The nature of these contaminants can be organic/inorganic, water-insoluble/soluble and their toxic aspects towards humans and the environment also vary accordingly. According to an estimate, more than 700 different types of chemical entities belonging to 20 different classes are present in water. The commonly found contaminants are pesticides, nitrate, uranium, halogenated aromatic substances, and selenium [7, 8].
(b) Biological pollutants: The development of next-gen. sequencing has offered a convenient and affordable understanding of microbial communities in the presented sample or habitat. It is well known that a large number of microbes are present in water solutions, and the pyro-sequencing method showed the availability of firmicutes, cyanobacteria, bacteroidetes, and proteobacteria in the groundwater sample [9].
(c) Suspended solids: Total suspended solids include fine particulate particles comprising various insoluble substances such as sand, dust, clay, animal materials, decaying plants, and silt. The standard size of SS is less than 62 μm. The suspended particles make the water non-potable and cloudy and have a negative impact on invertebrates and fish [10].

16.3 2D Nanomaterial-Based Water Purification Membranes

The water treatment upholds the prime function of eliminating the unwanted composites from water, so the functional membranes provide a barrier (physical) to remove these composites based on permeability or size from water. It has been noted that acquiring selectivity and permeability at the same time is a bit difficult, but the functional membrane offers both selectivity and permeability at the same time. Also, the fact that there is an inverse relation between membrane thickness and permeability allows for

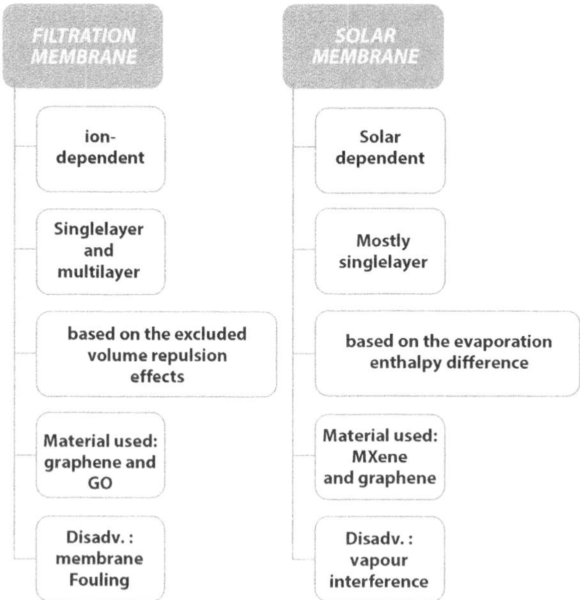

Figure 16.2 2D nanomaterial-based water purification membranes.

more significant use of 2D nanomaterials as filtration membranes, as the thickness of 2DM ranges from one atom to a few atoms. The incorporation of 2DM into a functional membrane forms an atomic-thin separation unit with the facility of selective water separation. On the basis of the water phase that passes through the membrane (either liquid or gas phase), separation (functional) membranes can be classified into membrane filtration units for liquid and water phases and solar membrane-desalination units for gas and water phases [1]. Figure 16.2 discusses that the filtration membrane specifically depends on the hydrodynamic size design of the pores for size-dependent separation, while the solar desalination membrane design involves various factors like hydrodynamic transport of water, light-matter reaction, and evaporation enthalpy of water.

16.4 Solar Desalination Membrane

The direct use of sunlight to produce desalinated water has emerged as a promising solution to positively address the increased shortage of global freshwater. The working of the desalination membrane is based on the evaporation enthalpy difference between the evaporation enthalpy of contaminants and water. A standard solar-desalination device consists of four main components: a reservoir of sea water, a solar-desalination membrane, a condenser, and a freshwater collector. From all the discussed components, the solar-desalination membrane plays the most important role in generating steam by converting solar energy into heat. Black 2DM, like MXene and graphene, have been studied for applications in solar-water evaporation owing to their excellent photon-harvesting ability, adjustable thermal conductivity, and lower molar specific heat *via* chemical doping [11, 12]. Solar desalination offers vertically aligned 2DM designs with high porosity that promote water transportation. For instance, the vertically aligned membrane sheet of graphene (VA-GSM) was developed for the conversion of solar energy into heat energy to produce desalinated water. VA-GSM with random structure and lamellar alignment when compared to graphene demonstrated explicit advantages for water transportation and thereby revealed higher solar-desalination performance. The VA-GSM suggested a higher rate of vapor generation (1.6 kg m^{-2} h^{-1} under 1 sun illumination and 6.3 kgm^{-2} h^{-1} under 4 sun illumination) with a better efficiency of up to 86.5% for 1 sun illumination and 94.2% for 4 sun illumination, respectively. To increase the employment of solar energy and decrease thermal loss, effective management of heat and excellent solar absorption have been suggested as the two primary factors for carrying out the solar desalination process. An ideal absorber of sunlight decreases the effects of transmittance or reflectance over a broad range of wavelengths. To date, various strategic designs have been examined to fabricate efficient anti-reflection with effective absorbers. Higher density for optical modes and strong light coupling for excellent absorption [13]. Owing to the eco-friendly and cost-effective properties of carbon-based 2DM, they are extensively studied for solar steam and membrane desalination. The porous nanostructures of graphene overcome the limitations of standard graphene structures as they depict 5%–10% reflection at the interface of air-dielectric. The 2DM unit's on-demand can preferably confine sunlight in small areas *via* resonant recirculation, elevating light interaction inside the materials. Further, the plasmon examination revealed that over the metal surface, the light energy can initiate electron excitation altogether if its frequency is relative to the oscillation frequency exhibited by delocalized electrons [14]. During this phenomenon, the hot electron carriers finish re-distribution *via* electron–electron scattering, producing a significant quantity of heat, which can be transferred to the surrounding material *via* thermal conduction. Several plasmonic components incorporated with 2DM have been investigated on membrane substrates. For instance, Wang *et al.* constructed a solar-desalination membrane by using graphene-oxide and a multiscale plasmonic nanostructures, with Ag/Au nanocubes utilized as blocks of assembly to develop plasmonic colloidosomes. The plasmonic coupling within metal nanoparticles resulted in good efficiency (92%) at 10 kWm^{-2} with a rate of steam generation of about 12.96 $kg\ m^{-2}\ h^{-1}$ for the solar-desalination membrane. In spite of all advantages discussed, solar desalination exhibits disadvantages such as lower long-term stability and interference of vapors during the collection of light [1, 15]. Thus, the introduction of enhanced membranes or integrated solar-steam generating systems with standard membrane distillation phenomena (advanced distillation geometries) can maximize the large-scale application of solar-desalination membranes.

16.5 Filtration Membrane

In size-dependent separation, membrane-based water purification depends on the excluded volume repulsion effects between the unwanted contaminant entities and membrane pores. Thereby, two strategies (i.e., single-layer membranes and stacked-multilayer membranes) have been investigated to develop filtration membranes based on 2DM with an optimized pore size. 2DM demonstrates lower atomic thickness, lower energy consumption, and high water permeability, thereby promoting the purification process. The introduction of nanopores with an optimized size *via* irradiation, electrochemical oxidation, and ion/plasma etching. The 2DM single-layer membranes are a potential candidate to replace standard separation membranes. Experimental studies have revealed that nanoporous single layers are highly effective in water purification. Carried *via* pressure difference conditions, the graphene nanoporous single layer develops a physical mesh design of mesh that allows selective penetration of water molecules *via* the membrane. The calculations of molecular dynamics revealed that nanoporous graphene can possibly achieve a good water flux (66 $Lcm^{-2}\ day^{-1}MPa^{-1}$) with a salt elimination efficiency of about 99%. The graphene nanopores in a single layer play an important role in the desalination of water, permitting selective passage of water [16]. By optimizing the functional groups and size of pores present on a porous 2D-nanomaterial single layer, a higher rate of water penetration can be achieved in comparison to conventional RO membranes. Experimental work on graphene has revealed that the electron beam can be utilized to produce nanopore structures over the 2D sheets with nanometer accuracy. A nanoporous graphene single layer was fabricated by O'Hern *et al.* through CVD graphene transfer on porous polycarbonate. The presence of nanopores over graphene offered the selective transfer of various molecules, involving tetramethylammonium chloride, tetramethylrhodamine, and potassium chloride [17]. The composites of graphene allowed tetramethylammonium chloride and KCl to pass through the membrane but exempted tetramethylrhodamine. A scanning electron microscope suggested that about 83% of the pore structures were comparatively small, with a diameter of less than 10 nm. The ion bombardment method has been used to fabricate nanoporous graphene filtration membranes. In comparison to standard pore production strategies, which specifically depend on the direct oxidative method, ion bombardment-dependent methods (like argon-ion irradiation) can produce pores with relatively optimized density and size. Moreover, the graphene-oxide (GO) nanosheets (multilayered) are inherently hydrophilic in nature. The GO nanosheets are mainly vacuum-filtered, so they lack the required bonding between graphene-oxide nanosheets and thereby may undergo structural deterioration under the conditions of cross-flow commonly witnessed in industrial operations [18]. Thus, to improve the stability of structure for stacked 2DM membranes, great efforts have been made in the layer-by-layer assembly approach to present inter-layer bonding *via* covalent linkage and/or electrostatic interactions during the fabrication of the membrane. Multiple molecular interactions, consisting of covalent bonding, coordinate interactions, biologically specific interactions, and charge-transfer interactions, have been largely studied and investigated to provide layer-by-layer assembly to stabilize functional multilayers. The spacing between the layers of stacked nanosheets is an important parameter that determines the performance of the multilayered membranes [19, 20]. However, despite remarkable advancements in 2DM-based filtration membranes, specific challenges like large-scale development of membranes with long-span stability remain unattended. Fouling of the membranes is the standard problem, which abstains the long-span function of the membranes as it elevates the consumption of energy and decreases the recovery of water from the desalination process [1]. Thereby, membrane units can be self-fixing and self-cleaning designs with the incorporation of anti-bacterial nanoparticles (like silver-nanoparticles), which can facilitate the commercial utilization of 2DM nanomaterial membranes for the purification of water.

16.6 Properties of Widely Used 2DM for Water Purification

The significance of commonly used 2DMs namely, MXene, g-C_3N_4, black phosphorus, graphene, and hexagonal-boron nitride has been discussed in Table 16.1.

16.6.1 MXene

In the year 2011, a novel class of 2DM, transition metal nitride, carbonitride, and carbide (MXene), was developed and has progressively become an efficient nanomaterial (2DM). Basically, MXene (2DM) is synthesized *via* selective etching of particular atomic layers (like Si, Ge, and Al) from their original MAX phase in the presence of mixed fluoride salts (e.g., LiF, NH_4F, and NaF) and strong acids like HCl or HF. $M_{n+1}X_nT_x$ is the structural formula of MXene, where M stands for initial transition metals (like Zr, Nb, V, and Ti), X represents N and/or C elements, and T specifies the functional groups on the surface (like -F, -O, and -OH). Due to the various number of atomic layers within a unit cell, n can range from one to three, leading to the formation of the standard structures of M_2XT_x, $M_3X_2T_x$, and $M_4X_3T_x$, respectively [21, 22]. Initially, extensive studies on MXene were based on storage systems and energy conversion (batteries) because of its better electrical conductivity and high capacity for theoretical gravimetrics. The photocatalytic activity of MXene has also been acknowledged over the past few years. MXene has ample functional groups as a result of the wet-chemical etching method, which is advantageous for the development of intimate connection interfaces between semiconductors and MXene. The alignment of the band gap for MXene could be tuned by modulating the chemistry of the surface. Also, the layered structure of MXene with a conductive metal core has efficient metallic conductivity with electron acceptance capability. Thus, MXene is considered as a potential candidate in comparison to other 2DMs and has particularly been investigated for several water purification operations [21, 23, 24]. To realize the antibacterial activities of MXene, Rasool *et al.* prepared a 2DM to purify water. The developed MXene membrane was assayed against *B. subtilis* and *E. coli* by utilizing bacterial growth curves based on colonies growth over nutritive agar plates and optical density. MXene showed excellent antibacterial activity, as about 98% of the bacterial cell viability was lost at 200 μg/mL MXene for both bacterial colonies in about 4 h of exposure, as assured by the regrowth curve and Colony Forming Unit (CFU) [25]. Thus, MXene

Table 16.1 Significance of various 2DM structures.

Sr. no.	2DM	Structure	Chemical composition	Properties	Advantage	Disadvantage	Ref.
1.	MXene	M: transition metal X: N or C elements T:Functional groups	transition metal nitride, carbonitride, and carbide	• Abundant functional groups • Tunable surface chemistry • Layered structure with conductive metal-core	• better electrical conductivity • high capacity for theoretical gravimetric	• Relatively less dispersible • Less hydrophilic and • less stable	[21, 23, 24]
2.	g-C_3N_4	2D layered sheets comprising of carbon (C) and nitrogen (N)	nitrogen and carbon atoms form a pie–pie conjugated electronic system (sp^2)	• metal-free • photo catalyst • easily forms composites with other materials	• higher thermal and chemical stability • stable • cost-effective	• Low disparity in many solvents resulting in declined filtration	[26, 28, 29]

(Continued)

Table 16.1 Significance of various 2DM structures. (*Continued*)

Sr. no.	2DM	Structure	Chemical composition	Properties	Advantage	Disadvantage	Ref.
3.	BP	exfoliated into few layered-BP from bulk BP.	Allotrope of phosphorus	• thermodynamically stable • tunable bandgap • in-plane anisotropy	• ultra-high charge mobility • wide-light absorption Anti-microbial actions	• Stability issues • Relatively low conductivity • Medium catalytic functions	[31, 32, 33]
4.	Graphene	is a single layer of graphite	carbon allotrope designed in the honeycomb lattice model	• semi-metal • zero-bandgap • ambi-polar electrical properties	• high thermal conductivity • high Young's modulus excellent specific surface area of	• potential risk towards ecosystem and human beings low chemical stability	[10, 38, 39]
5.	h-BN	2D hexagonal-boron nitride (h-BN) is a boron (B) and nitrogen (N) layered nanostructure	B-N interactions are covalent with partial ionic bonds	• tunable of surface chemistry • decreased pie-pie forces • selective functionalization	• outstanding mechanical strength • enhanced catalytic functions • increased water-dispensability	• Improvisation required in: • surface chemistry • thickness-dependent properties • charge-transfer phenomenon	[40, 42, 43]

membranes can be effectively utilized to purify water. However, more dispersible, hydrophilic, and stable MXene membranes can be developed to improve the working efficiency of membranes with eco-friendly synthesis and large-scale fabrication.

16.6.2 g-C_3N_4

Graphite-carbon nitride, i.e., g-C_3N_4, is a photocatalyst (metal-free) with various benefits, such as cost-effective and eco-friendly properties, convenient synthesis, and excellent stability. It can be considered a potential 2DM material in water purification applications. The g-C_3N_4-based composites include both organic and inorganic characteristics, which makes them a competent material to purify water. The structure of g-C_3N_4 is made up of 2D layered sheets in which nitrogen and carbon atoms form a pie–pie conjugated electronic system *via* sp^2 hybridization and are then repeatedly stacked to make a 3D crystal unit [26, 27]. The standard g-C_3N_4 units exist in two different forms, consisting of tri-s-triazine (C_6N_7) and s-triazine (C_3N_3). According to density-functional theory, the former is relatively more stable and is hence considered a theoretical design in most experimental studies. Consisting of earth-ample elements, g-C_3N_4 in bulk can be developed *via* the direct thermal condensation process of nitrogen-containing organic molecules such as urea, dicyandiamide, melamine, and cyanamide. The kind of precursor demonstrates an important impact on the outcome, adsorption ability, photocatalytic activity, and microstructure of g-C_3N_4. In general, g-C_3N_4 acquired from thiourea demonstrates a low negative zeta potential while g-C_3N_4 acquired from urea demonstrates a more precise surface area, but the outcome is low, ranging from 1% to 5%. This establishes the preferable functionality of urea-acquired g-C_3N_4 for photocatalytic interactions like degradation of organic contaminants and hydrogen generation. The structural characteristics of g-C_3N_4 endow outstanding properties, comprising: (1) higher thermal and chemical stability permits g-C_3N_4 to function under various conditions; (2) the triazine ring and amino group present over the surface increases the possibility of making composites with other units; and (3) the nonmetal composition of g-C_3N_4 abstains from causing contamination towards the environment [28, 29]. To analyze the g-C_3N_4 unit, Arumugham *et al.* prepared a g-C_3N_4-CuO nanosheets-based ultra-filtration membrane through a facile calcination pathway as a one-pot fabrication process to treat the water solution. The membrane displayed notable penetration flux for protein solution (131 $Lm^{-2}h^{-1}$) and pure water (202 $Lm^{-2}h^{-1}$), with a higher rejection rate for proteins of about 96%. Thus, the constructed g-C_3N_4 filtration unit revealed comprehensive membrane performance with applicable potential to treat water [30]. Despite the advantages offered by g-C_3N_4, a few drawbacks are to be noted and worked upon, such as the fact that the disparity of g-C_3N_4 is low for many solvents, leading to decreased interface dispersion and subsequently an uneven coating of the layer with declined filtration performance. Also, studies based on the multifunctionality of g-C_3N_4 can be worked on to increase its commercial utility.

16.6.3 Black Phosphorus

Black phosphorus is a favorable option that can possibly fill the existing gap between 2D transition metal dichalcogenides and graphene to act as a nearly perfect 2DM. Phosphorus (P) is abundantly present on the earth's surface, occupying 0.1% of the volume of the earth's surface. Phosphorus comprises four standard allotropes, which are black phosphorus (BP), red phosphorus, white phosphorus, and violet phosphorus. Out of these, BP is the most thermodynamically stable allotrope in comparison to the others. BP gained the required attention after when the bulk BP was exfoliated into a few layers of BP (phosphorene) *via* the Scotch tape micro-cleavage process due to feeble Van der Waals bonds in between the layers. The experimental and theoretical studies suggested that bandgap energy for BP is dependent on thickness, with values of 0.3 eV for bulk BP and 2.0 eV for monolayer phosphorene. This signifies that BP can carry out wide-light absorption ranging from ultra-violet light to visible light, and also involving near-infrared light, permitting PBP-based units to work fascinatingly for photoremediation and related processes [31, 32, 33]. Moreover, BP possesses ultra-high charge mobility and a tunable bandgap, along with excellent in-plane anisotropy and relatively high thermal stability, thereby optimizing its use as an excellent 2DM to treat water [31]. Further, BP-based quantum dots have demonstrated remarkable functionality in the purification of water. Tang *et al.* prepared the BP-based quantum dots through a standard solution-dependent top-down approach. The quantum dots depict a notable reduction in friction and anti-wear characteristics even at quite low concentrations (0.005 wt%), which decreased the coefficient of friction by 32.3% and the wear volume of the liquid base by 56.4%. While load-supporting ability for the liquid base improved to 300 N from 200 N. The BP-based quantum dots suggested even and small particle sizes, high reactivity over frictional surfaces, and better dispersion stability [34]. BP also exhibits antimicrobial properties that can be utilized to remove microbial contamination in water [35]. Though BP has considerable catalytic applications, its inherent properties (of an electrical and optical nature) do not fulfill the required practical functions. Various properties have been worked upon, such as conductivity elevation, enhanced stability, accelerated carry mobility, and recombination of photo-induced charge carriers to positively evaluate the catalytic properties of BP. Thus, modifications (edge designs, carbonaceous materials, plasmonic metals, and morphology) can increase the applicability of BP-based materials [31]. BP is a significant part of the post-graphene era, holding the potential to remarkably affect the water purification industries in a better way. As a result, BP can be effectively used as a water purification material, even on a large scale.

16.6.4 Graphene

Graphene is a single layer of graphite, is the first 2DM with one atom thickness to be extracted in nature. For the last five decades, substantial efforts have been conducted to obtain a single layer of graphene sheets, aiming to access the predicted characteristics of a compactly bonded, one atom-thick sheet of carbon (sp^2). Initially, the work was based on developing graphite intercalation units. In these units, the host atom or molecule gets 'sandwiched, in between the layers of graphite, leading to weakened inter-planar forces and enabling layer separation [10, 36]. The first single layer of graphite was extracted by Andre Geim *et al.* in 2004. The presented technique for acquiring actual one-atom-thick layers having macroscopic lateral dimensions was a convenient method. The process involved the rubbing of pyrolitic graphite against other surfaces, leaving behind several flakes over the surface. The obtained flakes consisted of more than 10 layers, along with a few single-layered structures [37]. Graphene, the carbon allotrope designed in the honeycomb lattice model, has gathered interest owing to its exotic and fascinating characteristics. It is a semi-metal, zero-bandgap, having ambi-polar electrical properties and exhibiting efficient mechanical parameters, with a high thermal conductivity of $3 * 10^3$ W, a high Young's modulus of 1.1 TPa, and an excellent specific surface area of 2,630 m^2g^{-1}. Also, graphene demonstrates extraordinary optical characteristics with 97.7% transmittance within the visible spectrum (high transparency), and commendable electrical characteristics having a higher electrical conductivity of approximately about 104 Ω^{-1} cm^{-1}. Moreover, nanosheets developed from graphene are automatically 2D thin materials, with efficient chemical and physical characteristics in terms of convenient modification, optimized structure, and tunable surface parameters. Isolated graphene nanosheets are regarded as efficient adsorbers of contaminants due to their strong pie bond interactions, ample functional groups, and larger surface area. Graphene-oxide and reduced graphene-oxide have also shown desirable properties (structural wrinkles, specific edges, and surface defects) to treat contaminated water and eliminate metallic ions or/and organic pollutants from the water. However, direct utilization of nanoscale graphene can inevitable release graphene into nature, leading to a potential risk to ecosystems and human beings [38, 39]. Therefore, the safe application of graphene can be carried out by incorporating graphene into specifically designed devices, for example, the integration of graphene nanosheets with two-dimensional membrane units. As a result, the graphene composites can be preferably utilized to purify water solutions and relatively remove contaminants from the solution to give clean water.

16.6.5 h-BN

2D hexagonal-boron nitride (h-BN) is a boron (B) and nitrogen (N) layered nanostructure in a sp^2 hybridization honeycomb-designed lattice structure. The van der Waals bonds establish the vertical assembly of layers, resembling graphite structure, while the in-plane B-N interactions are covalent with partial ionic bonds owing to the electronegativity difference between boron and nitrogen. The bond strength between boron and nitrogen offers outstanding mechanical strength; however, the absence of pie electrons on each plane surface decreases the surface attraction as a cause of pie–pie forces. This decrease in pie forces increases the chemical stability of h-BN in comparison to other 2D materials like transition metal dichalcogenides and graphite, where pie–pie conjugation overpowers the surface interaction process. The partially charged regions over the surface, i.e., B (+) and N (−), can act as anchoring areas for Lewis acid–base attractions [40, 41]. Also, the corresponding empty orbitals of boron and a lone pair on nitrogen demonstrate a favorable route for selective operations of h-BN. The observed functionalization of h-BN permits the tuning of surface chemistry to establish desirable characteristics along with the preservation of intrinsic properties. Hydroxylation of h-BN increases its water-dispersibility, edge amination increases its acid–base interaction on the surface, point defects enhance its catalytic functions, and fluorination improves its photoluminescent emission. Thus signifying that surface modifications are an important step to harnessing the full ability of h-BN nanomaterial [40, 42, 43]. The h-BN nanosheets have been utilized recently to purify water. For instance, Ivanova *et al.* constructed a purifying composite comprising exfoliated h-BN and cluster complexes of hexamolybdenum. The cluster was deposited on the h-BN nanosheets through impregnation of the matrix by solvents. A very few concentrations of water molecules in the acetone solution-initiated hydrolysis process of the molybdenum-cluster precursor with labile ligands of NO_3^-, which were not present in the resultant final state. The intermediate hydrolyzed cluster formed an anchor over the h-BN surface and thereby promoted the growth of insoluble compounds as the final resultant hydrolysis product [44]. The h-BN-based unit showed reasonable filtration and action on the contaminants present in the water. Though h-BN exhibits various advantages, parameters like surface chemistry, thickness-dependent properties of h-BN layers, and charge-transfer phenomena of the h-BN nanosheets can be improved in order to attain extraordinary 2DM [40]. The integration of h-BN into other membranes or its utilization as a core shell can possibly increase the purification efficiency of h-BN. As a result, the h-BN-based composites depicted the relevant elimination and filtration of contaminants from the water solution.

16.7 Synthesis of 2DM

16.7.1 Top-Down Approach

Top-down approaches can be applied to single- or few-layered 2DM by eliminating van der Waals attraction in-between the stacked layers of the bulk layered crystals. The commonly used top-down approaches comprise liquid exfoliation *via* oxidation/ion intercalation/mechanical force, mechanical cleavage, ion exchange, and selective etching [45].

16.7.1.1 Liquid Exfoliation via Oxidation/Ion Intercalation/Mechanical Force

Liquid exfoliation (LE) *via* the oxidation method (Figure 16.3) is commonly utilized to separate the layers of graphite through a modified Hummers' process. Hummers' method employs strong oxidizing agents, chiefly a combined mixture of H_2SO_4 and $KMnO_4$, to specifically oxidize graphite layers to develop graphite oxide. The executed oxidation phenomenon generates a generous amount of oxygen-comprising functional groups, like, carboxyl, epoxy, and hydroxyl groups, present over the graphene layers, which can possibly expand the inter-layer spacing and consequently weaken the van der Waals attraction between adjoining bulk-graphite layers. Following successive sonication treatments, the expanded graphite-oxide can be

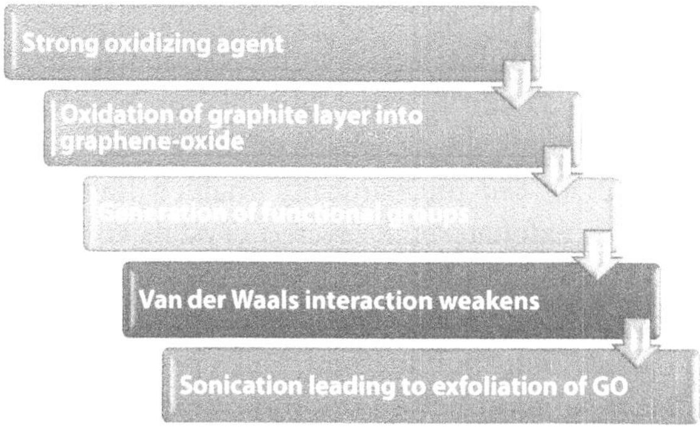

Figure 16.3 Liquid exfoliation *via* oxidation.

conveniently exfoliated into single-layer graphene-oxide nanosheets [46, 47]. Single layer graphene-oxide nanosheets with a higher yield and high concentration in the aqueous form can be attained *via* this method. However, this method is noxious because it uses strong oxidizing agents and favorable extension of the method to other layered materials remains difficult.

The liquid exfoliation process of 2DM *via* ion intercalation technique is based on the prime principle of cationic intercalation, such as Na^+, K^+, Li^+, etc., within the inter-layers of the bulk crystals (Figure 16.4). The intercalation technique weakens the van der Waals attraction in between adjoining layers of bulk crystals. The intercalated substances can be further exfoliated into single-layer or few-layer sheets through the mild sonication method in water. Generally, intercalated ions may interact with water solutions and produce hydrogen gas, leading to the separation of adjoining layers at the time of sonication [48]. High-yielding nanosheet layers can be acquired

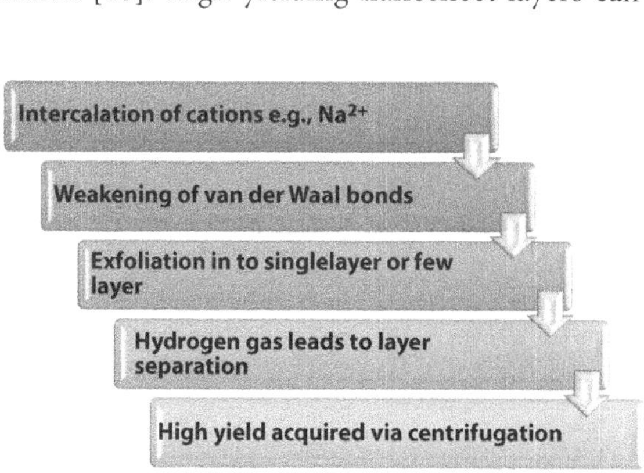

Figure 16.4 Liquid exfoliation *via* ion intercalation.

following the centrifugation step. The single layer 2DM procured *via* the ion intercalation method develops a tinier lateral size. However, a combined chemical ion and hydrothermal intercalation-enabled liquid exfoliation process can overcome the drawback of lateral size.

Generation of 2DM *via* mechanically enabled liquid exfoliation is a progressive route to exfoliate bulk layered crystals (Figure 16.5). Several liquid exfoliation techniques have been suggested by efficient functions of mechanical strength in aqueous solutions based on shear force and sonication. In the standard sonication method, bulk crystals are evenly dispersed within the solvent solution before the sonication process, and following sonication, the formed suspension is purified through centrifugation. Sonication is a convenient method that transmits mechanical forces in the aqueous form, which consequently disintegrates the van der Waals interaction between the layers without effecting the covalent bonds within the layers [49, 50]. Additionally, the exfoliation efficiency can be elevated by relating the surface properties of layered bulk crystals with the solvent.

16.7.1.2 Mechanical Cleavage

Mechanical cleavage, also known as the Scotch tape process, is a standard pathway to develop thin nanosheets of 2DM through the peeling-off step from the layered bulk crystals *via* weakening of van der Waals attraction within layers without disturbing in-plane covalent forces (Figure 16.6). This kind of exfoliation process was primarily established to acquire a single layer of graphene from bulk graphite and was eventually utilized to cleave other 2DMs like BP, h-BN, and transition metal chalcogenides (TaS_2, WS_2, MOS_2, Re_2, etc.). The cleavage process includes the attachment of the bulk crystal (surface) on the Scotch tape, followed by the peeling-off step *via*

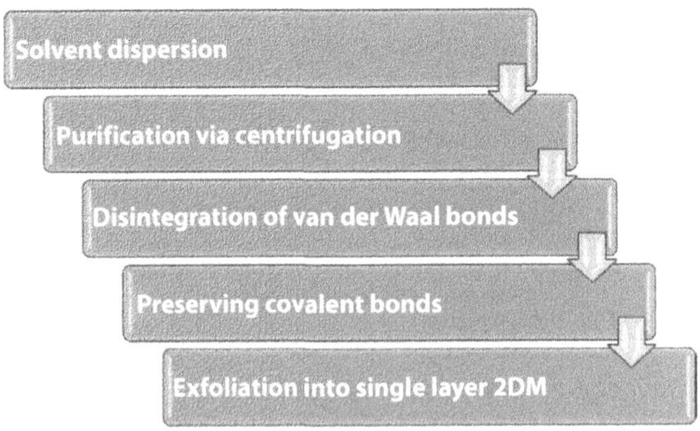

Figure 16.5 Liquid exfoliation *via* mechanical force.

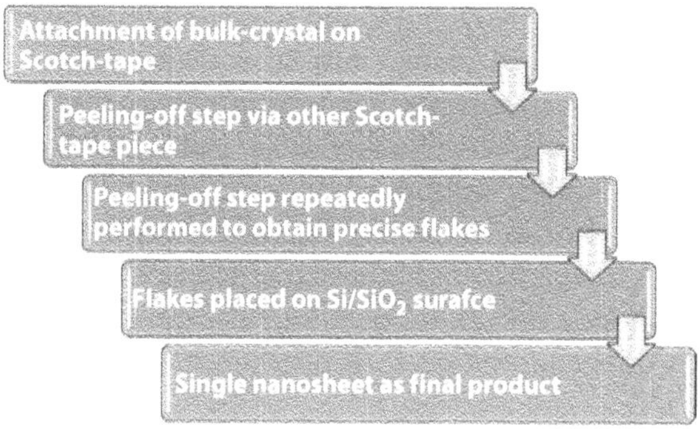

Figure 16.6 Mechanical cleavages.

another Scotch tape piece, and the process is repeatedly performed to procure a precise thin flake. The obtained flake is safely transferred to the Si or SiO2 target surface to fabricate single-layered or few-layered nanosheet units [51, 52]. Thus, the developed nanosheets *via* cleavage demonstrated minimum defects, better crystal quality, and a clean surface. However, the production rate and yield *via* the cleavage method are low, and its manual operations make it difficult to control the production parameters.

16.7.1.3 Ion Exchange

Exfoliation *via* anionic or cationic exchange has been widely reported for the fabrication of 2DM from bulk crystals (Figure 16.7). The cationic exchange method is established primarily for the extraction of layered

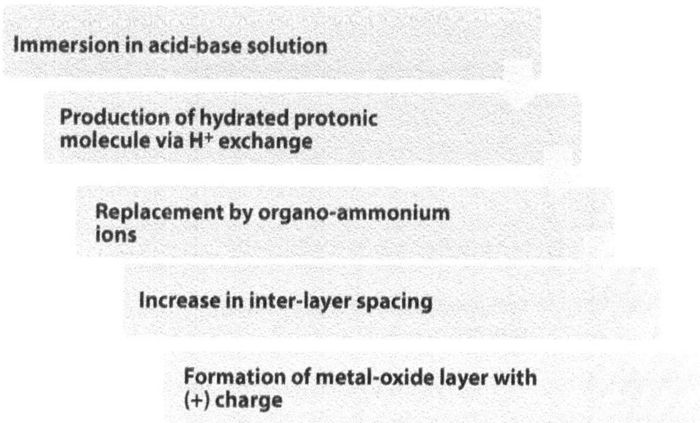

Figure 16.7 Ion exchange.

metal phosphorus-trichalcogenides and metal oxides. During the immersion step of layered metal oxides into an aqueous solution of acid and base, hydrated protonic molecules are produced *via* H^+ (cation) exchange. The inter-layer proton in the further steps can be replaced by organo-ammonium ions, thereby leading to increased inter-layer spacing in the layers of bulk metal-oxide crystals owing to the organo-ammonium ion's large radius. This increase leads to the extraction of metal-oxide layers with a positive (+) charge over the surface [53]. The ion-exchange method can be successfully utilized for large-scale operations to achieve high yields.

16.7.1.4 Selective Etching

The selective etching process is mainly applied to exfoliate MXene nanosheets from MAX (bulk phase) (Figure 16.8). The standard etching process involves the dispersion of MAX phase (crystal powders) in 50% conc. HF solution for the etching of layers. The etched result is then cleaned with water and sonicated to attain 2DM sheets. To obtain better results, the parameters, like time, reaction temperature, and HF concentration, can be modulated accordingly. The etching technique is convenient and offers more MXene generation with higher yields [54]. The technique offers the availability of functional groups (F, H, OH, O, etc.) over the surface, making the developed nanosheets suitable for a few applications.

16.7.2 Bottom-Up Approach

The development of 2DM from bottom-up approaches is mainly *via* direct chemical production methods, such as the Chemical Vapor Deposition (CVD) method and the wet-chemical method [45].

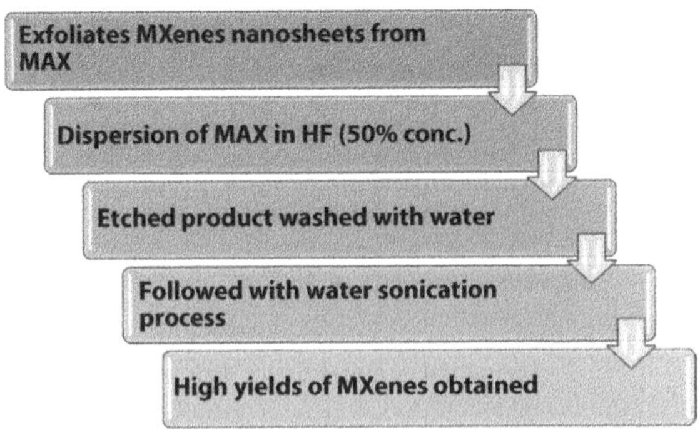

Figure 16.8 Selective etching.

16.7.2.1 *Chemical Vapor Deposition*

CVD technique is the most common method to develop thin films of 2DM over solid substrates (Figure 16.9). This technique involves a precursor, vapor, or gas that can decompose or react on the pre-selected substrate in a vacuum chamber at a high temperature. The nanosheets can grow over the substrate surface with/without the use of a catalyst. The CVD technique was primarily developed to extensively fabricate several 2DM nanosheets like silicone, h-BN, graphene metal carbides, and borophenes. With the optimization of CVD parameters like catalyst, substrate type, temperature, and precursor, the graphene layer growth can be modulated. CVD offers a high production rate for 2DM with excellent purity and crystal quality and low defects over the substrate [55, 56]. But CVD is a cost-demanding process with nanosheet transfer with every deposition, making it a laborious task.

16.7.2.2 *Wet-Chemical (WC) Synthesis*

The WC pathway involves chemical reactions in the aqueous phase by utilizing precursors under optimized experimental conditions. This synthesis pathway gives a high degree of reproducibility and controllability for the fabrication of nanosheets. Template synthesis, hot injection, interface-mediated synthesis, template synthesis, self-assembly, and solvothermal synthesis are the primary WC synthesis routes for 2DM. The solvothermal/hydrothermal (Figure 16.10) method is a scalable and convenient approach that is widely utilized to prepare 2DM. In a standard solvothermal method, an organic solvent or water acts as a medium of reaction in the closed

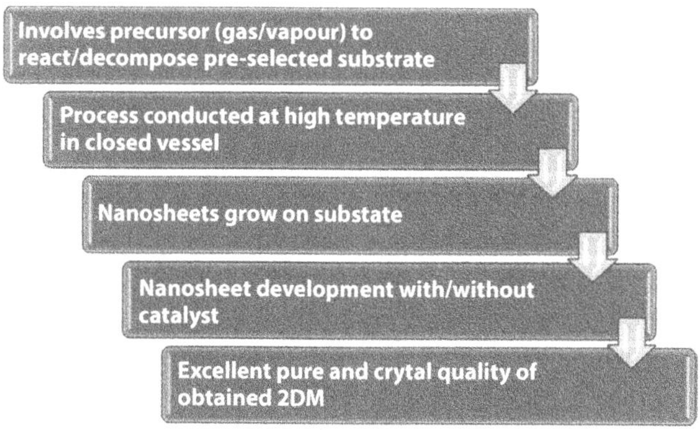

Figure 16.9 Chemical vapor deposition.

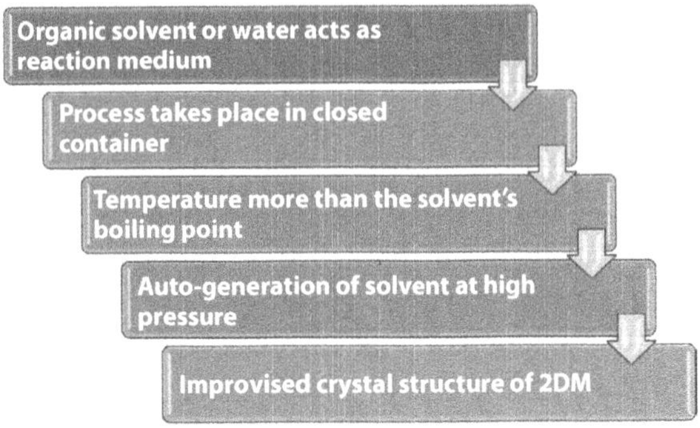

Figure 16.10 Wet-chemical synthesis (solvothermal method).

container. The process has a higher reaction temperature, i.e., more than the solvent boiling point, which results in the auto-generation of solvent at high pressure conditions, improving the crystal structure of the synthesized nanomaterial. WC offers cost-effective and high-yield production methods [57, 58]. But the development of single-layer nanosheets is rather difficult to acquire through WC synthesis, while few-layered nanosheets can be conveniently obtained by WC synthesis.

16.8 Adsorption of Contaminants From Water

The potential applications and working of 2DM composites to adsorb various pollutants have been discussed in Table 16.2.

16.8.1 Removal of Ions

Several noxious elements, including anions/oxyanions, have a negative charge over the surface, so the electrostatic interactions of the 2DM adsorbents must be redesigned accordingly. For a long time, arsenic has been considered a poisonous element with lethal traits. In addition, F^-, $C_2O_4^{2-}$, and NO_2^-/NO_3^- also exhibit detrimental impacts at massive concentrations. The presence of anions also changes the taste of water. Thus, to remove such anionic elements from water, anion intercalation, ion exchange, and other similar procedures have been reported to develop and design 2DM-based adsorbents. Carbon-based 2D composites have received massive attention due to their versatile characteristics. Venkatswarlu *et al.* presented a graphene-oxide 2D carbon flake by utilizing natural resources (discarded onion sheathing). The 2D composite was augmented with crystalline

Table 16.2 Potential applications and working of 2DM composites.

Sr. no.	2DM	Substrate	Synthesis	Removal process	Adsorption capacity	Adsorbed compound	Ref.
1.	GO	Fe_2O_3	Facile synthesis from discarded onion peel	Adsorption(pseudo-second-order reaction)	57.4 mg/g	As^{3-}	[59]
2.	LDH	MoS_4^{2}	Anion-exchange reaction	Adsorption (pseudo-second-order rate)	130 mg/g, 56 mg/g, and 99 mg/g, respectively	As (III), Cr (VI), As (V)	[60]
3.	MXene	PDDA	etching	Adsorption	363 mg/g	Re (VII)	[61]
4.	GO	MgAl-LDH	coagulation	Adsorption (Langmuir and Freundlich)	2.435 mmol/g	Sr^{2-} ion and SeO_4^{2-} ions	[62]
5.	K10 montmorillonite	Fe^{3+}	Ion-exchange	Adsorption (Langmuir,freundlich model)	0.0919–1.211 mg/L	phosphate	[63]
6.	MXene	Ti_3C_2	intercalation	Adsorption (Langmuir)	9.3 mg/g	Barium	[64]
7.	MXene	Ti_3C_2	etching	Adsorption (Langmuir/ pseudo-second-order)	104 mg/g	chromium	[65]
8.	Polyaniline nanofibrils	Rice-husk nanoparticles	Facile synthesis	Adsorption (Langmuir)	95% and 92%, respectively	Zinc, chromium	[66]

(Continued)

Table 16.2 Potential applications and working of 2DM composites. (*Continued*)

Sr. no.	2DM	Substrate	Synthesis	Removal process	Adsorption capacity	Adsorbed compound	Ref.
9.	MXene	Ti3C2Tx	ultra sonication	Adsorption (pseudo-second-order/ Freundlich)	78.45 mg/g	copper	[67]
10.	MXene/LDH	Titanium carbide	intercalation	Adsorption (Redlich-Peterson isotherms/ pseudo- second-order kinetics)	97.35%	nickel	[68]
11.	RUB-15	silicate	Ion-exchange	Adsorption (Langmuir)	338 mg/g and 190 mg/g, respectively.	Lead. cadmium	[69]
12.	COF	1,3,5-tris(4-aminophenyl) benzene (TAPB), and 2,5-bis(methylthio) terephthalaldehyde (DMTTPA,	Solvothermal	Adsorption	197 mg/g	mercury	[70]
13.	graphene	Melamine sponge and mineral wool	centrifugation-assisted dip coating	Photocatalysis	94.6%	Oil spill	[71]

(*Continued*)

Table 16.2 Potential applications and working of 2DM composites. (*Continued*)

Sr. no.	2DM	Substrate	Synthesis	Removal process	Adsorption capacity	Adsorbed compound	Ref.
14.	Carbon nano hybrid	Iron nanoparticles	ultrasonic spray pyrolysis	Photocatalysis	25 mg/g	Cr(VI)	[72]
15.	Graphene-oxide	TiO_2	Calcination treatment	Photocatalysis	10 mg/g	dye	[73]
16.	g-C_3N_4	Nb_2O_5	Facile (heating)	Photocatalysis	–	Antibiotic contaminant	[74]
17.	MXene	Polyimide	Freeze dry	adsorption	5.8 g g^{-1}	pump oil	[75]
18.	WS_2	1T	Hydrothermal	adsorption	210.2 mg/g	Methylene blue	[76]
19.	g-C_3N_4	silver	*in-situ* implanting approach	Photocatalysis	31%	microbes	[77]
20.	graphene	NiO	*In-situ* growth	adsorption	198 mg/g	chromium	[78]

nanoparticles of Fe_2O_3. The Fe_2O_3/carbon unit effectively adsorbed 57.4 mg/g of As^{3-} ions. The synergetic interaction of graphene/carbon with Fe_2O_3 had a positive impact on As^{3-} adsorption, and a state of equilibrium was attained in almost 60 min by following the pseudo-second-order reaction model. Excellent reusability with recovery in a short time span of 28 s was also reported, thereby indicating practical applications [59]. On the same line, Ma *et al.* eliminated As (III), Cr (VI), and As (V) from water solutions *via* MoS_4^{2-}/Layered-double-hydroxide (LDH) nanocomposite. In the absence of NO_3^-, Cl^-, and SO_4^{2-} non-toxic ions, the toxic CrO_4^{2-}, $HAsO_4^{2-}$, and $HAsO_3^{2-}$ anions were selectively eliminated. The estimated adsorption ability for oxyanions Cr (VI), As (V), and As (III) were about 130 mg/g, 56 mg/g, and 99 mg/g. The Langmuir model suggested single-layer adsorption over the sorbent. Oxyanions followed a pseudo-second-order rate model and showed the ability to reach equilibrium in less than 5 min. The interaction of oxoanions with sulfur and chromium reduction processes effectively increased the sorption process [60]. With the use of MXene, Wang *et al.* presented a nanocomposite made from polyelectrolyte and MXene with the augmentation of poly (diallyldimethylammonium chloride)/(PDDA) with improved stability and regulated surface charge. The composite effectively removed Re (VII) (up to 363 mg/g) from the contaminated water. Also, selective adsorption of ReO^{4-} was observed even in the presence of other competing anions (SO_4^{2-}, Cl^-). The adsorption process was pH-dependent, as when an increase in pH was observed, the native environment relatively changed from Re (IV) (octahedra-coordinated) to Re (VII) (tetrahedral-coordinated). Thus, the MXene-based composite significantly contains Re ions and, in a few cases, technetium ions [61]. Co-sorption of cationic and anionic radioactive nuclides is favorable for eliminating radioactive pollutants from water. Koilraj *et al.* fabricated a GO/MgAl-LDH nanocomposite to absorb Sr^{2-} ions and SeO_4^{2-} ions from the water. Both ions synergistically facilitated the adsorption process of the contaminants, with a removal efficiency of about 2.435 mmol/g of graphene-oxide. The elimination of SeO_4^{2-} occurred *via* an inter-change reaction between LDH and NO^{3-}, while Sr^{2+} was eliminated via carboxyl/alkoxy groups in GO through the ring-opening of epoxides. The composite adsorbed the contaminants using Langmuir and Freundlich isotherms [62]. Leite *et al.* created a Fe^{3+}-modified K10 montmorillonite to adsorb the phosphate. The unit adsorbed 0.0919–1.211 mg/L of phosphate and followed the Langmuir-Freundlich model to adsorb the contaminants. The capacity of adsorption increased with the increase in ionic strength, thereby depicting inner-sphere complexes sustaining electrostatic attraction as an adsorption phenomenon [63].

16.8.2 Removal of Heavy Metals

Heavy metal ions are toxic contaminants specifically found in industrial and poorly maintained water resources. Lead, chromium, cadmium, and mercury are the well-known heavy metals present in water. The consumption of such heavy metals can cause nausea, stomachache, vomiting, and carcinoma. Optimization of the surface with enhancement in bond strength can bring about efficient removal of heavy metals [1]. To eliminate barium ions, Fard *et al.* presented a composite involving nanosheets of Titanium (III) Carbide (II) (Ti3C2) incorporated with MXene. A maximum adsorption ability of about 9.3 mg/g for a 55 ppm concentration of barium, which was comparatively more than most 2DM competitors [64]. Similarly, Karthikeyan *et al.* used MXene/Ti3C2Tx nanocomposite to eliminate chromium ions from water. The adsorption process followed Langmuir isotherms and was kinetically driven by pseudo-second-order rates. The uptake of chromium ions was based on surface interactions, chemical adsorption, complexation, and electrostatic adsorption. A maximum of 104 mg/g of chromium was adsorbed from the water [65]. Diab *et al.* reported polyaniline nanofibrils augmented with rice-husk silica nanoparticles to eliminate zinc and chromium ions from water. The unit exhibited excellent adsorption capacities of about 95% for zinc and 92% for chromium. The combined composite had better adsorbing ability in comparison to only polyaniline. The adsorption process best fitted the Langmuir model between the Langmuir model and the Freundlich model [66]. Shahzad *et al.* developed MXene/Ti3C2Tx nanosheets to effectively remove copper from water solutions. The nanosheets depicted positive aspects such as explicit surface functional characteristics, hydrophilicity, and a more specific surface area. A total of 78.45 mg/g of copper was eliminated from the water solution [67]. Thus, 2DM materials can be efficaciously removed from the water.

The 2DM-based nanomaterials can efficiently remove several contaminants from the water, thereby purifying it in an effective way. Some of the removed elements are discussed below.

16.9 Photocatalytic Purification of Water

To resolve environment-related problems with minimal input of energy, photocatalysis by utilizing solar energy to purify water solutions has been widely studied and regarded as a sustainable approach. In general, photocatalysis includes single or various photoreaction phenomena catalyzed *via* a photocatalyst, where free electrons undergo excitation to reach the conduction band and holes (n) are left behind in the valence band. The electron/hole pair plays a crucial role in photocatalytic operations, including pollutant degradation, carbon dioxide (CO2) reduction, and water splitting. The charge carriers produced *via* excited photo-generation lack stability and can easily re-combine, which decreases the efficiency of conversion for photocatalysis. Owing to the distinctive structure of 2DM and their nanosheets, 2DM-dependent photocatalyst units are promising candidates that can provide signature properties like ample surface-active regions, more specific surface areas, and a variety of host-guest species [1, 79]. Recently, various 2DM-dependent photocatalytic units like carbon nitride, chalcogenides, metal oxide, and graphene nanocomposites have been studied for several applications. The size/thickness of nanosheets, edge morphologies, and chemical structures of 2DM are crucial factors when using 2DM in photocatalytic methods. According to the kind of contaminant, the function of a 2DM photocatalyst can be primarily divided into three prime categories: organic decontamination, inorganic pollutant elimination, and biological disinfection. Some inorganic contaminants can be transformed into a less noxious or, preferably, non-toxic phase *via* oxidative/reductive photocatalysis. These phenomena may include the oxidative conversion of H_2S into S, NH_3 into N_2, and the reduction of Cr^{6+} into Cr^{3+}. The In_2S_2/CuS photocatalytic nanosheets were prepared by Praks *et al.*, which photocatalytically transformed H_2S (toxic) into $S_2O_3^{2-}$ (nontoxic) with the generation of H_2. Also, the single-layer of g-C_3N_4 depicted good photocatalytic operations for the elimination of total ammonia and nitrogen from the contaminated water [80]. The oxidation process of ammonia can be thermodynamically operated on the g-C_3N4 surface, as the corresponding redox levels for ammonia are completely covered through the g-C_3N_4 bandgap range. The most important point to be noted was that the nanosheets of 2DM showed better photocatalytic operations in comparison to bulk materials. Elimination of organic contaminants requires the generation of abundant reactive oxygen species (ROS), which can mineralize or destroy organic pollutants completely. To create an efficient photocatalytic unit, the band design of the 2DM structure needs to be carefully designed, as they play an important role in the redox potential for the photogenerated pair of electron-holes [81]. Compared to bulk, single- and few-layer 2DM nanocomposites have large extended bandgaps that considerably shift the edge position of the bandgap to produce a redox potential, which suits ROS. The 2D/2D nanocomposite units are heterostructures with explicit coupling of heterointerface, favoring the differentiation and transfer of photo-generated charge carriers, thus bringing about higher photocatalytic function. Furthermore, photocatalysts are also used for the bio-disinfection process. Bacteria and viruses are pathogens responsible for causing diseases in humans and other living species. By employing strategies like nanocomposting and bandgap engineering various 2DM nanocomposites to eliminate such biological pollutants from water. During a standard photocatalytic phenomenon, the photoexcited charge carriers aid in producing ROS, which can possibly deactivate unwanted bacterial species [1, 82, 83]. Generally, the 2D nanocomposites absorb the light range accordingly, rupture the cell, and consequently deactivate the undesirable microbes. As a result, the photocatalytic units have shown significant results in the elimination of contaminants from the water *via* ROS generation and other dependent reactions.

16.10 Conclusion and Future Prospects

The enhancements in water treatment technologies have leveraged the development of sophisticated nanocomposites for clean water availability. Nanotechnology offers a controllable, sustainable, scalable, and affordable mode of water purification. However, large-scale application of 2DM is still a matter of concern as the synthesis process of 2DM is laborious and cost dependent. Also, the fabrication process should offer production stability for large-scale employment. Focus should be given to the development of biocompatible 2DM to enhance the purification process. In the era of next-generation intelligent materials to purify water, it remains highly essential to form a basic structure-processing-performance relationship by using innovative empirical approaches (like machine learning). The artificial intelligence-based techniques aid in designing nanocomposites with excellent properties and purification behaviors to effectively treat water resources. An emphasis can be given on developing these composites with the integration of AI-based patterns with nanotechnology to develop extraordinary 2DM structures for water purification processes. Moreover, stress could be placed on fabricating multifunctional 2DM for simultaneous removal of contaminants. The multifunctionality of the 2DM can promote the simultaneous removal of various contaminants from the water, thereby promoting the purification process.

References

1. Zeng, M. *et al.*, Engineered two-dimensional nanomaterials: An emerging paradigm for water purification and monitoring. *Mater. Horiz.*, 8, 3, 758–802, 2021, doi: 10.1039/d0mh01358g.

2. Chimene, D., Alge, D.L., Gaharwar, A.K., Two-dimensional nanomaterials for biomedical applications: Emerging trends and future prospects. *Adv. Mater.*, 27, 45, 7261–7284, 2015, doi: 10.1002/adma.201502422.
3. Jin, H. *et al.*, Emerging two-dimensional nanomaterials for electrocatalysis. *Chem. Rev.*, 118, 13, 6337–6408, Jul. 2018. doi: 10.1021/acs.chemrev.7b00689.
4. Koski, K.J. and Cui, Y., The new skinny in two-dimensional nanomaterials. *ACS Nano*, 7, 5, 3739–3743, 2013, doi: 10.1021/nn4022422.
5. Pal, K. *et al.*, Cutting edge development on graphene derivatives modified by liquid crystal and CdS/TiO2 hybrid matrix: Optoelectronics and biotechnological aspects. *Crit. Rev. Solid State Mater. Sci.*, 46, 5, 385–449, 2021, doi: 10.1080/10408436.2020.1805295.
6. Khan, S.T. and Malik, A., Engineered nanomaterials for water decontamination and purification: From lab to products. *J. Hazard. Mater.*, 363, 295–308, 2019, doi: 10.1016/j.jhazmat.2018.09.091.
7. Bartram, J. and Cairncross, S., Hygiene, sanitation, and water: Forgotten foundations of health. *PloS Med.*, 7, 11, e1000367, Nov. 2010, doi: 10.1371/journal.pmed.1000367.
8. Allgeier, S.C. and Magnuson, M.L., Responding to threats and incidents of intentional drinking water contamination. *J. Contemp. Water Res. Educ.*, 129, 1, 13–17, 2009, doi: 10.1111/j.1936-704x.2004.mp129001004.x.
9. Parvez, K., Two-dimensional nanomaterials: Crystal structure and synthesis, in: *Biomedical Applications of Graphene and 2D Nanomaterials*, Elsevier Inc, London, pp. 1–5, 2019.
10. Mas-Ballesté, R., Gómez-Navarro, C., Gómez-Herrero, J., Zamora, F., 2D materials: To graphene and beyond. *Nanoscale*, 3, 1, 20–30, 2011, doi: 10.1039/c0nr00323a.
11. Xie, X., Chen, C., Zhang, N., Tang, Z.-R., Jiang, J., Xu, Y.-J., Microstructure and surface control of MXene films for water purification. *Nat. Sustain.*, 2, 9, 856–862, 2019, doi: 10.1038/s41893-019-0373-4.
12. Li, X. *et al.*, Graphene oxide-based efficient and scalable solar desalination under one sun with a confined 2D water path. *Proc. Natl. Acad. Sci.*, 113, 49, 13953–13958, 2016, doi: 10.1073/pnas.1613031113.
13. Lee, W., Ji, R., Gösele, U., Nielsch, K., Fast fabrication of long-range ordered porous alumina membranes by hard anodization. *Nat. Mater.*, 5, 9, 741–747, 2006, doi: 10.1038/nmat1717.
14. Li, W., Tekell, M.C., Huang, Y., Bertelsmann, K., Lau, M., Fan, D., Synergistic high-rate solar Steaming and mercury removal with mos2/c @ polyurethane composite sponges. *Adv. Energy Mater.*, 8, 32, 1–11, 2018, doi: 10.1002/aenm.201802108.
15. Guo, Z. *et al.*, Intensifying sustainable solar water production by steam heat internal circulation. *Mater. Adv.*, 2, 5, 1731–1738, 2021, doi: 10.1039/d0ma00911c.
16. Zodrow, K. *et al.*, Polysulfone ultrafiltration membranes impregnated with silver nanoparticles show improved biofouling resistance and virus removal. *Water Res.*, 43, 3, 715–723, Feb. 2009, doi: 10.1016/j.watres.2008.11.014.

17. Garaj, S., Liu, S., Golovchenko, J.A., Branton, D., Molecule-hugging graphene nanopores. *Proc. Natl. Acad. Sci.*, 110, 30, 12192–12196, 2013, doi: 10.1073/pnas.1220012110.
18. Hu, M. and Mi, B., Enabling graphene oxide nanosheets as water separation membranes. *Environ. Sci. Technol.*, 47, 8, 3715–3723, Apr. 2013, doi: 10.1021/es400571g.
19. Cohen-Tanugi, D., Lin, L.-C., Grossman, J.C., Multilayer nanoporous graphene membranes for water desalination. *Nano Lett.*, 16, 2, 1027–1033, Feb. 2016, doi: 10.1021/acs.nanolett.5b04089.
20. Russo, C.J. and Golovchenko, J.A., Atom-by-atom nucleation and growth of graphene nanopores. *Proc. Natl. Acad. Sci. U. S. A.*, 109, 16, 5953–5957, Apr. 2012, doi: 10.1073/pnas.1119827109.
21. Zhan, X., Si, C., Zhou, J., Sun, Z., MXene and MXene-based composites: Synthesis, properties and environment-related applications. *Nanoscale Horiz.*, 5, 2, 235–258, 2020, doi: 10.1039/c9nh00571d.
22. Jun, B.M. *et al.*, Applications of metal-organic framework based membranes in water purification: A review. *Sep. Purif. Technol.*, 247, 1–49, 2020, doi: 10.1016/j.seppur.2020.116947.
23. Kuang, P., Low, J., Cheng, B., Yu, J., Fan, J., MXene-based photocatalysts. *J. Mater. Sci. Technol.*, 56, 18–44, 2020, doi: 10.1016/j.jmst.2020.02.037.
24. Ihsanullah, I., MXenes (two-dimensional metal carbides) as emerging nanomaterials for water purification: Progress, challenges and prospects. *Chem. Eng. J.*, 388, 124–340, 2020.
25. Rasool, K., Helal, M., Ali, A., Ren, C.E., Gogotsi, Y., Mahmoud, K.A., Antibacterial activity of Ti3C2Tx MXene. *ACS Nano*, 10, 3, 3674–3684, 2016, doi: 10.1021/acsnano.6b00181.
26. Zhang, M., Yang, Y., An, X., Hou, L., A critical review of g-C3N4-based photocatalytic membrane for water purification. *Chem. Eng. J. Elsevier B.V.*, 412, 128663, 2021, doi: 10.1016/j.cej.2021.128663.
27. Truong, H.B., Bae, S., Cho, J., Hur, J., Advances in application of g–C3N4–based materials for treatment of polluted water and wastewater via activation of oxidants and photoelectrocatalysis: A comprehensive review. *Chemosphere*, 286, P2, 131737, 2022, doi: 10.1016/j.chemosphere.2021.131737.
28. Ren, Q., Liu, J., Yang, Q., Shen, W., A review: Photocatalysts based on biocl and g-c3n4 for water purification. *Catalysts*, 11, 9, 1–29, 2021, doi: 10.3390/catal11091084.
29. Zhang, Y., Yuan, J., Ding, Y., Liu, B., Zhao, L., Zhang, S., Research progress on g–C3N4–based photocatalysts for organic pollutants degradation in wastewater: From exciton and carrier perspectives. *Ceram. Int.*, 47, 22, 31005–31030, 2021, doi: 10.1016/j.ceramint.2021.08.063.
30. Arumugham, T., Amimodu, R.G., Kaleekkal, N.J., Rana, D., Nano CuO/g-C 3 N 4 sheets-based ultrafiltration membrane with enhanced interfacial affinity, antifouling and protein separation performances for water treatment application. *J. Environ. Sci. (China)*, 82, 57–69, 2019, doi: 10.1016/j.jes.2019.03.001.

31. Li, B. *et al.*, Black phosphorus, a rising star 2d nanomaterial in the post-graphene era: Synthesis, properties, modifications, and photocatalysis applications. *Small*, 15, 8, 1–30, 2019, doi: 10.1002/smll.201804565.
32. Gui, R., Jin, H., Sun, Y., Jiang, X., Sun, Z., Two-dimensional group-VA nanomaterials beyond black phosphorus: Synthetic methods, properties, functional nanostructures and applications. *J. Mater. Chem. A*, 7, 45, 25712–25771, 2019, doi: 10.1039/c9ta09582a.
33. Li, D. *et al.*, Filtration-based water treatment system embedded with black phosphorus for NIR-triggered disinfection. *Environ. Sci. Nano*, 6, 10, 2977–2985, 2019, doi: 10.1039/c9en00774a.
34. Tang, W., Jiang, Z., Wang, B., Li, Y., Black phosphorus quantum dots: A new-type of water-based high-efficiency lubricant additive. *Friction*, 9, 6, 1528–1542, 2021, doi: 10.1007/s40544-020-0434-8.
35. Liang, M. *et al.*, Silver-laden black phosphorus nanosheets for an efficient *in vivo* antimicrobial application. *Small*, 16, 13, 1–11, 2020, doi: 10.1002/smll.201905938.
36. Yu, W., Yu, T., Graham, N., Development of a stable cation modified graphene oxide membrane for water treatment. *2D Mater.*, 4, 4, 045006, 2017.
37. Novoselov, K.S. *et al.*, Two-dimensional atomic crystals. *Proc. Natl. Acad. Sci. U. S. A.*, 102, 30, 10451–10453, 2005, doi: 10.1073/pnas.0502848102.
38. Yang, K., Wang, J., Chen, X., Zhao, Q., Ghaffar, A., Chen, B., Application of graphene-based materials in water purification: From the nanoscale to specific devices. *Environ. Sci. Nano*, 5, 6, 1264–1297, 2018, doi: 10.1039/c8en00194d.
39. Singh, P., Shandilya, P., Raizada, P., Sudhaik, A., Rahmani-Sani, A., Hosseini-Bandegharaei, A., Review on various strategies for enhancing photocatalytic activity of graphene based nanocomposites for water purification. *Arab. J. Chem.*, 13, 1, 3498–3520, 2020, doi: 10.1016/j.arabjc.2018.12.001.
40. Angizi, S., Alem, S. A. A., Pakdel, A., Towards integration of two-dimensional hexagonal boron nitride (2D h-BN) in energy conversion and storage devices. *Energies*, 15, 3, 1162, 2022, https://doi.org/10.3390/en/15031162.
41. Angizi, S. *et al.*, Review—towards the two-dimensional hexagonal boron nitride (2D h-BN) electrochemical sensing platforms. *J. Electrochem. Soc.*, 167, 12, 126513, 2020, doi: 10.1149/1945-7111/abaf29.
42. Song, L., Jia, H., Zhang, H., Cao, J., Graphene-like h-BN/CdS 2D/3D heterostructure composite as an efficient photocatalyst for rapid removing rhodamine B and Cr(&xxx2165;) in water. *Ceram. Int.*, 46, 15, 24674–24681, 2020, doi: 10.1016/j.ceramint.2020.06.257.
43. Kostoglou, N., Polychronopoulou, K., Rebholz, C., Thermal and chemical stability of hexagonal boron nitride (h-BN) nanoplatelets. *Vacuum*, 112, 42–45, 2015, doi: 10.1016/j.vacuum.2014.11.009.
44. Ivanova, M.N. *et al.*, Hexamolybdenum clusters supported on exfoliated h-bn nanosheets for photocatalytic water purification. *Inorg. Chem.*, 59, 9, 6439–6448, 2020, doi: 10.1021/acs.inorgchem.0c00528.

45. Pottathara, Y.B., Grohens, Y., Kokol, V., Kalarikkal, N., Thomas, S., Synthesis and processing of emerging two-dimensional nanomaterials, in: *Nanomaterials Synthesis: Design, Fabrication and Applications*, Elsevier Inc., London, pp. 1–25, 2019.
46. Li, D., Müller, M.B., Gilje, S., Kaner, R.B., Wallace, G.G., Processable aqueous dispersions of graphene nanosheets. *Nat. Nanotechnol.*, 3, 2, 101–105, 2008, doi: 10.1038/nnano.2007.451.
47. Hummers, W.S. and Offeman, R.E., Preparation of graphitic oxide. *J. Am. Chem. Soc.*, 80, 6, 1339, Mar. 1958, doi: 10.1021/ja01539a017.
48. Zheng, J. *et al.*, High yield exfoliation of two-dimensional chalcogenides using sodium naphthalenide. *Nat. Commun.*, 5, 1, 2995, 2014, doi: 10.1038/ncomms3995.
49. Hamilton, C.E., Lomeda, J.R., Sun, Z., Tour, J.M., Barron, A.R., High-yield organic dispersions of unfunctionalized graphene. *Nano Lett.*, 9, 10, 3460–3462, Oct. 2009, doi: 10.1021/nl9016623.
50. Li, H. *et al.*, Fabrication of single- and multilayer MoS2 film-based field-effect transistors for sensing NO at room temperature. *Small*, 8, 1, 63–67, Jan. 2012, doi: 10.1002/smll.201101016.
51. Ciesielski, A. and Samorì, P., Graphene via sonication assisted liquid-phase exfoliation. *Chem. Soc. Rev.*, 43, 1, 381–398, 2014, doi: 10.1039/C3CS60217F.
52. Yi, M. and Shen, Z., A review on mechanical exfoliation for the scalable production of graphene. *J. Mater. Chem. A*, 3, 22, 11700–11715, 2015, doi: 10.1039/C5TA00252D.
53. Liang, J., Ma, R., Iyi, N., Ebina, Y., Takada, K., Sasaki, T., Topochemical synthesis, Anion exchange, and exfoliation of co–ni layered double hydroxides: A route to positively charged Co–Ni hydroxide nanosheets with tunable composition. *Chem. Mater.*, 22, 2, 371–378, Jan. 2010, doi: 10.1021/cm902787u.
54. Anasori, B. *et al.*, Two-dimensional, ordered, double transition metals carbides (MXenes). *ACS Nano*, 9, 10, 9507 9516, Oct. 2015, doi: 10.1021/acsnano.5b03591.
55. Yu, J., Li, J., Zhang, W., Chang, H., Synthesis of high quality two-dimensional materials via chemical vapor deposition. *Chem. Sci.*, 6, 12, 6705–6716, 2015, doi: 10.1039/C5SC01941A.
56. Zhan, Y., Liu, Z., Najmaei, S., Ajayan, P.M., Lou, J., Large-area vapor-phase growth and characterization of MoS 2 atomic layers on a SiO 2 substrate. *Small*, 8, 7, 966–971, 2012, doi: 10.1002/smll.201102654.
57. Sun, Y. *et al.*, Pits confined in ultrathin cerium(IV) oxide for studying catalytic centers in carbon monoxide oxidation. *Nat. Commun.*, 4, 1, 2899, 2013, doi: 10.1038/ncomms3899.
58. Fan, Z. and Zhang, H., template synthesis of noble metal nanocrystals with unusual crystal structures and their catalytic applications. *Acc. Chem. Res.*, 49, 12, 2841–2850, Dec. 2016, doi: 10.1021/acs.accounts.6b00527.

59. Venkateswarlu, S., Lee, D., Yoon, M., Bioinspired 2D-carbon flakes and fe3o4 nanoparticles composite for arsenite removal. *ACS Appl. Mater. Interfaces*, 8, 36, 23876–23885, 2016, doi: 10.1021/acsami.6b03583.
60. Ma, L. *et al.*, Selective and efficient removal of toxic oxoanions of As (III), As (V), and Cr (VI) by layered double hydroxide intercalated with MoS42-. *Chemistry of Materials ACS Chem. Mater.*, 29, 7, 3274–3284, 2017.
61. Wang, L. *et al.*, Effective removal of anionic re(vii) by surface-modified ti2ctx mxene nanocomposites: Implications for Tc(VII) sequestration. *Environ. Sci. Technol.*, 53, 7, 3739–3747, 2019, doi: 10.1021/acs.est.8b07083.
62. Koilraj, P., Kamura, Y., Sasaki, K., Cosorption characteristics of SeO42- and Sr2+ radioactive surrogates using 2D/2D graphene oxide-layered double hydroxide nanocomposites. *ACS Sustain. Chem. Eng.*, 6, 11, 13854–13866, 2018, doi: 10.1021/acssuschemeng.8b02056.
63. Leite, S.T., do Nascimento, F.H., Masini, J.C., Fe(III)-polyhydroxy cations supported onto K10 montmorillonite for removal of phosphate from waters. *Heliyon*, 6, 4, e03868, 2020, doi: 10.1016/j.heliyon.2020.e03868.
64. Fard, A.K., Mckay, G., Chamoun, R., Rhadfi, T., Preud'Homme, H., Atieh, M.A., Barium removal from synthetic natural and produced water using MXene as two dimensional (2-D) nanosheet adsorbent. *Chem. Eng. J.*, 317, 331–342, 2017, doi: 10.1016/j.cej.2017.02.090.
65. Karthikeyan, P., Ramkumar, K., Pandi, K., Fayyaz, A., Meenakshi, S., Park, C.M., Effective removal of Cr(VI) and methyl orange from the aqueous environment using two-dimensional (2D) Ti3C2Tx MXene nanosheets. *Ceram. Int.*, 47, 3, 3692–3698, 2021, doi: 10.1016/j.ceramint.2020.09.221.
66. Diab, M.A., Attia, N.F., Attia, A.S., El-Shahat, M.F., Green synthesis of cost-effective and efficient nanoadsorbents based on zero and two dimensional nanomaterials for Zn2+ and Cr3+ removal from aqueous solutions. *Synth. Met.*, 265, 116411, May 2020, doi: 10.1016/j.synthmet.2020.116411.
67. Shahzad, A. *et al.*, Two-dimensional Ti3C2Tx MXene nanosheets for efficient copper removal from water. *ACS Sustain. Chem. Eng.*, 5, 12, 11481–11488, 2017, doi: 10.1021/acssuschemeng.7b02695.
68. Feng, X. *et al.*, Self-assembling 2D/2D (MXene/LDH) materials achieve ultrahigh adsorption of heavy metals Ni2+ through terminal group modification. *Sep. Purif. Technol.*, 253, 117525, 2020, doi: 10.1016/j.seppur.2020.117525.
69. Chen, Z., Liang, Y., Jia, D., Chen, W., Cui, Z., Wang, X., Layered silicate RUB-15 for efficient removal of UO22+ and heavy metal ions by ion-exchange. *Environ. Sci. Nano*, 4, 9, 1851–1858, 2017, doi: 10.1039/c7en00366h.
70. Huang, N., Zhai, L., Xu, H., Jiang, D., Stable covalent organic frameworks for exceptional mercury removal from aqueous solutions. *J. Am. Chem. Soc.*, 139, 6, 2428–2434, 2017, doi: 10.1021/jacs.6b12328.
71. Ge, J. *et al.*, Joule-heated graphene-wrapped sponge enables fast clean-up of viscous crude-oil spill. *Nat. Nanotechnol.*, 12, 5, 434–440, 2017, doi: 10.1038/nnano.2017.33.

72. Masud, A., Cui, Y., Atkinson, J.D., Aich, N., Shape matters: Cr(VI) removal using iron nanoparticle impregnated 1-D vs 2-D carbon nanohybrids prepared by ultrasonic spray pyrolysis/. *J. Nanopart. Res.*, 20, 3, 1–12, 2018.
73. Jiang, G. *et al.*, TiO2 nanoparticles assembled on graphene oxide nanosheets with high photocatalytic activity for removal of pollutants. *Carbon N. Y.*, 49, 8, 2693–2701, 2011, doi: 10.1016/j.carbon.2011.02.059.
74. Hong, Y. *et al.*, Efficient and stable Nb2O5 modified g-C3N4 photocatalyst for removal of antibiotic pollutant. *Chem. Eng. J.*, 299, 74–84, 2016, doi: 10.1016/j.cej.2016.04.092.
75. Zhou, H. *et al.*, Study on contact angles and surface energy of MXene films. *RSC Adv.*, 11, 10, 5512–5520, 2021, doi: 10.1039/d0ra09125a.
76. Jo, Y.K., Lee, J.M., Son, S., Hwang, S.J., 2D inorganic nanosheet-based hybrid photocatalysts: Design, applications, and perspectives. *J. Photochem. Photobiol. C Photochem. Rev.*, 40, 150–190, 2019, doi: 10.1016/j.jphotochemrev.2018.03.002.
77. Sun, D. *et al.*, *In-situ* homodispersely immobilization of Ag@AgCl on chloridized g-C3N4 nanosheets as an ultrastable plasmonic photocatalyst. *Chem. Eng. J.*, 384, 1–33, 2020, doi: 10.1016/j.cej.2019.123259.
78. Liu, N. *et al.*, Removal mechanisms of aqueous Cr(VI) using apple wood biochar: A spectroscopic study. *J. Hazard. Mater.*, 384, 121371, 2020.
79. Ghanbari, F. *et al.*, Evaluation of peroxymonosulfate/O3/UV process on a real polluted water with landfill leachate: Feasibility and comparative study. *Korean J. Chem. Eng.*, 38, 7, 1416–1424, 2021, doi: 10.1007/s11814-021-0782-8.
80. Zhang, Y., Tang, Z.-R., Fu, X., Xu, Y.-J., Engineering the unique 2D Mat of graphene to achieve graphene-TiO2 nanocomposite for photocatalytic selective transformation: What advantage does graphene have over its forebear carbon nanotube? *ACS Nano*, 5, 9, 7426–7435, Sep. 2011, doi: 10.1021/nn202519j.
81. Yang, M.Q., Zhang, N., Xu, Y.J., Synthesis of fullerene-, carbon nanotube-, and graphene-TiO2 nanocomposite photocatalysts for selective oxidation: A comparative study. *ACS Appl. Mater. Interfaces*, 5, 3, 1156–1164, 2013, doi: 10.1021/am3029798.
82. Bai, S., Wang, L., Chen, X., Du, J., Xiong, Y., Chemically exfoliated metallic MoS2 nanosheets: A promising supporting co-catalyst for enhancing the photocatalytic performance of TiO2 nanocrystals. *Nano Res.*, 8, 1, 175–183, 2015, doi: 10.1007/s12274-014-0606-9.
83. Zhang, X.-L., Wang, L., Chen, L., Ma, X.-Y., Xu, H.-X., Ultrathin 2D conjugated polymer nanosheets for solar fuel generation. *Chin. J. Polym. Sci.*, 37, 2, 101–114, 2019, doi: 10.1007/s10118-019-2171-x.

17

Insights into the Exciton Dynamics of Functionalized 2D Nanomaterials for Robust Photoelectrochemical Sensing Applications

Dipyaman Mohanta[1]*, Koushik Barman[2†], Abhinandan Mahanta[3], Bishal Bhuyan[4] and Arpita Paul Chowdhury[5]

[1]*Department of Chemistry, Cachar College Silchar, Silchar, Assam, India*
[2]*Department of Chemistry and Biochemistry, Queens College-CUNY, Flushing, New York, USA*
[3]*Department of Chemistry, Assam University Silchar, Silchar, Assam, India*
[4]*Department of Chemistry, Majuli College, Majuli, Assam, India*
[5]*Department of Chemistry, Dayananda Sagar College of Engineering, Bangalore, Karnataka, India*

Abstract
Photoelectrochemical sensing is a robust analytical tool for rapid, selective, and sensitive monitoring of chemical and biochemical substances. In contrast to conventional sensing techniques, the photoelectrochemical sensing method employs two separate energy sources: light for excitation and current for detection, thereby enhancing the catalytic activity and sensitivity. The superiority of photoelectrochemical sensing depends greatly on the design and engineering of photoactive materials. Visible light-harvesting ability is a prerequisites for an efficient photoelectrochemical sensor. Furthermore, the charge separation of photogenerated excitons (electron-hole pairs) also plays a crucial role in the efficacy of photoactive materials. Therefore, the design and development of engineered materials with enhanced light-harvesting capacity and improved exciton lifespan are of prime importance for robust photoelectrochemical sensing applications.

**Corresponding author*: mdipyaman@gmail.com
†*Corresponding author*: koushik.barman2@gmail.com

Subhendu Chakroborty and Kaushik Pal (eds.) 2D Nanomaterials: Synthesis, Properties and Applications, (397–414) © 2024 Scrivener Publishing LLC

Two-dimensional nanomaterials are promising candidates for photoelectrochemical sensors owing to their intriguing properties, such as high surface-to-volume ratio, good stability, high electron mobility, and exciting mechanical and interfacial properties. Recently, 2D materials such as graphene, reduced graphene oxide (rGO), graphitic carbon nitride (gC_3N_4), molybdenum disulfide (MoS_2), and boron nitride (BN) have been extensively utilized as photoactive materials for the photoelectrochemical sensing of target molecules with enhanced sensitivity and selectivity. Semiconducting 2D nanomaterials can generate excitons (electron-hole pairs) under visible light excitation owing to their compatible narrow-band structures. In addition, some 2D nanomaterials can enhance the life span of photogenerated excitons by lowering the charge recombination owing to the extensive π-architecture. Moreover, charge confinement in one dimension leads to extraordinary exciton dynamics in 2D nanostructures, which can be fine-tuned by controlled structural and morphological modifications. Surface modification of 2D nanomaterials with element doping, hybridization with noble metals, nanoheterojunction with semiconducting metal oxide/metal sulfide nanomaterials, surface functionalization with organic ligands, and metal-organic frameworks, have also been reported to have immense potential to synergistically alter exciton dynamics and enhance photoelectrochemical sensing performance.

Therefore, this chapter attempts to summarize and compare the recent progress in the use of 2D nanomaterials in the field of photoelectrochemical sensors and their applications, starting from the features of their synthesis and concluding remarks on the perspectives of photoelectrochemical sensing applications.

Keywords: 2D nanomaterials, photoelectrochemical sensing, exciton dynamics

17.1 Introduction

Photoelectrochemical (PEC) analysis is an emerging and highly innovative detection technique because of its strong distinctive qualities, such as cost-effectiveness, ease of operation, low background current, quick response, high selectivity, and sensitivity [1, 2]. PEC technology provides significantly higher detection sensitivity and a lower background signal compared to electrochemical techniques. This methodology is less expensive than conventional sensing methods, such as fluorescence, surface-enhanced Raman scattering techniques, and electrochemiluminescence (ECL), making real-time detection possible. With light serving as both the source of excitation and the detection signal in the photoelectrochemical approach, the background noise is greatly minimized, and the sensitivity is increased. The photoactive material is activated by the incident light to generate electron-hole pairs, which in turn are then transmitted to the electrode surface, resulting in a photocurrent [3, 4]. Notably, this method is simply ECL in reverse. Photoelectrochemical sensors and biosensors including metal ions, bacteria, and small organic molecules, have been developed for a variety of targets [5–8].

As a novel sensing platform, photoelectrodes have a significant impact on the detection abilities of photoelectrochemical assays. Because of the suitable bandgap of semiconductors for the transmission of photogenerated electrons, the majority of photoelectrodes are made of these materials [9]. Among them, two-dimensional nanomaterials are fascinating prospects for promising photoelectrochemical sensors owing to their unique characteristics such as high surface-to-volume ratio, strong stability, ease of preparation and exfoliation, good biocompatibility, high electron mobility, and intriguing mechanical and interfacial properties [10, 11]. Interestingly, several 2D stacked semiconductor nanomaterials have been created and used in a variety of applications, including graphene, graphitic carbon nitride, 2D boron nitride, doped graphene, graphene oxides, and transition metal dichalcogenides [12, 13].

Although most of the two-dimensional nanomaterials listed above have been used in PEC assays, there is still plenty of room for the use of 2D semiconducting nanostructured materials in photoelectrochemical sensing technology. Further use of various engineered nanostructures has great potential for improving the light-harvesting nature and generation of electron-hole pairs with increased charge separation, thereby enhancing the efficiency of photoelectrochemical sensing. The applicability of bulk 2D semiconductor nanomaterials is constrained by their drawbacks, which include slow electron migration, difficult dispersion in solution, and rapid recombination of photogenerated electron-hole pairs. To overcome this, 2D semiconductor nanomaterials typically require additional alterations, functionalization, or other treatment processes, such as element doping, hybridization with noble metal nanoparticles, or combination with other semiconductors to form nanoheterojunction structures [14–16]. The visible light activity, electron transfer rate, PEC response, and detection sensitivity can all be significantly enhanced by these modifications. Other sensing devices, including gas sensors [17], microfluidic sensors [18], electrochemical sensors [19], fluorescence sensors [20], and plasmonic sensors have also confirmed the extensive use of 2D nanomaterials [21, 22]. It is pertinent to mention that compared with other types of nanomaterials, the signal responsiveness of the sensor may be significantly improved by a large surface area and quick electron movement.

This chapter provides a brief account of recent advances in the design and development of engineered 2D nanomaterials for efficient photoelectrochemical monitoring of both chemical and biochemical target species. The immense potential of engineered 2D nanomaterials and the control over exciton dynamics by tailoring structural, morphological, and interfacial attributes have motivated us to summarize the structure–property relationship of functionalized 2D nanostructures for photoelectrochemical sensing applications.

17.2 Basic Theory and Working Principle of Photoelectrochemical Sensing

Photoelectrochemical (PEC) detection is an attractive technique because of several advantages, such as small background current, low cost, easy handling, quick response, and high sensitivity [23–29], compared with other detection procedures, (e.g., as optosensors, surface plasmon resonance (SPR) sensors, electrochemical sensors, and electrochemiluminescence sensors). As a new and important method of detection, photoelectrochemical sensing has great potential for the sensitive and selective detection of target molecules owing to appropriate combination of photochemistry and electrochemistry [30]. In photoelectrochemical sensing, the light is used to excite the photoactive material, and the generated electric signal is considered as the detection signal. Under illumination, the photoactive sensing material generates electron-hole pairs. The charges when moving to the electrode surface produce a photocurrent and thus, photo-energy is converted to electrical energy [31–34]. In the PEC method, oxidation or reduction reactions occur between the analyte in solution and the photoexcited species generated at the electrode/electrolyte interface during light illumination [35, 36], which has been broadly applied to photocatalysis, photovoltaics, and solar cell. A schematic representation of the working principle of typical photoelectrochemical cell has been represented in Figure 17.1.

Photoactive sensing materials play an important role in photoelectrochemical sensing. In the photo-to-electric transformation procedure, photoactive sensing materials are used to harvest light photons which leads to the generation of charge carriers that provide the required energy for redox reactions [37–40]. In photocatalysis, an electron-hole pair is directly utilized to drive a redox reaction. However, rapid recombination is a major

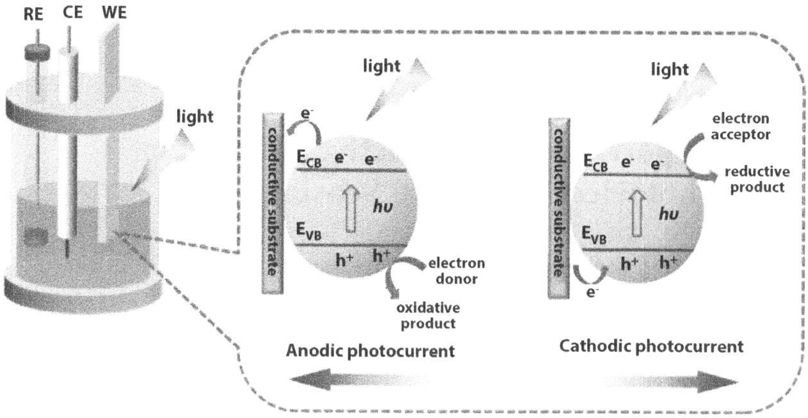

Figure 17.1 Schematic representation of the working principle of typical photoelectrochemical cell [5].

problem in photocatalysis. However, in photoelectrocatalysis, the applied potential helps to reduce the number of recombination processes between the electrons and generated holes, which leads to an increase in the efficiency of the conversion of light to chemical energy. To date, several nanostructured materials have shown PEC activity and have attained immense improvements, and 2D nanomaterial-based photoelectrochemical sensing techniques are becoming a promising method for environmental monitoring and biomedical applications [41–45].

17.3 Experimental Setup of Photoelectrochemical Cell

Semiconductors (photoactive materials) are usually used in a PEC cell; either the semiconductor is directly used as an electrode or it is deposited on the surface of the working electrode (e.g., carbon, platinum), and a potential is applied between the working and reference electrodes. The current between the working and counter electrodes was measured. The working media contains the supporting electrolyte (inert species) with a solvent, redox species, or analyte that needs to be detected by means of oxidation or reduction reactions. A UV–vis lamp is usually used as a light source to illuminate the working electrode, which is a semiconductor or a conducting substrate containing the semiconductor material on its surface. Usually, a quartz cell is used for PEC measurements because it is photoinactive. A monochromator can be used to illuminate the working electrode surface with light of a specific wavelength.

17.4 Importance of Photoactive Material in Photoelectrochemical Sensing

In PEC sensing, the photoactive material plays a vital role in initiating the catalytic process. The harvesting of light by the photoactive material leads to the generation of an electron-hole pair. The photogenerated charge-pair then transfers to the electrode or electrolyte to participate in the redox processes involved in the catalytic process. When a PEC process is considered in terms of its high sensitivity, low detection limit, and durability, the ease of initiation and propagation of the process is unavoidable. The process is always more feasible and favorable when the photoactive material is highly responsive to light harvesting and sufficiently efficient to generate an electron-hole pair. Once the pair is formed, another important factor to be considered is the prevention of recombination of the electron-hole pair, thus, addressing the rapidity and durability of the catalytic process [46, 47]. Hence, the basic role of the photoactive counterpart of the PEC sensor is to generate electron-hole pairs by harvesting energy from the encountered photon and providing the photogenerated charge pairs with sufficient time by suppressing the recombination so that the charge pairs may transfer to the analyte media and eventually participate in the PEC catalytic process.

Owing to the light harvesting and electron-hole pair generation ability due to the presence of a suitable bandgap between the valance and conduction bands, semiconductors such as metal oxides, metal chalcogenides and different carbon-based materials such as g-C_3N_4 and other polymer systems are often used as photoactive materials in PEC systems [48–51]. The low efficiency of photocurrent conversion and high photocorrosion limits their widespread applicability. To address these issues, several strategies have been adopted, such as, mixing different elements with a semiconductor as dopants or constructing different heterojunctions by combining different compatible semiconductors. This strategy provides a system with new dynamics in charge separation and recombination of charge carriers by introducing localized electronic levels between the valence and conduction bands [52]. The combination of two different semiconductors with different bandgaps can improve the conversion efficiency by providing sufficient time for the charge carriers to migrate towards catalytic participation prior to recombination [53]. Carbon materials are known for their high electrical conductivity, which accelerates photogenerated charges and suppresses recombination, high surface area to adsorb electroactive species, and stability. The combination of carbon materials such as graphene and reduced graphene oxide with inorganic semiconductor materials can produce photoelectrocatalysts with excellent analytical properties such as rapid sensing with high sensitivity and low detections limits [54].

17.5 2D Nanomaterials in Photoelectrochemical Sensing

Among the various photoactive materials used for photoelectrochemical sensing, graphene has received immense attention. Graphene is a thin sheet of graphite with a carbon architecture and possesses great advantages such as large specific surface area-to-volume ratio, high electronic conductivity, and simple and low-cost preparation, thereby making it an attractive material for photoelectrochemical sensors. However, because it is a non-photoactive material with a zero bandgap, graphene has been extensively used as support material for photoelectrochemical sensors as an electron transport promoter. Various graphene-based nanocomposites, such as Au nanoparticles decorated on graphene [55], nitrogen-doped graphene [56], and TiO_2-gC_3N_4-graphene [57], have been successfully fabricated and applied as photoelectrochemical sensors.

Hexagonal graphitic carbon nitride as a metal-free 2D semiconductor nanomaterial, exhibits great advantages such as a narrow band structure, visible light harvesting capacity, low fabrication cost, low toxicity, and unique optoelectronic properties, which make graphitic carbon nitride one of the most noteworthy nanostructures for photoelectrochemical analysis. Yanyan Niu and co-workers [58] have reported g-C_3N_4-based photoelectrochemical sensor for detection of Cd (II) ions, Dong Liang *et al.* [59] fabricated g-C_3N_4-based regenerative photoelectrochemical sensor for the detection of Cu (II) ions and Amir Hatamie and colleagues [60] have also developed g-C_3N_4-based fast and ultra-sensitive PEC sensor for efficient detection of Ph (II) ions. g-C_3N_4 based PEC sensors have also been employed for the detection of toxic heavy metals (arsenic, lead, cadmium, etc.) toxic chemical compounds (aromatic nitro compounds, phenolic compounds, etc.), pesticides, insecticides, antibiotics, neurotransmitters as well as pathogens [61].

Among 2D transition metal dichalcogenides, MoS_2 has received the most attention in photoelectrochemical sensing applications owing to its simple and low-cost synthetic strategy, easy exfoliation, suitable band structure, and visible light activity. Li *et al.* [62] fabricated single-layer MoS_2

for efficient photoelectrochemical detection of methionine. Similarly, Hun *et al.* [63] used single-layer MoS_2 for fast and ultrasensitive photoelectrochemical detection of dopamine.

However, the utilization of pristine 2D nanomaterials in photoelectrochemical sensing applications is restricted owing to the high rate of the recombination of the photogenerated electron hole pairs. To address this limitation, several modifications have been employed to engineer 2D nanostructure-based photoelectrochemical sensing materials.

The most common method for the modification of 2D nanomaterials is hybridization with noble nanomaterials. The incorporation of noble metal nanoparticles in semiconducting 2D nanosheets can enhance the photoelectrochemical performance by many fold, essentially due to the surface plasmon resonance of noble metal nanoparticles and a decrease in charge recombination due to the potential barrier at the semiconductor/metal interface. With the surface plasmon resonance effect, noble metal nanoparticles, such as Au and Ag can harvest visible light to produce photogenerated excitons. The photogenerated electrons may be transferred to the conduction band of the semiconducting 2D nanosheets, hybridized with the noble nanoparticles in the composite, depending on the band alignment of the hybrid. Thus, the localized noble metal nanoparticles on 2D nanomaterials can improve the visible light absorption owing to the SPR effect and the excellent electron mobility of 2D nanomaterials, and the potential barrier across the metal/semiconductor interface can enhance the charge separation, thereby, synergistically improving the photoelectochemical sensing efficiency. Xu and colleagues [64] used plasmonic Au/gC_3N_4 composites for efficient photoelectrochemical detection of 4-chlorophenol. The photocurrent density of Au/g-C_3N_4 is approximately 3.4 times increase in that of g-C_3N_4 alone. Wang *et al.* [65] investigated the effect of dodecahedral gold nanocrystals decorated over MoS_2 nanosheets for photoelectrochemical immunosensing of tetrabromobisphenol A. It was reported that the integration of dodecahedral gold nanocrystals into MoS_2 nanosheets significantly enhanced the photocurrent response owing to the charge mobility and surface plasmon resonance effects of Au nanoparticles. A schematic representation of photocurrent generation mechanism of Au/gC_3N_4/ITO electrode has been given in Figure 17.2. Similarly, Dai *et al.* [66] reported that the incorporation of multifunctional Au nanorods into hexagonal graphitic carbon nitride nanosheets resulted in hybrid localized surface plasmonic resonance and a Schottky barrier, which significantly lowered the charge recombination rate, resulting in an increased photoelectrochemical performance.

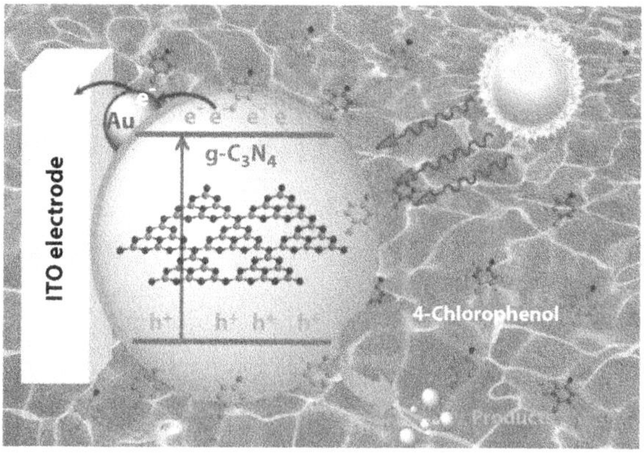

Figure 17.2 Schematic representation of photocurrent generation mechanism of Au/g-C_3N_4/ITO electrode and photoelectrochemical sensing of 4-chlorophenol. Reprinted from Ref. [65] with permission from Elsevier.

Heterogeneous structures of 2D nanomaterials and other semiconductors have also been effectively exploited to improve the performance of various PEC sensors. Various binary heterojunctions involving semiconducting metal oxides, metal sulfides with 2D nanosheets with an appropriate electronic structure, and band alignment can synergistically enhance the photoelectrochemical response of the composite. Semiconductors with suitable bandgaps can utilize incident radiation to generate electron-hole pairs. The photogenerated electrons may be transferred from the conduction band of one semiconductor to that of the other, thereby increasing the charge mobility. In addition, the potential barrier across the interfacial contact of the semiconductors may lower the charge recombination rate, thereby enhancing the life span of the photogenerated excitons and improving the PEC sensing performance. Wang *et al.* [67] fabricated graphene-TiO_2 nanohybrids for the ultrasensitive photoelectrochemical sensing of nicotinamide adenine dinucleotide under solar irradiation. Li *et al.* [68] reported g-C_3N_4/α-Fe_2O_3 heterostructures as efficient photoelectrochemical glucose sensor exhibiting synergistic photoelectrochemical response compared to the pristine counterparts. Yang *et al.* [69] demonstrated that incorporation of copper oxide nanoparticles to graphitic carbon nitride results into highly sensitive photoelectrochemical aptasensing response towards bisphenol A. Co_3O_4/g-C_3N_4 heterojunction has also been investigated for efficient photoelectrochemical aptasensing of oxytetracycline [70]. Schematic representation of photoelectrochemical sensing mechanism using Co_3O_4 nanoparticle/graphitic carbon nitride heterojunction has been represented in Figure 17.3.

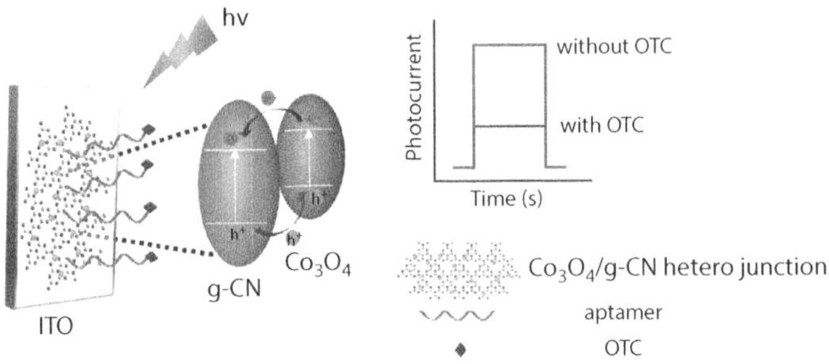

Figure 17.3 Schematic diagram of PEC sensing of oxytetracycline using Co_3O_4 nanoparticle/graphitic carbon nitride heterojunction and charge separation-charge transfer process of the Co_3O_4/g-CN heterojunction under light irradiation. Reprinted from Ref. [70] with permission from Elsevier.

Various 2D–2D nanoheterojunctions have also been fabricated owing to their enhanced stability and regeneration ability and their photoelectrichemical sensing performance has been investigated. Yuan *et al.* [71] fabricated 2D-2D g-C_3N_4/Ti_3C_2 composites for selective and efficacious photoelectrochemical sensing of ciprofloxacin. Schematic representation of the photoelectrochemical sensing mechanism of 2D/2D heterojunction is given in Figure 17.4. Yan *et al.* [72] constructed MoS_2/C_3N_4 heterojunction with enhanced visible-light photoelectrochemical activity towards ultrasensitive detection of Cu (II) ions.

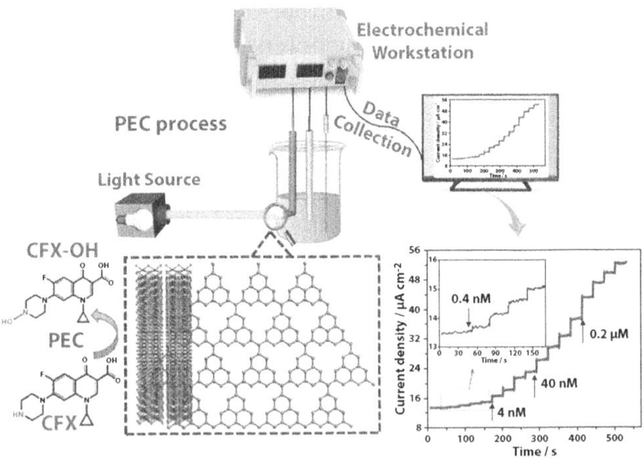

Figure 17.4 Schematic illustration of the photoelectrochemical sensing mechanism of 2D/2D g-C_3N_4/Ti_3C_2 composites for photoelectrochemical detection of ciprofloxacin. Reprinted from Ref. [71], with permission from Elsevier.

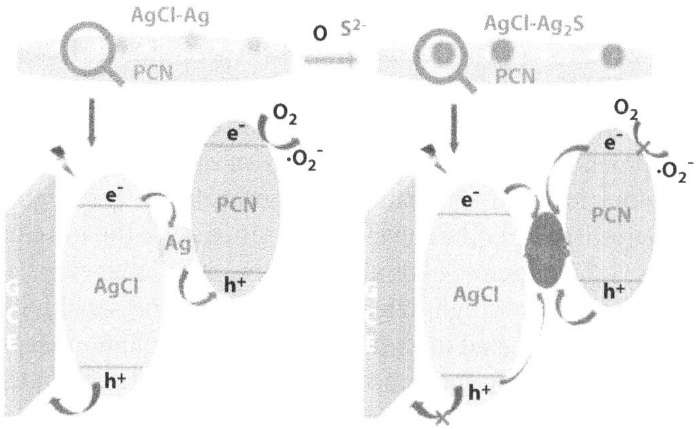

Figure 17.5 Schematic representation of the transfer mechanism and tuning of photogenerated electrons in AgCl-Ag-porous carbon nitride upon the addition of S^{2-}. Reprinted from Ref. [75], with permission from Elsevier.

In addition to the above-mentioned binary heterojunction structures, multicomponent heterojunction systems have also been employed to construct efficient PEC sensors. For example, Ahmadi *et al.* [73] constructed a TiO_2-CeO_2-graphene quantum dot nanocomposite for photoelectrochemical detection of dopamine. Owing to the synergistic effect of the nanocomposite components, an outstanding photocurrent response was observed in the PEC sensor. Sui *et al.* [74] fabricated ternary $ZnO/MoS_2/C_3N_4$ heterojunction for photoelectrochemical detection of 5-hydroxymethylcytosine. Similarly, Liang *et al.* [75] fabricated AgCl-Ag coupled with a porous carbon nitride ternary nanocomposite for the Z-scheme cathodic photoelectrochemical detection of hydrogen sulfide. The dynamics between the band alignment and bandgap of the components of the heterojunction and the interfacial potential barrier to prevent charge recombination play a crucial role in determining the heterojunction structures and the efficiency with which the synergistic photoelectrochemical performance may result. Figure 17.5 represents the transfer mechanism and tuning of photogenerated electrons in AgCl-Ag-porous carbon nitride.

17.6 Current Challenges and Future Prospects

Considering the current scenario, the practical utility of the sensor requires proper attention considering the obstacles that the PEC sensor may encounter in the future. The current prospects of various nanomaterial-based photoelectrochemical sensors are primarily based on laboratory research.

Moreover, little attention has been paid to the advancement of PEC-related equipment, which is a major constraint in the overall development of the photoelectrochemical sensing field. Although the photoelectrochemical approach has advantages such as high sensitivity, selectivity, reasonably priced equipment, quick response, and ease of operation, it still has some serious weaknesses such as reliability, reproducibility, and high throughout. One major issue that has a significant impact on the development of PEC sensors is the scarcity of commercial PEC sensing devices. Therefore, rather than focusing on high detection selectivity and sensitivity, further research should be directed toward the development of photoelectrochemical sensing devices. Therefore, photoelectrochemical detection techniques should be coupled with intelligent technology to accomplish real-time detection and monitoring of target molecules.

17.7 Conclusion

Photoelectrochemical sensing is an emerging analytical technique with immense potential for the detection of environmental pollutants such as toxic heavy metals (such as arsenic, lead, cadmium, and chromium), organic compounds (such as aromatic nitro compounds, and phenolic compounds), pharmaceuticals, insecticides, pesticides, antibiotics, and simple organic molecules, as well as great prospects for biosensing applications. Photoactive nanoparticles play a significant role in the photoelectrode and have a significant impact on the detection capabilities of PEC sensor owing to their advantages of low background noise, affordable instrumentation, quick response, and high selectivity and sensitivity. 2D nanomaterials because of their large surface area, high electron mobility, low cost, easy fabrication and easy structural and morphological modification capabilities, they have attracted significant interest for the development of novel and efficient photoelectrochemical sensors.

It is worth mentioning that many 2D nanomaterials require modification or exfoliation owing to their fast recombination rates of photogenerated electron-hole pairs, which results in difficulty in material preparation repeatability and the complexity of experimental operations. Moreover, compared with other types of nanomaterials, the signal responsiveness of the sensor may be significantly improved by the large surface area and quick electron movement. The use of 2D nanomaterials in plasmonic sensing is still in its early stages, even though they have been thoroughly studied and applied in plasmonic sensors because of their advantages over conventional materials, including a large surface area to volume ratio, high thermal and chemical stability, and biocompatibility. However, strategically induced structural and morphological modifications in 2D nanomaterials by noble techniques such as hierarchical architecture and bandgap engineering of 2D nanomaterials with suitable modifications such as doping, integration with noble metals, and heterojunctions with other semiconductors may overcome all the limitations of developing highly efficient photoelectrochemical sensors for environmental and biosensing applications.

References

1. Gaudin, V., Advances in biosensor development for the screening of antibiotic residues in food products of animal origin-a comprehensive review. *Biosens. Bioelectron.*, 90, 363–377, 2017.
2. Lv, S., Zhang, K., Zhu, L., Tang, D., ZIF-8-assisted NaYF4: Yb, Tm@ZnO converter with exonuclease III-powered DNA walker for near-infrared light responsive biosensor. *Anal. Chem.*, 92, 1470–1476, 2020.
3. Long, D., Li, M., Wang, H. *et al.*, A photoelectrochemical biosensor based on fullerene with methylene blue as a sensitizer for ultrasensitive DNA detection. *Biosens. Bioelectron.*, 142, 111579, 2019.
4. Wang, Y., Li, X., Waterhouse, G.I.N. *et al.*, Photoelectrochemical biosensor for protein kinase Adetection based on carbon microspheres, peptide functionalized Au-ZIF-8 and TiO2/g-C3N4. *Talanta*, 196, 197–203, 2019.
5. Qiu, Z. and Tang, D., Nanostructure-based photoelectrochemical sensing platforms for biomedical applications. *J. Mater. Chem. B*, 8, 2541–2561, 2020.
6. Zhou, Q. and Tang, D., Recent advances in photoelectrochemical biosensors for analysis of mycotoxins in food. *TrAC-Trend. Anal. Chem.*, 124, 115814, 2020.
7. Zhou, Y., Yin, H., Zhao, W.W., Electrochemical, electrochemiluminescentand photoelectrochemical bioanalysis of epigenetic modifiers. *A Compr. review. Coord. Chem. Rev.*, 424, 213519, 2020.
8. Zhao, W.W., Xu, J.J., Chen, H.Y., Photoelectrochemical immunoassays. *Anal. Chem.*, 90, 615–627, 2018.
9. Guo, L., Yin, H., Xu, M. *et al.*, *In situ* generated plasmonic silver nanoparticle-sensitized amorphous titanium dioxide for ultrasensitive photoelectrochemical sensing of formaldehyde. *ACS Sens.*, 4, 2724–2729, 2019.
10. Dai, H., Zhang, S., Hong, Z., Lin, Y., A potentiometric addressable photoelectrochemical biosensor for sensitive detection of two biomarkers. *Anal. Chem.*, 88, 9532–9538, 2016.
11. Li, R., Liu, Y., Cheng, L. *et al.*, Photoelectrochemical aptasensing of kanamycin using visible light-activated carbon nitride and graphene oxide nanocomposites. *Anal. Chem.*, 86, 9372–9375, 2014.

12. Huang, C., Li, Y., Wang, N. *et al.*, Progress in research into 2D graphdiyne-based materials. *Chem. Rev.*, 118, 7744–7803, 2018.
13. Zhu, C., Du, D., Lin, Y., Graphene-like 2D nanomaterial-based biointerfaces for biosensing applications. *Biosens. Bioelectron.*, 89, 43–55, 2017.
14. Wu, E., Xie, Y., Liu, Q. *et al.*, Photoinduced doping to enable tunable and high-performance anti-ambipolar $MoTe_2/MoS_2$heterotransistors. *ACS Nano*, 13, 5430–5438, 2019.
15. Jiang, D., Li, J., Xing, C. *et al.*, Two-dimensional $CaIn_2S_4/g\text{-}C_3N_4$ heterojunction nanocomposite with enhanced visible-light photocatalytic activities: Interfacial engineering and mechanism insight. *ACS Appl. Mater. Interfaces*, 7, 19234–19242, 2015.
16. Sui, C., Li, F., Wu, H. *et al.*, Photoelectrochemical biosensor for 5hmC detection based on the photocurrent inhibition effect of ZnO on MoS_2/C_3N_4 heterojunction. *Biosens. Bioelectron.*, 142, 111516, 2019.
17. Yang, S., Liu, Y., Chen, W. *et al.*, High sensitivity and good selectivity of ultralong MoO_3nanobelts for trimethylamine gas. *Sens. Actuators B Chem.*, 226, 478–485, 2016.
18. Liu, J., Jiang, X., Zhang, R. *et al.*, MXeneenabled electrochemical microfluidic biosensor: Applications toward multicomponent continuous monitoring in whole blood. *Adv. Funct. Mater.*, 29, 1807326, 2019.
19. Yin, H., Cui, L., Ai, S. *et al.*, Electrochemical determination of bisphenolA at Mg–Al–CO_3 layered double hydroxide modified glassy carbon electrode. *Electrochim. Acta*, 55, 603–610, 2010.
20. Yin, F., Gu, B., Lin, Y. *et al.*, Functionalized 2D nanomaterials for gene delivery applications. *Coord. Chem. Rev.*, 347, 77–97, 2017.
21. Chai, L.X., Fan, X.X., Zuo, Y.H. *et al.*, Low-dimensional nanomaterials enabled autoimmune disease treatments: Recent advances, strategies, and future challenges. *Coord. Chem. Rev.*, 432, 213697, 2021.
22. Gui, R., Jin, H., Bu, X. *et al.*, Recent advances in dual-emission ratiometric fluorescence probes for chemo/biosensing and bioimaging of biomarkers. *Coord. Chem. Rev.*, 383, 82–103, 2019.
23. Gaudin, V., Advances in biosensor development for the screening of antibiotic residues in food products of animal origin–a comprehensive review. *Biosens. Bioelectron.*, 90, 363–377, 2017.
24. Lv, S., Zhang, K., Zhu, L., Tang, D., ZIF-8-assisted NaYF4: Yb, Tm@ ZnO converter with exonuclease III-powered DNA walker for near-infrared light responsive biosensor. *Anal. Chem.*, 92, 1, 1470–1476, 2019.
25. Luo, Z., Qi, Q., Zhang, L., Zeng, R., Su, L., Tang, D., Branched polyethylenimine-modified upconversion nanohybrid-mediated photoelectrochemical immunoassay with synergistic effect of dual-purpose copper ions. *Anal. Chem.*, 91, 6, 4149–4156, 2019.
26. Zhou, Q., Li, G., Chen, K., Yang, H., Yang, M., Zhang, Y., Zhang, Y., Simultaneous unlocking optoelectronic and interfacial properties of C60

for ultrasensitive immunosensing by coupling to metal–organic framework. *Anal. Chem.*, 92, 1, 983–990, 2019.
27. Zhao, L., Ji, J., Shen, Y., Wu, K., Zhao, T., Yang, H., Zhang, Y., Exfoliation and sensitization of 2D carbon nitride for photoelectrochemical biosensing under red light. *Chem. Eur. J.*, 25, 68, 15680–15686, 2019.
28. Dashtian, K., Hajati, S., Ghaedi, M., L-phenylalanine-imprinted polydopamine-coated CdS/CdSe nn type II heterojunction as an ultrasensitive photoelectrochemical biosensor for the PKU monitoring. *Biosens. Bioelectron.*, 165, 112346, 2020.
29. Chen, Y., Zhou, Y., Yin, H., Li, F., Li, H., Guo, R., Ai, S., Photoelectrochemical biosensor for histone acetyltransferase detection based on ZnO quantum dots inhibited photoactivity of BiOI nanoflower. *Sens. Actuators B Chem.*, 307, 127633, 2020.
30. Qiu, Z. and Tang, D., Nanostructure-based photoelectrochemical sensing platforms for biomedical applications. *J. Mater. Chem. B*, 8, 13, 2541–2561, 2020.
31. Long, D., Li, M., Wang, H., Wang, H., Chai, Y., Yuan, R., A photoelectrochemical biosensor based on fullerene with methylene blue as a sensitizer for ultrasensitive DNA detection. *Biosens. Bioelectron.*, 142, 111579, 2019.
32. Wang, Y., Li, X., Waterhouse, G., II, Zhou, Y., Yin, H., Ai, S., Photoelectrochemical biosensor for protein kinase A detection based on carbon microspheres, peptide functionalized Au-ZIF-8 and TiO2/g-C3N4. *Talanta*, 196, 197–203, 2019.
33. Sui, C., Wang, T., Zhou, Y., Yin, H., Meng, X., Zhang, S., Ai, S., Photoelectrochemical biosensor for hydroxymethylated DNA detection and T4-β-glucosyltransferase activity assay based on WS2 nanosheets and carbon dots. *Biosens. Bioelectron.*, 127, 38–44, 2019.
34. Çakıroğlu, B. and Özacar, M., A self-powered photoelectrochemical glucose biosensor based on supercapacitor Co3O4-CNT hybrid on TiO2. *Biosens. Bioelectron.*, 119, 34–41, 2018.
35. Zhao, W.W., Xu, J.J., Chen, H.Y., Photoelectrochemical bioanalysis: The state of the art. *Chem. Soc. Rev.*, 44, 3, 729–741, 2015.
36. Zhou, H., Liu, J., Zhang, S., Quantum dot-based photoelectric conversion for biosensing applications. *TrAC Trends Anal. Chem.*, 67, 56–73, 2015.
37. Zeng, R., Luo, Z., Su, L., Zhang, L., Tang, D., Niessner, R., Knopp, D., Palindromic molecular beacon based Z-scheme BiOCl-Au-CdS photoelectrochemical biodetection. *Anal. Chem.*, 91, 3, 2447–2454, 2019.
38. Li, M., Zheng, Y., Liang, W., Yuan, Y., Chai, Y., Yuan, R., An ultrasensitive "on–off–on" photoelectrochemical aptasensor based on signal amplification of a fullerene/CdTe quantum dots sensitized structure and efficient quenching by manganese porphyrin. *Chem. Commun.*, 52, 52, 8138–8141, 2016.
39. Wang, B., Mei, L.P., Ma, Y., Xu, Y.T., Ren, S.W., Cao, J.T., Zhao, W.W., Photoelectrochemical-chemical-chemical redox cycling for advanced signal

amplification: Proof-of-concept toward ultrasensitive photoelectrochemical bioanalysis. *Anal. Chem.*, 90, 21, 12347–12351, 2018.
40. Zeng, R., Zhang, L., Su, L., Luo, Z., Zhou, Q., Tang, D., Photoelectrochemical bioanalysis of antibiotics on rGO-Bi2WO6-Au based on branched hybridization chain reaction. *Biosens. Bioelectron.*, 133, 100–106, 2019.
41. Qiu, Z. and Tang, D., Nanostructure-based photoelectrochemical sensing platforms for biomedical applications. *J. Mater. Chem. B*, 8, 13, 2541–2561, 2020.
42. Han, F., Song, Z., Nawaz, M.H., Dai, M., Han, D., Han, L., Niu, L., MoS2/ZnO-heterostructures-based label-free, visible-light-excited photoelectrochemical sensor for sensitive and selective determination of synthetic antioxidant propyl gallate. *Anal. Chem.*, 91, 16, 10657–10662, 2019.
43. Wang, C., Wang, Y., Zhang, H., Deng, H., Xiong, X., Li, C., Li, W., Molecularly imprinted photoelectrochemical sensor for carcinoembryonic antigen based on polymerized ionic liquid hydrogel and hollow gold nanoballs/MoSe2 nanosheets. *Anal. Chim. Acta*, 1090, 64–71, 2019.
44. Gao, H., Jia, J., Guo, F., Li, B., Dai, D., Deng, X., Liu, G., The electronic structure and photoactivity of TiO2 modified by hybridization with monolayer g-C3N4. *J. Photochem. Photobiol. A Chem.*, 364, 328–335, 2018.
45. Xue, J., Gao, C., Zhang, L., Cui, K., He, W., Yu, J., A single-interface photoelectrochemical sensor based on branched TiO 2 nanorods@ strontium titanate for the detection of two biomarkers. *J. Mater. Chem. B*, 6, 28, 4697–4703, 2018.
46. Shu, J. and Tang, D., Recent advances in photoelectrochemical sensing: From engineered photoactive materials to sensing devices and detection modes. *Anal. Chem.*, 92, 1, 363–377, 2019.
47. Devadoss, A., Sudhagar, P., Terashima, C., Nakata, K., Fujishima, A., Photoelectrochemical biosensors: New insights into promising photoelectrodes and signal amplification strategies. *J. Photochem. Photobiol. C Photochem. Rev.*, 24, 43–63, 2015.
48. Hoang, S., Berglund, S.P., Hahn, N.T., Bard, A.J., Mullins, C.B., Enhancing visible light photo-oxidation of water with TiO2 nanowire arrays via cotreatment with H2 and NH3: Synergistic effects between Ti3+ and N. *J. Am. Chem. Soc.*, 134, 8, 3659–3662, 2012.
49. Zhang, K., Lv, S., Lin, Z., Li, M., Tang, D., Bio-bar-code-based photoelectrochemical immunoassay for sensitive detection of prostate-specific antigen using rolling circle amplification and enzymatic biocatalytic precipitation. *Biosens. Bioelectron.*, 101, 159–166, 2018.
50. Miao, H., Zhang, G., Hu, X., Mu, J., Han, T., Fan, J., Hou, X., A novel strategy to prepare 2D g-C3N4 nanosheets and their photoelectrochemical properties. *J. Alloys Compd.*, 690, 669–676, 2017.
51. Li, Y., Zhang, N., Zhao, W.W., Jiang, D.C., Xu, J.J., Chen, H.Y., Polymer dots for photoelectrochemical bioanalysis. *Anal. Chem.*, 89, 9, 4945–4950, 2017.

52. Teoh, W.Y., Scott, J.A., Amal, R., Progress in heterogeneous photocatalysis: from classical radical chemistry to engineering nanomaterials and solar reactors. *J. Phys. Chem. Lett.*, 3, 5, 629–639, 2012.
53. Low, J., Yu, J., Jaroniec, M., Wageh, S., Al-Ghamdi, A., Heterojunction photocatalysts. *Adv. Mater.*, 29, 1601694, 2017.
54. Zhou, Q. and Tang, D., Recent advances in photoelectrochemical biosensors for analysis of mycotoxins in food. *TrAC Trends Anal. Chem.*, 124, 115814, 2020.
55. Hu, Y., Xue, Z., He, H., Ai, R., Liu, X., Lu, X., Photoelectrochemical sensing for hydroquinone based on porphyrin-functionalized Au nanoparticles on graphene. *Biosens. Bioelectron.*, 47, 45–49, 2013.
56. Liu, Y., Yan, K., Okoth, O.K., Zhang, J., A label-free photoelectrochemical aptasensor based on nitrogen-doped graphene quantum dots for chloramphenicol determination. *Biosens. Bioelectron.*, 74, 1016–1021, 2015.
57. Cai, Z., Rong, M., Zhao, T., Zhao, L., Wang, Y., Chen, X., Solar-induced photoelectrochemical sensing for dopamine based on TiO2 nanoparticles on g-C3N4 decorated graphene nanosheets. *J. Electroanal. Chem.*, 759, 32–37, 2015.
58. Niu, Y., Chen, Y., Zhang, X., Xie, H., Luo, G., Sun, W., Target-enhanced photoelectrochemical aptasensor for Cd (II) detection using graphite-like carbon nitride as sensitizer with high sensitivity. *Microchem. J.*, 168, 106394, 2021.
59. Liang, D., Liang, X., Zhang, Z., Wang, H., Zhang, N., Wang, J., Qiu, X., A regenerative photoelectrochemical sensor based on functional porous carbon nitride for Cu2+ detection. *Microchem. J.*, 156, 104922, 2020.
60. Hatamie, A., Jalilian, P., Rezvani, E., Kakavand, A., Simchi, A., Fast and ultra-sensitive voltammetric detection of lead ions by two-dimensional graphitic carbon nitride (g-C3N4) nanolayers as glassy carbon electrode modifier. *Measurement*, 134, 679–687, 2019.
61. Vinoth, S., Devi, K.S., Pandikumar, A., A comprehensive review on graphitic carbon nitride based electrochemical and biosensors for environmental and healthcare applications. *TrAC Trends Anal. Chem.*, 140, 116274, 2021.
62. Li, Y., Mei, S., Liu, S., Hun, X., A photoelectrochemical sensing strategy based on single-layer MoS2 modified electrode for methionine detection. *J. Pharm. Biomed. Anal.*, 165, 94–100, 2019.
63. Hun, X., Wang, S., Wang, S., Zhao, J., Luo, X., A photoelectrochemical sensor for ultrasensitive dopamine detection based on single-layer NanoMoS2 modified gold electrode. *Sens. Actuators B Chem.*, 249, 83–89, 2017.
64. Xu, L., Ling, S., Li, H., Yan, P., Xia, J., Qiu, J., Yuan, S., Photoelectrochemical monitoring of 4-chlorophenol by plasmonic Au/graphitic carbon nitride composites. *Sens. Actuators B Chem.*, 240, 308–314, 2017.
65. Wang, Y., Chen, F., Ye, X., Wu, T., Wu, K., Li, C., Photoelectrochemical immunosensing of tetrabromobisphenol A based on the enhanced effect of

dodecahedral gold nanocrystals/MoS2 nanosheets. *Sens. Actuators B Chem.*, 245, 205–212, 2017.
66. Dai, H., Zhang, S., Li, Y., Lin, Y., Excellent graphitic carbon nitride nanosheets-based photoelectrochemical platform motivated by Schottky barrier and LSPR effect and its sensing application. *Analyst*, 140, 10, 3514–3520, 2015.
67. Wang, K., Wu, J., Liu, Q., Jin, Y., Yan, J., Cai, J., Ultrasensitive photoelectrochemical sensing of nicotinamide adenine dinucleotide based on graphene-TiO2 nanohybrids under visible irradiation. *Anal. Chim. Acta*, 745, 131–136, 2012.
68. Li, W., Jiang, D., Yan, P., Dong, J., Qian, J., Chen, J., Xu, L., Graphitic carbon nitride/α-Fe2O3 heterostructures for sensitive photoelectrochemical non-enzymatic glucose sensor. *Inorg. Chem. Commun.*, 106, 211–216, 2019.
69. Yang, L., Zhao, Z., Hu, J., Wang, H., Dong, J., Wan, X., Li, M., Copper oxide nanoparticles with graphitic carbon nitride for ultrasensitive photoelectrochemical aptasensor of bisphenol A. *Electroanalysis*, 32, 7, 1651–1658, 2020.
70. Chen, Y., Wang, Y., Yan, P., Ouyang, Q., Dong, J., Qian, J., Li, H., Co3O4 nanoparticles/graphitic carbon nitride heterojunction for photoelectrochemical aptasensor of oxytetracycline. *Anal. Chim. Acta*, 1125, 299–307, 2020.
71. Yuan, C., He, Z., Chen, Q., Wang, X., Zhai, C., Zhu, M., Selective and efficacious photoelectrochemical detection of ciprofloxacin based on the self-assembly of 2D/2D g-C3N4/Ti3C2 composites. *Appl. Surf. Sci.*, 539, 148241, 2021.
72. Yan, J., Chen, Z., Ji, H., Liu, Z., Wang, X., Xu, Y., Li, H., Construction of a 2D graphene-like MoS2/C3N4 heterojunction with enhanced visible-light photocatalytic activity and photoelectrochemical activity. *Chem. Eur. J.*, 22, 14, 4764–4773, 2016.
73. Ahmadi, N., Bagherzadeh, M., Nemati, A., Comparison between electrochemical and photoelectrochemical detection of dopamine based on titania-ceria-graphene quantum dots nanocomposite. *Biosens. Bioelectron.*, 151, 111977, 2020.
74. Sui, C., Li, F., Wu, H., Yin, H., Zhang, S., Waterhouse, G., II, Ai, S., Photoelectrochemical biosensor for 5hmC detection based on the photocurrent inhibition effect of ZnO on MoS2/C3N4 heterojunction. *Biosens. Bioelectron.*, 142, 111516, 2019.
75. Liang, D., Luo, J., Huang, Y., Liang, X., Zhou, Z., Wang, J., Qiu, X., Z□Scheme cathodic photoelectrochemical sensors for detection of hydrogen sulfide based on AgCl-Ag coupled with porous carbon nitride. *Appl. Surf. Sci.*, 532, 147424, 2020.

18

Fabrication of 2D Nanomaterials-Based Biosensor

Arpita Paul Chowdhury*, M. Dinamani and K. S. Anantharaju

Department of Chemistry, Dayananda Sagar College of Engineering, Bengaluru, India

Abstract

The development of graphene has paved the way for the discovery of more two-dimensional (2D) nanomaterials, and new 2D materials derived from graphene are more versatile, functional, and flexible. Owing to the attractive properties of graphene and its derivatives, they can be used in a wide range of applications. These nanomaterials lead to biomolecular recognition, which can be used as a diagnostic platform. In this appraisal, we provide insights into the emerging 2D based biosensors. The known 2D materials are *g*-C_3N_4, graphene, black phosphorus, transition metal dichalcogenides, and Mxenes, which are cost-effective and possess rapid, and sensitive detection tools. Different types of biosensors, namely electrochemical, fluorescence, colorimetric, field effect transistor (FET), and SERS biosensors, are highlighted. Important aspects and physicochemical properties of the fabrication techniques are discussed. Finally, the challenges and thoughts on future perspectives in this field are also addressed.

Keywords: 2D nanomaterials, electochemical biosensor, fluorescence, field effect transistor, SERS biosensor

18.1 Introduction

With developments in science and technology, nanomaterials have aroused today's scientific community due to their unique nanostructure with outstanding mechanical, electronic, and biological performances [1].

**Corresponding author*: chemarpitachowdhury@gmail.com

Subhendu Chakroborty and Kaushik Pal (eds.) 2D Nanomaterials: Synthesis, Properties and Applications, (415–434) © 2024 Scrivener Publishing LLC

Nanomaterials are classified based on their dimensional structures, and their properties vary according to their classification. The crystal structures were classified as 0D, 1D, 2D, or 3D. A fascinating feature of 2D material is the atoms of the materials which are exposed to the surface, making it comparatively more comfortable to change the structure and accordingly the properties of the materials. The inherent properties of 2D materials, such as electron mobility, optical properties, and facile synthesis, make them the first choice. The 2D material such as graphitic C_3N_4 (g-C_3N_4) [2], graphene [3], black phosphorus (BP) [4], transition metal dichalcogenides (TMDCs) [5], MXenes [6], boron nitrides (BN) [7], and phosphorus chalcogenides (MPX_3) [8] have been developed by chemists, physicists, and materials worldwide. These 2D materials possess potential because of their 2D flexible layered structures, strong interaction with biomolecules, and distinctive physicochemical properties, and have proven to be potential materials for the technological needs of today [9]. Currently, sensing devices are highly efficient for the rapid testing of specific analytes. For these sensing applications, 2D materials are primarily chosen because of their wide practical applicability [10]. Biosensors are rapidly gaining importance in various fields, such as industries, agriculture, and medicine [11]. Leland C. Clark is known as the father of biosensors [12]. With progress in science and technology, great developments in biosensors have been achieved in environmental, medical, and military fields [13]. A biosensor comprises an analyte to be detected, a transducer, and a signal processor [14, 15]. Figure 18.1 represents a schematic representation of the various components of the biosensor.

Since the discovery of 2D materials, they have not only attracted considerable academic interest but also become potential materials, especially in sensing applications. Figure 18.2 illustrates the application of 2D nanomaterials in a biosensor.

Thus, ample work on the rapid synthesis, characterization, and applications of 2D materials has been carried out by the scientific community. The incorporation of emerging technologies in 2D material synthesis is expected to provide more in-depth insights into the physicochemical properties of materials. In the following section, we discuss the introduction and synthesis strategies for 2D nanomaterials. Next, the important role of 2D based biosensor is highlighted. Finally, various 2D nanomaterial-based biosensors are discusssed.

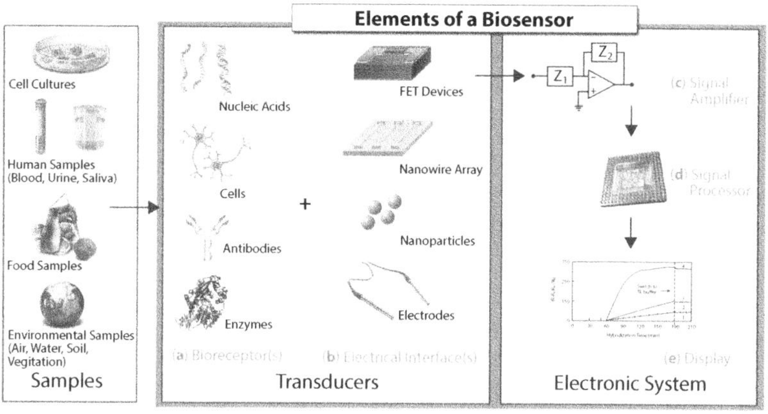

Figure 18.1 Schematic representation of a biosensor. Reproduced with permission from Reference [16]. Copyright 2008 MDPI.

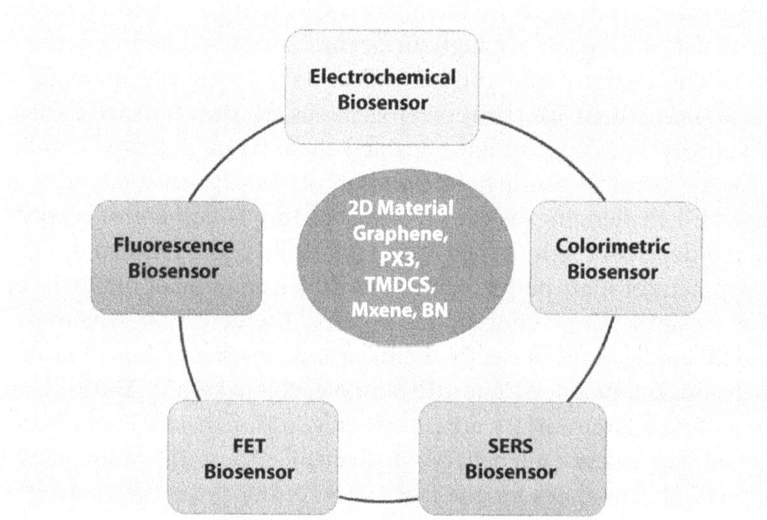

Figure 18.2 Prospective applications of 2D material as biosensor.

18.2 2D Nanomaterial Synthesis Strategies

It is evident that the rapid development of ultrathin 2D materials in various disciplines of science is due to their novel properties. Consequently, thickness and size control in the synthesis of 2D materials has emerged as a challenging area for scientists. The two synthetic strategies for 2D materials are top-down and bottom-up methods. Atomic precursors are allowed to react in the bottom-up approach for nanoscale 2D synthesis, whereas macrostructures are broken down into nanostructures using the top-down method of synthesis. The most common top-down methods include mechanical milling, sputtering, laser ablation, etching, chemical deposition (CVD) vapor sol–gel method, and bottom-up synthesis routes includes solvothermal, hydrothermal, and reverse micelle methods [17].

18.3 Role of 2D Materials in Biosensor

Many fields have witnessed the rapid development of sensors owing to their optical, magnetic, thermal, and electrical properties. Electrochemical detection by sensors incorporating a 2D material improves the sensitivity and detection range of the sensors [23]. Two-dimensional nanomaterials are a good choice for the fabrication of electrode materials. These materials are of great interest in electrochemistry owing to their enhanced flexibility, high electrical conductivity, high surface area, presence of more active sites, thermal conductivity, and good chemical stability. Special characteristics, such as surface area, excellent heterogeneous electron transfer ability, and good adsorption capacity have enabled them to be potential candidates for electrochemical sensing. Preparation of two-dimensional nanomaterials is still challenging, as the layered structure is vulnerable to aggregation and deprives them of their vast applications. The rational design and development of high-performance two-dimensional nanomaterials are of major concern. These efforts have improve the detection sensitivity and speed for applications in electrochemical and biosensing. Single-layer two-dimensional nanomaterials adsorb biomolecules in large quantities because they possess a large surface area. Further, two-dimensional nanomaterials are good fluorescence quenchers, which enables these 2D nanomaterials to be potential candidates for the fabrication of fluorescent biosensors [24]. Some different types of biosensors are discussed in the following sections.

18.3.1 Electrochemical Biosensors

Electrochemical biosensors are important because of their various advantages. Electrochemical sensors operate at ambient temperatures and require very low power for operation. Electrochemical biosensors are widely used in the promising fields of clinical diagnosis, food-processing quality control, and environmental monitoring [25, 26]. The 2D material carbon is widely used in sensing applications because of its availability, and chemical inertness compared to other types of electrode materials. Thus, graphene is considered an excellent material for electrochemical sensing applications. Disha *et al.* synthesized a magnetic GO biosensor. The sensor exhibited excellent progesterone detection with a good detection limit. The current response is inversely proportional to the progesterone concentration under optimal experimental conditions [27]. Hao *et al.* constructed a GO-based biosensor for high glucose detection. The glucose detection sensitivity was 3.80×10^{-9} mol cm^{-2} [28]. Wu *et al.* prepared a biosensor for tyrosinase detection using GDY. The biosensor proved to be a prominent platform for BPA detection with a sensitivity of 2990.8 mA cm^{-2} M^{-1} [29]. Recently developed 2D layered nanomaterials MXenes have been used in bioanalytical applications. Liu *et al.* prepared a Mxene-$Ti_3C_2T_x$ composite for multicomponent analysis of whole blood [30]. Lei *et al.* reported a stretchable multifunctional biosensor composite MXene/Prussian blue [31]. The electrochemical sensitivity for biomolecules glucose (35.3 μA mM^{-1} cm^{-2}) and lactate (11.4 μA mM^{-1} cm^{-2}) are achieved using artificial sweat. Shahzad *et al.* demonstrated the $Ti_3C_2T_x$ Mxenes stabilized with Nafion for dopamine sensing. The biosensor showed good reproducibility, stability, and recovery for DA detection in real samples [32]. AuNPs/Ti_3C_2 Mxene biosensor was fabricated by Yang *et al.* for sensitive miRNA-155 detection. The LOD of 0.35 fM was achieved with a linear range from 1.0 fM to 10 nM [33]. Wang *et al.* reported an electrochemical DNA biosensor based on MXene (Ti_3C_2) nanosheets for the detection of gliotoxin. The biosensor exhibited a detection range from 5 pM to 10 nM with an LOD of 5 Pm [34]. Figure 18.3 represents the working principle of the sensor. TDN probes can easily capture the signal probes. The streptavidin-targeting aptamer (green loop) pulled horseradish peroxidase and produced a current signal. Transition-metal dichalcogenides exhibit excellent metallic, fluorescent, electrical, and electron transfer properties [35]. Hamami *et al.* fabricated an electrochemical flexible (Au/MoS_2/Au) biosensor on a polyethylene terephthalate substrate for the detection of glycoprotein GP120 [36]. Kim *et al.* reported a polyimide substrate-based MoS_2 biosensor. The parathyroid hormone (PTH), triiodothyronine (T3), and thyroxine (T4) were detected by the MoS_2 based sensor [37]. Cui *et al.* reported MoS_2 nanosheets biosensor for detection of circulating tumor DNA (ctDNA) with high sensitivity [38]. WS_2 nanosheet/graphite microfiber material fabricated by Zhang *et al.* and the electrocatalytic oxidation of adenine and guanine was observed with high sensitivity and good stability [39]. Boron nitride nanosheets are a class of 2D material which consists of alternating boron and nitrogen atoms with a honeycomb lattice structure and possess numerous advantages, including such as low cytotoxicity, high, good thermal conductivity,

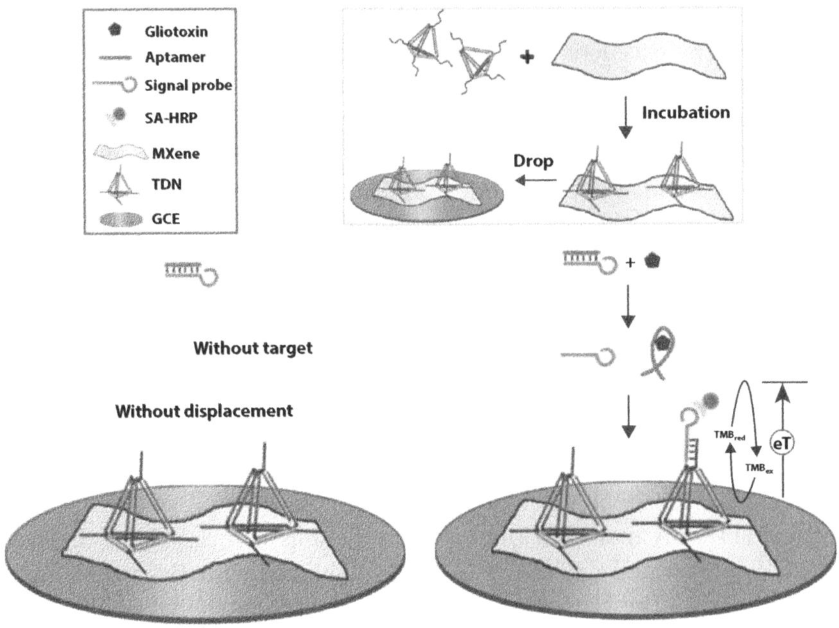

Figure 18.3 Scheme of preparation of TDN/MXene-modified electrode with gliotoxin detection principle. Reproduced with permission from Reference [34]. Copyright 2019 Elsevier.

and mechanical strength. A new modified boron nitride nanosheet was developed by Ondes *et al.* for the detection of cancer antigen 125 (CA125). The immunosensor demonstrated good detection properties for CA125, with a detection limit of 1.18 U/mL in the linear range of 5–100 U [40]. Ouyang *et al.* reported a C-hBN/GCE electrode for sensing the central neurotransmitter dopamine (DA) and for diagnosing DA-linked diseases owing to the remarkable tunable band gaps, current switching ratios, and physicochemical properties of hBN. Moreover, the sensor showed good ability to detect real samples with good efficiency [41]. Black phosphorus (LDBP) materials have emerged as potential candidates for sensor properties [42]. Paweł *et al.* fabricated functionalized black phosphorus electrode (f-FLBP-DMF-T) and demonstrated the biosensing application for the bacterial protein (*Haemophilus influenzae*) detection [43]. Kumar *et al.* designed a myoglobin (Mb) sensor based on BPNSs functionalized with PLL. A sensitivity of 51 cm^{-2} $\mu A\ pg^{-1}$ Ml and LOD 0.137 $pg\ mL^{-1}$ have been reported [44].

18.3.2 Fluorescence Biosensors

The fluorogenic probe detection of biomolecules has gained popularity. Fluorescence biosensors are widely used in medicine, chemistry, and biology. Intriguing properties such as binding kinetics and ease of automation are the advantages of biosensors [45]. Some of the advantages of fluorescence biosensors are their simple operation, sensitivity, selectivity, fast response, simple instrumentation, and multiple analyses [46–48]. Transition and heavy metal ions can be detected by fluorescence spectroscopy with high sensitivity and simplicity [49]. The changes in the fluorophore properties, such as intensity, quantum yield, and lifetime, are sensed using a fluorescence-based sensor [50]. Recently, in FRET-based fluorescent biosensors 2D materials received significant interest because of their high quenching efficiencies and large surface areas [51]. The most common 2D nanomaterial is graphene, which has attracted considerable attention. Graphene consists of layered hexagonally arranged carbon atoms with interesting properties such as electron mobility at room temperature, mechanical properties, and high thermal stability [52]. Interestingly, GO acts as a fluorescence quencher to quench the fluorescence intensity. Fan *et al.* demonstrated that GO sensors could detect various DNA targets in the same solution [53]. A GO-based fluorescent sensor reported by Zhang *et al.* Here, the researchers reported an LOD of up to 0.5 nM for Hg^{2+} detection with a poly(dT) DNA duplexes-GO biosensor [54].

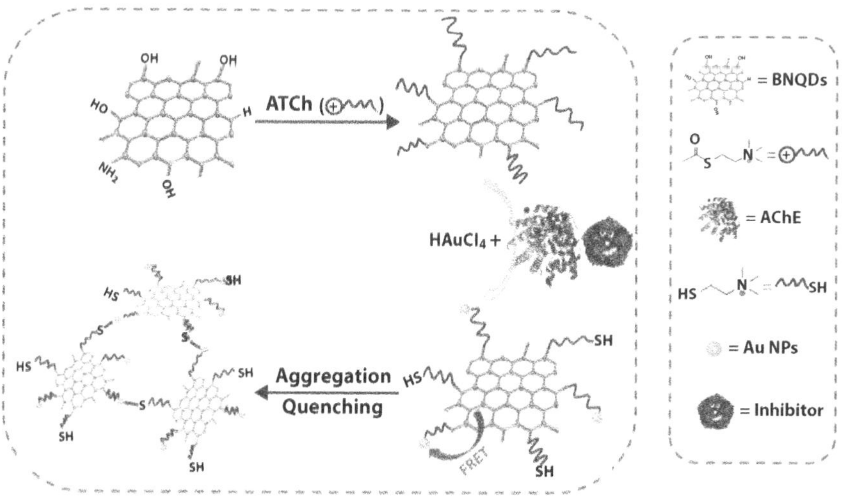

Figure 18.4 Schematic illustration of AchE detection activity using the fabricated BNQDs. Reproduced with permission from Reference [62]. Copyright 2019 Elsevier.

Li *et al.* reported the detection of thrombin using a GO-based FRET biosensor with high specificity and sensitivity for both buffer and blood serum detection [55]. Gao *et al.* synthesized a new fluorescent nanosensor for the detection of AA. Researchers have reported fluorescence quenching by the formation of the SQA–iron (III) complex [56]. The fluorescence quenching between g-C_3N_4 nanosheets and a DNA probe demonstrated by Wang *et al.* was explained by static quenching [57]. Molybdenum disulfide (MoS_2) is a 2D material with a direct energy bandgap. Li *et al.* developed MoS_2-based cellular PDA-PEG-Peptide biosensor for the determination of caspase-3 [58]. Recently, a fluorescence-quenching biosensor based on WS_2 2D materials was described by Yuan *et al.* [59]. Xiao *et al.* reported a water-soluble MoS_2 nanosheet fluorescence biosensor with a DNA detection limit of 140 pM [60]. Shanmugaraj *et al.* reported bilirubin sensing in real samples by MoS_2 QDs [61]. Boron nitride (BN) is a 2D material analogous to graphene with B and N atoms and better known as white graphene due to its ivory white texture. The two crystalline forms of BN are hexagonal (h-BN) and cubic (c-BN). Zhan *et al.* developed a BN-based composites (TCh-BNQDs/AuNPs) for detection of AChE activity *via* FRET mechanism [62]. The LOD was 0.0212 mU/mL, with a linear range from 0.05 to 6.0 mU/mL. The schematic diagram for the preparation, sensor, and detection mechanism is presented in Figure 18.4. The rapid detection of DNA molecules by BNNS colloids reported by Zhang *et al.* The high quenching efficiency and large surface area of BNNS result in rapid and selective detection of analytes [63]. A recent potential member in 2D material is black phosphorus (BP) which possess technological applications [64]. Zhou *et al.* described RNA detection using BP nanosheets. BP nanosheets act as fluorescent-quenching materials. The biosensor can exhibit an RNA response ranging from 10 to 1,000 nm [65].

18.3.3 Colorimetric Biosensor

Colorimetric biosensors work on the principle of color change in the analyte, which is detected using simple portable optical detectors for the rapid detection of viruses, DNA, small molecules, proteins, and metallic cation analytes [66, 67]. Lan *et al.* reported the use of WS_2 nanosheets for DNA detection [68]. Reduced GO QDs with high sensitivity were reported by Hwang *et al.* and can be applied for glucose sensing based on portable smartphones [69]. Tong *et al.* fabricated a PDDA-rGO-POMOF composite. This work demonstrates the prospects for application of colorimetric sensing systems for food quality control and safety [70]. Fan *et al.* reported a colorimetric method with good sensitivity and selectivity for ascorbic acid detection. Co_3O_4 nanoparticle-graphene microspheres were synthesized *via* an annealing process [71]. Tang *et al.* developed a facile colorimetric aptasensor WS_2 nanosheet for the first time developed for kanamycin detection. The peroxidase-mimicking activity was significantly high [72]. A new application of GO for miRNA detection was established by Lee *et al.* based on a NET strategy mediated by a colorimetric assay that occurs in two steps [73]. The activation of Dz with miRNA elicited a deeper color in the presence of colorimetric reagents. The control experiment without GONET showed an LOD of 12.92 Nm while in the presence of GONET, the target concentration increased, which was visible by a sharp color change (Figure 18.5).

Wu and co-workers reported *S. typhimurium* colorimetric detection with the $ZnFe_2O_4$–rGO hybrid nanostructures. In this study, the $ZnFe_2O_4$–rGO nanoenzyme showed activity towards the catalysis of TMB. A clear change in the color was observed with the detection range of 11–1.10 × 105 CFU/mL for *S. typhimurium* in buffer solution [74]. The detection of dopamine (DA) and glutathione (GSH) in human serum samples using a colorimetric method with graphene nanoribbon/silver nanoparticle was reported by Roshtami *et al.* [75]. The researchers observed a color change from green to red in the plasmonic band. Sharma *et al.* introduced a CuO:GNS nanocomposite as a colorimetric sensor for the dual detection of cholesterol and H_2O_2 with LOD of 78 μM and 6.8 μM, respectively [76]. Li *et al.* developed a simple and efficient MXene@NiFe-LDH. The sensor displayed a low detection limit of 84nM for glutathione (GSH) [77].

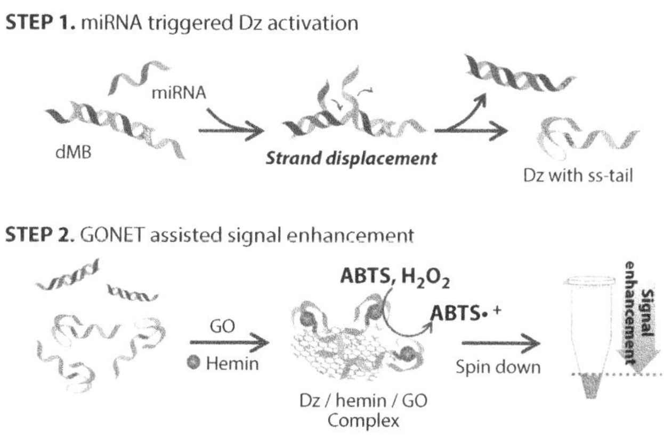

Figure 18.5 Colorimetric miRNA detection strategies. Reproduced with permission from Reference [73]. Copyright 2019, Elsevier.

18.3.4 Field-Effect Transistor Biosensor

Field-effect transistor (FET) biosensors are among the various biosensors with several advantages. In FET biosensors, sensing elements are immobilized on sensing channels. There two types of FET biosensors: n- and p-type sensors [78]. The GFET biosensor virus was immobilized on the surface of graphene, which was detected at the output [79]. Ono *et al.* reported an influenza virus sensor, glycan-functionalized G-FET. The sensor exhibited high sensitivity for the detection of lectins [80]. Seo *et al.* developed a graphene-based COVID-19 biosensor for SARS-CoV-2 virus detection. The FET sensor detected the target SARS-CoV-2 antigen protein with an LOD of 1 fg/mL [81]. Hwang *et al.* reported FET-based biosensors based on graphene for detection of DNA/RNA molecules with 600 zM LOD [82].

A MoS_2-FET-based biosensor device was reported by Lee *et al.* for the quantitative detection of PSA with a minimum detectable concentration of 1 pg/mL [83]. Xu *et al.* fabricated an effective Ti_3C_2–MXene FET biosensor for neural activity and dopamine detection [84]. The MXene FET biosensor synthesized by selective etching was found to be compatible with long-term neuronal culture with electrical measurements up to a temporal resolution of 50 ms. Li and his coworkers developed MXene–graphene FET sensor. The sensor showed good results in the detection of H_1N_1 virus and SARS-CoV-2 with an LOD of 125 copies/mL^{-1} and 1 fg mL^{-1} respectively

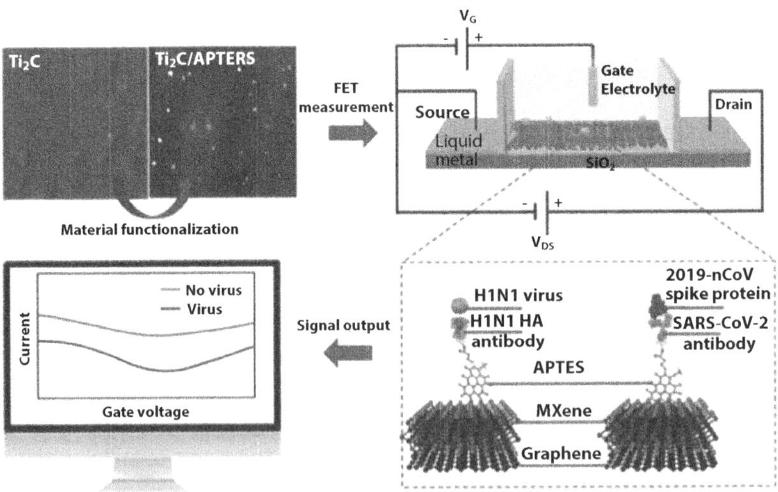

Figure 18.6 Illustration of fabrication of MXene–graphene FET sensor and detection mechanism of H_1N_1 Virus. Reproduced with permission from Reference [85]. Copyright 2021, ACS.

(Figure 18.6) [85]. Rahman *et al.* fabricated surface plasmon resonance-induced graphene-tungsten disulfide (WS_2) biosensors for the detection of DNA hybridization [86]. Meanwhile, a black phosphorous FET biosensor labeled with gold nanoparticle-antibody conjugates for human immunoglobulin G was reported by Chen *et al.* [87].

18.3.5 Surface-Enhanced Raman Spectroscopy (SERS)

SERS is a useful technique for bioanalyte detection using fingerprint recognition. Recently, two-dimensional (2D) materials have exhibited outstanding optical properties and good biocompatibility with SARS-CoV-2 [88]. Quantitative determination of 2-Thiouracil using SERS biosensor with substrates of silver nanoparticles seeded on graphene nanosheets (Ag/G) was reported by Saleh *et al.* for the detection of 2-Thiouracil at ultra-low concentrations and proved to be promising [89]. Wang *et al.* reported sensitive detection of neurotransmitters (dopamine and serotonin). The researchers demonstrated dopamine and serotonin detection using a graphene–Au nanopyramid heterostructure with a spectroscopic enhancement factor of ~10^{10} [90]. Ouyang *et al.* proposed a new graphene-Ag array for the sensitive detection of methylated DNA. Sensitivity and reproducibility were highly achieved with LOD around 1.8 pmol L^{-1}. The SERS substrate was highly reusable even after five cycles [91]. Liu *et al.* reported ($Ti_3C_2T_x$)-Ag nanoparticles for the detection of 4-MBA molecules [92]. A schematic illustration of the MXene-PDDA-Ag preparation is shown in Figure 18.7. The $Ti_3C_2T_x$ flakes have a zeta potential of −33.7 mV, and after modification with PDDA, the potential increased to 56.4 mV. The change in potential indicates the adsorption of PDDA molecules on the surface of the $Ti_3C_2T_x$ nanosheets. Two-dimensional Ag/Ti_3C_2Tx composite was fabricated by Yang *et al.* for the detection of beta-human chorionic

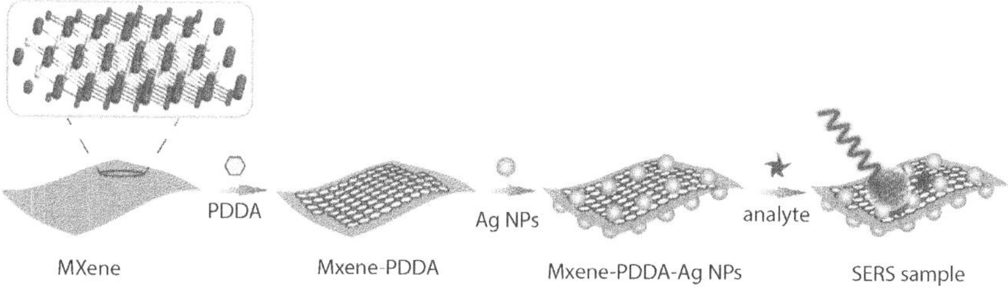

Figure 18.7 Schematic illustration of the preparation of MXene-PDDA-Ag NPs substrate and SERS detection. Reproduced with permission from Reference [92]. Copyright 2021 Elsevier.

gonadotropin with satisfactory recovery for the detection of actual serum samples [93].

18.4 Conclusions and Future Prospective

Since the discovery of the first 2D nanomaterial, graphene, the scientific community has been fascinated by its unique physicochemical properties. The exceptional properties of these 2D nanomaterials allow them to be used in different fields, such as imaging, antimicrobial agents, drug delivery, tissue engineering, and cancer therapy. Recently, biosensors have been fabricated with significant achievements in the biological, medical, and industrial fields. Advanced nanomaterials have provided new opportunities for the fabrication of biosensors. In this chapter, we first highlighted the intriguing properties of 2D materials, such as graphene, graphitic C_3N_4 (g-C_3N_4), black metal dichalcogenides, phosphorus (BP), transition metals,and metal-organic framework boron nitrides (BN). Methods for active material synthesis are also required to realize practical and functional applications. Therefore, recent research on 2D nanomaterial-based biosensors has been focused on. Biosensors provide quantitative information about the composition of the sensor-situated environment. The robustness and accuracy of biosensors are necessary for rapid field analysis compared to existing tools. Further exploration of this research area will create several possibilities. The long-term stability of the biosensor is a key challenging task in a biosensor fabrication. In addition, the quantity, quality, and production rate of the 2DMs are limited. Moreover, the biosensing ability of the biosensor is affected by the presence of impurities detected during the preparation methods. Modifications in nanomaterial-based biosensor synthesis are expected to speed up the diagnosis, treatment, and monitoring process. Thus, a better understanding of biosensors will significantly assist in the design of new and improved biosensors. Nanostructured materials can be further coupled with newly emerging 2D materials to explore a wide range of applications in a broad area.

References

1. Shuang, L., Ma, L., Zhou, M., Yuanqi, L., Xia, Y., Fan, X., Cheng, C., Luo, H., New opportunities for emerging 2D materials in bioelectronics and biosensors. *Curr. Opin. Biomed. Eng.*, 13, 32, 2020.

2. Zhang, X., Wang, H., Wang, H., Zhang, Q., Xie, J., Tian, Y., Wang, J., Xie, Y., Single-layered graphitic-C_3N_4 quantum dots for two-photon fluorescence imaging of cellular nucleus. *Adv. Mater.*, 26, 4438, 2014.
3. Yildiz, G., Warberg, M.B., Awaja, F., Graphene and graphene oxide for bio-sensing: General properties and the effects of graphene ripples. *Acta Biomater.*, 131, 62, 2021.
4. Yi, Y., Sun, Z., Li, J., Chu, P.K., Yu, X.F., Optical and optoelectronic properties of black phosphorus and recent photonic and optoelectronic applications. *Small Methods*, 3, 1900165, 2019.
5. Rodenas, T., Luz, I., Prieto, G., Seoane, B., Miro, H., Corma, A., Kapteijn, F., Xamena, F.X.L., Gascon, J., Metal–organic framework nanosheets in polymer composite materials for gas separation. *Nat. Mater.*, 14, 48, 2015.
6. Molle, A., Goldberger, J., Houssa, M., Xu, Y., Zhang, S.-C., Akinwande, D., Buckled two-dimensional Xene sheets. *Nat. Mater.*, 16, 163, 2017.
7. Yankowitz, M., Ma, Q., J-H, P., Leroy, B.J., van der Waals heterostructures combining graphene and hexagonal boron nitride. *Nat. Rev. Phys.*, 1, 112, 2019.
8. Gusmão, R., Sofer, Z., Pumera, N., Metal phosphorous trichalcogenides ($mpch_3$): From synthesis to contemporary energy challenges. *Angew. Chem. Int. Ed.*, 58, 9326, 2019.
9. Tan, C., Cao, X., Wu, X.J., He, Q., Yang, J., Zhang, X., Chen, J., Zhao, W., Han, S., Nam, G.H., Sindoro, M., Zhang, H., Recent advances in ultrathin two-dimensional nanomaterials. *Chem. Rev.*, 117, 6225, 2017.
10. Vaidyanathan, A., Mathew, M., Radhakrishnan, S., Rout, C.S., Chakraborty, B., Theoretical insight on the biosensing applications of 2D materials. *J. Phys. Chem. B*, 124, 11098, 2020.
11. Ravindra, N.M., Prodan, C., Fnu, S., Padronl, I., Sikha, S.K., Advances in the manufacturing, types and applications of biosensors. *JOM*, 59, 37, 2007.
12. Clark Jr., L.C., Wolf, R., Granger, D., Taylor, Z., Continuous recording of blood oxygen tensions by polarography. *J. Appl. Physiol.*, 6, 189, 1953.
13. Lei, Z.L. and Guo, B., 2D material-based optical biosensor: Status and prospect. *Adv. Sci.*, 9, 2102924, 2022.
14. Filice, M., Marchal, J.A., Gamiz, F., Chapter 7 - Biosensors based on two-dimensional materials. in: *2D Materials for Nanophotonics*, pp. 245–312, Elsevier, 2021.
15. Chauhan, N., Saxena, K., Tikadar, M., Jain, U., Recent advances in the design of biosensors based on novel nanomaterials: An insight. *Nanotechnol. Precis. Eng.*, 4, 045003, 2021.
16. Grieshaber, D., MacKenzie, R., Vörös, J., Reimhult, E., Electrochemical biosensors-sensor principles and architectures. *Sensors*, 8, 1400, 2008.
17. Jiang, H., Zhang, P., Wang, X. *et al.*, Synthesis of magnetic two-dimensional materials by chemical vapor deposition. *Nano Res.*, 14, 1789, 2021.
18. Shi, Y., Yuan, S., Ovanessian, B.T., Hermange, K., Huo, Y., Normand, B., Enhancing the barrier effect of sol-gel derived inorganic coating by doping h-BN nanosheet. *Appl. Surf. Sci.*, 544, 148849, 2021.

19. Duraisamy, S., Ganguly, A., Sharma, P.K., Benson, J., Davis, J., Papakonstantinou, P., One-step hydrothermal synthesis of phase-engineered MoS2/MoO3 electrocatalysts for hydrogen evolution reaction. *ACS Appl. Nano Mater.*, 4, 2642, 2021.
20. Choi, S.H., Yun, S.J., Won, Y.S. *et al.*, Large-scale synthesis of graphene and other 2D materials towards industrialization. *Nat. Commun.*, 13, 1484, 2022.
21. Yu, L., Yap, P.L., Tran, D.N.H., Santos, A.M.C., Losic, D., High-yield preparation of edge-functionalized and water dispersible few-layers of hexagonal boron nitride (hBN) by direct wet chemical exfoliation. *Nanotechnology*, 32, 405601, 2021.
22. Valiev, R., Nanomaterial advantage. *Nature*, 419, 887, 2002.
23. Yang, Y., Liu, T., Cheng, L., Song, G.S., Liu, Z., Chen, M.W., MoS_2-Based Nanoprobes for Detection of Silver Ions in Aqueous Solutions and Bacteria. *ACS Appl. Mater. Interfaces*, 7, 7526, 2015.
24. Abdulbari, H.A. and Basheer, E.A.M., Electrochemical biosensors: Electrode development, materials, design, and fabrication. *ChemBioEng Rev.*, 4, 92, 2017.
25. Choudhary, M. and Arora, K., Electrochemical biosensors for early detection of cancer, in: *Biosensor Based Advanced Cancer Diagnostics*, pp. 123–151, Elsevier, 2021.
26. Yang, Y., Yang, X., Yang, Y., Yuan, Q., Aptamer-functionalized carbon nanomaterials electrochemical sensors for detecting cancer relevant biomolecules. *Carbon*, 129, 380, 2018.
27. Disha, D., Kumari, P., Nayak, M.K., Kumar, P., An electrochemical biosensing platform for progesterone hormone detection using magnetic graphene oxide. *J. Mater Chem. B.*, 9, 5264, 2021.
28. Fang, Y.H.M., Xua, C., Ying, Z., Wang, H., Zhang, R., Cheng, H., Zeng, Y., A graphene-laminated electrode with high glucose oxidase loading for highly-sensitive glucose detection. *J. Mater. Sci. Technol.*, 66, 57, 2021.
29. Wu, L., Gao, J., Lu, X., Huang, C., Chen, D.J., Graphdiyne: A new promising member of 2D all-carbon nanomaterial as robust electrochemical enzyme biosensor platform. *Carbon*, 156, 568, 2020.
30. Liu, J., Jiang, X., Zhang, R., Zhang, Y., Wu, L., Lu, W., Li, J., Li, Y., Zhang, H., MXene-enabled electrochemical microfluidic biosensor: Applications toward multicomponent continuous monitoring in whole blood. *Adv. Funct. Mater.*, 29, 1807326, 2019.
31. Lei, Y., Zhao, W., Zhang, Y., Jiang, Q., He, J.H., Baeumner, A.J., Wolfbeis, S., Wang, Z.L., Salama, K.N., Alshareef, H.N., A MXene-based wearable biosensor system for high-performance *in vitro* perspiration analysis. *Small*, 15, 1901190, 2019.
32. Shahzad, F., Iqbal, A., Zaidi, S.A., Hwang, S.W., Koo, C.M., Nafion-stabilized two-dimensional transition metal carbide (Ti_3C_2Tx MXene) as a high-performance electrochemical sensor for neurotransmitter. *J. Ind. Eng. Chem.*, 79, 338, 2019.

33. Yang, X., Feng, M., Xia, J., Zhang, F., Wang, Z., An electrochemical biosensor based on AuNPs/Ti_3C_2 MXene three-dimensional nanocomposite for microRNA-155 detection by exonuclease III-aided cascade target recycling. *J. Electroanal. Chem.*, 878, 114669, 2020.
34. Wang, H., Li, H., Huang, Y., Xiong, M., Wang, F., Li, C., A label-free electrochemical biosensor for highly sensitive detection of gliotoxin based on DNA nanostructure/MXene nanocomplexes. *Biosens. Bioelectron.*, 142, 111531, 2019.
35. Pumera, M. and Loo, A.H., Layered transition-metal dichalcogenides (MoS_2 and WS_2) for sensing and biosensing. *Trends Analyt. Chem.*, 61, 49, 2014.
36. Shin, M., Yoon, J., Yi, C., Lee, T., Choi, J.W., Flexible HIV-1 biosensor based on the Au/MoS_2 nanoparticles/Au nanolayer on the PET substrate. *Nanomaterials*, 9, 1076, 2019.
37. Kim, H., Kim, H.Y., Seok, H., Kanade, V., Yoo, H., Park, K.Y., Lee, J.H., Lee, M.H., Kim, T., Flexible MoS2–polyimide electrode for electrochemical biosensors and their applications for the highly sensitive quantification of endocrine hormones: PTH, T3, and T4. *Anal. Chem.*, 92, 6327, 2020.
38. Cui, Z., Li, D., Yang, W., Fan, K., Liu, H., Wen, F., Li, L., Dong, L., Wang, G., Wua, W., An electrochemical biosensor based on few-layer MoS_2 nanosheets for highly sensitive detection of tumor marker ctDNA. *Anal. Methods*, 14, 1956, 2022.
39. Zhang, J., Han, D., Wang, S., Zhang, X., Yang, R., Ji, Y., Yu, X., Electrochemical detection of adenine and guanine using a three-dimensional WS_2 nanosheet/graphite microfiber hybrid electrode. *Electrochem. Commun.*, 99, 75, 2019.
40. Öndeş, B., Evli, S., Uygun, M., Uygun, D.A., Boron nitride nanosheet modified label-free electrochemical immunosensor for cancer antigen 125 detection. *Biosens. Bioelectron.*, 191, 113454, 2021.
41. Ouyang, H., Li, W., Long, Y., Carbon-doped h-BN for the enhanced electrochemical determination of dopamine. *Electrochim. Acta*, 369, 137682, 2021.
42. An, D., Zhang, X., Bi, Z., Shan, W., Zhang, H., Xia, S., Qui, M., Low-dimensional black phosphorus in sensor applications: Advances and challenges. *Adv. Funct. Mater.*, 31, 2106484, 2021.
43. Jakóbczyk, P., Kowalski, M., Brodowski, M., Dettlaff, A., Dec, B., Nidzworski, D., Ryl, J., Ossowski, T., Bogdanowicz, R., Low-powermicrowave-induced fabrication of functionalised few-layer black phosphorus electrodes: A novel route towards haemophilus influenzae pathogen biosensing devices. *Appl. Surf. Sci.*, 539, 148286, 2021.
44. Kumar, V., Brent, J.R., Shorie, M., Kaur, H., Nanostructured aptamer-functionalized black phosphorus sensing platform for label-free detection of myoglobin, a cardiovascular disease biomarker anostructured aptamer-functionalized black phosphorus sensing platform for label-free detection of myoglobin, a cardiovascular disease biomarker. *ACS Appl. Mater. Interfaces*, 8, 22860, 2016.

45. Zhu, C., Du, D., Lin, Y., Graphene-like 2D nanomaterial-based biointerfaces for biosensing applications. *Biosens. Bioelectron.*, 89, 43, 2017.
46. Eun Park, T. and Lee, S.H., A micellized fluorescence sensor based on amplified quenching for highly sensitive detection of non-transferrin-bound iron in serum. *Dalton Trans.*, 49, 4660, 2020.
47. Wu, D., Chen, L., Lee, W., Ko, G., Yin, J., Yoon, J., Recent progress in the development of organic dye based near-infrared fluorescence probes for metal ions. *Coord. Chem. Rev.*, 354, 74, 2018.
48. Liu, Z., Mo., Z., Niu, X., Yang, X., Jiang, Y., Zhao, P., Liu, N., Guo, R., Highly sensitive fluorescence sensor for mercury(II) based on boron- and nitrogen-co-doped graphene quantum dots. *J. Colloid Interface Sci.*, 566, 357, 2020.
49. Girigoswami, K. and Akhtar, N., Nanobiosensors and fluorescence based biosensors: An overview. *Int. J. Nano Dimens.*, 10, 1, 2019.
50. Carter, K.P., Young, A.M., Palmer, A.E., Fluorescent Sensors for Measuring Metal Ions in Living Systems. *Chem. Rev.*, 114, 4564, 2014.
51. Zhang, L., Guo, S., Dong, S., Wang, E., Pd nanowires as new biosensing materials for magnified fluorescent detection of nucleic acid. *Anal. Chem.*, 84, 3568, 2012.
52. Pham, H., Hoseini Soflaee, M., Karginov, A.V. *et al.*, Förster resonance energy transfer biosensors for fluorescence and time-gated luminescence analysis of rac1 activity. *Sci. Rep.*, 12, 5291, 2022.
53. He, S., Song, B., Li, D., Zhu, C., Qi, W., Wen, Y., Wang, L., Song, S., Fang, H., Fan, C., A graphene nanoprobe for rapid, sensitive, and multicolor fluorescent DNA analysis. *Adv. Funct. Mater.*, 20, 453, 2010.
54. Zhang, J.R., Huang, W.T., Xie, W.Y., Wen, T., Luo, H.Q., Li, N.B., Highly sensitive, selective, and rapid fluorescence Hg^{2+} sensor based on DNA duplexes of poly(dt) and graphene oxide. *Analyst*, 137, 3300, 2012.
55. Chang, H., Tang, L., Wang, Y., Jiang, J., Li, J., Graphene fluorescence resonance energy transfer aptasensor for the thrombin detection. *Anal. Chem.*, 82, 2341, 2010.
56. Gao, Y., Yan, X., Li, M., Gao, H., Sun, J., Zhu, S., Han, S., Jia, L.N., Zhao, X.N., Wang, H., A "turn-on" fluorescence sensor for ascorbic acid based on graphene quantum dots via fluorescence resonance energy transfer. *Anal. Methods*, 10, 611, 2018.
57. Wang, Q., Wang, W., Lei, J., Xu, N., Gao, F., Ju, H., Fluorescence quenching of carbon nitride nanosheet through its interaction with DNA for versatile fluorescence sensing. *Anal. Chem.*, 85, 12182, 2013.
58. Li, X., Li, Y., Qiu, Q., Wen, Q., Zhang, Q., Yang, W., Yuwen, L., Weng, L., Wang, L., Efficient biofunctionalization of MoS2 nanosheets with peptides as intracellular fluorescent biosensor for sensitive detection of caspase-3 activity. *J. Colloid Interface Sci.*, 543, 96, 2019.
59. Yuan, Y., Li, R., Liu, Z., Establishing water-soluble layered WS2 nanosheet as a platform for biosensing. *Anal. Chem.*, 86, 3610, 2014.

60. Xiao., L., Xu, L., Gao, C., Zhang, Y., Yao, Q., Zhang, G.J., A MoS_2 nanosheet-based fluorescence biosensor for simple and quantitative analysis of DNA methylation. *Sensors*, 16, 1561, 2016.
61. Shanmugaraj, K. and John, S.A., Water-soluble MoS2 quantum dots as effective fluorescence probe for the determination of bilirubin in human fluids. *Spectrochim. Acta A Mol. Biomol. Spectrosc.*, 215, 290, 2019.
62. Zhan, Y., Yang, J., Guo, L., Luo, F., Qiu, B., Hong, G., Lin, Z., Targets regulated formation of boron nitride quantum dots-Gold nanoparticles nanocomposites for ultrasensitive detection of acetylcholinesterase activity and its inhibitors. *Sens. Actuators B Chem.*, 279, 61, 2019.
63. Zhang, X., An, L., Bai, C., Chen, L., Yu, Y., Hexagonal boron nitride quantum dots: Properties, preparation and applications. *Mater. Today Chem.*, 20, 100425, 2021.
64. Sun, Z., Xie, H., Tang, S., Yu, X., Guo, Z., Shao, J., Zhang, H., Huang, H., Wang, H., Chu, P.K., Ultrasmall black phosphorus quantum dots: Synthesis and use as photothermal agents. *Angew. Chem.*, 127, 11688, 2015.
65. Li, A.Z., Ying, M., Liu, M., Wang, X., Wang, X., Cao, L., Zhag, H., Xu, G., Black phosphorus nanosheets for rapid microRNA detection. *Nanoscale*, 10, 5060, 2018.
66. Zhao, V.X.T., Wong, T., Zheng, X.T., Tan, Y.N., Zhou, X., Colorimetric biosensors for point-of-care virus detections. *J. Mater. Sci. Technol.*, 3, 237, 2020.
67. Liu, B., Zhuang, J.Y., Wei, G., Recent advances in the design of colorimetric sensors for environmental monitoring. *Environ. Sci. Nano*, 7, 2195, 2020.
68. Lan, L.Y., Yao, Y., Ping, J.F., Ying, Y.B., Ultrathin transition-metal dichalcogenide nanosheet-based colorimetric sensor for sensitive and label-free detection of DNA. *Sens. Actuators B Chem.*, 290, 565, 2020.
69. Hwang, J., Le., A.D.D., Trinh, C.T., Le, Q.T., Lee, K.G., Kim, J., Green synthesis of reduced-graphene oxide quantum dots and application for colorimetric biosensor. *Sens. Actuator A Phys.*, 318, 2021, 112495.
70. Tong, Z., Xu, M., Li, Q., Liu, C., Wang, Y., Sha, J., Polyelectrolyte-functionalized reduced graphene oxide wrapped helical POMOF nanocomposites for bioenzyme-free colorimetric biosensing. *Talanta*, 220, 121373, 2020.
71. Fan, S., Zhao, M., Ding, L., Li, H., Chen, S., Preparation of Co3O4/crumpled graphene microsphere as peroxidase mimetic for colorimetric assay of ascorbic acid. *Biosens. Bioelectron.*, 89, 846, 2017.
72. Tang, Y., Hu, Y., Zhou, P., Wang, C., Tao, H., Wu, Y., Colorimetric detection of kanamycin residue in foods based on the aptamer-enhanced peroxidase-mimicking activity of layered WS_2 nanosheets. *J. Agric. Food Chem.*, 69, 2884, 2021.
73. Lee, J., Kim, Y.K., Lee, S., Yoon, S., Kim, W.K., Graphene oxide-based NET strategy for enhanced colorimetric sensing of miRNA. *Sens. Actuators B Chem.*, 282, 861, 2019.
74. Wu, S.J., Duan, N., Qiu, Y.T., Li, J.H., Wang, Z.P., Colorimetric aptasensor for the detection of Salmonella enterica serovar typhimurium using

$ZnFe_2O_4$-reduced graphene oxide nanostructures as an effective peroxidase mimetics. *Int. J. Food Microbiol.*, 261, 42, 2017.
75. Rostami, S., Mehdinia, A., Niroumand, R., Jabbari, A., Enhanced LSPR performance of graphene nanoribbons-silver nanoparticles hybrid as a colorimetric sensor for sequential detection of dopamine and glutathione. *Anal. Chim. Acta*, 1120, 11, 2020.
76. Sharma, V. and Mobin, S.M., Cytocompatible peroxidase mimic CuO:graphene nanosphere composite as colorimetric dual sensor for hydrogen peroxide and cholesterol with its logic gate implementation. *Sens. Actuators B Chem.*, 240, 338, 2017.
77. Li, H., Wen, Y., Zhu, X., Wang, J., Zhang, L., Sun, B., Novel heterostructure of a MXene@NiFe-LDH nanohybrid with superior peroxidase-like activity for sensitive colorimetric detection of glutathione. *ACS Sustain. Chem. Eng.*, 8, 520, 2020.
78. Vu, C.A. and Chen, W.U., Field-effect transistor biosensors for biomedical applications: Recent advances and future prospects. *Sensors*, 19, 4214, 2019.
79. Sengupta, J. and Hussain, C.M., Graphene-based field-effect transistor biosensors for the rapid detection and analysis of viruses: A perspective in view of COVID-19. *Carbon Trends*, 2, 100011, 2021.
80. Ono, T., Oe, T., Kanai, Y., Ikuta, T., Ohno, Y., Maehashi, K., Inoue, K., Watanabe, Y., Nakakita, S., Suzuki, Y., Kawahara, T., Matsumoto, K., Glycan-functionalized graphene-FETs toward selective detection of human-infectious avian influenza virus. *Jpn. J. Appl. Phys.*, 56, 030302, 2017.
81. Seo, G., Lee, G., Kim, M.J., Baek, S.H., Choi, M., Ku, K.B., Lee, C.S., Jun, S., Park, D., GiKim, H., Kim, S., Lee, J., Kim, B.T., Park, E.C., IlKim, S., Rapid detection of COVID-19 causative virus (SARS-CoV-2) in human nasopin human-effect transistor based biosensor. *ACS Nano*, 14, 5135, 2020.
82. Hwang, M.T., Heiranian, M., Kim, Y. *et al.*, Ultrasensitive detection of nucleic acids using deformed graphene channel field effect biosensors. *Nat. Commun.*, 11, 1543, 2020.
83. Lee, J., Dak, P., Lee, Y. *et al.*, Two-dimensional layered MoS_2 biosensors enable highly sensitive detection of biomolecules. *Sci. Rep.*, 4, 7352, 2014.
84. Xu, B., Zhu, B., Zhang, W., Zhen, X., Pei, Z., Xue, Q., Zhi, C., Shi, P., Ultrathin MXene-micropattern-based field-effect transistor for probing neural activity. *Adv. Mater.*, 28, 3333, 2016.
85. Li, Y., Peng, Z., Holl, N.J., Hassan, M.R., Pappas, J.M., Wei, C., Izadi, O.H., Wang, Y., Dong, X., Wang, C., Huang, Y.W., Kim, D., Wu, C., T1-MXene–graphene field-effect transistor sensing of influenza virus and SARS-CoV-2. *ACS Omega*, 6, 6643, 2021.
86. Rahman, M.S., Hasan, M.S., Rikta, K.A., Anower, M.S., A novel graphene coated surface plasmon resonance biosensor with tungsten disulfide (WS_2) for sensing DNA hybridization. *Opt. Mater.*, 75, 567, 2018.

87. Chen, Y., Ren, R., Pu, H., Chang, J., Mao, S., Chen, J., Field-effect transistor biosensors with two-dimensional black phosphorus nanosheets. *Biosens. Bioelectron.*, 89, 505, 2017.
88. Das, G.M., Managò, S., Mangini, M., DeLuca, A.C., Biosensing using SERS active gold nanostructures. *Nanomaterials*, 12, 2679, 2021.
89. Saleh, T.A., Al-Shalalfeh, M.M., Al-Saadi, A., Silver loaded graphene as a substrate for sensing 2-thiouracil using surface-enhanced Raman scattering. *Sens. Actuators B Chem.*, 254, 1110, 2018.
90. MingXia, P.W., Liang, O., Sun, K., Cipriano, A.F., Schroeder, T., Liu, H., Xie, Y.H., Label-free SERS selective detection of dopamine and serotonin using graphene-au nanopyramid heterostructure. *Anal. Chem.*, 87, 10255, 2015.
91. Ouyang, L., Hu, Y., Zhu, L., Cheng, G.J., Irudayaraj, J., A reusable laser wrapped graphene-Ag array based SERS sensor for trace detection of genomic DNA methylation. *Biosens. Bioelectron.*, 92, 755, 2017.
92. Liu, R., Jiang, L., Yu, Z., Jing., X., Liang, X., Wang, D., Yang, B., Lu, C., Zhou, W., Jin, S., MXene (Ti_3C_2Tx)-Ag nanocomplex as efficient and quantitative SERS biosensor platform by *in-situ* PDDA electrostatic self-assembly synthesis strategy. *Sens. Actuators B Chem.*, 333, 129581, 2021.
93. Yang, J., Xu, C., Yang, Q., Wei, W., Wang, C., Ag nanoparticle *in situ* decorated on Ti3C2Tx with excellent SERS and EIS immunoassay performance for beta-human chorionic gonadotropin. *Microchim. Acta*, 189, 348, 2022.

19

Transition Metal Dichalcogenide (TMD)-Based 2D Nanomaterials for Various Kinds of Rechargeable Batteries

Periyakaruppan Karuppasamy[1]* and Varatharaj Rajapanian[2]

[1]*Department of Chemistry, Dayananda Sagar College of Engineering, Bengaluru, India*
[2]*Department of Chemistry, Sri Ramakrishna Mission Vidyalaya College of Arts and Science, Coimbatore, India*

Abstract

Transition metal di-chalcogenides (TMDCs) have the common formula MX_2, where M represents the transition metal atom (Ni, Mo, Co, Mn, W, etc.), and X represents the chalcogenide atoms (S, Se, Te, etc.). These MX_2 types of TMDCs have received great interest as electrode materials for energy storage applications in rechargeable batteries because of their atomically layered structure, high surface area, and enhanced electrochemical characteristics. The aforementioned remarkable properties of these materials have made it possible for researchers to discover new materials in all fields over the past two decades. The weak van der Waals forces of attraction between the two layers are mainly responsible for their covalently bound nature. Various synthesizing methods, rechargeable batteries such as Li, Na, K, Mg, Zn, Al, and Li-S ions energy storage applications using various 2D-TMDCs were discussed in this chapter.

Keywords: Transition metal dichalcogenides (TMDCs), 2D-nanomaterials, heterostructures, energy storage devices, rechargeable batteries

**Corresponding author*: periyakaruppankaruppasamy@gmail.com
ORCID ID: https://orcid.org/0000-0002-9503-1728

Subhendu Chakroborty and Kaushik Pal (eds.) 2D Nanomaterials: Synthesis, Properties and Applications, (435–474) © 2024 Scrivener Publishing LLC

19.1 Introduction

In the past decade, nanotechnology has been one of the most important and pioneering fields for numerous applications. This emerging technology explains nanomaterial synthesis, structural insights, and applications in various scientific fields. Many dimensions (0D, 1D, 2D, and 3D) of TMDCs are currently available, among those dimensions 2D-TMDCs exhibit excellent characteristics and are useful for various applications in electronics, sensors, energy storage devices, photocatalysts, electrocatalysts, photonics, dye sensitized solar cells (DSSCs), and biomedical fields [1–6]. In particular, 2D nanostructure TMDCs can provide benefits such as exceptional active sites, larger surface area, enhanced functionalities, better compatibility with device integration, band gap, and layered structure. These materials can be used to fabricate flexible electronic devices in integrated circuits (ICs), supercapacitors, rechargeable batteries, and solar cells [7–15]. Among all all TMDC synthesis methods, many researchers have utilized bottom-up techniques over top-down approaches. Many researchers have employed CVD methods for synthesizing 2D-TMDCs layered materials owing to the synthesis of good crystalline products, well-defined material structures, phases, morphologies, and extensive lateral layer sizes of TMDCs. However, these CVD methods also have shortcomings such as high cost and low commercial scalability. TMDC-layered materials contain atomic layers of transition elements, such as molybdenum (Mo) and tungsten (W), which are sandwiched between two chalcogenide atomic layers, such as sulfur (S), selenium (Se), and tellurium (Te), which stacked by van der Waals interactions between the layers. These stacking patterns of layers are responsible for the formation of various polymorphs/phases (Figure 19.1), which can alter the properties of TMDCs. Generally, the determination of properties of electrical, conductive, and semiconductive materials mainly depends on the combination of the d-electron configuration in the transition element and their coordination nature. TMDCs phases/polymorphs are represented as 1T, 2H, 3R, and distorted 1T0, and they are denoted as X-M-X in a unit cell. 2D-TMDCs have excellent properties, such as tunable band gap, structure, electrical properties, and semiconducting properties, and their tunable bandgap depends on the number of monolayers. Semiconducting TMDCs (MoS_2 and WS_2) layered structure materials are used in the sensor and optical fields owing to their luminescent properties in the visible region. Generally, TMDCs have either rhombohedral or hexagonal symmetry and the transition metal atoms have trigonal prismatic or octahedral coordination. Based on the combination of X and M, TMDCs can provide two types of geometries such as trigonal prismatic (2H) and octahedral (1T). As can be seen in the trigonal-prismatic crystal structure, M coordinate to 6X-atoms providing a stable polymorph, whereas an octahedral crystal structure of 6-X atoms coordinates with the M atom and creates a distorted (metastable) octahedral polymorph (Figure 19.1). However, both polymorphs exhibited distinct electronic characteristics. In trigonal-prismatic TMDCs, the d-orbital is split into dz^2, dx^2-y^2, d_{xy}, d_{xz}, and d_{yz} and possesses significant band gap energy, whereas in octahedral TMDCs, the d-orbital is split into two degenerated dz^2, dx^2-y^2, d_{xy}, and dxz, yz (Figure 19.2). This orbital split-up tendency provides electronic characteristics that are dissimilar to those of TMDCs used as semiconductors. For instance, 2H–$TaSe_2$ (*d*1) is metallic in nature owing to the partially filled d_z^2 orbital, whereas 2H–WSe_2 (d^2) is a semiconducting material owing to its fully occupied d_z^2 orbital. Currently, more than 60 TMDCs materials have been synthesized, out of 60 more than 40 compounds have 2D-layer structures. Examples include, tungsten disulfide (W-S_2), molybdenum disulfide (Mo-S_2), titanium disulfide (Ti-S_2), niobium disulfide (Nb-S_2), tantalum disulfide (Ta-S_2), zirconium disulfide (Zr-S_2), niobium diselenide (Nb-Se_2), tantalum diselenide (Ta-Se_2), titanium diselenide (Ti-Se_2), and tungsten diselenide (W-Se_2) [16–20]. Many studies have proved that 2D-TMDCs are potential materials for various applications such as electronics, magnetics, photocatalysis, electrocatalysis, energy storage devices, solar cells, lithium-ion batteries, sensors, and water-splitting [21–35]. Molybdenum disulfide (Mo-S_2) is an important and attractive 2D-TMDCs materials owing to its distinctive chemical, optical, electrical, and electrochemical properties, and it has been used to fabricate active electrode materials in lithium, sodium, and magnesium ion batteries [36]. MoS_2 is in bulk form it has a 1.2 eV indirect band gap whereas it changes to a layer structure band gap (1.9 eV) changes to semiconducting material [37–39]. MoS_2 is a distinctive three-layered TMDCs, where the Mo atoms of a transition metal are positioned between two S-atom layers to create an excellent sandwich structure (S–Mo–S). The three-layered structure of MoS_2 has three possible geometries such as hexagonal (2H, MoS_2), trigonal (1T, MoS_2) and rhombohedral (3R, MoS_2) geometries. In particular, 2H-MoS_2 was the most stable state compared to the other two geometries. MoS_2 monolayer is showed two polymorphs/phases owing to coordination types between Mo atom and the nearest six S atoms and it gives an octahedral (D_{3d}), trigonal-prismatic (D_{3h}) coordination that commonly called as 1T and 1H (Figure 19.1) [40]. Tungsten disulfide (WS_2) is also an important semiconductor TMDC material that possesses the same structure as MoS_2. It exhibits a bandgap energy (1.3 eV) for the bulk and a bandgap (2.1 eV) for the monolayer form, and the mobility of the MoS_2 layer structure at 298 K was measured to be 19 $cm^2/V/s$. The unusually high anisotropy of WS_2 provides excellent distinct properties such as morphology, crystalline size, weak van der Waals interactions between the layers, and the orientation of the crystals, making WS_2 an important material for many applications [41–43]. Titanium disulfide (TiS_2) is an essential sequential layered TMDC material with repeating units of Ti and S, and it has a hexagonal crystal structure analogous to that of CdI_2. The weak van der Waals interaction dominates the two S–S bond layers and the strong bond between the Ti atom layer and the S layer [44]. These van der Waals interactions provide excellent stability to the [001] surface of the layer and allow for additional atoms in the material. There are six S-atoms coordinated to Ti atoms that provide an octahedral geometry [44]. TiS_2 exhibits a range of band-gap energy between 0.5 eV and 2.5 eV as a semiconducting material with a Bohr radius of ~6.43 nm and the lattice parameter values a ($a = b$) and c of TiS_2 were found to be 3.40 A°, 5.96 A°, respectively. Different TiS_2 morphologies are synthesized successfully still now like nano-disc, nanotube, nano-cluster, thin film, fullerene, and whisker and used for various applications [45–49]. Many researchers

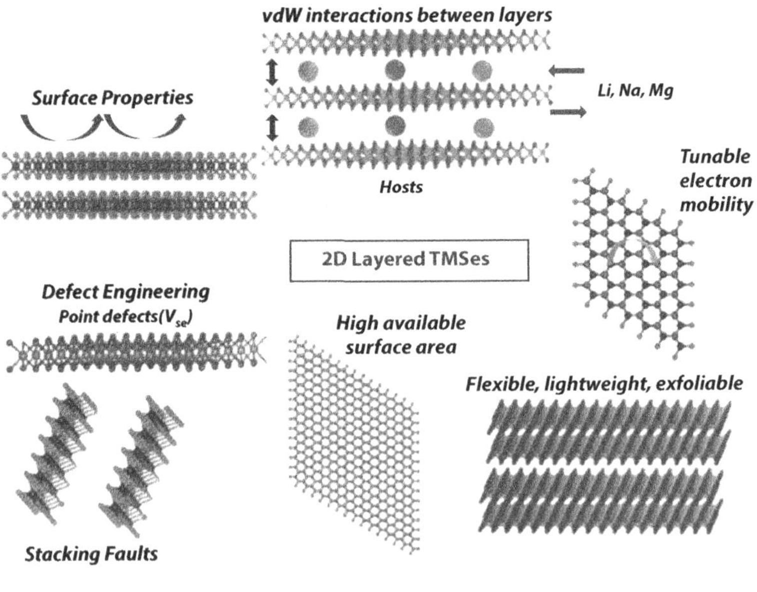

Figure 19.1 Structural properties of 2D-TMDC materials for enhancing the performance of energy storage applications.

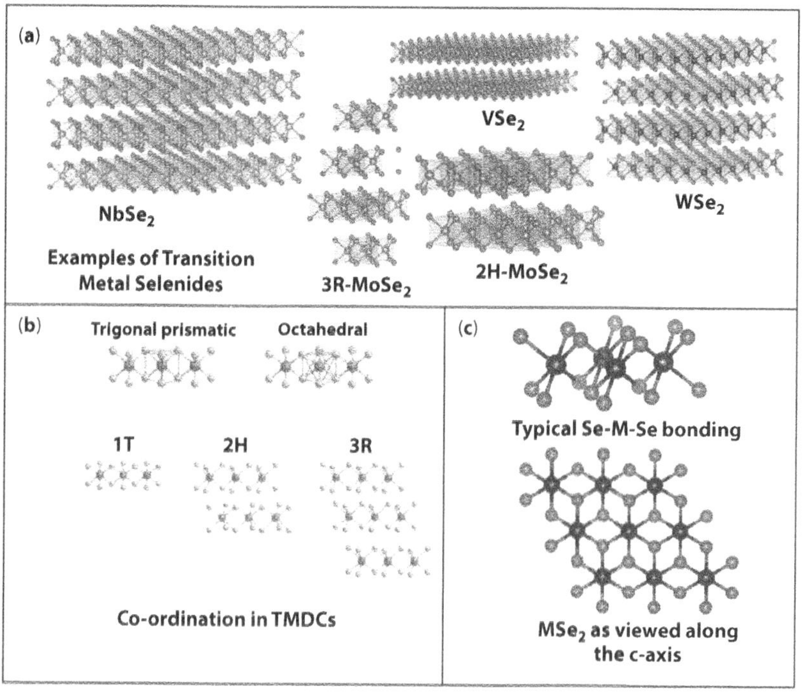

Figure 19.2 (a) Metal selenides (TM-Se) utilized in different battery applications ((LIBs), (NIBs), and (MIBs)); (b) Coordination geometries in TMSes. Replicated with consent from Ref. [52]; and (c) Visualizing TMSes bonding sandwiched between transition metal and selenium in different axes.

have used 2D-TMDCs for making devices such as energy conversion and storage, solar cells, fuel cells, light-emitting diodes, supercapacitors, and Li/Na-ion battery devices. For instance, few 2D-TMDCs are MoS_2, $MoSe_2$, $MoTe_2$, WS_2, WSe_2, and WTe_2 materials are potential aspirants for fabricating anode/cathode electrodes in the field of supercapacitors and rechargeable Li-ion batteries [50, 51].

This chapter presents an overall interpretation of the synthesis, characterization, and uses of 2D-transition metal di-chalcogenides (TMDCs) in various rechargeable batteries. It provides basic information about the structure of 2D-transition metal di-chalcogenides and their structural insights. It includes various methods for synthesizing 2D-TMDCs and discusses rechargeable battery applications.

19.2 Synthesis of 2D-TMDCs

To synthesize semiconductor-based TMDCs electrode materials for fabricating new transportable flexible devices for achieving enhanced performance using various synthetic methods such as layer deposition, exfoliation, hydrothermal, one-pot solvothermal, chemical vapor deposition, and electrospinning. Various synthesis approaches for synthesizing 2D-TMDCs, characteristic size and thickness of 2D-TMDCs material, and their electrochemical applications are shown in Figure 19.3.

Chemical vapor deposition (CVD) is a vacuum deposition approach to deposit a solid material and then form a film like thin-layer at high temperature that have ~1 μM thickness and divergent properties as compared to bulk material. The volatile gaseous starting material decomposes over the

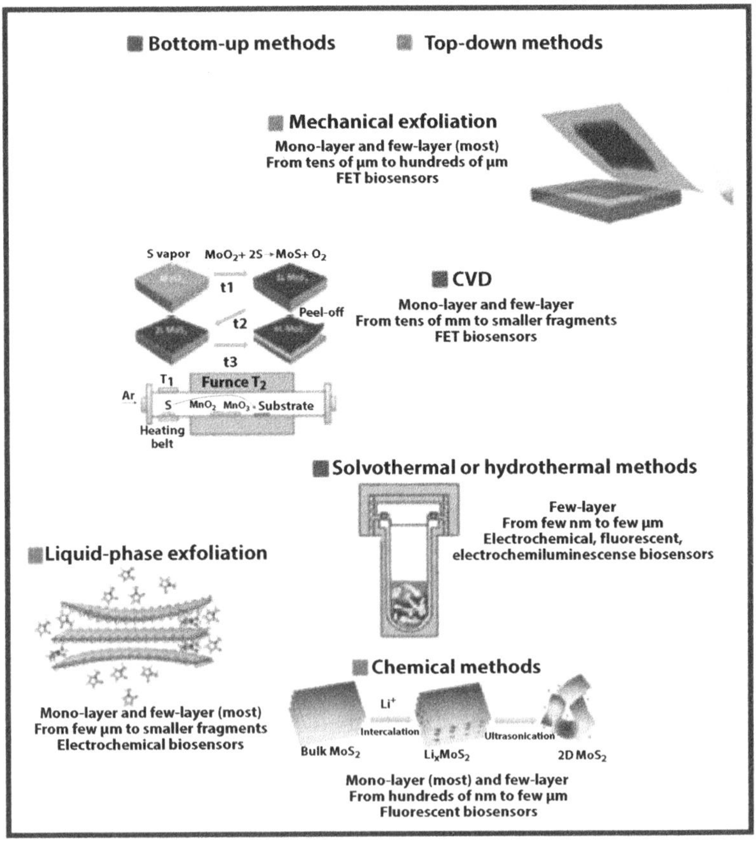

Figure 19.3 Various synthesis approaches for synthesizing 2D-TMDCs, characteristic size and thickness of 2D-TMDCs material, and their electrochemical applications. Reproduced with permission from Ref. [53].

surface of the substrate because of the chemical reaction. This CVD process is a reliable method for preparing homogeneous film growth, crystal arrays, and better reproducibility, excellent relative surface area, and high-quality atomic thin-layer material. Hydro/solvothermal methods involve the synthesis of materials under closed conditions at high pressure and temperature using aqueous and non-aqueous solvents in a stainless vessel, Zhan *et al.* synthesized a nanocrystalline 2H-MoS_2 semiconductor with 100 nm using pyridine, MoO_3, and sulfur precursors *via* a solvothermal approach [54]. This approach was used to obtain an average 100 nm diameter and semiconducting materials of different shapes. Fan *et al.* prepared nano-crystallites of MoS_2 and $MoSe_2$ by hydrothermal process with Na_2MoO_4 and $Na_2S_2O_3$/Na_2SeSO_3 starting materials resulting crystallite sizes are found to be 4 and 7 nm respectively that annealed at 350 °C about 9 h resulting the formation of 2H phase-MoS_2 and $MoSe_2$ semiconducting materials [55]. Wang *et al.* synthesized self-assembled monolayer MoS_2 by a solvothermal method using octylamine and ethanol as the solvent and additive, respectively [56]. The rapid injection method has been utilized to synthesize a variety of quantum dots (QDs) such as cadmium sulfide (CdS), selenide (CdSe), and telluride (CdTe) in containers at higher temperatures with coordination ligands and solvents [57, 58]. Recently, researchers have developed a hot-injection method to synthesize monodisperse TMDCs by varying the stabilizing ligands, precursors, reaction times, and temperatures to obtain TMDCs with different morphologies, sizes, and polymorphs. Yin *et al.* prepared blue-light-emitting WS_2 QDs *via* the hot injection method and it exhibited a higher quantum yield than reported materials [59]. Savjani and co-workers synthesized oleylamine (OAm)-stabilized ^{1}H-Mo-S_2 material *via* a hot injection method used a precursor of $Mo_2O_2S_2(S_2COEt)$ and lateral sizes between 100 and 1,000 nm. The reaction time was minimal (3 min) for growing nanosheets at 200 °C, and the obtained nanosheet showed an average lateral dimension of 4.78 nm than those materials (11.29 nm) prepared at 325 °C [60]. Jung *et al.* produced size dependent Mo-Se_2 and W-Se_2 nanosheet materials with various capping ligands (oleylamine, oleyl alcohol, and oleic acid) with different binding energies for improving their lateral growth (200–400 nm) of the nanosheets [61]. Zhou *et al.* prepared 2H-WS_2 layered semiconducting materials using capping ligands such as oleic acid (OA), oleylamine (OAm), trioctylphosphine oxide (TOPO), and squalane ligands [62]. Geisenhoff *et al.*, discussed the effect of ligands on the structure of WSe_2 nanocrystalline materials synthesized by the hot injection method [63] using Ph_2Se_2, $W(CO)_6$, OA, and TOPO, respectively. The mixing ratio of OA and $W(CO)_6$ changed the reactivity, phases, and morphologies of the WSe_2 material under various conditions. Researchers have adopted the shape conversion approach to prepare solution-based ultrathin sheets of WS_2 through sulfidation of $W_{18}O_{49}$ nanorods using a hot injection technique [64, 65]. Seo *et al.* synthesized lateral-size 2 H-WS_2 nano-sheets by converting $W_{18}O_{49}$ into WS_2 through sulfidation using carbon disulfide, hexadecylamine and $W_{18}O_{49}$ nano-rods at 250 °C. This process, in which hexadecylamine acted as a solvent and a stabilizing agent, was confirmed by TEM analysis [64]. 2D-WS_2 nanosheet material formation via shape-transformation technique. (a) Schematic representation of synthesizing of 2D-WS_2 nanosheets. (b) TEM image of W18O49 nanorods. TEM image of WS_2 for (c) 10 min

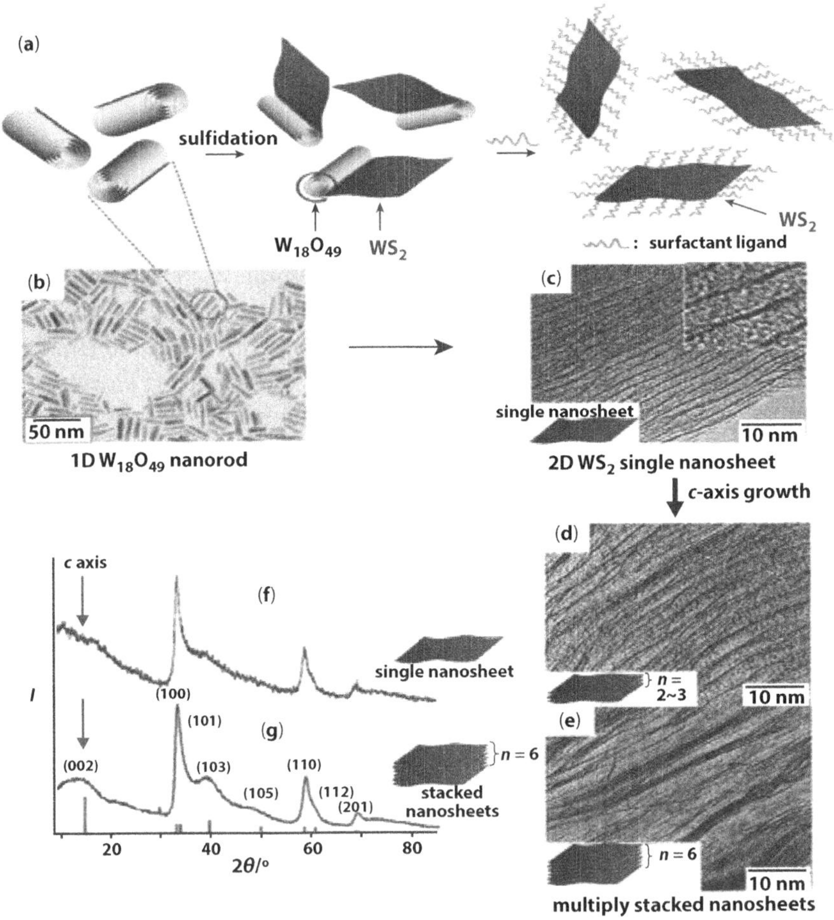

Figure 19.4 2D-WS_2 nanosheet material formation *via* shape-transformation technique. (a) Schematic representation of synthesizing of 2D-WS_2 nanosheets. (b) TEM image of $W_{18}O_{49}$ nanorods. TEM image of WS_2 for (c) 10 min (d) 30 min, and (e) 1 h, respectively. (f) X-ray diffraction pattern of WS_2, (g) WS_2 nano-sheet materials. Reproduced with permission from Ref. [64].

(d) 30 min, and (e) 1 h, respectively. (f) X-ray diffraction pattern of WS_2, (g) WS_2 nano-sheet materials are depicted in Figure 19.4.

Ding and his coworkers have prepared variety of TMDCs QDs such as 2H-WS_2, MoS_2, $MoTe_2$, $MoSe_2$, RuS_2, $RuSe_2$, and WSe_2 by single step synthesis technique using D_2O as solvent, gluconate, polyarginine, cysteine, and bovine serum albumin as surfactants resulting the formation of quantum dots at ~3.9 nm [66]. Sharma *et al.* used multistep bottom-up approach to produce colloidal WS_3 which is calcinated at 850 °C transforming to WS_2 nano-sheets [67]. As a result, the bottom-up approach is a versatile technique for synthesizing high-quality 2D-TMDCs with different sizes, phases, morphologies, relative surface-to-volume ratios, and electrical properties are presented (Table 19.1).

19.3 Applications of 2D-Transition Metal Di-Chalcogenides (2D-TMDCs) in Various Categories of Rechargeable Batteries

Gaston Plante (a French physicist) made a lead acid rechargeable battery for the storage of electrical energy in the 1859 [90]. MoS_2 was patented as a cathode material by Haering *et al.* for rechargeable LIBs applications [91]. In 1991, Sony developed and commercialized first LIB using layered $LiCoO_2$ as a cathode [92].

Recently, many scientific communities and research groups have shown interest in developing TMDCs for use in rechargeable batteries owing to their attractive characteristics, such as an enhanced electrical conductivity, larger relative surface area, and multivalent redox activities. Recently, lithium-ion batteries have exhibited better theoretical capacitances than sodium-ion (Na-ion), potassium-ion (K-ion), and magnesium-ion (Mg), and zinc-ion (Zn-ion) batteries owing to larger ionic radii and short operational lifespans, respectively. Significant efforts are required to improve electrochemical performance of rechargeable batteries to achieve high specific capacity, low cost, and safety. Recently, the cycle capability of lithium (Li)/sulfur (S) batteries has been improved by incorporating glass ceramics as solid electrolytes. The electrochemical cell (In/$LiCoO_2$) with a glass ceramic as a solid electrolyte ($80Li_2S{\cdot}20P_2S_5$) preserved an excellent cyclic capacity of approximately 1,000 mAh/g with an efficiency of 100% used for 200 cycles, and an increasing trend of ionic conductivity was observed with increasing Li-ion concentrations [93–95]. The 2D-TMDC applications for rechargeable batteries and energy storage devices and their mechanisms are depicted in Figures 19.5 and 19.6, respectively.

Table 19.1 Comparison of dissimilar synthetic approaches used for preparing numerous 2D-TMDCs.

Methods	2D-TMDCs	Morphology	Phase	Advantage	Disadvantage	Ref.
Hydrothermal	MoS_2, $MoSe_2$, WS_2, WSe_2	Microrods, micro/nanoparticles, microspheres, nano-sheets, nano-flowers, laterally confined monolayers (QDs)	Amorphous, 2H, 1T/1T0	Simple, one-step process, aqueous phase synthesis	High temperatures and pressures required	[55, 68–75]
Solvothermal	MoS_2, WS_2	Platelets, nano-sheets, 3D self-assembled tubes	2H	Simple, one-step process	Requires solvents	[76–79]
Hot injection	MoS_2, $MoSe_2$, WS_2, WSe_2	Laterally confined monolayers (QDs), nano-flowers, monolayer and few-layer nano-sheets	1H, 2H, 1T0	High level of controllability, Mono disperse products	High temperatures, solvents	[80–89]

(Continued)

Table 19.1 Comparison of dissimilar synthetic approaches used for preparing numerous 2D-TMDCs. (*Continued*)

Methods	2D-TMDCs	Morphology	Phase	Advantage	Disadvantage	Ref.
One-pot synthesis	Quantum dots of MoS_2, RuS_2, $RuSe_2$, WS_2, WSe_2, $MoSe_2$, $MoTe_2$, etc.	Restrained monolayers of quantum dots	2 H	Extremely fast response time, even, mono-dispersed Layer product, extensive range of TMDCs conceivable	–	[42]
Shape transformation	WS_2 ultrathin Nanosheets	Ultrathin nano-sheets, stacked nano-sheets	1T0, 2H	Highly mono disperse Product	needs highly pure precursors and time, more than one step process	[64, 65]
Calcination method	WS_2	Nano-sheets	3 R, 2 H	Extremely scalable, no need of Surfactants, to be produced larger quantity of products	Several step Process, high temperature is needed.	[67]

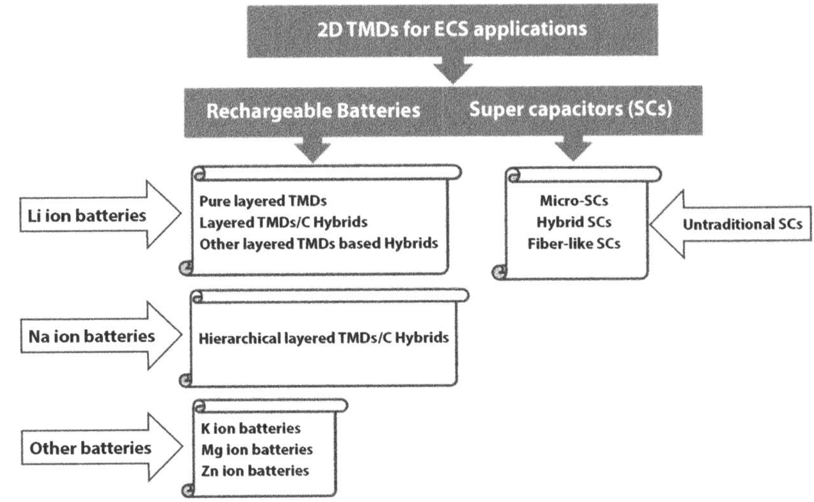

Figure 19.5 Flowchart for various applications of TMDCs in batteries.

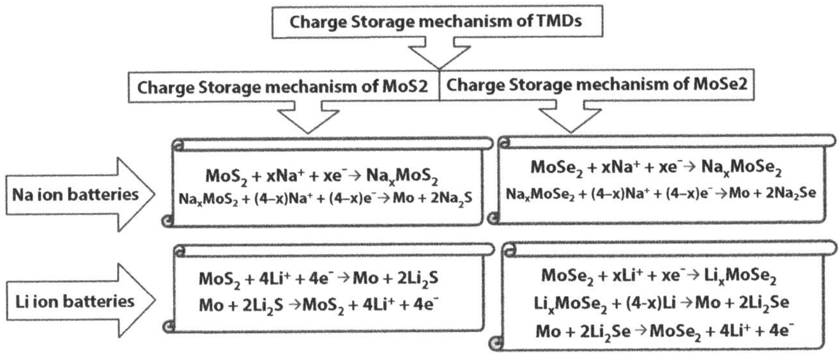

Figure 19.6 Flowchart for charge storage mechanism of TMDCs in batteries.

19.3.1 Li-Ion Batteries (LIBs)

Li-ion batteries (LIBs) have been extensively studied and successfully applied in commercial devices owing to their desirable properties. However, the cell is comprised of the anode and cathode which are immersed in an electrolyte containing Li ions, and the electrodes are separated using a porous polymer membrane. The membrane exchanges Li ions from one electrode to another. Because of the external electrical supply to the two connected electrodes, electrons are transferred from the cathode to the anode through the electrolyte medium by charging. At the same time, Li ions move from the cathode to the anode through an electrolyte, resulting in the production of energy, which is stored in the battery in chemical energy form by means of different chemical potentials. During discharging, electron transfer occurs between the anode and cathode *via* an external load for external applications, where Li ions move from the anode to the cathode through the electrolyte medium. Schematic diagram of lithium-ion batteries (a) Charging (b) Discharging cycles is given in Figure 19.7.

Generally, 2D-graphite material prefers as anode materials for LIBs because of their excellent conductivity due to π-electron delocalization and suitable structure for intercalation and Li-ion diffusion [96]. However, the theoretical specific capacitance of graphite in Li-ion batteries is inadequate; for example, individual Li ions intercalate to six C-atoms in the hexagonal sheet structure of graphite and the intercalation of one Li atom into graphite is occupied by six C-atoms, which reduces the specific capacity of graphite anode substances ranging from 280 to 330 mAh/g [97]. Recently, TMDCs have received considerable attention among the different kinds of reported 2D materials. Li and his workers [98] have deliberated an adsorption and diffusion environment of Li-species at two dissimilar morphologies of MoS_2 (Nano-sheets, Zig-Zag nanoribbons) theoretically and to match up to bulk structure material. The obtained MoS_2 nanosheet showed a noticeable Li-species intercalation owing to the high inter-layer (~6.37 Å) distance as match up to the inter-layer (6.18 Å) distance of bulk MoS_2. However, the binding energy of the Li species in the MoS_2 nanosheets is smaller than that in bulk MoS_2. Hence, MoS_2 nanosheets are unsuitable for cathode applications. Shu *et al.* [99] exhibited Li ion induced physical changes of MoS_2 nanosheets during discharging which pursues a layer-by-layer dissociation reaction pathway. Interim modification of the structure of MoS_2 nanosheets from the 2H-phase to the 1T-phase owing to ultra-low

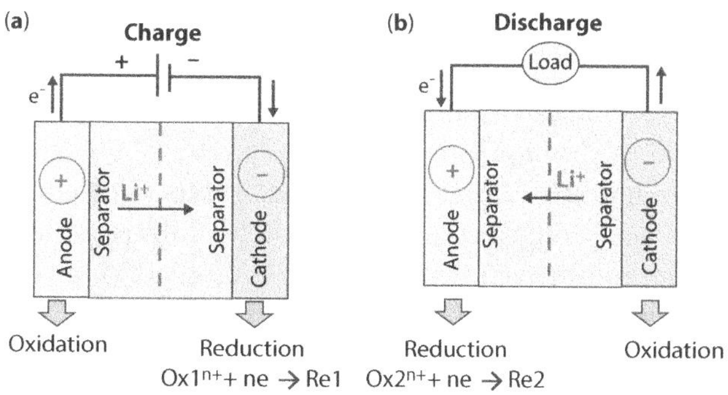

Figure 19.7 Schematic diagram of lithium-ion batteries (a) Charging, (b) Discharging cycles.

transition barriers results in a more rapid capacity loss. Lian *et al.* [100] synthesized high-quality graphene nanosheets for electrochemical performance in LIBs with a high reversible capacitance. Hwan *et al.* [101] studied the structure of layer structured Mo-S_2 nanoplates of approximately 30 nm synthesized by a one-pot hydrothermal approach, and the resultant material had a larger (0.69 nm) inter-layer distance than the interlayer distance (0.62 nm) of the bulk material. The obtained larger disordered interlayer structure led to a better rate capability at 50 °C (53.1 A/g) and exhibited a reversible capacity of 700 mAh/g. Additionally, the $LiCoO_2$/MoS_2 cell exhibited a superb capacity retention of 60 cycles, and the interface structure of graphene/MoS_2 was differentiated using the lattice fringe diameter *via* transmission electron microscopy. The composite graphene/MoS_2 aerogel-based anode material exhibited a steady specific charge capacitance compared with that of the MoS_2 anode in LIBs for several cycles [102]. Zhang *et al.* [103] prepared morphology proscribed MoS_2 nanosheets by means of graphene/acid *via* hydrothermal synthesis and synthesized Mo-S_2 nanosheets are self-arranged like cockscomb with (100) lattice plane lying on graphene sheets. A larger aggregation of MoS_2 nanoplates was grown without restraint on the graphene sheets without acetic acid. This graphene-coated MoS_2 shows potential as an anode in LIBs because of its better conductivity and reduced capacity loss throughout the charging/discharging process [104]. Zhang *et al.* synthesized MoS_2 nanotube material through an ionic liquid-assisted hydrothermal reaction and exhibited its first discharging/charging capacities of 2,065/1,334 mAh/g with a columbic efficiency of 64%, while the second and third discharging capacities were found to be 1,467 mAh/g and 1,473 mAh/g with efficiencies of 93 and 95%, respectively [105]. Ren *et al.* synthesized TMDCs using graphene foam coated by carbon nanotube (CNT) and decorated with MoS_2 nanomaterial and the fabricated electrode assembly (GF@CNT/MoS_2) was integrated by hydrothermal process and followed by nickel foam prototype *via* CVD technique. The fabricated battery was incapable of carrying a specific capacitance of 935 mAh/g at 0.1 A/g of a reversible capacity of 606 mAh/g for more than 200 cycles at 0.2 A/g current density for flexible anode for LIBs [106]. Yu *et al.* synthesized graphene-based MoS_2 TMDCs *via* a controlled CVD method for LIBs and they obtained a 1,260 mAh/cm^3 volumetric capacity for 1.31 V of working voltage than a layered graphene electrode (1.46 V) for energy storage [107]. Wang *et al.* prepared MoS_2-coated vertical graphene nanosheet (MoS_2/VGNS) hybrid electrodes *via* a solvothermal process followed by plasma-assisted CVD for LIBs. The as-prepared TMDCs delivered a specific capacity of 1,277 mAh/g, current density of 100 mA/g and columbic efficiency of 76.65%.

The prepared heterostructures retained a specific capacity of 1,109 mAh/g at a current density of 200 mA/g after 100 cycles [108]. Nguyen *et al.* studied the electrochemical properties of as-synthesized Ag-decorated 1T-MoS_2 nanosheets *via* liquid exfoliation and reported a specific capacity of 510 mAh/g after 100 cycles. The obtained value was ~73% times higher than the initial specific capacity, revealing that Ag-decorated 1T-MoS_2 nanosheets exhibit higher Li-ion storage performance [109]. Wang *et al.* synthesized a MoS_2 nanosheet material in an ethanol and octylamine solution mixture by the solvothermal method and used it as a portable electrode material for LIBs. The attained discharge capacity, and current density were 960 mAh/g and 0.1 A/g, respectively. The MoS_2 nanosheet material retains a capacity of 606 mAh/g after 50 cycles [110]. Kato *et al.* prepared VS_2 material by hydrothermal method for storage of Li ion and indicated that the lithium intercalation reaction follows 2-electron transfer mechanism, and a charge capacity is found to be 400 mAh/g at 0.1A/g of current density [111]. The results suggest that the superior electrochemical performance of VS_2 is due to its 2D metallic structure and ionic conductivity. Ju *et al.* [112] reported hybrid TMDCs with carbon allotropes were synthesized by hydrothermal method for achieving a steady power production. The current density and discharge capacity of the for as-synthesized Fe_2O_3/MoS_2/rGO material were 906 mAh/ g and 0.2 A/g, respectively for 100 cycles. The increased current density trend retained a discharging capacity of 711 mAh/g after 500 cycles. The obtained experimental data indicate that Fe_2O_3/MoS_2/rGO is a suitable anode material for LIBs. Zhu *et al.* studied a MoS_2/TiO_2 nanohybrid composite TMDC material synthesized by a hydrothermal technique and found the discharging capacity and current density to be 604 mAh/g and 0.1 A/g for 100 cycles, respectively [113]. Bai *et al.* prepared a composite TMDC (MoS_2/Co) material as an anode for LIBs by a hydrothermal route followed by ball-milling, and they were found to be 0.1 A/g and 1,130 mAh/g values of current density and discharge capacity, respectively. The electrochemical capacity increased to 1,340 mAh/g after 50 cycles [114].

The structure of the composite (MoS_2/Co) material reduced the path length of the Li+ ion and accelerated the electron transfer kinetics to enhance the electrochemical properties. Rui *et al.* [115] also reported MoS_2/Co composite material as anodes for LIBs was synthesized by hydrothermal method followed by carbonization. They obtained a high specific capacity (568.5 mAh/g) after a current density (0.2 A/g) value for 500 cycles. Graphene-decorated 2D-TMDCs energy storage materials are used as anode material for LIBs. They displayed fast ion diffusion and electron transport, which enhanced the electrochemical functionality of LIBs.

The columbic efficiency and specific capacity were found to be 70% and 1,217 mAh/g, respectively, for initial charging/discharging cycles at 0.2 C using expanded 1T-rich $MoSe_2$ (e-$MoSe_2$) with NH_4^+ intercalation synthesized *via* the hydrothermal method by Zhou *et al.* [116]. Li *et al.* used CoSe/NC anode material for LIBs and found to be 1,244 and 310.11 mAh/g reversible capacity values behind 190 and 500 cycles which corresponds to current density values between 0.1 and 1.0 A/g [117]. Panda *et al.* [118] utilized $MoTe_2$ material and they found to be 432 mAh/g of initial specific capacity at current density of 1.0 A/g and a value of 291 mAh/g reversible specific capacity after 250 cycles for LIBs. In addition, they also used 1T phase-$MoTe_2$/LCO material and reported a value of 388.4 mAh/g reversible capacity of 0.1 A/g after 100 cycles with 74% for that one former capacity and Columbic efficiency of ~96%. Zhang *et al.* synthesized single-walled carbon nanotube gallium chalcogenide (Ga-X-NS/SWCNT, where X = S/Se) to be used as anodes for LIBs, and they displayed specific capacities of 838 and 713 mAh/g for GaS-NS and GaSe-NS, respectively [119]. Xiao *et al.* [120] reported MoS_2-PEO composite (ratio:0.05) and PEO is utilized as stabilizing agent. The reported PEO-MoS_2 composite exhibited a specific capacity value of 1,000 mAh/g and MoS_2-based Li-ion anode and its combination, for example, CNTs-MoS_2 [121, 122] and CMK-3-MoS_2 [123] revealed *via* various synthetic methods and demonstrated superior performance. Jiang and his coworkers [124] displayed distinctive mesoporous layered carbon for synthesizing 2D-MoS_2/m-C hybrids consist of monolayer Mo-S_2 and meso-C sandwiched in a discontinuous order. The oleic-acid-secluded monolayer MoS_2 nanosheet with dopamine was self-polymerized in the interlayer of MoS_2. Finally, it is clear that the MoS_2 nanosheet significantly enhances the LIB performance compared to the reported Mo-S_2-C and g-MoS_2 hybrids. George and his coworkers [125] studied the behavior of carbon (C) as a stabilizer and to compared with standard powdered form of carbon and CNTs. The CNTs provide ~2-fold enhancement in specific capacity and a 45% reduction in resistance by means of 90% columbic efficiency. Further, WSe_2-composites such as carbon-coated WSe_2, were synthesized using mechanical milling followed by annealing with carbon black by Zhang *et al.* [126]. The Figure 19.8 exhibits (a) CVs of 2D-WSe_2 vs. Li (2 cycles) at 0.1 mV/s of scan rate, (b) Galvanostatic profile diagram (potential vs. specific capacity) for two cycles of 2D-WSe_2. (c) Discharge capacity vs. cyclic numbers graph, (d) electrochemical cycling performance graph between specific capacity and cyclic numbers for discharging and charging capacities, (e) illustration of synthesis of bulk 2D-WSe_2 *via* CVD, (f, g) HR-SEM images of 2D-WSe_2 (h); (i) AFM image of 2D-WSe_2 (thickness is approximately 1.8–0.3 nm).

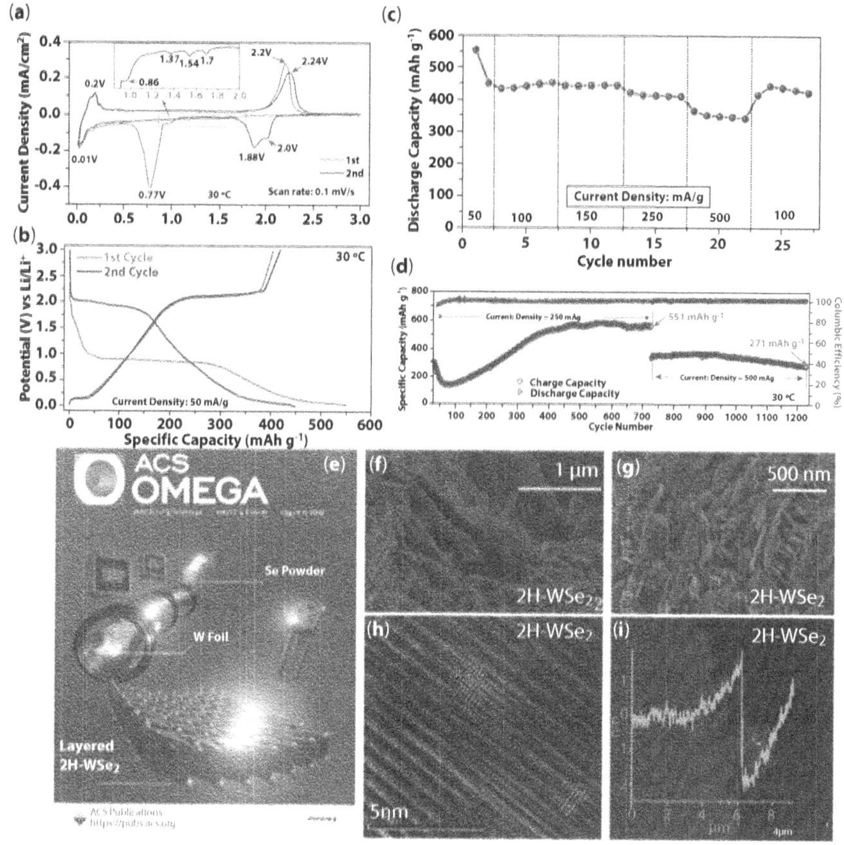

Figure 19.8 (a) CVs of 2D-WSe_2 vs. Li (2 cycles) at 0.1 mV/s of scan rate, (b) Galvanostatic profile diagram (potential vs. specific capacity) for two cycles of 2D-WSe_2. (c) Discharge capacity vs. cyclic numbers graph, (d) electrochemical cycling performance graph between specific capacity and cyclic numbers for discharging and charging capacities, (e) illustration of synthesis of bulk 2D-WSe_2 *via* CVD, (f, g) HR-SEM images of 2D-WSe_2 (h); (i) AFM image of 2D-WSe_2 (thickness is approximately 1.8–0.3 nm). (Figure subpanels a–g were reproduced from ACS Omega, Ref. [127]. Figure subpanels h and i were reproduced from Elsevier, Ref. [128].

The prepared powder forms of WSe_2-C material, which is directly utilized as an electrode and showed 467 mAh/g of discharge capacity at 200 mAh/g current density due to the C-matrix and an improved electronic conductance of the WSe_2-C material in LIBs. The redox process of LIBs indicated that the increased capacity was equal to 500 cycles, and 250 mAh/g discharge capacity to the 50th cycle for the as-synthesized substance showed a constant value of 550 mAh/g for extended cycling. Subsequently, 500 cycles of a specific capacity (271 mAh/g) with a retention of 91% were recorded at

500 mAh/g. The results of this study of 2H-WSe_2 indicated that the layered 2D microstructure has a prismatic rate and stable cycling ability for energy storage devices.

19.3.2 Sodium-Ion Batteries

Sodium-ion batteries are an alternative and are competent for LIBs owing to their richness and cost effectiveness compared to lithium in rechargeable batteries for obtaining higher energy and power densities. Zhou *et al.* prepared WS_2 nanosheet TMDCs using a sodium chloride template-assisted *in situ* method for Na-ion anode material and obtained 453.2 mAh/g value of reversible capacity of 0.1 A/g current density value [129]. A layered $MoSe_2$ nanoplate TMDCs material was used as the anode material for SIBs, and the material was synthesized by a thermal decomposition process. $MoSe_2$ nanoplate material established that the 440/513 mAh/g values were found for first charging/discharging capacities. The result found *via* Density functional theory (DFT) studies corroborated the Na-ion diffusion reaction and revealed that the inherent electrical conductivity of the $MoSe_2$ nanoplate material is responsible for the enhanced electrochemical properties [130]. Mukherjee *et al.* fabricated $MoSe_2$ nanoflakes TMDCs electrode materials for SIBs, and the results were found to be 399.10 mAh/g of stable initial dissociation capacity by means of ~100% columbic efficiency [131]. Sun *et al.* established a new anode using Cu-Ge chalcogenide for SIBs, it exhibited 463.3 mAh/g a reversible capacity [132]. Various TMDCs have been used as anode materials, such as TiS_2, ZrS_2, NbS_2, and MoS_2, to calculate their theoretical capacities for SIBs ranging from 260 to 339 mAh/g, as reported by Yang *et al.* [133]. David *et al.* [134] used the similar kind of MoS_2-graphene material in NIBs and the fabricated electrode exhibited better cycling ability and charge capacity ~230 mAh/g relating to weight of electrode by means of ~99% Columbic efficiency. Other studies have reported that MoS_2-graphene composite materials and metal nanoclusters (Ag1) intercalate in graphene/MoS_2 nanosheets in Na-ion batteries for vast charge loss issues [135–137]. Liu *et al.* prepared a 2D-Ti_2PTe_2 layered material with metallic behavior that was used as an anodic material for SIBs and exhibited a high theoretical capacity of 280.72 mAh/g with a high power density and rapid charging/discharging [138]. MoSeTe TMDCs were synthesized as anode materials and characterized by XRD, FE-SEM, TEM, and Raman spectroscopy.

Hexagonal geometry by means of a space group (p63/mmc) and the crystal structure of MoSeTe were confirmed. A half-cell electrochemical study of the as-synthesized MoSeTe anode material for SIBs with potential

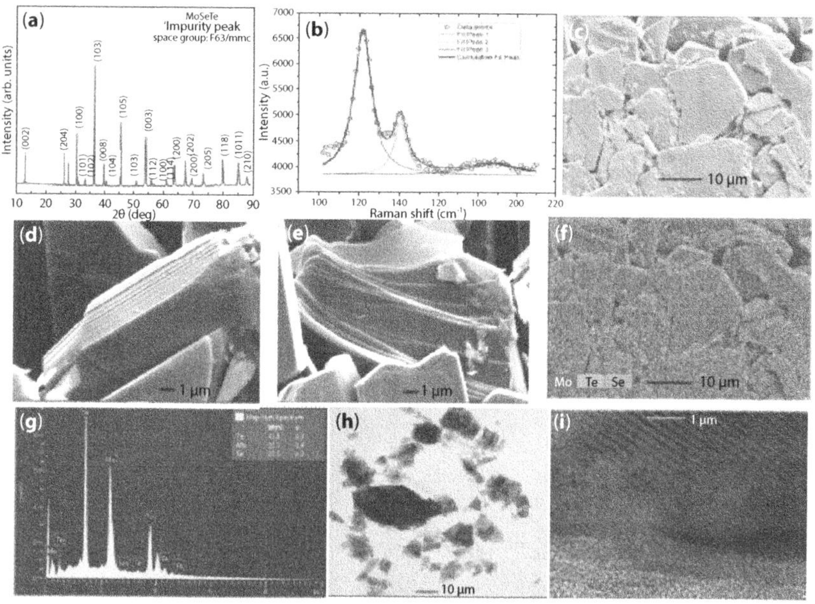

Figure 19.9 Structural depictions of the MoSeTe material: (a) XRD pattern, (b) Raman spectra, (c–e) FE-SEM images of MoSeTe, (f, g) EDX spectrum, (h) TEM morphology of MoSeTe, and (i) magnified TEM exhibiting fringes [139, 140].

between 0.01 and 3.0 V and metallic Na used as auxiliary or reference electrodes resulting considerable discharging capacities 475–355 mAh/g at 50–100 mA/g of current densities (as shown in Figure 19.9). However, a discharging capacity of 200 mAh/g in the second cycle with 100% columbic efficiency confirmed the enhanced stability. The cyclic voltammetry (CV) technique revealed the reversible nature of MoSeTe after 1st cycle and there was no change in the peak positions in the obtained CV of the material. The EIS study confirmed that the low charge-transfer resistance of the new cell compared to other cells after the 10th cycle revealed a diffusion-controlled process, and the coefficient values were found to be 10–14 cm^2s^{-1} [141]. The electrochemical capacities of MoSeTe: (a) CVs of cell at 0.05 mV/s of scan rate, (b) charging-discharging curves of the cell at 100 mA/g current density, (c) charging-discharging curves of the cell at 50 mA/g for 10 cycles, (d) cycle performance, and columbic efficiency of the cell at 100 mA/g current density, and (e) at 50 mA/g of current density is presented in Figure 19.10.

The fabrication of metal-rich Cu_2Se electrodes by selenization was examined as a feasible material for SIBs [142]. The fabricated Cu_2Se demonstrated excellent reversible capacity (250 mAh/g), excellent cyclic

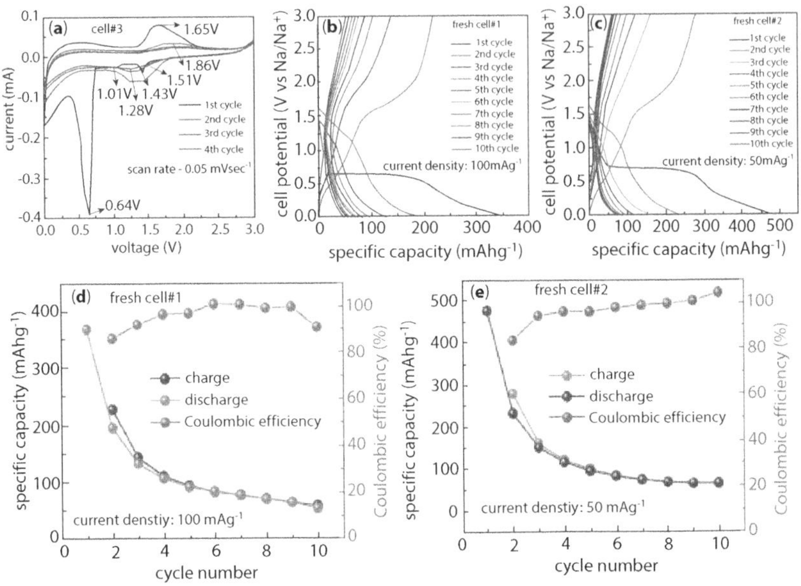

Figure 19.10 Electrochemical capacities of MoSeTe: (a) CVs of cell at 0.05 mV/s of scan rate, (b) charging-discharging curves of the cell at 100 mA/g current density, (c) charging-discharging curves of the cell at 50 mA/g for 10 cycles, (d) cycle performance, and columbic efficiency of the cell at 100 mA/g current density, and (e) at 50 mA/g of current density [141].

permanence, and low polarization for SIBs. Share *et al.* reported a WSe_2 pristine electrode for NIBs, which exhibited a reversible capacity of 200 mAh/g at a current density of 20 mAh/g [142]. The electrochemical performance of (a) CVs curve for the first three cycles (WSe_2 devices with CMC, and EC/DEC). (b) CVs of the chemical storage in WSe_2 and WS_2 in 3rd cycle. (c) Graph between dQ/dE of WSe_2 electrode and voltage (V vs. Na^+/Na) at a current density of 10 mA/g, (d) galvanostatic graph at five different current densities (20, 40, 100, 200, and 400 mAh/g), (e) EDS map of the WSe_2 electrode before cycling (scale bar 1/4 100 nm). (f) EDS of the WSe_2-battery electrode after the 5th discharge, (g) EDS spectra, (f) Na-added (after 5th discharge). (h) Representation of the Na storage with WSe_2 is shown in Figure 19.11.

Wang *et al.* synthesized a SnS/$SnSe_2$ composite TMDC material *via* the CVD method, followed by template-assisted calcination and found 624.9/1,131.3 mAh/g values of preliminary charging/discharging capacities correspond to 0.1 A/g current density. Subsequently, they reached a stable capacity of 580 mAh/g after 100 cycles with 90.2% capacity retention [143].

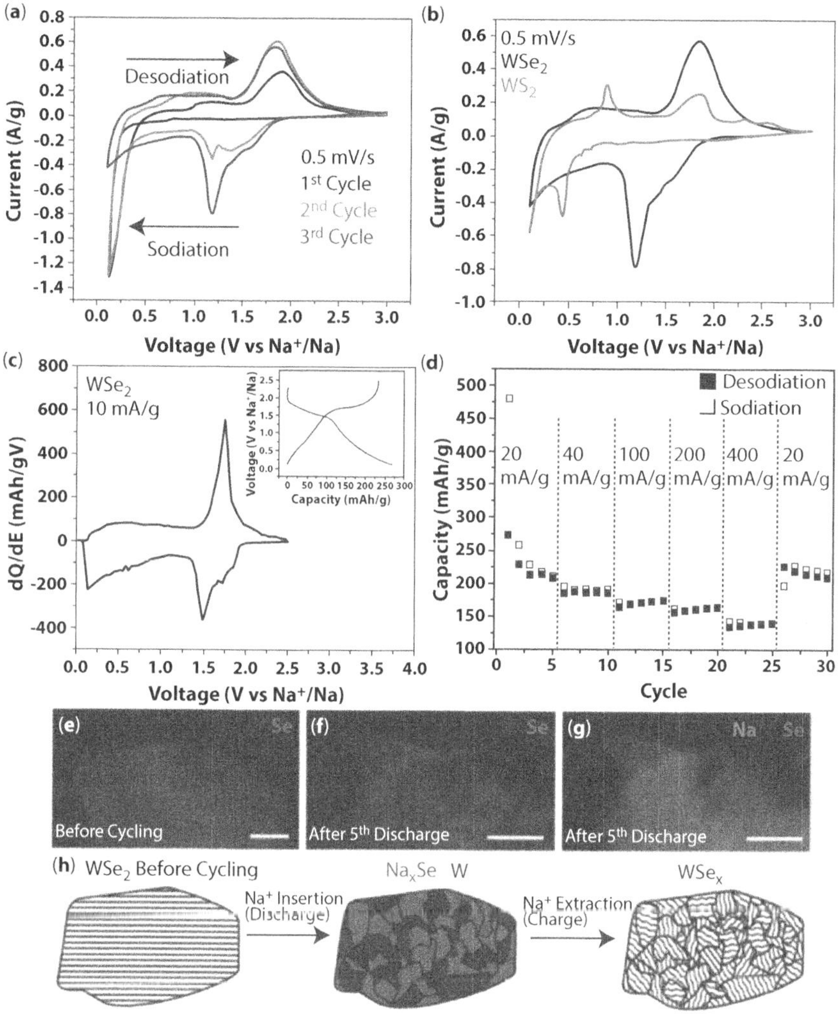

Figure 19.11 (a) CVs curve for the first three cycles (WSe_2 devices with CMC, and EC/DEC). (b) CVs of the chemical storage in WSe_2 and WS_2 in 3rd cycle. (c) Graph between dQ/dE of WSe_2 electrode and voltage (V vs. Na^+/Na) at a current density of 10 mA/g, (d) galvanostatic graph at five different current densities (20, 40, 100, 200, and 400 mAh/g), (e) EDS map of the WSe_2 electrode before cycling (scale bar 1/4 100 nm). (f) EDS of the WSe_2-battery electrode after the 5th discharge, (g) EDS spectra, (f) Na-added (after 5th discharge). (h) Representation of the Na storage with WSe_2. All figures were reproduced from *RSC Adv.*, Ref. [142].

19.3.3 Potassium Ion Batteries (PIBs)

Wu *et al.* demonstrated that N-doped carbon-containing FeS_2 and $FeSe_2$ materials (FeS_2@NC) can be utilized as anodes for potassium-ion batteries (PIBs) after 5,000 cycles and they obtained values 525.5 and 154.7 mAh/g values of reversible capacity which corresponds to 0.1 and 10 A/g of current densities. They also studied a full-cell setup using FeS_2@NC as the anode and found 88% specific capacity retention after 120 cycles with 99.9% columbic efficiency [144]. Diffusion obstructions for K ions over the TMDCs layers were low (<0.05 eV), permitting K ions to migrate spontaneously. VS_2 and TiS_2 materials are promising anodes that exhibit unexpected characteristics with respect to the electronic conductivity, K-ion flow, optimum open circuit voltage (OCV), and higher theoretical K-ion storage capacities [145]. MoS_2 nanoflowers with extended interlayer spacings and basal plane defects were used as anodes for KIBs at voltage ranges from 0.5 to 2.5 V. In this study, intercalation occurred during the conversion

Table 19.2 Summarized data for the various morphologies and rechargeable battery applications of TMSes.

Transition metal selenides	Morphology	Batteries application	Rate capability (mAh/g)	Ref.
WSe_2 NWs on W foil	Nanowire	Mg-ions	103/3,000	[149]
Cu_2Se on Cu grids	Nano thin film	Na-ions	250	[150]
WSe_2		Na-ions	200	[151]
WSe_2	Nanosheets	Li- ions	550	[152]
$MoSe_2$ from MoO_3	Yolk shell	Na-ions	527	[153]
Metal organic framework-derived SnSe nano-platelets in nitrogen-doped graphene cages	SnSe-MOF Nano platelets wrapped with Nitrogen doped graphene cages	Na-ions	650	[154]

to store K ions in-between the van der-Waals crevices. MoS_2 nanoflowers showed improved K-storage compared to their pristine interlayer spacing counterparts [146, 147]. Layer-2D-TMDCs MoS_2 exhibited a capacity of 118 mAh/g, with a cyclability of 85% after 100 cycles and a rate capability of 41 mAh/g at a current density of 2 A/g in a low-crystalline state. However, the higher crystalline state of MoS_2 provided an intercrystalline capacity of 330 mAh/g at a current density of 1 A/g and retained an intercrystalline capacity of 161 mAh/g at a current density of 20 A/g [148].

19.3.4 Magnesium Ion Batteries (MIBs)

The nanostructures of TMSe show potential for use as cathode materials for magnesium batteries. Liu *et al.* prepared W-Se_2 nanowires as promising binder-free cathodes in MIBs [149]. WSe_2 nanowires exhibited better performance in Mg-ion batteries without changing the morphology, and the fabricated battery exhibited a reversible capacity of ~203 mAh/g at a current density of 50 mA/g, while the W-Se_2 exhibited a capacity retention rate (CRR) of 10% after 100 cycles. The electrodes exhibited reversible capacities of 142, 120, and 103 mAh/g at 800, 1500, and 3,000 mA/g used for 50 cycles. The columbic efficiency (CE) was found to be 98.5% for 160 cycles as compared to the CE of 93.8% for 50 cycles for the bulk material, and this result indicated the cycle stability and rate capacity. We have provided a summarized table for all reported transition metal selenides for energy storage capabilities as presented in Table 19.2.

19.3.5 Zinc Ion Batteries (ZIBs)

Li *et al.* [155] synthesized an expanded (E-MoS_2) TMDCs material by glucose-assisted hydrothermal method for Zn-ion battery (ZIBs). The discharging capacity was found to be 202.6 mAh/g at a current density of 0.1 A/g. The consequent values of the capacity retention ratio and energy density were found to be 98.6 % and 148.2 Wh/kg, respectively, whereas cycle stability was established for 600 cycles [30]. Lui *et al.* [156] studied the hydrated MoS_2 nanosheet TMDCs was prepared *via* carbon fiber-assisted hydrothermal route and showed the values of reversible capacitance of 182 mAh/g and 0.1 A/g of current density, respectively. The corresponding capacity retention ratio was 94% greater than 1,500 cycles at 2 A/g for the ZIBs. They revealed that the increased capacity was due to existence of water, and the water molecules enlarged the inter-planar spacing and accelerated the charge-transfer (CT) route in the charge storage process. Cai *et al.* prepared metallic nature of 1T-MoS_2 by means of an interlayer

Table 19.3 Comparison table of different synthesized TMDCs in different electrolyte and their energy storage performances.

TMDCs	Electrolyte	Performance	Ref.
MoS_2 (exfoliation)	1M $NaClO_4$ mixed in PC/EC (1:1)	386 mAh/g after 100 cycles at 0.04 A/g	[158]
MoS_2 (exfoliation)	1M $NaClO_4$ mixed in FEC/PC	161 mAh/g after 50 cycles at 0.02 A/g	[161]
MoS_2 (CVD)	1M $NaClO_4$ mixed in PC/FEC	520 mAh/g after 100 cycles at 1 A/g	[157]
MoS_2 (hydrothermal)	1M $NaClO_4$ mixed in EC/DMC (1:1)	370 mAh/g after 200 cycles at 2 A/g	[158]
$MoSe_2$ (colloidal)	1M $NaClO_4$ in EC/DMC (1:1)	345 mAh/g after 200 cycles at 42.2 mA/g	[161]
$MoSe_2$	1M $NaClO_4$ mixed in EC/PC mixed with 5% FEC (1:1)	450 mAh/g after 90 cycles at 200 mA/g	[162]
$MoTe_2$	1M $NaClO_4$ mixed in PC/EC (1:1) with 3% FEC	270 mAh/g after 200 cycles at 1 A/g	[163]
VS_2 (solvothermal)	1M $NaClO_4$ in DEC/EC (1:1) with 6% FEC	700 mAh/g after 100 cycles at 0.1 A/g	[164]
VS_2 (hydrothermal)	1M $NaSO_3CF_3$ mixed in DGM	565 mAh/g after 1200 cycle at 2.0 A/g	[165]
$NbSe_2$ (Sintering technique)	1M $NaClO_4$ mixed in DEC/EC (1:1)	98.1 mAh/g after 100 cycles at 0.1 A/g	[166]
NbS_2 (Exfoliation Method)	1M $NaPF_6$ mixed in EC/DMC (1:1)	157 mAh/g after 100 cycles at 0.5 A/g	[167]
WTe_2 (one-pot hydrothermal)	1M $NaClO_4$ mixed in DEC/EC (1:1) Blend with 5% FEC	221 mAh/g after 100 cycles at 0.1 A/g	[168]
WTe_2 (CVD)	1M $NaClO_4$ mixed in DEC/EC (1:1) Blend with 5% FEC	260 mAh/g after 40 cycles at 0.1 A/g	[168]
MoSeTe as anode material	1M $NaClO_4$ mixed with EC and DEC mixed in 1:1 ratio as electrolyte	200 mAh/g in 2nd cycle at 50 and 100 mA/g	[169]

spacing of 1.025 nm *via* a hydrothermal process and provided a specific capacity value (125 mAh/g) at a current density of 2 A/g surrounded by a 100% capacity retention rate (CRR) of more than 500 cycles [157]. Guo *et al.* [158] prepared an inter-layered extended MoS_2/N-doped carbon with a hierarchical structure by a hydrothermal approach followed by calcinations and the as-synthesized electrochemical data revealed that a superb cycling ability and elevated discharge capacity due to structure, large specific surface area of N-doped carbon inter-layer spacing of MoS_2. Aqueous ZIBs are divalent cation (Zn^{2+})-based rechargeable batteries that have very strong interactions with water, which increases the complexity of Zn^{2+} ion diffusion and intercalation, hindering the progress of ZIBs. To overcome this issue, many researchers have utilized MoS_2 as an electrode material, to enhance the electrochemical properties of ZIBs by improving the interlayer spacing *via* doping with a heteroatom source. The different TMDCs synthesized in different electrolytes and their energy storage performances are shown in Table 19.3. He *et al.* reported VS_2 nanosheets with a high interlayer distance (0.576 nm) as electrodes for ZIBs. The high interlayer capacity facilitates zinc ion diffusion, and the discharge capacity was 190.3 mAh/g at a current density of 50 mA/g with 98.0% cyclic stability over 200 cycles [159]. Wu *et al.* studied ultra-thin VSe_2 nanosheets used as an electrode material for ZIBs and they provided enhanced discharging capacities of 79.5 mAh/g, 93.9 mAh/g, 105.2 mAh/g, 114.6 mAh/g, and 118.4 mAh/g at current densities of 2,000 mA/g, 1,000 mAh/g, 500 mAh/g, 200 mAh/g, and 100 mAh/g, respectively [160].

19.3.6 Lithium Sulfur Batteries (LSBs)

Liu *et al.* [170] have reported MoS_2 nanosheets an oxygen plasma treatment (OPT) for creating vacancies/defects and incorporation of heteroatom through oxygen plasma engineering and the inimitable hybrid oxygen-plasma mediated MoS_2 nano-sheet for enhancing the better electrochemical properties with suitable reversible capacities, cyclic stability, and rate capabilities. Recently, many attempts have been made to develop Li-S batteries (LSBs). However, the insulating property of S and the polymerization nature of sulfur make it challenging to fabricate S as a cathode. In addition, LSBs are possessed capacity degradation problem initiates from the electrolyte *via* the reduced electrical conductivity of S and disbanding Li-polysulfide development, which lowers battery performance. Choi *et al.* [171] demonstrated a successful technique for coating Li-anode over 2D-MoS_2 for achieving better performance LSBs. The MoS_2-protected anode with S-CNTs as the cathode resulted in an energy density of

589 Wh/kg for 1,200 cycles at 0.5 °C. Guo *et al.* [172] established interlayer functionalities consist of composite 2D-MoS_2 with graphene like a polysulfide barrier for LSBs and exhibited greater rate capacity and enhanced cycling capacity. Derlim *et al.* [173] described that the synthesis and used poly(S-r-DIB)-MoS_2 composite enriched with S-matrix and Mo-S_2 insertion for Li-S cathode material. The as-synthesized composite material facilitated the disbanding of Li-polysulfide redox yields through Mo-S_2 insertions, performing as "polysulfide anchors" and the stability of the composite was confirmed by elemental analysis. The fabricated cathodes exhibited larger cycle lifetimes of 1,000 cycles, along with a minimum capacity decay for Li-S and a reversible specific capacity of 500 mAh/g at 5 °C. Other researchers have also reported MoS_2/graphene composite materials as polysulfide barriers in LSBs [174].

19.3.7 Aluminum Ion Batteries (AIBs)

MoS_2 and its composites have been used as cathodes in aluminum-ion batteries (AIBs). Lin *et al.* [175] prepared MoS_2 microsphere material by hydrothermal process using precursors like $(NH_4)_6Mo_7O_{24}{\cdot}4H_2O$ and $(NH_4)_2CS$ and the MoS_2 microsphere was obtained an electrochemical performance of AIBs with MoS_2 microsphere used as cathode material. However, the intrinsic defects of MoS_2 microspheres affect the electrochemical properties of MoS_2 electrode materials, and future research will focus on enhancing their electrochemical performance [176]. Yang *et al.* [176] examined flexible self-supporting MoS_2/CNFs cathode material for rechargeable AIBs and the MoS_2/CNFs are synthesized *via* electro-spinning followed by annealing. MoS_2/CNFs exhibited improved cycling stability and elevated rate capacity compared to bulk MoS_2 microspheres.

Wu *et al.* reported the use of a VS_2 nanosheet material as a cathode for AIBs. To increase stability and conductivity, the synthesized VS_2 nanosheets were modified using graphene (G-VS_2). The graphene-modified (G-VS_2) cathode provided a higher discharge capacity (186 mA h/g) at 100 mA/g with 100% coulombic efficiency over 50 cycles [177]. Thomas Nann's groups have utilized $MoSe_2$, MoS_2, and MoS-Se nanomaterials as cathodes for AIBs voltages discharge ranging from 0.5 V to 1.6 V. Interestingly $MoSe_2$ achieved better performance than MoS_2. The MoS-Se material showed an excellent specific capacity compared to MoS_2 and $MoSe_2$ materials; however, the energy density was found to be lower at a current density of 40 mA/g. The fabricated cells were found to be more stable at 100 mA/g after 200 cycles with 90% coulombic efficacy [178].

19.4 Conclusion

In summary, we have discussed the fundamentals, properties, various synthesis approaches, and electrochemical performance of 2D-TMDCs. In particular, advances in 2D-TMDCs for rechargeable batteries, such as sodium, potassium, lithium, magnesium, zinc, aluminum, and lithium-sulfur applications, are well discussed. Detailed structural insights, properties, characterization, and electrochemical mechanistic aspects were also discussed. Numerous methods for the bottom-up synthesis of 2D-TMDCs and their properties facilitate the production of materials with changeable shapes, sizes, morphologies, and phases. As a result, the 2D-structure of TMDCs with exceptional properties such as semiconding state, conductivity, flexibility, cost effectiveness, scalability, transparency, and large surface area makes them an important material for developing a novel wearable, flexible device for battery applications. Thus, this chapter will be helpful to researchers in designing and enhancing the electrochemical performance of 2D-TMDCs electrode materials for rechargeable battery applications.

References

1. Chou, S.S., Sai, N., Lu, P., Coker, E.N., Liu, S., Artyushkova, K. *et al.*, Understanding catalysis in a multiphasic two-dimensional transition metal dichalcogenide. *Nat. Commun.*, 6, 8311, 2015.
2. Chia, X. and Pumera, M., Layered transition metal dichalcogenide electrochemistry: Journey across the periodic table. *Chem. Soc. Rev.*, 47, 5602, 2018.
3. An, S., Kim, Y.H., Lee, C., Park, D.Y., Jeong, M.S., Exfoliation of transition metal dichalcogenides by a high-power femtosecond laser. *Sci. Rep.*, 8, 12957, 2018.
4. Tian, Y., Zeng, G., A. Rutt, T., Kim, H., Wang, J., Koettgen, J., Sun, Y., Ouyang, B., Chen, T., Lun, Z., Rong, Z., Persson, K., Ceder, G., Promises and challenges of next-generation "beyond Li-ion" batteries for electric vehicles and grid decarbonization. *Chem. Rev.*, 121, 1623, 2021.
5. Maiti, S., Konar, R., Sclar, H., Grinblat, J., Talianker, M., Tkachev, M., Wu, X., Kondrakov, A., Nessim, G.D., Aurbach, D., Stabilizing high-voltage lithium-ion battery cathodes using functional coatings of 2D tungsten diselenide. *ACS Energy Lett.*, 7, 1383, 2022.
6. Sun, W., Zhang, Y., Kang, W., Deng, N., Wang, X., Kang, X., Yan, Z., Pan, Y., Ni, J., Synthesis of MoS2-based nanostructures and their applications in rechargeable ion batteries, catalysts and gas sensors: A review. *RSC Adv.*, 12, 19512, 2022.

7. Kannan, P.K., Late, D.J., Morgan, H., Rout, C.S., Recent developments in 2Dlayered inorganic nano materials for sensing. *Nanoscale.*, 7, 13293, 2015.
8. Yang, W., Gan, L., Li, H., Zhai, T., Two-dimensional layered nanomaterials for gas-sensing applications. *Inorg. Chem. Front.*, 3, 433, 2016.
9. Lin, L., Zhang, S., Allwood, D.A., Transition metal dichalcogenides for energy storage applications, in: *Two-Dimensional Transition Metal Dichalcogenides Synthesis, Properties, and Applications*, N.S. Arul, and V.D. Nithya, (Eds.), pp. 173–201, Springer, Singapore, 2019.
10. Li, B.L., Wang, J., Zou, H.L., Garaj, S., Lim, C.T., Xie, J., Li, N.B., Leong, D.T., Low-dimensional transition metal dichalcogenide nanostructures based sensors. *Adv. Funct. Mater.*, 26, 7034, 2016.
11. Yang, S., Jiang, C., Wei, S.H., Gas sensing in 2D materials. *Appl. Phys. Rev.*, 4, 021304, 2017.
12. Guo, S. and Dong, S., Graphene nanosheet: Synthesis, molecular engineering, thin film, hybrids, and energy and analytical applications. *Chem. Soc. Rev.*, 40, 2644, 2011.
13. Lu, Q., Yu, Y., Ma, Q., Chen, B., Zhang, H., 2D transition-metal-dichalcogenide nanosheet based composites for photocatalytic and electrocatalytic hydrogen evolution reactions. *Adv. Mater.*, 28, 1917, 2016.
14. Cao, X., Tan, C., Zhang, X., Zhao, W., Zhang, H., Solution-processed two-dimensional metal dichalcogenide-based nanomaterials for energy storage and conversion. *Adv. Mater.*, 28, 6167, 2016.
15. Butler, S.Z., Hollen, S.M., Cao, L., Cui, Y., Gupta, J.A., Gutiérrez, H.R., Heinz, T.F., Hong, S.S., Huang., J., Ismach, A.F., Johnston-Halperin, E., Progress, challenges, and opportunities in two dimensional materials beyond graphene. *ACS Nano*, 7, 2898, 2013.
16. Wang, Q.H., Kalantar-Zadeh, K., Kis, A., Coleman, J.N., Strano, M.S., Electronics and optoelectronics of two-dimensional transition metal dichalcogenides. *Nat. Nanotechnol.*, 7, 699, 2012.
17. Kumar, A. and Ahluwalia, P.K., Electronic structure of transition metal dichalcogenides monolayers 1H-MX2 (M= Mo, W; X= S, Se, Te) from ab-initio theory: New direct bandgap semiconductors. *Eur. Phys. J. B*, 85, 186, 2012.
18. Chia, X., Ambrosi, A., Sofer, Z., Luxa, J., Pumera, M., Catalytic and charge transfer properties of transition metal dichalcogenides arising from electrochemical pretreatment. *ACS Nano*, 9, 5164, 2015.
19. Das, S., Prakash, A., Salazar, R., Appenzeller, J., Toward low-power electronics: Tunneling phenomena in transition metal dichalcogenides. *ACS Nano*, 8, 1681, 2014.
20. Lv, R., Robinson, J.A., Schaak, R.E., Sun, D., Sun, Y., Mallouk, T.E., Terrones, M., Transitionmetal dichalcogenides and beyond: Synthesis, properties, and applications of single-and few layer nanosheets. *Acc. Chem. Res.*, 48, 56, 2014.
21. Pospischil, A., Furchi, M.M., Mueller, T., Solar-energy conversion andlight emission in an atomic monolayer p–n diode. *Nat. Nanotechnol.*, 9, 257, 2014.

22. Rao, C.N.R., Ramakrishna Matte, H.S.S., Subrahmanyam, K.S., Maitra, U., Unusual magneticproperties of graphene and related materials. *Chem. Sci.*, 3, 45, 2012.
23. Zhou, Y., Yang, C., Xiang, X., Zu, X., Remarkable magnetism and ferromagnetic coupling insemi-sulfuretted transition-metal dichalcogenides. *Phys. Chem. Chem. Phys.*, 15, 14202, 2013.
24. Pumera, M., Sofer, Z., Ambrosi, A., Layered transition metal dichalcogenides for electrochemical energy generation and storage. *J. Mater. Chem. A*, 2, 8981, 2014.
25. Tributsch, H., Photo-electrochemical behaviour of layer-type transition metal dichalcogenides. *Faraday Discuss. Chem. Soc.*, 70, 189, 1980.
26. Chhowalla, M., Shin, H.S., Eda, G., Li, L.J., Loh, K.P., Zhang, H., The chemistry of two dimensional layered transition metal dichalcogenide nanosheets. *Nat. Chem.*, 5, 263, 2013.
27. Wang, H., Feng, H., Li, J., Graphene and graphene-like layered transition metal dichalcogenides in energy conversion and storage. *Small*, 10, 2165, 2014.
28. Huang, K.J., Liu, Y.J., Wang, H.B., Liu, Y.M., Wang, L.L., Synthesis of polyaniline/2-dimensional graphene analog MoS2 composites for high-performance supercapacitor. *Electrochim. Acta*, 109, 587, 2013.
29. Muller, G.A., Cook, J.B., Kim, H.S., Tolbert, S.H., Dunn, B., High performance pseudocapacitor based on 2D layered metal chalcogenide nanocrystals. *Nano Lett.*, 15, 1911, 2015.
30. Ellmer, K., Preparation routes based on magnetron sputtering for tungsten disulfide (WS2) films for thin-film solar cells. *Phys. Status Solidi.*, 245, 1745, 2008.
31. Shanmugam, M., Bansal, T., Durcan, C.A., Bin, Y., Molybdenum disulphide/titanium dioxide nanocomposite-poly 3-hexylthiophene bulk heterojunction solar cell. *Appl. Phys. Lett.*, 100, 153901, 2012.
32. Bhandavat, R., David, L., Singh, G., Synthesis of surface-functionalized WS2 nanosheets and performance as Li-ion battery anodes. *J. Phys. Chem. Lett.*, 3, 1523, 2012.
33. Xiao, J., Choi, D., Cosimbescu, L., Koech, P., Liu, J., Lemmon, J.P., Exfoliated MoS2 nanocomposite as an anode material for lithium ion batteries. *Chem. Mater.*, 22, 4522, 2010.
34. Kong, D., Cha, J.J., Wang, H., Lee, H.R., Cui, Y., First-row transition metal dichalcogenide catalysts for hydrogen evolution reaction. *Energy Environ. Sci.*, 6, 3553, 2013.
35. Chen, T.Y., Chang, Y.H., Hsu, C.L., Wei, K.H., Chiang, C.Y., Li, L.J., Comparative study on MoS_2 and WS_2 for electrocatalytic water splitting. *Int. J. Hydrog. Energy*, 38, 12302, 2013.
36. Yang, L., Wang, S., Mao, J., Deng, J., Gao, Q., Tang, Y., Schmidt, O.G., Hierarchical MoS_2/polyaniline nanowires with excellent electrochemical performance for lithium-ion batteries. *Adv. Mater.*, 25, 1180, 2013.

37. Kam, K.K. and Parkinson, B.A., Detailed photocurrent spectroscopy of the semiconducting group VIB transition metal dichalcogenides. *J. Phys. Chem.*, 86, 463, 1982.
38. Ho, W., Yu, J.C., Lin, J., Jiaguo, Y., Li, Puishan, Preparation and photocatalytic behavior of MoS_2 and WS_2 nanocluster sensitized TiO_2. *Langmuir*, 20, 5865, 2004.
39. Mak, K.F., Lee, C., Hone, J., Shan, J., Heinz, T.F., Atomically thin MoS2: A new direct-gap semiconductor. *Phys. Rev. Lett.*, 105, 136805, 2010.
40. Chhowalla, M., Shin, H.S., Eda, G., Li, L.J., Loh, K.P., Zhang, H., The chemistry of two dimensional layered transition metal dichalcogenide nanosheets. *Nat. Chem.*, 5, 263, 2013.
41. Wang, H., Yuan, H., Hong, S.S., Li, Y., Cui, Y., Physical and chemical tuning of two-dimensional transition metal dichalcogenides. *Chem. Soc. Rev.*, 44, 2664, 2015.
42. Rao, C.N.R., Maitra, U., Waghmare, U.V., Extraordinary attributes of 2-dimensional MoS2 nanosheets. *Chem. Phys. Lett.*, 609, 172, 2014.
43. Terrones, H., López-Urías, F., Terrones, M., Novel hetero-layered materials with tunable direct band gaps by sandwiching different metal disulfides and diselenides. *Sci. Rep.*, 3, 1, 2013.
44. Wilson, J.A. and Yoffe, A.D., The transition metal dichalcogenides discussion and interpretation of the observed optical, electrical and structural properties. *Adv. Phys.*, 18, 193, 1969.
45. Gatensby, R., McEvoy, N., Lee, K., Hallam, T., Berner, N.C., Rezvani, E., Winters, S., O'Brien, M., Duesberg, G.S., Controlled synthesis of transition metal dichalcogenide thin films forelectronic applications. *Appl. Surf. Sci.*, 297, 139, 2014.
46. Zhu, Z., Cheng, Y., Schwingenschlogl, U., Topological phase diagrams of bulk and monolayer TiS2–xTex. *Phys. Rev. Lett.*, 110, 077202, 2013.
47. Zhang, Y., Li, Z., Jia, H., Luo, X., Xu, J., Zhang, X., Yu, D., TiS_2 whisker growth by a simple vapor-deposition method. *J. Cryst. Growth*, 293, 124, 2006.
48. Laua, K.T., Edwards, S., Diamond, D., Nanostructured semiconductor oxides for the next generation of electronics and functional devices. *Sens. Actuators B.*, 98, 12, 2004.
49. Tao, Z.-L., Xu, L.-N., Gou, X.-L., Chen, J., Yuan, H.-T., TiS_2 nanotubes as the cathode materials of Mg-ion batteries. *Chem. Commun.*, 18, 2080, 2004.
50. Du, G., Guo, Z., Wang, S., Zeng, R., Chen, Z., Liu, H., Superior stability and high capacity of restacked molybdenum disulfide as anode material for lithium-ion batteries. *Chem. Commun.*, 46, 1106, 2010.
51. Chang, K., Chen, W., Ma, L., Li, H., Li, H., Huang, F., Xu, Z., Zhang, Q., Lee, J.Y., Graphene-like MoS2/amorphous carbon composites with high capacity and excellent stability as anode materials for lithium-ion batteries. *J. Mater. Chem.*, 21, 6251, 2011.

52. Coogan, A. and Gunko, Y.K., Solution-based "bottom-up" synthesis of group VI transition metal dichalcogenides and their applications. *Mater. Adv.*, 2, 146, 2021.
53. Gan, X.R., Zhao, H.M., Quan, X., Two-dimensional MoS2: A promising building block for biosensors. *Biosens. Bioelectron.*, 89, 56, 2017.
54. Zhan, J.H., Zhang, Z.D., Qian, X.F., Wang, C., Xie, Y., Qian, Y.T., Solvothermal synthesis of nanocrystalline MoS_2 from MoO_3 and elemental sulfur. *J. Solid State Chem.*, 141, 270, 1998.
55. Fan, R. and Chen, Z., A novel route to obtain molybdenum dichalcogenides by hydrothermal reaction. *Chem. Lett.*, 29, 920, 2000.
56. Wang, P.P., Sun, H.X., Ji, Y., Li, W., Wang, X., Three-dimensional assembly of single-layered MoS_2. *Adv. Mater.*, 26, 964, 2014.
57. Murray, C.B., Kagan, C.R., Bawendi, M.G., Synthesis and characterization of monodisperse nanocrystals and close-packed nanocrystal assemblies. *Annu. Rev. Mater. Sci.*, 30, 545, 2000.
58. Bai, X., Purcell-Milton, F., Gunko, Y., Optical properties, synthesis, and potential applications of Cu-based ternary or quaternary anisotropic quantum dots, polytypic nanocrystals, and core/shell heterostructures. *Nanomaterials*, 9, 85, 2019.
59. Yin, W., Bai, X., Chen, P., Zhang, X., Su, L., Ji, C., Gao, H., Song, H., Yu, W.W., Rational control of size and photoluminescence of WS_2 quantum dots for white light-emitting diodes. *ACS Appl. Mater. Interfaces*, 10, 43824, 2018.
60. Savjani, N., Lewis, E.A., Bissett, M.A., Brent, J.R., Dryfe, R.A.W., Haigh, S.J., O'Brien, P., Synthesis of lateral size-controlled monolayer 1*H*-MoS_2@Oleylamine as supercapacitor electrodes. *Chem. Mater.*, 28, 657, 2016.
61. Jung, W., Lee, S., Yoo, D., Jeong, S., Miro, P., Kuc, A., Heine, T., Cheon, J., Colloidal synthesis of single-layer MSe_2 (M = Mo, W) nanosheets via anisotropic solution-phase growth approach. *J. Am. Chem. Soc.*, 137, 7266, 2015.
62. Zhou, P., Tanghe, I., Schiettecatte, P., van Thourhout, D., Hens, Z., Geiregat, P., Ultrafast carrier dynamics in colloidal WS_2 nanosheets obtained through a hot injection synthesis. *J. Chem. Phys.*, 151, 164701, 2019.
63. Geisenhoff, J.Q., Tamura, A.K., Schimpf, A.M., Using ligands to control reactivity, size and phase in the colloidal synthesis of WSe_2 nanocrystals. *Chem. Commun.*, 55, 8856, 2019.
64. Seo, J.W., Jun, Y.W., Park, S.W., Nah, H., Moon, T., Park, B., Kim, J.G., Kim, Y.J., Cheon, J., Two-dimensional nanosheet crystals. *Angew. Chem. Int. Ed.*, 46, 8828, 2007.
65. Mastria, R., Scarfiello, R., Altamura, D., Giannini, C., Liscio, A., Kovtun, A., Bianco, G.V., Bruno, G., Grillo, V., Tavabi, A.H., Dunin-Borkowski, R.E., Nobile, C., Cola, A., Cozzoli, P.D., Gambino, S., Rizzo, A., An insight into chemistry and structure of colloidal 2D-WS_2 nanoflakes: Combined XPS and XRD study. *Sci. Rep.*, 9, 1, 2019.

66. Ding, X., Peng, F., Zhou, J., Gong, W., Slaven, G., Loh, K.P., Lim, C.T., Leong, D.T., Defect engineered bioactive transition metals dichalcogenides quantum dots. *Nat. Commun.*, 10, 1, 2019.
67. Sharma, P., Kumar, A., Bankuru, S., Chakraborty, J., Puravankara, S., Large-scale surfactant-free synthesis of WS_2 nanosheets: An investigation into the detailed reaction chemistry of colloidal precipitation and their application as an anode material for lithium-ion and sodium-ion batteries. *New J. Chem.*, 44, 1594, 2020.
68. Therese, H.A., Li, J., Kolb, U., Tremel, W., Synthesis of fullerene-and nanotube-like SnS_2 nanoparticles and Sn/S/carbon nanocomposites. *Solid State Sci.*, 7, 67, 2005.
69. Chakravarty, D. and Late, D.J., Microwave and hydrothermal syntheses of WSe_2 micro/nanorods and their application in supercapacitors. *RSC Adv.*, 5, 21700, 2015.
70. Kumar, N., George, B.P.A., Abrahamse, H., Parashar, V., Ngila, J.C., Sustainable one-step synthesis of hierarchical microspheres of PEGylated MoS_2 nanosheets and MoO_3 nanorods: Their cytotoxicity towards lung and breast cancer cells. *Appl. Surf. Sci.*, 396, 8, 2017.
71. Masimukku, S., Hu, Y.C., Lin, Z.H., Chan, S.W., Chou, T.M., Wu, J.M., High efficient degradation of dye molecules by PDMS embedded abundant single-layer tungsten disulfide and their antibacterial performance. *Nano Energy*, 46, 338, 2018.
72. Askari, M.B., Kalourazi, A.F., Seifi, M., Shahangian, S.S., Askari, N., Manjili, T.J., Hydrothermal synthesis of molybdenum disulfide (MoS_2) and study of structure, optical, electrical and high antibacterial properties. *Optik*, 174, 154, 2018.
73. Wang, J., Wang, N., Guo, Y., Yang, J., Wang, J., Wang, F., Sun, J., Xu, H., Liu, Z.H., Jiang, R., Metallic-phase MoS_2 nanopetals with enhanced electrocatalytic activity for hydrogen evolution. *ACS Sustain. Chem. Eng.*, 6, 13435, 2018.
74. Xie, J., Zhang, H., Li, S., Wang, R., Sun, X., Zhou, M., Zhou, J., Lou, X.W., Xie, Y., Defect-rich MoS_2 ultrathin nanosheets with additional active edge sites for enhanced electrocatalytic hydrogen evolution. *Adv. Mater.*, 25, 5807, 2013.
75. Chu, H., Liu, X., Liu, B., Zhu, G., Lei, W., Du, H., Liu, J., Li, J., Li, C., Sun, C., Hexagonal 2H-$MoSe_2$ broad spectrum active photocatalyst for Cr (VI) reduction. *Sci. Rep.*, 6, 1, 2016.
76. Lu, Y., Yao, X., Yin, J., Peng, G., Cui, P., Xu, X., MoS_2 nanoflowers consisting of nanosheets with a controllable interlayer distance as high-performance lithium ion battery anodes. *RSC Adv.*, 5, 7938, 2015.
77. Giubileo, F., Grillo, A., Passacantando, M., Urban, F., Iemmo, L., Luongo, G., Pelella, A., Loveridge, M., Lozzi, L., di Bartolomeo, A., Field emission characterization of MoS_2 nanoflowers. *Nanomaterials*, 9, 717, 2019.

78. Thang, N.T., Hong, L.T., Thoan, N.H., Hung, C.M., van Duy, N., van Hieu, N., Hoa, N.D., Controlled synthesis of ultrathin MoS_2 nanoflowers for highly enhanced NO_2 sensing at room temperature. *RSC Adv.*, 10, 12759, 2020.
79. Sun, Y., He, W., Sun, X., Liu, B., MoS_2 quantum dots as a specific fluorescence sensor for selection of rutin and for temperature sensing. *Luminescence*, 35, 1416, 2020.
80. Sun, Y., Xu, J., Zhu, Z., Wang, Y., Xia, H., You, Z., Lee, C., Tu, C., Comparison of MoS_2 nanosheets and hierarchical nanospheres in the application of pulsed solid-state lasers. *Opt. Mater. Express*, 5, 2924, 2015.
81. Zhan, J.H., Zhang, Z.D., Qian, X.F., Wang, C., Xie, Y., Qian, Y.T., Synthesis of nanocrystalline cobalt selenide in nonaqueous solvent. *J. Solid State Chem.*, 141, 270, 1998.
82. Liu, Y., Zhang, N., Kang, H., Shang, M., Jiao, L., Chen, J., Design, synthesis, and energy-related applications of metal sulfides. *Chem.-Eur. J.*, 21, 11878, 2015.
83. Wang, P.P., Sun, H., Ji, Y., Li, W., Wang, X., Three-dimensional assembly of single-layered MoS_2. *Adv. Mater.*, 26, 964, 2014.
84. Wang, W.T., Legut, D., Fan, Y., Qin, J., Li, X., Zhang, Q., Building fast diffusion channel by constructing metal sulfide/metal selenide heterostructures for high-performance sodium ion batteries anode. *Nano Lett.*, 20, 6199, 2020.
85. Sokolikova, M.S., Sherrell, P.C., Palczynski, P., Bemmer, V.L., Mattevi, C., Direct solution-phase synthesis of 1T-WSe_2 nanosheets. *Nat. Commun.*, 10, 1, 2019.
86. Zhou, P., Tanghe, I., Schiettecatte, P., van Thourhout, D., Hens, Z., Geiregat, P., Ultrafast carrier dynamics in colloidal WS_2 nanosheets obtained through a hot injection synthesis. *J. Chem. Phys.*, 151, 164701, 2019.
87. Wang, Y., Zhang, F., Wang, Q., Yang, P., Lin, H., Schiettecatte, P., Geiregat, P., Hens, Z., Ultrafast carrier dynamics in few-layer colloidal molybdenum disulfide probed by broadband transient absorption spectroscopy. *J. Phys. Chem. C*, 123, 10571, 2019.
88. Ding, X., Peng, F., Zhou, J., Gong, W., Slaven, G., Loh, K.P., Lim, C.T., Leong, D.T., Defect engineered bioactive transition metals dichalcogenides quantum dots. *Nat. Commun.*, 10, 1, 2019.
89. Kurzweil, P., Gaston planté and his invention of the lead–acid battery-the genesis of the first practical rechargeable battery. *J. Power Sources*, 195, 4424, 2010.
90. R.R. Haering, J.A.R. Stiles, K. Brandt, Lithium molybdenum disulphide battery cathode. US Patent US4224390A, 1980.
91. Nagaura, T. and Tozawa, K., Lithium-ion rechargeable battery. *Prog. Batteries Sol. Cells*, 9, 209, 1990.
92. Pumera, M., Graphene-based nanomaterials for energy storage. *Energy Environ. Sci.*, 4, 668, 2011.

93. Mahmood, N., Zhang, C., Yin, H., Hou, Y., Graphene-based nanocomposites for energy storage and conversion in lithium batteries, supercapacitors and fuel cells. *J. Mater. Chem. A*, 2, 15, 2014.
94. Srivastava, M., Singh, J., Kuila, T., Layek, R.K., Kim, N.H., Lee, J.H., Recent Advances in graphene and its metal-oxide hybrid nanostructures for lithium-ion batteries. *Nanoscale*, 7, 4820, 2015.
95. Goodenough, J.B. and Park, K.S., The Li-ion rechargeable battery: A perspective. *J. Am. Chem. Soc.*, 135, 1167, 2013.
96. Yang, Z., Zhang, J., Kintner-Meyer, M.C.W., Lu, X., Choi, D., Lemmon, J.P. *et al.*, Electrochemical energy storage for green grid. *Chem. Rev.*, 111, 3577, 2011.
97. Li, Y., Wu, D., Zhou, Z., Cabrera, C.R., Chen, Z., Enhanced Li adsorption and diffusion on MoS_2 Zigzag nanoribbons by edge effects: A computational study. *J. Phys. Chem. Lett.*, 3, 2221, 2012.
98. Shu, H., Li, F., Hu, C., Liang, P., Cao, D., Chen, X., The capacity fading mechanism and improvement of cycling stability in MoS_2-based anode materials for lithium-ion batteries. *Nanoscale*, 8, 2918, 2016.
99. Lian, P., Zhu, X., Liang, S., Li, Z., Yang, W., Wang, H., Large reversible capacity of high quality graphene sheets as an anode material for lithium-ion batteries. *Electrochim. Acta*, 55, 3909, 2010.
100. Hwang, H., Kim, H., Cho, J., MoS_2 nanoplates consisting of disordered graphene-like layers for high rate lithium battery anode materials. *Nano Lett.*, 11, 4826, 2011.
101. Zhong, Y., Shi, T., Huang, Y., Cheng, S., Chen, C., Liao, G. *et al.*, Three-dimensional MoS_2/graphene aerogel as binder-free electrode for Li-ion battery. *Nanoscale Res. Lett.*, 14, 85, 2019.
102. Zhang, K., Kim, H.J., Shi, X., Lee, J.T., Choi, J.M., Song, M.S. *et al.*, Graphene/acid coassisted synthesis of ultrathin MoS_2 nanosheets with outstanding rate capability for a lithium battery anode. *Inorg. Chem.*, 52, 9807, 2013.
103. Wang, Y., Chen, B., Seo, D.H., Han, Z.J., Wong, J.I., Ostrikov, K. *et al.*, MoS_2-coated vertical graphene nanosheet for high-performance rechargeable lithium-ion batteries and hydrogen production. *NPG Asia Mater.*, 8, e268, 2016.
104. Zhang, G., Feng, H., Ma, C., Chen, J., Wang, Z., Zheng, W., MoS_2 nanotubes via ionic-liquid-assisted assembly of MoS_2 nanosheets for lithium storage. *ACS Appl. Nano Mater.*, 4, 3397, 2021.
105. Ren, J., Ren, R.P., Lv, Y.K., A flexible 3D graphene@ CNT@ MoS_2 hybrid foam anode for high-performance lithium-ion battery. *Chem. Eng. J.*, 353, 419, 2018.
106. Yu, S., Jung, J.W., Kim, I.D., Single layers of WS_2 Nanoplates embedded in nitrogen-doped carbon nanofibers as anode materials for lithium-ion batteries. *Nanoscale*, 7, 11945, 2015.

107. Wang, Z., Chen, T., Chen, W., Chang, K., Ma, L., Huang, G., Chen, D., Lee, J.Y., CTAB-assisted synthesis of single-layer MoS_2-graphene composites as anode materials of Li-ion batteries. *J. Mater. Chem. A*, 1, 2202, 2013.
108. Nguyen, T.P. and Kim, I.T., Ag nanoparticle-decorated MoS_2 nanosheets for enhancing electrochemical performance in lithium storage. *Nanomaterials*, 11, 626, 2021.
109. Wang, P.P., Sun, H., Ji, Y., Li, W., Wang, X., Three-dimensional assembly of single-layered MoS_2. *Adv. Mater.*, 26, 964, 2014.
110. Kato, K., Sayed, F.N., Babu, G., Ajayan, P.M., All 2D materials as electrodes for high power hybrid energy storage applications. *2D Mater.*, 5, 025016, 2018.
111. Ju, W., Dong, C., Jin, B., Zhu, Y., Wen, Z., Jiang, Q., Composites of reduced graphene oxide and Fe_2O_3 nanoparticles anchored on MoS_2 nanosheets for lithium storage. *ACS Appl. Nano Mater.*, 3, 9009, 2020.
112. Zhu, X., Yang, C., Xiao, F., Wang, J., Su, X., Synthesis of nano-TiO_2-decorated MoS_2 nanosheets for lithium-ion batteries. *New J. Chem.*, 39, 683, 2015.
113. Bai, J., Zhao, B., Zhou, J., Fang, Z., Li, K., Ma, H., Dai, J., Zhu, X., Sun, Y., Improved electrochemical performance of ultrathin MoS_2 nanosheet/Co composites for lithium-ion battery anodes. *ChemElectroChem*, 6, 1930, 2019.
114. Rui, B., Li, J., Chang, L., Wang, H., Lin, L., Guo, Y., Nie, P., Engineering MoS_2 nanosheets anchored on metal organic frameworks derived carbon polyhedra for superior lithium and potassium storage. *Front. Energy Res.*, 7, 142, 2019.
115. Zhou, R., Wang, H., Chang, J., Yu, C., Dai, H., Chen, Q., Zhou, J., Yu, H., Sun, G., Huang, W., Ammonium intercalation induced expanded 1T-rich molybdenum diselenides for improved lithium-ion storage. *ACS Appl. Mater. Interfaces*, 13, 17459, 2021.
116. Li, Z., Zhang, L.Y., Zhang, L., Huang, J., Liu, H., ZIF-67-derived CoSe/NC composites as anode materials for lithium-ion batteries. *Nanoscale Res. Lett.*, 14, 1, 2019.
117. Panda, M.R., Gangwar, R., Muthuraj, D., Sau, S., Pandey, D., Banerjee, A., Chakrabarti, A., Sagdeo, A., Weyland, M., Majumder, M., Bao, Q., Mitra, S., High performance lithium-ion batteries using layered 2H-$MoTe_2$ as anode. *Small*, 16, 2002669, 2020.
118. Zhang, C., Park, S.H., Ronan, O., Harvey, A., Ascaso, A.S., Lin, Z., McEvoy, N., Boland, C.S., Berner, N.C., Duesberg, G.S., Rozier, P., Coleman, J.N., Nicolosi, V., Enabling flexible het-erostructures for Li-ion battery anodes based on nanotube and liquid-phase exfoliated 2D gallium chalcogenide nanosheet colloidal solutions. *Small*, 13, 1701677, 2017.
119. Xiao, J., Choi, D., Cosimbescu, L., Koech, P., Liu, J., Lemmon, J.P., Exfoliated MoS_2 nanocomposite as ananode material for lithium ion batteries. *Chem. Mater.*, 22, 4522, 2010.

120. Chen, Y.M., Yu, X.Y., Li, Z., Paik, U., Lou, X.W., Hierarchical MoS_2 tubular structures internally wired bycarbon nanotubes as a highly stable anode material for lithium-ion batteries. *Sci. Adv.*, 2, e1600021, 2016.
121. Shi, Y., Wang, Y., Wong, J.I., Tan, A.Y.S., Hsu, C.L., Li, L.J. *et al.*, Self-assembly of hierarchical MoS_x/CNT nanocomposites (2,x,3): Towards high performance anode materials for lithium ion batteries. *Sci. Rep.*, 3, 2169, 2013.
122. Zhou, X., Wan, L.J., Guo, Y.G., Facile synthesis of MoS_2@CMK-3 nanocomposite as an improved anode material for lithium-ion batteries. *Nanoscale*, 4, 5868, 2012.
123. Jiang, H., Ren, D., Wang, H., Hu, Y., Guo, S., Yuan, H. *et al.*, 2D monolayer MoS_2-carbon inter overlapped superstructure: Engineering ideal atomic interface for lithium ion storage. *Adv. Mater.*, 27, 3687, 2015.
124. George, C., Morris, A.J., Modarres, M.H., De Volder, M., Structural evolution of electrochemically lithiated MoS_2 nanosheets and the role of carbon additive in Li-ion batteries. *Chem. Mater.*, 28, 7304, 2016.
125. Zhang, Z., Yang, X., Fu, Y., Nanostructured WSe_2/C composites as anode materials for sodium-ion batteries. *RSC Adv.*, 6, 12726, 2016.
126. Konar, R., Rosy, R.S., Perelshtein, I., Teblum, E., Telkhozhayeva, M., Tkachev, M., Richter, J.J., Cattaruzza, E., PietropolliCharmet, A., Stoppa, P., Noked, M., Nessim, G.D., Scalable synthesis of few-layered 2D tungsten diselenide (2H-WSe_2) nanosheets directly grown on tungsten (W) foil using ambient-pressure chemical vapor deposition for reversible li-ion storage. *ACS Omega*, 5, 19409, 2020.
127. Konar, R. and Nessim, G.D., A mini-review focusing on ambient-pressure chemical vapor deposition (AP-CVD) based synthesis of layered transition metal selenides for energy storage applications *Mater. Adv.*, 3, 4471, 2022.
128. Zhou, J., Qin, J., Guo, L., Zhao, N., Shi, C., Liu, E., He, F., Ma, L., Lia, J., He, C., Scalable synthesis of high-quality transition metal dichalcogenide nanosheets and their application as sodium-ion battery anodes. *J. Mater. Chem. A*, 4, 17370, 2016.
129. Lan, H.W.X., Jiang, D., Zhang, Y., Zhong, H., Zhang, Z., Jiang, Y., Sodium storage and transport properties in pyrolysis synthesized MoSe2 nanoplates for high performance sodium-ion batteries. *J. Power Sources*, 283, 187, 2015.
130. Mukherjee, S., Turnley, J., Mansfield, E., Holm, J., Soares, D., David, L., Singh, G., Exfoliated transition metal dichalcogenide nanosheets for supercapacitor and sodium ion battery applications. *R. Soc. Open Sci.*, 6, 190437, 2019.
131. Sun, Q., Fu, L., Shang, C., A novel open-framework Cu-Ge-based chalcogenide anode material for sodium-ion battery. *Scanning*, 27, 3876525, 2017.
132. Yang, E., Ji, H., Jung, Y., Two-dimensional transition metal dichalcogenide monolayers as promising sodium ion battery anodes. *J. Phys. Chem. C*, 119, 2637, 2015.
133. David, L., Bhandavat, R., Singh, G., MoS_2/graphene composite paper for sodium-ion battery electrodes. *ACS Nano*, 8, 1759, 2014.

134. Sun, D., Ye, D., Liu, P., Tang, Y., Guo, J., Wang, L. *et al.*, MoS_2/graphene nanosheets from commercialbulky MoS_2 and graphite as anode materials for high rate sodium-ion batteries. *Adv. Energy Mater.*, 8, 1702383, 2018.
135. Sahu, T.S. and Mitra, S., Exfoliated MoS_2 sheets and reduced graphene oxide-an excellent and fast anode forsodium-ion battery. *Sci. Rep.*, 5, 12571, 2015.
136. Ji, G., Yu, Y., Yao, Q., Qu, B., Chen, D., Chen, W. *et al.*, Promotion of reversible Li1 storage in transition metal dichalcogenides by Ag nanoclusters. *NPG Asia Mater.*, 8, e247, 2016.
137. Liu, H., Huang, Z., Wu, G., Wu, Y., Yuan, G., He, C., Qi, X., Zhong, J., A novel $WS_2/NbSe_2$ 2d-heterostructure as an ultrafast charging and discharging anode material for lithium-ion batteries. *Mater. Chem. A*, 6, 17040, 2018.
138. Roy, A., Movva, H.C.P., Satpati, B., Kim, K., Dey, R., Rai, A., Pramanik, T., Guchhait, S., Tutuc, E., Banerjee, S.K., Structural and electrical properties of $MoTe_2$ and $MoSe_2$ grown by molecular beam epitaxy. *ACS Appl. Mater. Interfaces*, 8, 7396, 2016.
139. Zhang, J., Kang, W., Jiang, M., You, Y., Cao, Y., Ng, T.W., Yu, D.Y.W., Lee, C.-S., Xu, J., Conversion of 1T-$MoSe_2$ to 2H-$MoS_{2x}Se_{2-2x}$ mesoporous nanospheres for superior sodium storage performance. *Nanoscale*, 9, 1484, 2017.
140. Mudgal, P., Arora, H., Pati, J. *et al.*, Electrochemical investigation of MoSeTe as an anode for sodium-ion batteries. *Proc. Indian Natl. Sci. Acad.*, 88, 430, 2022.
141. Yue, J.L., Sun, Q., Fu, Z.W., Cu_2Se with facile synthesis as a cathode material for rechargeable sodium batteries. *Chem. Commun.*, 49, 5868, 2013.
142. Share, K., Lewis, J., Oakes, L., Carter, R.E., Cohn, A.P., Pint, C.L., Tungsten diselenide (WSe_2) as a high capacity, low overpotential conversion electrode for sodium ion batteries. *RSC Adv.*, 5, 101262, 2015.
143. Wang, T., Legut, D., Fan, Y., Qin, J., Li, X., Zhang, Q., Building fast diffusion channel by constructing metal sulfide/metal selenide heterostructures for high-performance sodium ion batteries anode. *Nano Lett.*, 20, 6199, 2020.
144. Wu, C., Zhao, G., Yu, X., Liu, C., Lyu, P., Maurin, G., Le, S., Sun, K., Zhang, N., MoS_2/graphene heterostructure with facilitated Mg-diffusion kinetics for high-performance rechargeable magnesium batteries. *Chem. Eng. J.*, 412, 128736, 2021.
145. Zhang, Z., Yang, M., Zhao, N., Wang, L., Li, Y., Two-dimensional transition metal dichalcogenides as promising anodes for potassium ion batteries from first-principles prediction. *Phys. Chem. Chem. Phys.*, 21, 23441, 2019.
146. Wu, Y., Xu, Y., Li, Y., Lyu., P., Wen, J., Zhang, C., Zhou, M., Fang, Y., Zhao, H., Kaiser, U., Lei, Y., Unexpected intercalation-dominated potassium storage in WS_2 as a potassium-ion battery anode. *Nano. Res.*, 12, 2997, 2019.
147. Xu, Y., Bahmani, F., Zhou, M., Li, Y., Liang, F., Kazemi, S.H., Kaiser, U., Meng, G., Lei, Y., Enhancing potassium-ion battery performance by defect and interlayer engineering. *Nanoscale Horiz.*, 4, 202, 2019.

148. Dong, Y., Xu, Y., Li, W., Fu, Q., Wu, M., Manske, E., Kröger, J., Lei., Y., Insights into the crystallinity of layer-structured transition metal dichalcogenides on potassium ion battery performance: A case study of molybdenum disulfide. *Nanomicro Small*, 15, 1900497, 2019.
149. Liu, B., Luo, T., Mu, G., Wang, X., Chen, D., Shen, G., Rechargeable Mg-Ion batteries based on WSe_2 nanowire cathodes. *ACS Nano*, 7, 8051, 2013.
150. Yue, J.L., Sun, Q., Fu, Z.W., Cu_2Se with facile synthesis as a cathode material for rechargeable sodium batteries. *Chem. Commun.*, 49, 5868, 2013.
151. Wang, Q.Z., Yichun, W., Xin, Z., Wang, G., Ji, P., Yin, F., WSe_2/reduced graphene oxide nanocomposite with superfast sodium ion storage ability as anode for sodium ion capacitors. *J. Electrochem. Soc.*, 165, A3642, 2018.
152. Konar, R., Tamari, R., Teblum, E., Nessim, G.D., Meshi, L., In-depth characterization of stacking faults forming during the growth of transition-metal Di-chalcogenides (TMDCs) by ambient pressure-CVD. *Mater. Charact.*, 184, 111666, 2021.
153. Ko, Y.N., Choi, S.H., Park, S.B., Kang, Y.C., Hierarchical MoSe2 yolk–shell microspheres with superior Na-ion storage properties. *Nanoscale*, 6, 10511, 2014.
154. Lu, C., Li, Z., Xia, Z., Ci, H., Cai, J., Song, Y., Yu, L., Yin, W., Dou, S., Sun, J., Liu, Z., Confining MOF-derived SnSe nanoplatelets in nitrogen-doped graphene cages via direct CVD for durable sodium ion storage. *Nano Res.*, 12, 3051, 2019.
155. Li, H., Yang, Q., Mo, F., Liang, G., Liu, Z., Tang, Z., Ma, L., Liu, J., Shi, Z., Zhi, C., MoS_2 nanosheets with expanded interlayer spacing for rechargeable aqueous Zn-ion batteries. *Energy Storage Mater.*, 19, 94, 2021.
156. Liu, H., Wang, J.G., Hua, W., You, Z., Hou, Z., Yang, J., Wei, C., Kang, F., Boosting zinc-ion intercalation in hydrated MoS_2 nanosheets toward substantially improved performance. *Energy Storage Mater.*, 35, 731, 2021.
157. Cai, C., Tao, Z., Zhu, Y., Tan, Y., Wang, A., Zhou, H., Yang, Y., A nano interlayer spacing and rich defect 1T-MoS_2 as cathode for superior performance aqueous zinc-ion batteries. *Nanoscale Adv.*, 3, 3780, 2021.
158. Gao, Z., Yu, X., Zhao, J., Zhao, W., Xu, R., Liu, Y., Shen, H., Synthesis of long hierarchical MoS_2 nanofibers assembled from nanosheets with an expanded interlayer distance for achieving superb Na-ion storage performance. *Nanoscale.*, 9, 15558, 2017.
159. He, P., Yan, M., Zhang, G., Sun, R., Chen, L., An, Q., Mai, L., Layered VS_2 nanosheet-based aqueous Zn ion battery cathode. *Adv. Energy Mater.*, 7, 1601920, 2017.
160. Wu, Z., Lu, C., Wang, Y., Zhang, L., Jiang, L., Tian, W., Cai, C., Gu, Q., Sun, Z., Hu, L., Ultrathin VSe_2 nanosheets with fast ion diffusion and robust structural stability for rechargeable zinc-ion battery cathode. *Small*, 16, 2000698, 2020.
161. Wang, H., Wang, L., Wang, X., Quan, J., Mi, L., Yuan, L., Li, L., Zhang, B., Zhong, H., Jiang, Y., High quality $MoSe_2$ nanosphereswith superior

electrochemical properties forsodium batteries. *J. Electrochem. Soc.*, 163, A1627, 2016.

162. Zhang, Z., Yang, X., Fu, Y., Du, K., Ultrathin molybdenum diselenide nanosheets anchored on multi-walled carbon nanotubes as anode composites for high performance sodium-ion batteries. *J. Power Sources*, 296, 2, 2015.
163. Panda, M.R., Raj K, A., Ghosh, A., Kumar, A., Muthuraj, D., Sau, S., Yu, W., Zhang, Y., Sinha, A.K., Weyland, M., Bao, Q., Mitra, S., Blocks of molybdenum ditelluride: A high rate anode for sodium-ion battery and full cell prototype study. *Nano Energy*, 64, 103951, 2019.
164. Zhou, J., Wang, L., Yang, M., Wu, J., Chen, F., Huang, W., Han, N., Ye, H., Zhao, F., Li, Y., Li, Y., Hierarchical VS_2 nanosheet assemblies: A universal host material for the reversible storage of alkali metal ions. *Adv. Mater.*, 29, 1702061, 2017.
165. Wang, J., Luo, N., Wu, J., Huang, S., Yu, L., Wei, M., Hierarchical spheres constructed by ultrathin VS_2 nanosheets for sodium-ion batteries. *J. Mater. Chem.*, 7, 3691, 2019.
166. Xu, B., Ma, X., Tian, J., Zhao, F., Liu, Y., Wang, B., Yang, H., Xia, Y., Layer-structured $NbSe_2$ anode material for sodium-ion and potassium-ion batteries. *Ionics*, 25, 4171–4177, 2019.
167. Ou, X., Xiong, X., Zheng, F., Yang, C., Lin, Z., Hu, R., Jin, C., Chen, Y., Liu, M., *In-situ* x-ray diffraction characterization of NbS_2 nanosheets as the anode material for sodium-ion batteries. *J. Power Sources*, 325, 410, 2016.
168. Hong, M., Li, J., Zhang, W., Liu, S., Chang, H., Semimetallic 1t0 WTe2 nanorods as anode material for the sodium ion battery. *Energy Fuels*, 32, 6371, 2018.
169. Cho, J.S., Ju, H.S., Lee, J.K., Kang, Y.C., Carbon/two-dimensional $MoTe_2$ core/shell-structured microspheres as an anode material for Na-ion batteries. *Nanoscale*, 9, 1942, 2017.
170. Liu, Y., Zhang, L., Zhao, Y., Shen, T., Yan, X., Yu, C. *et al.*, Novel plasma-engineered MoS_2 nanosheets for superior lithium-ion batteries. *J. Alloys Compd.*, 787, 996, 2019.
171. Cha, A., Patel, M.D., Park, J., Hwang, J., Prasad, V., Cho, K. *et al.*, 2D MoS_2 as an efficient protective layer for lithium metal anodes in high-performance Li-S batteries. *Nat. Nanotechnol.*, 13, 337, 2018.
172. Guo, P., Liu, D., Liu, Z., Shang, X., Liu, Q., He, D., Dual functional MoS_2/graphene interlayer as an efficient polysulfide barrier for advanced lithium-sulfur batteries. *Electrochim. Acta*, 256, 28, 2017.
173. Dirlam, P.T., Park, J., Simmonds, A.G., Domanik, K., Arrington, C.B., Schaefer, J.L. *et al.*, Elemental sulfur and molybdenum disulfide composites for Li-S batteries with long cycle life and high-rate capability. *ACS Appl. Mater. Interfaces*, 8, 13437, 2016.

174. Lin, H., Yang, L., Jiang, X., Li, G., Zhang, T., Yao, Q. *et al.*, Electrocatalysis of polysulfide conversion by sulfur-deficient MoS_2 nanoflakes for lithium-sulfur batteries. *Energy Environ. Sci.*, 10, 1476, 2017.
175. Li, Z., Niu, B., Liu, J., Li, J., Kang, F., Rechargeable aluminum-ion battery based on mos_2 microsphere cathode. *ACS Appl. Mater. Interfaces*, 10, 9451, 2018.
176. Qiao, J., Zhou, H., Liu, Z., Wen, H., Yang, J., Defect-free soft carbon as cathode material for Al-ion batteries. *Ionics*, 25, 1235, 2019.
177. Wu, L., Sun, R., Xiong, F., Pei, C., Han, K., Peng, C., Fan, Y., Yang, W., An, Q., Mai, L., A rechargeable aluminum-ion battery based on a VS_2 nanosheet cathode. *Phys. Chem. Chem. Phys.*, 20, 22563, 2018.
178. Divya, S., Johnston, J.H., Nann, T., Molybdenum dichalcogenide cathodes for aluminum-ion batteries. *Energy Technol.*, 8, 2000038, 2020.

20

Effect of 2D Nanomaterial Addition to Performance and Emission Characteristics of Diesel Engine

Geetesh Goga[1]* **and M. V. B. Unnamatla**[2,3]

[1]*Department of Mechanical Engineering, Bharat Group of Colleges, Sardulgarh, Punjab, India*
[2]*Centro Conjunto de Investigación en Química Sustentable, UAEM-UNAM, Universidad Autónoma del Estado de México, Toluca, Estado de México, Mexico*
[3]*Facultad de Química, Universidad Autónoma del Estado de México, Unidad San Cayetano, Toluca, Estado de México, Mexico*

Abstract

Biodiesel is a superior alternative fuel designed to extend the value of fossil fuels, while maintaining the longevity and cleanliness of diesel engines. Owing to its closed carbon cycle, it reduces greenhouse gas emissions and dependence on foreign fuels. The disadvantages of biodiesel include an increase in nitrogen oxide emissions, its incompatibility with cold weather, and the necessity of replacing engine parts such as fuel filters, fuel tanks, and fuel lines due to clogging on a regular basis. By using 2D nanomaterials as fuel additives, it is possible to improve the properties of fuels and overcome their shortcomings. Recent investigations on fuel-added substances showed the consideration of 2D nano-sized particles (metallic, non-metallic, oxygenated, natural, and blended) with diesel-biodiesel fuel emulsions. The thermophysical properties were improved, the heat transfer rate was increased, and the fuel mixtures were stabilized. In addition, depending on the dosage of the 2D nanofluid additives, the engine performance parameters increased, and the exhaust emissions decreased. The dispersion of 2D nanoparticles in liquid fuel and the effect of adding these particles on the performance and emission parameters of diesel engines are covered in this chapter.

Keywords: 2D nanomaterials, diesel engine, performance, emission, fuel

**Corresponding author*: geeteshgoga@gmail.com

Subhendu Chakroborty and Kaushik Pal (eds.) 2D Nanomaterials: Synthesis, Properties and Applications, (475–488) © 2024 Scrivener Publishing LLC

20.1 Introduction

The swift variation in energy requirements, the effect of climate change on the environment, and vice-versa and expanding prices of energy sources have made researchers desirous of clean and non-conventional sources of energy that are effortlessly accessible and agreeable worldwide [1]. Although the use of electric vehicles is rapidly progressing for transportation systems, petrodiesel will act as the principal fuel for bulky engines in many financial sectors. The exhaustion of fossil fuels at very high speeds is also a major challenge. The transport sector is one of the key sectors that intensifies fuel use and emits noxious emissions [2]. Diesel engines are robust, dependable and consume less fuel than petrol engines, but higher toxic emissions from their tailpipes result in global warming. To overcome this drawback of diesel engine, researchers have tested numerous alternative fuels in the engine [3] and concluded that fuels with high oxygen content are amenable to minimizing fatal gases from the engine as they aid in the complete burning of fuel [4]. Among oxygenated liquid fuels, biodiesel and alcohols are considered to be most suitable fuel for diesel engines because they have the same physicochemical properties same as diesel fuel [5]. However, these fuels have some drawbacks because they cannot be used in diesel engines without blending or additives [6]. The combustion performance of conventional fuels can be enhanced by the addition of a few micrometer-sized metallic particles. However, these particles cause engine knocks owing to an augmented ignition lag [7]. When compared to metallic nanoparticle fuel additives, the lower surface-area-to-volume ratio is thought to be the cause of the ignition lag [8]. Because heat exchangers have a high heat transfer rate and better thermophysical properties, fuels with nanoparticle additives, known as nanofluids, are thought to be a new type of fuel [9].

2D nanomaterials are acknowledged for their exceptional physical properties and extensive engineering applications. It has been found from the latest experiments that the utilization of 2D nanomaterials in liquid fuels as additives aids in enhancing their properties.

The idea that nanoparticle addition to fuel improves the properties of fuel has been recognized by most researchers. Nanoparticles are used to improve fuel properties, allowing for lower pollutant exhaust emissions, improved engine performance, and the use of biofuels. Nanoparticles are added to commercially available fuels to achieve certain properties, including lowering exhaust pollution, increasing the quantity of oxygen in the engine, maintaining fluid stability, reducing ignition lag, and increasing the flashpoint and lubricity index. In order to reduce wear, contact between metal to metal should be minimized and contact between chemical to chemical should be maximized [10].

Researchers have also investigated whether the addition of 2D nanomaterials to liquid fuel aids in increasing the performance of engines and minimizing the pollution content [11]. The addition of 2D nanoparticles to fuel aids in enhancing the physicochemical properties of the fuel, such as the for abridged ignition lag, evaporation rate, and thermal conductivity. Another advantage of utilizing 2D nanoparticles as fuel additives is that they offer a desirable heat-release rate in engines [12].

The cetane number of the blended fuel decreased when alcohol was added to the diesel. Noise and vibration may arise because of the lower cetane number of the alcohol–diesel fuel blend [13]. The addition of 2D nanoparticles significantly reduces the number of pollution particles and enhances the performance of the engine. The increase in oxygen content and catalytic effect of the graphine oxide nano-additive resulted in a shorter ignition delay, reducing the likelihood of knocking. Nano-additives were more effective when the alcohol content is lower, but the difference is not significant when the alcohol content is higher [14]. Research into renewable fuels, particularly for diesel engines, is motivated by the exhaustion of fuel based on petroleum and the dangers posed by engine emissions. The addition of 2D nanomaterials holds promise for lowering diesel engine exhaust emissions and improving performance properties, such as BP, BSFC, and BTE. The purpose of this chapter is to study the impact of accumulating 2D nanomaterials to diesel and blended fuels on diesel engine emissions and performance. It provides an idea of the current trends in 2D nanomaterial research and is taken from research journals published up until 2022.

20.2 Performance Characteristics

The most intriguing alternative fuels, which are anticipated to eventually substitute diesel, are biofuels with characteristics comparable to those of diesel. However, diesel engines do not use biofuel to its full potential because of issues such as lower mixture stability, higher BSFC, lower power, and slower engine speed. Numerous researchers are looking for various technologies and approaches to address these issues. Utilizing nanoadditives is a promising strategy to enhance the properties of biofuel. A review of the performance parameters of BSFC, BP, and BTE for various biofuels containing various nanoparticles has been conducted. When nanoparticles are added to blends of diesel and biodiesel, it speeds up and completes the process of combustion owing to an enhanced heat transfer property, which increases the efficiency of combustion [15]. This topic has been the subject of numerous studies. Fuels containing metallic-based additives have lower BSFC than pure diesel [16]. In a diesel engine, the effects of Al2O3 nanoadditives on emission characteristics and performance enhancement were investigated. Two distinct amounts of Al2O3 nanoparticles, 0.1 and 0.5% by weight, were used as fuel additives. The results of the performance parameters were related to neat diesel. According to the findings, a fuel with a nanoparticle concentration of half than 1% performs better than a fuel with a nanoparticle concentration of 1%, and the addition of Al2O3 led to a 31% decrease in fuel consumption at full load *vis-a-vis* diesel [17]. The accumulation of nanoparticles lowers the BSFC because it slightly increases the calorific value of fuels. The high surface area of metallic additives reduces BSFC and improves the combustion quality [14] of an engine's characteristics by the addition of 60 ppm GQD nanoparticles to bioethanol containing one-tenth biodiesel and volumes of 2, 4, and 6%, respectively, of ethanol at three distinct engine speed settings. The accumulation of GQD nanoparticles usually results in increased engine performance and decreased emissions from the exhaust. The fuel chosen was 10% biodiesel, along with 4% ethanol and 60 ppm GQD as the best fuel at all paces. This composition of fuel enhances the power of the engine and torque by one-fourth and one-fifth, respectively, when associated with pure diesel fuel [18]. The experiments emphasize the effect of varying the injection timing on a mixture of 70% diesel, 20% biodiesel, and 10% ethanol (called BDE). The addition of Al_2O_3 results in reduced heat release and cylinder pressure away from the TDC, which improves the engine performance and combustion [19]. In a diesel engine with a single cylinder and a mixture of ethanol, diesel, and biodiesel, the impact of nanoparticles, injection timing, and combustion chamber geometry was checked as a whole. The fuel was checked at various injection times and combustion chamber geometries, including 21°–24° bTDC. The outcome showed a BTE of 33.8%, which was the highest. The lowest BSFC was attained at 22° bTDC, with an 11.51% increase in BTE over that of neat diesel [20]. The performance and emissions of an engine using blends of ethanol and biodiesel were investigated in relation to the effects of GQD nanoparticles. For the fuel blend, the GQD nanoparticles were used at a concentration of 30 ppm. The study was done at 1,800, 2,100, and 2,400 rpm engine speeds. The results of the investigation of the performance characteristics, such as specific fuel consumption, torque, and power, are presented here. When related to the D100 fuel, the addition of nanoparticles to the fuel enhanced power by approximately 28% and torque by 14.5% [21]. A study on the addition of TiO2 to the emission and performance parameters of pure diesel fuel was conducted. The study demonstrated that mixing the nanoparticles improves the viscidity, density, energy content, and fire point of petroleum diesel. Engine performance test results demonstrated a 22% reduction in BSFC and an enhancement in BTE at full load when nanoparticles were added to the fuel [22]. The impact of MoO3 nanoparticles on numerous parameters of diesel engines was investigated. 100CNT and 100MoO3 were the fuels being tested. Compared with natural diesel, the BSFCs of 100CNT and 100MoO3 decreased by 5.1% and 2.3%, respectively, and the BTE of 100CNT was enhanced by approximately 5% [23]. The performance and exhalation parameters of the two fuel blends (B20) were contrasted with those of pure diesel. At the maximum load, the new fuel blend had a higher BTE than that of natural diesel. At the maximum load, the BTE of the B20 and D80SBD15E4S1 alumina fuel blends increased by about15% and 18%, respectively. For B20 and D80SBD15E4S1's alumina fuel blend, complete combustion occurred because of the accessibility of ample oxygen in the fuel. In addition, the advanced evaporation amount contributed to an additional increase in efficiency. At the three-fourth maximum load, the BSFC was negligible in comparison to natural diesel. The presence of alumina nanoparticles increases the rate of atomization, which improves combustion and boosts the transfer of heat from combustion. This lowers the temperature and improves the engine performance. The specific fuel consumption for D80SBD15E4S1 alumina was 3.53, for B20 it is 3.7, and for neat diesel it is 4.33. This demonstrates that the full-load conditions decreased viscidity at advanced temperatures, overcoming the high viscidity of the fuel [24]. The performance and exhaust emissions of a diesel engine were compared and contrasted using three distinct nanoparticles, Al2O3, CNT, and silicon oxide, as additives. Pure diesel was mixed with nanoparticles at different concentrations (ppm). The test results showed that the BTE was improved by 18.8% and the BSFC of the CNT blend was reduced by up to 20%. Silicon oxide blends perform better than Al2O3 blends [25]. The effects of studying *Calophyllum inophyllum* biodiesel nanoemulsions at concentrations of 50 and 100 parts per million (ppm) were compared to those of 100% biodiesel and pure diesel fuels. The results showed that the highest BTE was approximately 26% for unadulterated biodiesel, 27% for biodiesel-Z100, 28% for biodiesel-Z50, 30% for biodieselT50, 31% for biodiesel-T100, and 33% for diesel. At all loadings, the CIME nanoemulsions and pure *C. inophyllum* biodiesel had an advanced BSFC compared with diesel. At full engine load, the BSFC of CIMET100 was 0.24 kg/kWh, while the BSFC of the CIME-Z100 is 0.258 kg/kWh.

The researchers concluded that because biodiesel has a lower calorific value, more fuel is required to produce the same amount of power. All CIME nanoemulsions, CIME-Z50, CIME-Z100, CIME-T50, and CIME-T100 had lower BSFC than CIME when nanoparticles were not added [26]. In dual-fuel systems, the impacts of numerous nanoparticles and the hydrogen mix on the performance and emission parameters of the engine were tested. The nanoparticles used for the review were CuO, CNT, CeO2, TiO2, and Al2O3. Nanoparticles were added at a concentration of 100 ppm with 20% hydrogen fuel. The test was performed at 1,800 rpm with engine loads of 0, 25, 50, 75, and 100%. The total increase in BTE is approximately 5% for CeO2 and 2.5% for Al2O3. In the interim, for CeO2 the BSFC is seen to be immaterial. CNT and TiO2 nanoparticles showed a reduction in the BSFC for different blends [27]. Owing to the increased surface area to volume ratio of the nanoparticles, which results in an advancement in BTE, the addition of nanoparticles increases fuel properties such as energy content and fuel viscidity, both of which increase the cetane number. Nanoparticles possess a catalytic property that speeds up ignition by enhancing mixing with air and promoting the microexplosion of primary droplets, which can vaporize the fuel. The enhancement of heat transfer, which facilitates combustion and the rate at which heat is released, is another beneficial effect of the nanoparticle additives. Consequently, by increasing the likelihood of complete combustion, the addition of nanoparticles to fuel improves engine performance [26–28].

A diesel engine running on a fuel blend of 30% sesame oil (SO) and 70% diesel fuel was the subject of this experimental study, which sought to ascertain the impact of varying quantities of graphene oxide (GO) on engine parameters. The assessment of the density of various fuels with diesel is depicted in Figure 20.1 Following that, an optimization using response surface methodology (RSM) was carried out to ascertain the ideal operating conditions for numerous engine loads. According to the experimental findings, the addition of GO nanoparticle to blends of diesel and biodiesel enhanced performance and reduced emissions. The optimal GO concentration for the performance characteristics was between 75 and 100 ppm. Power requires 75 parts per million (ppm) of GO, whereas BSFC and EGT require 100 ppm. Moreover, the highest GO measure of 100 ppm was optimal for CO and HC, and the 75 ppm GO sum was the best for NOx. In contrast, the optimization results demonstrated that the optimal condition for all responses was 100 ppm GO at a 1,950 W load. The study indicates that the maximum error between the optimal and experimental outcomes is around 5%. [29].

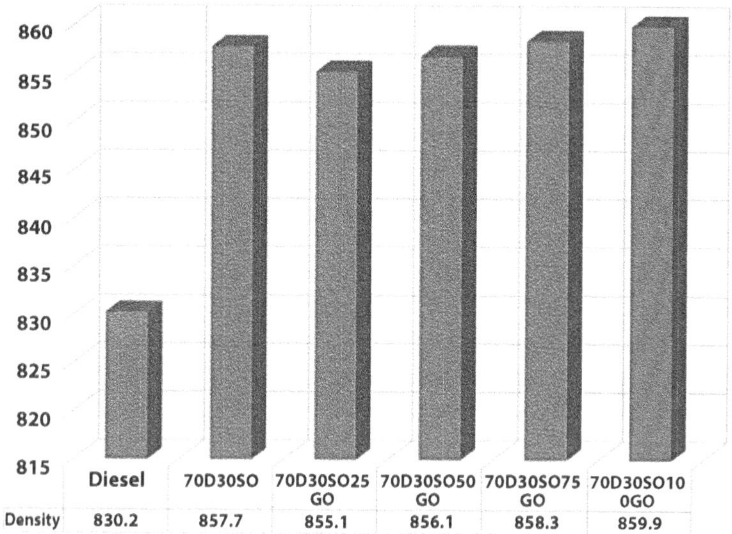

Figure 20.1 Comparison of density values of test fuels with diesel [29].

20.3 Emission Characteristics

According to previous research, the usual exposure to diesel exhaust may cause health harm in densely populated municipal areas. Despite their advantages, harmful gas emissions from diesel engines, such as CO2, CO, NOx, and sulfur dioxide [28], are significant contributors to air pollution. The environmentally friendly and health-conscious community has been focusing on diesel engine exhaust emissions and their impact on health of human health. Pollution from diesel engines has a significant impact on human health. Additionally, diesel engine emissions contain a wide range of harmful substances, including mutagens, NOx, SOx, CO, and CO_2. In addition, both humans and animals are frequently exposed to lung cancer, posing a health risk. According to a study conducted on road workers and drivers exposed to diesel engine exhaust because of their jobs, the risk of lung cancer increased with increasing years of exposure [30]. Numerous studies have been conducted on fuel additives to enhance the emission characteristics of diesel engines. The outcomes of numerous experiments on the effect of metallic nanoparticles have been reviewed. When associated with petroleum-based fuels, the use of aluminum oxide (Al2O3) nanoparticles as an additive decreases the emissions of NOx, CO2, and

HC [17]. When 60 ppm GQD nanoparticles were added to bioethanol fuel to check the performance and emission parameters of a diesel engine, a significant decrease in emission was attained with 10, 6, and 4% ethanol at various engine speeds. In comparison with diesel, the particular fuel B10 + E4 + GQD60 decreased the UHC and CO emissions by approximately 43% and 36%, respectively [22]. An experiment was conducted to determine how changing the injection timing affects the impact of alumina (Al2O3) nanoparticles on a mixture of 20% biodiesel, 70% diesel, and 10% ethanol (BDE). In a diesel engine, the addition of 25 ppm Al2O3 to a BDE blend at 19° bTDC IT reduced NOx, HC, CO, and smoke opacity [19]. Venu *et al.* studied the effects of nanoparticles as an additive, the design of the combustion chamber, and the timing of injection in a diesel engine with one cylinder. After alumina nanoparticles were added to diesel, biodiesel, and ethanol blends, which are referred to as ternary fuels, they referred to the fuel as a high-performance fuel. Various combustion chamber shapes and injection timings (21°, 22°, 23°, and 24° bTDC) were used to test the high-performance fuel. The experimental outcome showed decreased HC and CO discharges of approximately 9% and 17%, respectively, as opposed to 22° bTDC, and reduced NOx emissions by 22.53% in an examination with net diesel [20]. The performance and emissions of graphene quantum dot nanoparticles have been investigated in a diesel engine that uses a mixture of ethanol and biodiesel. The GQD nanoparticles had a concentration of 30 ppm. An experiment was conducted at various engine speeds. The investigation focused on emission parameters, such as CO, CO2, UHC, and NOx. CO and UHC decreased by 29.54% and 31.12%, respectively, and NOx discharge brought about a 1.29% addition, in contrast to net diesel fuel [21]. With and without the addition of TiO2 nanoparticles, an experiment was conducted on how the additive for TiO2 nanoparticles affects the emission characteristics of CI engine. With the development of nanoparticles, the discharge of hydrocarbons and carbon monoxide decreased by approximately 17% and 24%, respectively [22]. Deqing *et al.* examined how diesel emissions parameters in a diesel engine are affected by carbon nanotube (CNT) and nano molybdenum trioxide (MoO3). Studies on associating the engine emission parameters of nanofuels and diesel were performed using a single-cylinder common-rail diesel engine. The test energizes the incorporate 100CNT (diesel + 100 mg/L CTAB + 100 mg/L CNT) and 100MoO3 (diesel + 100 mg/L CTAB + 100 mg/L MoO3). The experiments revealed that when CNT and MoO3 were added to pure diesel, the levels of HC, CO, NOx, and soot emissions decreased. MoO3 was able to reduce HC, CO, NOx, and smoke emissions by 7.5%, 6.8%, 5.2%, and 8.3%, respectively, compared to neat diesel [23].

CNT were better able to reduce all emissions than neat diesel, lowering HC by 11.4%, CO by 9.6%, NOx by 8.9%, and smoke by 15.2%. The emission characteristics and engine performance of the two modified fuel blends, B20 (80% diesel and 20% soybean biodiesel) and D80SBD15E4S1, alumina (4% ethanol, 1% isopropanol as an emulsifier, and alumina nanoparticles of 100 mg/L) were examined by Shaafi and Velraj in comparison to pure diesel. The emission parameters were also the subject of encouraging findings. The inherent oxygen of soybean biodiesel and its enhanced utilization in the presence of alumina nanoparticles account for the emission reduction. However, when compared to pure diesel, the NOx emissions of the alumina fuel blend at full load were 9.9% higher for D80SBD15E4S1 [24]. Chena *et al.*, studied the parameters of aluminum oxide, silicon oxide, and carbon nanotube nanoparticle exhaust emissions from a four-stroke single-cylinder engine. The experiment revealed that the CNT nanoparticles resulted in 4.48% lower NOx emissions and higher CO, CO2, and HC emissions. Carbon nanotubes have the potential to be the subject of additional research as an additive for diesel fuel owing to their substantial reduction in NOx emissions during combustion [25, 30]. The CNT blends with diesel demonstrated improved performance in advancing the emission parameters. Nanthagopal *et al.* examined the performance and emission parameters of two-cylinder CI engines powered by *C. inophyllum* biodiesel (CIME) nanoemulsions. Owing to the oxidation impact of TiO2 and ZnO nanoparticles, all CIME nanoemulsions had lower unburned HC and CO emissions than pure CIME or diesel. Compared to all other fuels, the CIME-T100 fuel provided the greatest reduction in HC emission measurements. At full load, the CIME fuel process emitted 24.5% less CO than the pure diesel process. In comparison to the CIME-Z100 nanoemulsion fuel, the CIME-T100 nanoemulsion emitted relatively less CO. At full engine load, all CIME nanoemulsion fuels reduce NOx emissions by 7%–29% when compared to pure *C, inophyllum* biodiesel. This is due to the catalytic effect of nanoparticles [26, 30]. In a study by Manigandan *et al.*, the test was conducted under various engine load conditions to examine the effects of hydrogen mixed with Al2O3, CNT, TiO2, CeO2, and CuO nanoparticles on the emission characteristics. The results of the exhaust emissions showed that adding hydrogen and nanoparticles significantly reduced the emissions of carbon dioxide, carbon monoxide, and hydrocarbons. CNT, on the other hand, was the only blend that showed a reduction in NOx emissions compared to other blends. From the outcomes, they reasoned that the expansion of hydrogen and nanoparticles in unadulterated diesel decreased the gas discharge of fumes [27, 30]. Owing to its catalytic property and the increase in cetane number, the increase in nanoparticles in the fuel increases the total combustion. The reduced CO and HC concentrations result from complete combustion [28, 29]. The oxidation of N2 at the peak combustion temperature generates NOx in diesel motors. When diesel was mixed with nanoparticles, the top temperature of combustion increased, causing more N2 in the combustion chamber to be converted into NOx during combustion [30]. However, NOx outflow levels were lower when nanoparticle-emulsified biodiesel fuel was blended with diesel than when diesel fuel was used alone. The reason for this was the emulsified fuel heat sink, which dampened the gas temperatures in the cylinder during combustion.

20.4 Conclusion

The purpose of this review is to determine how various nanoparticles, when incorporated into conventional diesel and a variety of biofuels, affect CI engines. The performance and emission parameters of various nanoparticles, including CNT, MoO3, Al2O3, TiO2 ZnO, CeO2, and GQD, are discussed. Pure diesel, diesel biodiesel, and ethanol emulsion, ethanol–biodiesel blends, biodiesel, and diesel with 20% hydrogen are the fuels discussed in this chapter. The following conclusions were drawn from the studies conducted by combining the aforementioned nanoparticles and fuels. The additives CNT and Al2O3, which are distributed in diesel, biodiesel, and ethanol emulsions, respectively, have comparatively demonstrated the maximum BTE. The increased energy content and reduced ignition lag time because of the improved fuel combustion property account for the increase in BTE. The BSFC as a Hall decreased significantly when all nanoparticles were added. Al2O3 added to pure diesel has been shown to reduce the BSFC by 31%, while MoO3 in combination with pure diesel has been shown to reduce the BSFC by 2.3%. Concerning the power and brake force, despite the lack of information on the surveyed data, the accessible information for GQD in bioethanol with 10% (B10) and ethanol–biodiesel mixes energize demonstrates huge augmentation. When we look at CO and HC emissions, 90% of the studies show some kind of reduction. Nanoparticles made of Al2O3, GQD, and TiO2 significantly reduced CO and HC emissions. These reports revealed inconsistent results regarding NOx emissions. Most studies suggest that NOx emissions have decreased. CNT in pure diesel, Al2O3 in diesel, biodiesel, and ethanol emulsion, and GQD in ethanol–biodiesel blends all presented an increase in NOx emissions when compared to the base fuel in some of the reports.

References

1. Soudagar, M.E.M., Ghazali, N.N.N., Kalam, M.A., Badruddin, I.A., Banapurmath, N., Akram, N., The effect of nano-additives in diesel-biodiesel fuel blends: A comprehensive review on stability, engine performance and emission characteristics. *Energy Convers. Manage.*, 178, 146–77, 2018.
2. Hasan, M.M. and Rahman, M.M., Performance and emission characteristics of biodiesel–diesel blend and environmental and economic impacts of biodiesel production: A review. *Renew. Sust. Energ. Rev.*, 74, 938–48, 2017.
3. Goga, G., Chauhan, B.S., Mahla, S.K., Cho, H.M., Performance and emission characteristics of diesel engine fueled with rice bran biodiesel and n-butanol. *Energy Rep.*, 5, 78–83, 2019.
4. Horikoshi, S. and Serpone, N., Introduction to nanoparticles, in: *Microwaves in Nanoparticle Synthesis*, First Edition, pp. 1–24, Wiley-VCH, Wiley-VCH, 2013.
5. Arockiasamy, P. and Anand, R.B., Performance, combustion and emission characteristics of a DI diesel engine fuelled with nanoparticle blended jatropha biodiesel. *Period. Polytech. Mech. Eng.*, 59, 2, 82–93, 2015.
6. Keskin, A., Guru, M., Altıparmak, D., Biodiesel production from tall oil with synthesized Mn and Ni based additives: Effects of the additives on fuel consumption and emissions. *Fuel*, 86, 7–8, 1139–43, 2007.
7. Dreizin, E.L. and Schoentitz, M., Correlating ignition mechanisms of aluminum-based reactive materials with thermoanalytical measurements. *Prog. Energy Combust. Sci.*, 50, 81–105, 2015.
8. Wong, K.V. and de Leon, O., Applications of nanofluids: Current and future. *Adv. Mech. Eng.*, 2, 519659, 2010.
9. Ribeiro, N.M. *et al.*, The role of additives for diesel and diesel blended (ethanol or biodiesel) fuels: A review. *Energy Fuels*, 21, 4, 2433–2445, 2007.
10. Tutak, W., Lukács, K., Szwaja, S., Bereczky, A., Alcohol–diesel fuel combustion in the compression ignition engine. *Fuel*, 154, 196, 2015.
11. Banapurmath, N.R., Budzianowski, W.M., Basavarajappa, Y.H., Hosmath, R.S., Yaliwal, V.S., Tewari, P.G., Effects of compression ratio, swirl augmentation techniques and ethanol addition on the combustion of CNG-biodiesel in a dual-fuel engine. *Int. J. Sustain. Eng.*, 7, 55–70, 2013.
12. Kumar, N., Goel, V., Chauhan, S.R., Evaluation of the effects of engine parameters on performance and emissions of diesel engine operating with biodiesel blend. *Int. J. Ambient Energy*, 37, 121–135, 2014.
13. Yasar, A., Keskin, A., Yıldızhan, S., Uludamar, E., Ocakoglu, K., Effects of titanium-based additive with blends of butanol and diesel fuel on engine characteristics. *Int. J. Glob. Warm.*, 15, 1, 38–53, 2018.
14. Soudagara, M.E.M., Nik-Ghazalia, N.N., Abul Kalam, M., Badruddin, I.A., Banapurmath, N.R., Akrama, N., The effect of nano-additives in diesel-biodiesel fuel blends: A comprehensive review on stability, engine performance and emission characteristics. *Energy Convers. Manage.*, 178, 146–177, 2018.

15. Qi, D., Chen, H., Geng, L., Bian, Y.Z., Experimental studies on the combustion characteristics and performance of a direct injection engine fueled with biodiesel/diesel blends. *Energy Convers. Manage.*, 51, 12, 2985–2992, 2010.
16. D'Silva, R., Binu, K.G., Bhat, T., Performance and emission characteristics of a CI engine fuelled with diesel and TiO2 nanoparticles as fuel additive. *Mater. Today Proc.*, 2, 3728–3735, 2015.
17. Heidari-Maleni, A., Gundoshmian, T.M., Karimi, B., Jahanbakhshi, A., Ghobadian, B., A novel fuel based on biocompatible nanoparticles and ethanol-biodiesel blends to improve diesel engines performance and reduce exhaust emissions. *Fuel*, 276, 118079, 2020.
18. Venu, H. and Madhavan, V., Effect of Al_2O_3 nanoparticles in biodiesel-diesel-ethanol blends at various injection strategies: Performance, combustion and emission characteristics. *Fuel*, 186, 176–189, 2016.
19. Venu, H., Dhana Raju, V., Subramani, L., Combined effect of influence of nano additives, combustion chamber geometry and injection timing in a DI diesel engine fuelled with ternary (diesel-biodiesel-ethanol) blends. *Energy*, 174, 386–406, 2019.
20. Heidari-Maleni, A., Gundoshmian, T.M., Jahanbakhshi, A., Ghobadian, B., Performance improvement and exhaust emissions reduction in diesel engine through the use of graphene quantum dot (GQD) nanoparticles and ethanol biodiesel blends. *Fuel*, 267, 117116, 2020.
21. Hoseini, S.S., Najafi, G., Ghobadian, B., Ebadi, M.T., Mamat, R., Yusaf, T., Biodiesels from three feedstock: The effect of graphene oxide (GO) nanoparticles diesel engine parameters fuelled with biodiesel. *Renew. Energ.*, 145, 190–201, 2020.
22. Mei, D., Zuo, L., Adu-Mensah, D., Li, X., Yuan, Y., Combustion characteristics and emissions of a common rail diesel engine using nanoparticle-diesel blends with carbon nanotube and molybdenum trioxide. *Appl. Therm. Eng.*, 162, 1142, 2019.
23. Shaafi, T. and Velraj, R., Influence of alumina nanoparticles, ethanol and isopropanol blend as additive with diesel soybean biodiesel blend fuel: Combustion, engine performance and emissions. *Renew. Energ.*, 80, 655, 2015.
24. Chena, A.F., Adzmia, M.A., Adamb, A., Othmana, M.F., Kamaruzzamana, M.K., Mrwana, A.G., Combustion characteristics, engine performances and emissions of a diesel engine using nanoparticle-diesel fuel blends with aluminium oxide, carbon nanotubes and silicon oxide. *Energy Convers. Manage.*, 171, 461–477, 2018.
25. Nanthagopal, K., Ashok, B., Tamilarasu, A., Johny, A., Mohan, A., Influence on the effect of zinc oxide and titanium dioxide nanoparticles as an additive with Calophyllum inophyllum methyl ester in a CI engine. *Energy Convers. Manage.*, 146, 8–19, 2017.
26. Manigandan, S., Sarweswaran, R., Booma Devi, P., Sohret, Y., Kondratiev, A., Venkatesh, S., Rakesh Vimal, M., Jensin Joshua, J., Comparative study

of nanoadditives TiO2, CNT, Al2O3, CuO and CeO2 on reduction of diesel engine emission operating on hydrogen fuel blends. *Fuel*, 262, 116336, 2020.
27. Prabu, A. and Anand, R.B., Emission control strategy by adding alumina and cerium oxide nano particle in biodiesel. *J. Energy Inst.*, 89, 366–372, 2016.
28. Aalam, C.S., Saravanan, C.G., Samath, C.M., Influence of iron (Ii, Iii) oxide nanoparticles fuel additive on exhaust emissions and combustion characteristics of CRDI system assisted diesel engine. *Int. J. Adv. Eng. Res. Sci.*, 2, 3, 23–28, 2015.
29. Suleyman, S., Samet, U., Hatice, S., Response surface methodology-based parameter optimization of single-cylinder diesel engine fueled with graphene oxide dosed sesame oil/ diesel fuel blend. *Energy AI*, 10, 100200, 2022.
30. Nallamothu, R.B., Nallamothu, A.K., Nallamothu, S.K., Niranjan, I.N., Appa, B.V., Experimental investigation on effect of injection timing in multiple injection on NOx and smoke from CRDI diesel engine fueled with biodiesel blend. *Ethiop. J. Sci. Sustain. Dev.*, 5, 2, 105–114, 2018.

Index

Printed and bound by CPI Group (UK) Ltd, Croydon, CR0 4YY
17/09/2026
14964277-0007